Nephrotoxic Mechanisms of Drugs and Environmental Toxins

Nephrotoxic Mechanisms of Drugs and Environmental Toxins

Edited by
GEORGE A. PORTER, M.D.
University of Oregon Health Science Center
Portland, Oregon

Springer Science+Business Media, LLC

Library of Congress Cataloging in Publication Data

Main entry under title:

Nephrotoxic mechanisms of drugs and environmental toxins.
  Includes bibliographical references and index.
    1. Renal insufficiency — Etiology. 2. Kidneys — Effect of drugs on. 3. Toxicology. I.
Porter, George A., 1931–   . [DNLM: 1. Drug therapy — Adverse effects. 2. En-
vironmental exposure. 3. Kidney diseases — Chemically induced. 4. Kidney — Drug ef-
fects. WJ 300 N942]
RC918.R4N46 1982   616.6′1071   82-13156

ISBN 978-1-4684-4216-8   ISBN 978-1-4684-4214-4 (eBook)
DOI 10.1007/978-1-4684-4214-4

Softcover reprint of the hardcover 1st edition 1982

Contributors

B. Albini • Departments of Microbiology, Pathology, and Medicine, State University of New York at Buffalo, Buffalo, New York 14214

Giuseppe A. Andres • Departments of Microbiology, Pathology, and Medicine, School of Medicine, State University of New York at Buffalo, Buffalo, New York 14214

J.P. Bendirdjian • Tissue Physiology Group, University of Rouen, Rouen, France

William M. Bennett • Division of Nephrology, Department of Medicine, Oregon Health Sciences University, Portland, Oregon 97201

William O. Berndt • Department of Pharmacology and Toxicology, University of Mississippi Medical Center, Jackson, Mississippi 39216. Present address: Department of Pharmacology, College of Medicine, University of Nebraska Medical Center, Omaha, Nebraska 68105

Atul K. Bhan • Department of Pathology, Massachusetts General Hospital and Harvard Medical School, Boston, Massachusetts 02117

Roland C. Blantz • Department of Medicine, University of California, San Diego, School of Medicine, and Veterans Administration Medical Center, San Diego, California 92161

Wayne A. Border • University of California at Los Angeles School of Medicine, Harbor–UCLA Medical Center, Torrance, California 90509

Esther M. Brown • Department of Veterinary Anatomy–Physiology, College of Veterinary Medicine, University of Missouri—Columbia, Columbia, Missouri 65211

Shreekant Chopra • Renal Section, Boston Veterans Administration Medical Center, Boston, Massachusetts 02130

v

Michel G. Côté • Départment de Pharmacologie, Faculté de Médicine, Université de Montréal, Montréal, Québec, Canada H3C 3J7

Ramzi S. Cotran • Department of Pathology, Brigham and Women's Hospital and Harvard Medical School, Boston, Massachusetts 02115

Malcolm Cox • Renal-Electrolyte Section, Department of Medicine, Philadelphia Veterans Administration Center and University of Pennsylvania School of Medicine, Philadelphia, Pennsylvania 19104

Elvira Druet • Laboratory of Morphology and Renal Immunopathology, Broussais Hospital, Paris, France

Philippe Druet • Laboratory of Morphology and Renal Immunopathology, Broussais Hospital, Paris, France

W. Clayton Elliott • Division of Nephrology, Department of Medicine, Oregon Health Sciences University, Portland, Oregon 97201

J.P. Fillastre • Tissue Physiology Group, University of Rouen, Rouen, France

Walter Flamenbaum • Nephrology Division, Beth Israel Medical Center, New York, New York 10003

Bruce A. Fowler • Laboratory of Pharmacology, National Institute of Environmental Health Sciences, Research Triangle Park, North Carolina 27709

Justine S. Garvey • Department of Biology, Syracuse University, Syracuse, New York 13210

Marc Gehr • Renal Section, Boston Veterans Administration Center, Boston, Massachusetts 02130

I. Glurich • Departments of Microbiology, Pathology, and Medicine, State University of New York at Buffalo, Buffalo, New York 14214

M. Godin • Tissue Physiology Group, University of Rouen, Rouen, France

Robert A. Goyer • National Institute of Environmental Health Sciences, Research Triangle Park, North Carolina 27709

Philip W. Hall, III • Department of Medicine, Case Western Reserve University and Cleveland Metropolitan General Hospital, Cleveland, Ohio 44109

Robert Hamburger • Renal Section, Boston Veterans Administration Center, Boston, Massachusetts 02130

P.B. Hammond • Department of Environmental Health, University of Cincinnati Medical Center, Cincinnati, Ohio 45267

I.B. Hanenson • Department of Medicine, University of Cincinnati Medical Center, Cincinnati, Ohio 45267

Michael J. Hanley • Department of Medicine, Division of Renal Diseases, University of Texas Health Science Center at San Antonio, San Antonio, Texas 78284

William R. Hewitt • Department of Veterinary Anatomy–Physiology, College of Veterinary Medicine, and Department of Pharmacology, School of Medicine, University of Missouri—Columbia, Columbia, Missouri 65211

Francois Hirsch • Laboratory of Morphology and Renal Immunolopathology, Broussais Hospital, Paris, France

C.D. Hong • Department of Medicine, University of Cincinnati Medical Center, Cincinnati, Ohio 45267

Jerry B. Hook • Center for Environmental Toxicology, Michigan State University, East Lansing, Michigan 48824

Donald C. Houghton • Division of Nephrology, Department of Medicine, Oregon Health Sciences University, Portland, Oregon 97201

Russell F. Husted • Department of Medicine, University of Connecticut School of Medicine, Farmington, Connecticut 06032

James Kaufman • Renal Section, Boston Veterans Administration Medical Center, Boston, Massachusetts 02130

William M. Kluwe • National Toxicology Program, Research Triangle Park, North Carolina 27709

Jeffrey I. Kreisberg • Departments of Pathology and Medicine, University of Texas Health Science Center at San Antonio, San Antonio, Texas 78284

N. Lameire • Renal Division of the Department of Medicine, State University of Gent, University Hospital, B-9000 Gent, Belgium

S.I. Lerner • Department of Environmental Health, University of Cincinnati Medical Center, Cincinnati 45267

Friedrich C. Luft • Department of Medicine, Nephrology Section, Indiana University School of Medicine, Indianapolis, Indiana 46223

Erve G.J. Matthys • Departments of Pathology and Medicine, University of Texas Health Science Center at San Antonio, San Antonio, Texas 78284

Robert T. McCluskey • Department of Pathology, Massachusetts General Hospital and Harvard Medical School, Boston, Massachusetts 02117

Felix Milgrom • Department of Microbiology, School of Medicine, State University of New York at Buffalo, Buffalo, New York 14214

Hiroaki Miyajima • Département de Pharmacologie, Faculté de Médicine, Université de Montréal, Montréal, Québec, Canada H3C 3J7

J.P. Morin • Tissue Physiology Group, University of Rouen, Rouen, France

Gilbert H. Mudge • Departments of Medicine, Pharmacology, and Toxicology, Dartmouth Medical School, Hanover, New Hampshire 03755

Thomas G. Murray (deceased) • Hospital of the University of Pennsylvania, Philadelphia, Pennsylvania 19104

Gunnar F. Nordberg • Department of Environmental Hygiene, University of Umea, Umea, Sweden

E.J. O'Flaherty • Department of Environmental Health, University of Cincinnati Medical Center, Cincinnati, Ohio 45267

B. Olier • Tissue Physiology Group, University of Rouen, Rouen, France

Timo Palosuo • National Public Health Institute, Helsinki, Finland; and Department of Microbiology, School of Medicine, State University of New York at Buffalo, Buffalo, New York 14214

P. Pattyn • Renal Division of the Department of Medicine, State University of Gent, University Hospital, B-9000 Gent, Belgium

Gabriel L. Plaa • Département de Pharmacologie, Faculté de Médicine, Université de Montréal, Montréal, Québec, Canada H3C 3J7

Joseph P. Portanova • Division of Rheumatic Diseases, Department of Medicine, School of Medicine, University of Colorado Health Sciences Center, Denver, Colorado 80262

George A. Porter • Department of Medicine, Oregon Health Sciences University, Portland, Oregon 97201

J. Quatacker • Division of Anatomo-Pathology, State University of Gent, University Hospital, B-9000 Gent, Belgium

S. Ringoir • Renal Division of the Department of Medicine, State University of Gent, University Hospital, B-9000 Gent, Belgium

Robert L. Rubin • Division of Rheumatic Diseases, Department of Medicine, School of Medicine, University of Colorado Health Sciences Center, Denver, Colorado 80262

Catherine Sapin • Laboratory of Morphology and Renal Immunopathology, Broussais Hospital, Paris, France

John H. Schwartz • Renal Section, Department of Medicine, Thorndike Memorial Laboratory, Boston City Hospital and Boston University School of Medicine, Boston, Massachusetts 02118

Victoria C. Serbiá • Center for Environmental Toxicology, Michigan State University, East Lansing, Michigan 48814

Jay Stein • Department of Medicine, University of Texas Health Science Center, San Antonio, Texas 78284

Philip R. Steinmetz • Department of Medicine, University of Connecticut School of Medicine, Farmington, Connecticut 06032

Eng M. Tan • Division of Rheumatic Diseases, Department of Medicine, School of Medicine, University of Colorado Health Sciences Center, Denver, Colorado 80262

F. Gary Toback • Department of Medicine, Pritzker School of Medicine, The University of Chicago, Chicago, Illinois 60637

Bruce M. Tune • Division of Nephrology, Department of Pediatrics, Stanford University School of Medicine, Stanford, California 94305

L. Vakaet • Renal Division of the Department of Medicine, State University of Gent, University Hospital, B-9000 Gent, Belgium

R. Vanholder • Renal Division of the Department of Medicine, State University of Gent, University Hospital, B-9000 Gent, Belgium

Manjeri A. Venkatachalam • Departments of Pathology and Medicine, University of Texas Health Science Center at San Antonio, San Antonio, Texas 78284

G. Viotte • Tissue Physiology Group, University of Rouen, Rouen, France

Richard P. Wedeen • Veterans Administration Medical Center, East Orange, New Jersey 07019; and College of Medicine and Dentistry of New Jersey, Newark, New Jersey 07103

Curtis B. Wilson • Department of Immunopathology, Research Institute of Scripps Clinic, La Jolla, California 92037

Preface

The majority of the offending toxicants to be reviewed in this volume were developed to help mankind, and it is only with prolonged or widespread application that their adverse effects have been recognized. Conversely, in the case of prescription drugs, there has been an attempt to identify the adverse effects in advance and incorporate these risks into the decision of approval for human consumption. Unfortunately, for those drugs in which recognized injury occurs only after prolonged use, such appraisals are made in retrospect. Despite this, most renal injury induced by drugs or toxicants can be either prevented by excluding drugs with unacceptable side effects or interrupted by eliminating the offending agent once damage is manifested. The fact that prevention, reversibility, or arrest of renal injury is possible provided a major impetus for this publication.

Since no international registry for nephrotoxic injury exists, estimates of incidence must rely on less than ideal sources. Recently I, together with Dr. William Bennett, summarized a survey of the frequency of various categories of nephrologic disease (Porter and Bennett, 1981). Based on this survey, we projected that in nearly one of ten patients seeking nephrologic consultation a nephrotoxic etiology may be involved. Of cases of end-stage renal disease, between 3 and 4% are due to drug nephrotoxicity, according to recent published results(European Dialysis and Transplant Association, 1979). For acute renal failure, antibiotics and contrast agents persist as major offending agents, while for chronic renal failure, analgesics remain a worldwide problem.

Originally, it was assumed that nephrotoxins induced their cellular injury by physical–chemical interactions which were unique for each agent. However, it is now appreciated that other mechanisms of cellular injury are common. Examples of immunologically mediated cell injury, cytologic damage due to metabolic product(s), enzyme induction leading to the accumulation of "reactive intermediates," and interference with the delivery of substrate to cells are discussed in this volume.

Certain characteristics of the kidney increase its vulnerability to injury by clinical agents. The charged, endothelial surface of the glomeruli is vast and, coupled with the generous blood flow, maximizes both physical and chemical interactions with blood-borne material. It is not surprising that, with the advent of animals suitable for direct assessment of glomerular permselectivity, the effect of nephrotoxins, e.g., gentamicin, has been shown to change glomerular membrane permeability (Baylis *et al.*, 1977). Because many drugs are either organic cations or anions, binding to serum proteins protects them against filtration but does allow transport by proximal tubular cells. Such transport may lead to either elevated intracellular concentration, as in the case of certain cephalosporins, or high intraluminal concentrations in proximal tubular segments, as noted for penicillins. Conversely, potential toxicants that undergo filtration can gain access to intracellular organelles through direct penetration of apical cell boundary, coupled transport, or pinocytosis. Once inside the renal tubular cell, injury may occur by mechanisms such as uncoupling of the mitochondrial respiratory chain or modifying of critical enzyme activity sufficient to restrict or interrupt cellular energetics. For lipophilic toxicants, interaction with the plasma membrane, with associated increased permeability, may be sufficient to interfere with osmoregulation, leading to cell swelling and subsequent rupture. An additional mechanism of cell injury may be indirect. Drugs, through accumulation, metabolic conversion, and elimination, may interfere with normal vital mechanisms. Such a process has been suggested in analgesic nephropathy when, because of accumulation of either the parent compound or metabolites, or both, by medullary countercurrent processes, cellular injury leads to papillary necrosis. Another example is furnished by agents that act like osmotic diuretics, causing fluid and electrolyte depletion by obligating renal water excretion and altering distal tubular flow pattern. For acidic drugs with pKa between 5 and 6, terminal urinary acidification plus water reabsorption in the distal nephron segments may result in intraluminal precipitation and acute renal failure due to obstruction, while others may interfere with intracellular hormone action. Finally, the depletion of critical minerals from the body, e.g., potassium, calcium, or magnesium, by diuretics can compromise normal renal function. Examples of many of these renal effects will be discussed in this volume.

As can be confirmed by scanning the list of authors, multiple disciplines and varied backgrounds are represented. The objectives of this volume include: (1) to summarize current information on the wide range of subjects included for discussion, (2) to delineate the unanswered questions, and (3) to stimulate the search for new explanations by formulating testable hypotheses.

The contributions of section editors Giuseppe Andres, Jerry Hook, Jay Stein, and Bruce Tune, who collectively developed the character of this publication, are generously acknowledged, as is the support provided by Nancy Cummings, Bob Goyer, and Jim Scherbenske.

George A. Porter

Portland, Oregon

References

Baylis, C., Rennke, G.H., and Brenner, B.M., 1977, Mechanisms of defects in glomerular ultrafiltration associated with gentamicin administration, *Kidney Int.* **12:**344.

European Dialysis and Transplant Association, 1979, *Proceedings,* Excerpta Medica Foundation, Amsterdam.

Porter, G. A. and Bennett, W. M., 1981, Toxic nephropathies, in: *The Kidney* (B. M. Brenner and F. C. Rector, Jr., eds.), Saunders, Philadelphia, Pennsylvania, p. 2047.

Contents

Section II. Renal Failure Due to Antimicrobial Agents
Bruce Tune, *Section Editor*

Section III. Tubulointerstitial Nephropathy Due to Drugs and Environmental Toxicants

George A. Porter, *Section Editor*

Section IV. Pathophysiologic Mechanisms of Toxicity Induced by Environmental Toxins

Jerry B. Hook, *Section Editor*

Section V. Immunologic Mechanisms and Toxic Nephropathies
Giuseppe A. Andres, *Section Editor*

I

PATHOPHYSIOLOGY OF ACUTE RENAL FAILURE

JAY STEIN, Section Editor

1

Overview of Pathophysiology of Acute Renal Failure

JAY STEIN

1. Introduction

Acute renal failure is a common and invariably serious clinical syndrome which can be caused by a multiplicity of etiologic factors. An operational definition might be an abrupt, frequently reversible reduction in renal function usually associated with oliguria. Yet, even this simplistic definition is not totally adequate, since a nonoliguric form of the syndrome is being more frequently recognized.

There is obviously a wide variety of clinical disorders which would fall in this category of diseases (Table I). The major emphasis of this discussion will involve those entities that have been commonly included under the term "acute tubular necrosis." This would include those entities listed in Table I under hemodynamically mediated and nephrotoxic renal failure. The differentiation of these two types of acute renal failure is important because it is felt that the pathogenesis of renal functional impairment may be quite different in these two groups of entities.

Although it has been quite straightforward to recognize and classify these various causes of acute renal failure, it has not been possible to readily dissect the pathophysiology of these conditions from clinical studies. Thus, animal models have been utilized to determine the mechanism of renal functional impairment in acute renal failure.

JAY STEIN • Department of Medicine, University of Texas Health Science Center, San Antonio, Texas 78284.

TABLE I. Classification of Acute Renal Failure

Pre-renal failure
  A. Hypovolemia, e.g., hemorrhage
    1. Gastrointestinal losses: diarrhea, vomiting
    2. Skin losses: burns, sweating, heat stroke
    3. Renal losses: diuretic abuse, salt-wasting nephropathy
  B. Cardiovascular failure: myocardial failure, pericardial tamponade, vascular pooling

Post-renal failure
  A. Urethral obstruction: stricture, bladder neck obstruction, prostatic hypertrophy, carcinoma of bladder
  B. Functional obstruction of bladder: neuropathy, ganglionic blocking agents
  C. Ureteral obstruction: retroperitoneal fibrosis, stones, sulfonamide and uric acid crystals, blood clots, accidental ligation of ureters during pelvic surgery

Parenchymal renal failure
  A. Intrinsic renal disease
    1. Glomerulonephritis and/or vasculitis
        a. Acute post-streptococcal glomerulonephritis
        b. Systemic lupus erythematosus
        c. Polyarteritis nodosa
        d. Schonlein-Henoch purpura
        e. Subacute bacterial endocarditis
        f. Rapidly progressive glomerulonephritis
        g. Goodpasture's syndrome
        h. Serum sickness
        i. Hemolytic-uremic syndrome
        j. Malignant hypertension
        k. Scleroderma
        l. Drug-related vasculitis
        m. Post-partum acute renal failure
        n. Renal transplant rejection
    2. Interstitial nephritis
        a. Secondary to infection
        b. Drug-related (Methicillin)
        c. Hypercalcemia
        d. Myeloma kidney
        e. Radiation
  B. Acute fulminating infection
    1. Papillary necrosis
    2. Pyelonephritis
  C. Renal vascular disease
    1. Renal artery occlusion: embolism, thrombosis, dissecting aneurysm
    2. Renal vein thrombosis
  D. Hemodynamically mediated acute renal failure
    1. Major surgery: aortic surgery
    2. Obstetrical: septic abortion, placenta previa, abruptio placenta
    3. Trauma: crush injury
    4. Pigment release: hemoglobin, myoglobin
  E. Acute nephrotoxic renal failure
    1. Heavy metals: mercury, cadmium, uranium
    2. Organic solvents: carbon tetrachloride
    3. Glycols
    4. Antibiotics
    5. Pesticides
    6. Miscellaneous: methoxvfluorane

2. Initiation of Acute Renal Failure

Various factors have been suggested to play a role in either the initiation or maintenance of acute renal failure. The initial insult may be related to some hemodynamic event (i.e., ischemia) or to a direct toxic effect of a particular drug or agent. In other words, hypoxia and/or the direct effect of an agent on the metabolic machinery will cause cellular injury. The magnitude of this insult and the status of a poorly defined group of modifying factors will ultimately determine whether fixed acute renal failure will occur (Fig. 1).

3. Maintenance of Acute Renal Failure

After the initiating event, other factors come into play to maintain the renal functional impairment in acute renal failure. Four factors have been implicated and will be discussed.

3.1 Persistent Renal Vasoconstriction

It has been shown in many clinical and experimental studies that renal blood flow is reduced during the maintenance phase of acute renal failure (Ayer *et al.,* 1971; Flanigan *et al.,* 1965; Hollenberg *et al.,* 1968). The basis for this increase in renal resistance is not clear. Although the renin–angiotensin system has been implicated by circumstantial evidence, more recent data using agents that block the action of the renin–angiotensin system have not confirmed an important protective effect of decreased renin activity. In addition, authors have readily been able to dissociate the development of renal functional impairment and intrarenal renin activity (Bidani *et al.,* 1979).

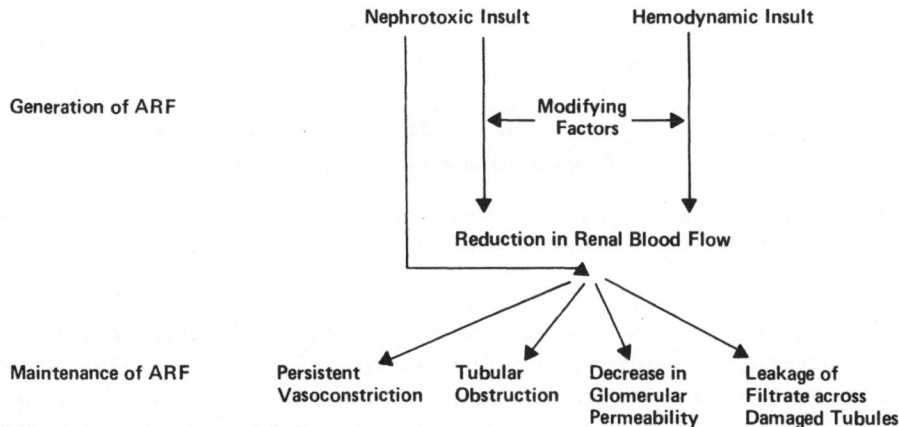

FIGURE 1. Summary of pathophysiologic events involved in the generation and maintenance of acute renal failure (ARF). [From Stein *et al.* (1978).]

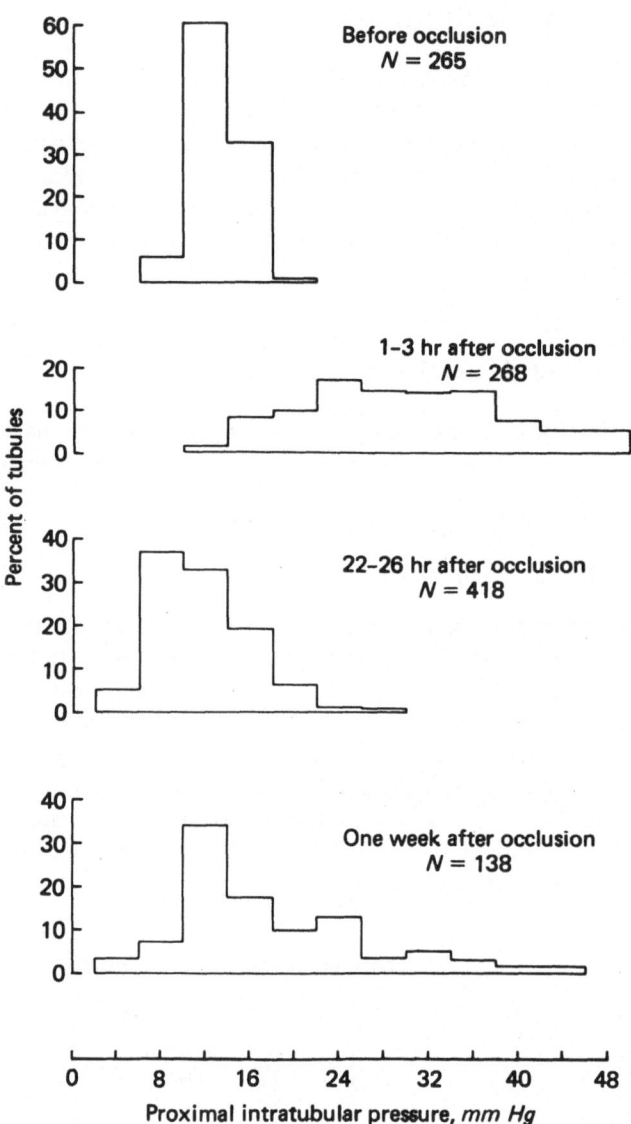

FIGURE 2. Proximal intratubular pressure prior to renal artery occlusion and at varying intervals after release of occlusion. [From Arendshorst *et al.* (1976).]

The prostaglandin system has also been evaluated and may be an important determinant of renal resistance in acute renal failure (Torres *et al.*, 1975). Finally, studies from this laboratory have demonstrated that the complement system may be involved in the increased renal resistance seen in experimental renal artery clamping (Barnes *et al.*, 1978). In this model, pretreatment of the animal with cobra venom factor markedly attenuated the renal hemodynamic changes seen in this model prior to and after volume expansion. Unfortunately, inulin clearance was not improved in the clamp model by complement depletion.

3.2 Tubular Obstruction

Recent studies have clearly demonstrated that tubular obstruction is a determinant of functional impairment in acute renal failure (Arendshorst *et al.*, 1975). As is shown in Fig. 2, there is a profound rise in intratubular pressure 1–3 hr after release of a renal artery clamp. Although intratubular pressure fell toward control levels 24 hr after clamping, this was associated with a marked increase in afferent arteriolar resistance. When volume expansion was induced, the characteristics of intratubular obstruction were again apparent. Sophisticated histologic studies have demonstrated that soon after release of renal artery clamping, desquamated blebs of proximal tubular brush borders became impacted in the pars recta, causing tubular obstruction (Venkatachalam *et al.*, 1978). These casts are eventually dissolved, but extensive case formation is then noted in the more distal portions of the nephron. There seems no doubt that tubular obstruction is a major determinant of renal functional impairment in acute renal failure.

3.3 Backleakage of Filtrate

A third pathogenic factor suggested to play a role in the pathogenesis of acute renal failure is leakage of filtrate across damaged tubular epithelium. This abnormality has been demonstrated by dye injections, microinjection studies (Fig. 3), and, more recently, by the demonstration of intravenously injected horseradish peroxidase in the renal interstitium of animals with renal artery clamping (Donohoe *et al.*, 1978). There is no doubt that leakage of filtrate across damaged tubular epithelium is of pathogenic importance, but the quantitative significance remains to be determined.

3.4 Decreased Glomerular Permeability

It has recently been suggested that a decrease in the glomerular capillary ultrafiltration coefficient K_f may also be a significant determinant of the functional impairment in acute renal failure (Cox *et al.*, 1974). This has been directly demonstrated in the gentamicin and uranyl nitrate model in the rat (Avasthi *et al.*, 1980). Whether this is the basis for the decrease in K_f in this model remains to be determined.

4. Prevention of Experimental Acute Renal Failure

In the past 5 years, a number of laboratories have evaluated various agents which may be useful in preventing or attenuating the development of acute renal failure. In studies performed in this laboratory, it has been found that prostaglandin E, bradykinin, furosemide, and mannitol all attenuate the fall in inulin clearance seen after intrarenal norepinephrine administration (Patak *et al.*,

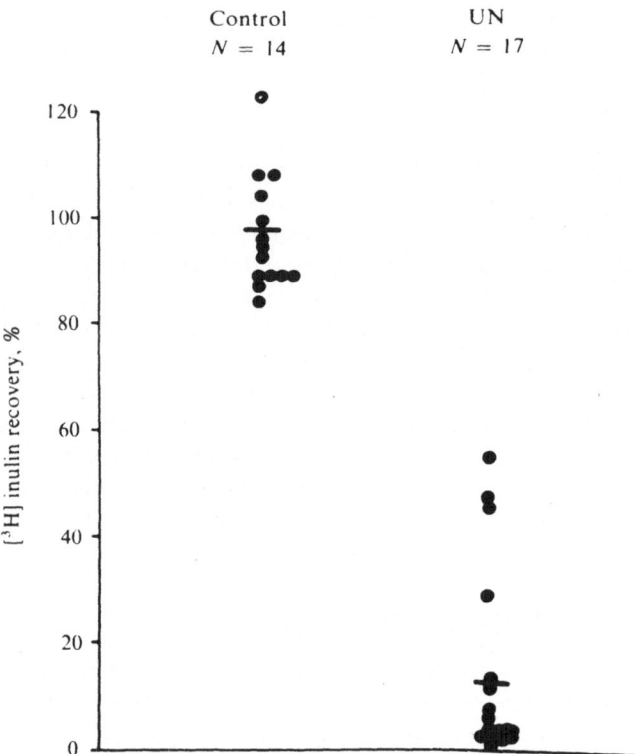

FIGURE 3. Comparison of the urinary recovery of [^{3}H] inulin injected into the proximal tubule of control and uranyl nitrate-treated dogs. [From Stein *et al.* (1975).]

1979). Histologic studies demonstrated that the administration of bradykinin markedly diminishes the tubular necrosis usually noted after norepinephrine.

Further studies have been performed in which total renal blood flow has been measured before, during, and after the administration of norepinephrine with and without concomitant infusions of these various protective agents (O'Neil *et al.*, 1981). Bradykinin significantly attenuated the degree of renal ischemia noted with norepinephrine alone, while furosemide and mannitol had no effect. All three protective agents blunted the reduction in the ATP charge ratio that occurred with norepinephrine administration. Thus, it is possible that these various agents act by altering cellular metabolism rather than by an effect on renal hemodynamics or solute clearance.

References

Arendshorst, W. J., Finn, W. F., and Gottschalk, C. W., 1975, Pathogenesis of acute renal failure following renal ischemia in the rat, *Circ. Res.* **37:**558.

Arendshorst, W. J., Finn, W. F., and Gottshalk, G. W., 1976, Micropuncture study of acute renal failure following temporary renal ischemia in the rat, *Kidney Int.* **10:**S100.

Avasthi, P., Evan, A. P., and Hay, D., 1980, Glomerular endothelial cells in uranyl nitrate-induced acute renal failure in rats, *J. Clin. Invest.* **65:**121.

Ayer, G., Grandchamp, A., Wyler, T., and Truniger, B., 1971, Intrarenal hemodynamics in glycerol-induced myohemoglobinuric acute renal failure in the rat, *Circ. Res.* **29:**128.

Barnes, J., Osgood, R. W., Pinckard, R. N., Reineck, H. J., and Stein, J. H., 1978, Effect of complement depletion on the post-ischemic model of acute renal failure, *Kidney Int.* **14:**721.

Bidani, A., Churchill, P., and Fleischman, L., 1979, Sodium-chloride-induced protection in nephrotoxic acute renal failure: Independence from renin, *Kidney Int.* **16:**481.

Cox, J. W., Baehler, R. W., Sharma, H., O'Dorisia, T., Osgood, R. W., Stein, J. H., and Ferris, T. F., 1974, Studies on the mechanism of oliguria in a model of unilateral acute renal failure, *J. Clin. Invest.* **53:**1546.

Donohoe, J. F., Venkatachalam, M. A., Bernard, D. B., and Levinsky, N. G., 1978, Tubular leakage and obstruction after renal ischemia: Structure-function correlations, *Kidney Int.* **13:**208.

Flanigan, W. J., and Oken, D. E., 1965, Renal micropuncture study of the development of anuria in the rat with mercury-induced acute renal failure, *J. Clin. Invest.* **44:**449.

Hollenberg, N. K., Epstein, M., Rosen, S. M., Basch, R. I., Oken, D. E., and Merrill, J. P., 1968, Acute oliguric renal failure in man: Evidence for preferential renal cortical ischemia, *Medicine* **47:**455.

O'Neil, T., Sinsteden, T., and Stein, J. H., 1981, Unpublished observations.

Patak, R. V., Fadem, S. Z., Lifschitz, M. D., and Stein, J. H., 1979, Study of factors which modify the development of norepinephrine-induced acute renal failure in the dog. *Kidney Int.* **15:**227.

Stein, J. H., Gottschall, J., Osgood, R. W. and Ferris, T. F., 1975, Pathophysiology of a nephrotoxic model of acute renal failure, *Kidney Int.* **8:**27.

Stein, J. H., Patak, R. V. and Lifschitz, M. D., 1978, Acute renal failure: Clinical aspects and pathophysiology, *Contrib. Nephrol.* **14:**118.

Torres, V. E., Strong, C. G., Romero, J. C., and Wilson, D. M., 1975, Indomethacin enhancement of glycerol-induced acute renal failure in rabbits, *Kidney Int.* **7:**170.

Venkatachalam, M. A., Bernard, D. B., Donohoe, J. F., and Levinsky, N. G., 1978, Ischemic damage and repair in the rat proximal tubule: Differences among the S_1, S_2 and S_3 segments, *Kidney Int.* **14:**31.

2

Pathology of Acute Renal Failure

JEFFREY I. KREISBERG, ERVE G. J. MATTHYS, and
MANJERI A. VENKATACHALAM

1. Introduction

Acute renal failure (ARF) can be defined as an abrupt decline in renal function sufficient to result in retention of nitrogenous waste. In this chapter we will describe and discuss, with an emphasis on structure–function correlations, a major subset of the clinical syndrome of ARF, that is, acute tubular necrosis (ATN). By definition, the term ATN includes all forms of acute urinary suppression in which tubular injury of acute onset is the primary basis for renal failure, and excludes all other causes of ARF, such as pre-renal azotemia, interstitial nephritis, glomerulone-phritis, disseminated intravascular coagulation with renal cortical necrosis, organic vascular disease, and obstruction of the lower urinary tract (Robbins and Cotran, 1979).

Although the clinical setting of ATN varies, the underlying etiologic factors have remained quite constant. It appears that prolonged renal ischemia is the most common pathogenetic factor (Anderson *et al.*, 1977; Bushinsky *et al.*, 1979; Miller *et al.*, 1978; Smolens and Stein, 1980, 1981). Furthermore, some degree of renal

JEFFREY I. KREISBERG, ERVE G. J. MATTHYS, and MANJERI A. VENKATACHALAM •
Departments of Pathology and Medicine, University of Texas Health Science Center at San Antonio,
San Antonio, Texas 78284.

11

ischemia may initiate all forms of ARF. Another etiologic factor commonly encountered in patients developing ATN is direct nephrotoxin exposure (Anderson and Schrier, 1980). Exposure to organic solvents, glycols, and heavy metals were once important causes of ATN (Schreiner and Maher, 1965). Now, however, aminoglycoside antimicrobial agents and radiographic contrast material are the most frequent inducers of direct nephrotoxicity (Cronin, 1978; Carvallo et al., 1978; Ansari and Baldwin, 1976; Swartz et al., 1978).

The major structural factors in ATN and the corresponding pathophysiologic mechanisms that they may underlie are: (1) tubular epithelial cell swelling and necrosis, the former leading to impaction of tubule lumina with plasma membrane blebs and the latter causing abnormal tubule permeability, (2) tubular obstruction of the distal nephron segments by casts, (3) structure abnormality in the filtering membrane, which may result in decreased glomerular filtration, and (4) contraction of arteries and arterioles, resulting in decreased glomerular blood flow. Studies in experimental ATN have provided evidence that all of these factors, to varying degrees, contribute to the overall impairment of renal function, but tubular damage appears to be the primary defect. It is also becoming clear that tubular damage in human ATN is commoner than had been previously thought and that tubular factors may play the primary role in the pathogenesis of human ATN as well.

2. Pathology of Human Acute Tubular Necrosis

2.1. Ischemic Type of Acute Tubular Necrosis

The ischemic type of ATN appears to result from one or more episodes of acute renal ischemia that occur in diverse clinical situations. Endogenous nephrotoxins liberated from necrotic tissue or bacterial products may also contribute to the development of renal injury in this type of ATN. Morphologically, the renal lesion is characterized by patchy tubular necrosis and cast formation (Oliver et al., 1951). Rents in the tubular epithelium may develop and are indicative of the dissolution of basement membrane at sites of necrosis (tubulorrhexis). There is a predilection for damage in the pars recta portion of the proximal tubule, but necrosis commonly involves the proximal convoluted tubules also and sometimes the distal tubules.

Based on histologic sections, tubular necrosis was once reported to be inconspicuous or absent in the ischemic form of ATN (Bohle et al., 1976; Olsen, 1976). In more recent studies, focal necrosis of single proximal tubule cells and isolated clusters of cells was observed in over 80% of cases (Solez et al., 1979). Many viable-appearing proximal tubule cells showed loss of brush border (Solez et al., 1979). Other changes included flattening of tubular epithelium and dilatation of tubule lumina. Many of these flattened tubule cells exhibited changes indicative of regeneration (i.e., mitoses and large hyperchromatic nuclei) following prior episodes of necrosis. Thus, regenerative changes and foci of necrosis can be observed in the same biopsy, implying that necrosis of tubular epithelium is an ongoing process (Solez et al., 1979) and may occur in episodic fashion (Oliver,

1951). This important concept may explain some of the differences between human ATN and experimental models.

In the ischemic type of ATN, distal tubules are characteristically occluded by urinary casts of the hyaline, granular, and pigmented types. The pathogenesis of cast formation is unknown. However, it is clear that the major component of the casts is the unique Tamm-Horsfall (TH) glycoprotein, a strongly anionic macromolecule of unknown function, normally secreted by the ascending thick limb of Henle and present in the urine. In monomeric form, the protein's molecular weight is 110,000, but it exhibits the singular tract of aggregation and complex polymer formation under conditions of dehydration and increasing Na^+, Ca^{++}, and H^+ ion concentration (Hoyer and Seiler, 1979). Polymerization of TH proteins within renal tubules results in cast formation. Coprecipitation of TH protein with Bence-Jones proteins, hemoglobin, and myoglobin (Kant et al., 1976; McQueen, 1962; Smolens et al., 1980) may be the basis for ARF in patients with myelomatosis, hemolysis, mismatched blood transfusion, and rhabdomyolysis.

2.2. Nephrotoxic Type of Acute Tubular Necrosis

The nephrotoxic type of ATN is caused by a variety of toxic chemicals, and the lesion consists primarily of necrosis in the proximal tubules and tubule lumen obstruction by casts. The tubule basement membranes appear to be intact and the location of necrosis in different segments of the proximal tubule follows patterns that are characteristic for some nephrotoxins (Oliver et al., 1951). In severe cases, necrosis of the distal tubule may occur along with tubulorrhexis. In contrast to the ischemic type of ATN, necrosis may be observed in all nephrons in ATN due to nephrotoxins. For an in-depth discussion of all types of toxic ATN the reader is referred to other sources (Heptinstall, 1974; Porter and Bennett, 1980).

3. *In Vivo* Experimental Models of Acute Tubular Necrosis

3.1. Hypotension, Arterial Occlusion, and Variants

The ischemic type of ATN may be induced in animals by (1) renal artery occlusion, (2) hemorrhagic shock, induced by controlled withdrawal of blood, and (3) reflex vasoconstriction of the renal vasculature, produced by hind limb trauma. In any of these models, infusion of hemolyzed blood can be used to enhance the lesion.

Since transient systemic hypotension is usually the most common antecedent to the circulatory form of acute renal failure, a useful animal model is hemorrhagic hypotension. Studies performed in hemorrhagic shock by Kreisberg et al. (1976) and later confirmed by Dobyan et al. (1977) showed that 1–2 hr of hypotension in rats does not result in ARF but induces necrosis of the straight portions of the proximal tubules. Morphologic alterations in the pars recta portion of the proximal tubule were noted after 60 and 90 min of hypotension. These alterations included dilated endoplasmic reticulum, clumped nuclear chromatin, swollen mitochondria with small flocculent densities, and internalization of microvilli. Cell death was observed extensively in the pars recta portions of the proximal tubule after 6 hr of reflow following hypotension.

The most evident functional response to 90 min of hypotension was a transient polyuria during the first 24 hr and normalization by 48 hr. The urine was hypoosmolar for 2 days following hypotension, and normal concentrating ability was restored by the third day. The 24-hr urine sodium excretion was depressed during the first and second days after hypotension, while potassium excretion was no different from controls, indicating that despite significant tubular necrosis, the kidney had retained its ability to reabsorb these essential electrolytes. Also, azotemia was not observed.

Another ischemic model that induces ARF is renal artery occlusion. In this model, proximal tubules, particularly the straight segments, are most vulnerable to ischemia. Selective vulnerability within the proximal tubule is based on segmentation of the proximal tubule by cell type (Venkatachalam et al., 1978). After 25 min of ischemia, the cells in the pars recta portion of the proximal tubule, but not the cells of the pars convoluta, undergo necrosis. Correspondingly, the renal failure that follows is also mild. Necrosis following 60 min of clamping is much more extensive and involves the entire proximal tubule, and the renal failure that occurs is irreversible.

The earliest morphologic alteration encountered following ischemia is plasma membrane bleb formation (Donohoe et al., 1978; Venkatachalam et al., 1978). Blebs appear to originate through ballooning of the microvilli or the intervening plasma membrane and many are shed into the tubule lumina. Blebs several micrometers in diameter are found in large numbers in the tubules. In their transit down the tubules, the blebs presumably continue to swell, and are impacted in large numbers in straight segments of the proximal tubules or more distally. Virtually every nephron is obstructed in this manner within 2 hr of blood reflow after 60 min of ischemia. Second, as was observed in the hypovolemia model, large masses of microvilli are incorporated into the cytoplasm of epithelial cells. These alterations reflect a profound defect in the plasma membrane. Whereas reversibly injured cells are able to reconstitute their lost brush border, in irreversibly injured cells, membrane damage is apparently a continuing process and is lethal to the cell. Between 2 and 4 hr of reflow after 60 min of ischemia, the membranous blebs that obstruct proximal straight tubules undergo lysis. Subsequently, well-organized casts develop in the distal nephron segments. These casts are known to contain TH glycoprotein (Hoyer and Venkatachalam, unpublished data).

Abnormal permeability of necrotic tubular epithelium can be demonstrated by microinjection of vital dyes or morphologically by using tracer methods (Arendshorst et al., 1975; Donohoe et al., 1978). For example, if horseradish peroxidase, a protein of 29 Å in molecular radius, is microinjected into proximal tubules from 1 to 2 hr following renal ischemia, there is backleak of the enzyme through necrotic single cells or isolated clusters of cells into the peritubular space. In control tubules not injured by ischemia, the protein is pinocytosed by the cells and segregated into lysosomes, and no backleak is seen (Donohoe et al., 1978).

Abnormalities in the glomerular filter that may interfere with glomerular filtration are also found relatively early (1–2 hr) following reflow of blood after ischemia (Barnes *et al.*, 1981). Significant numbers of glomeruli show flattening of epithelial foot processes, a defect which causes reduction in the area of filtration slits overlying the glomerular basement membrane.

Swelling of the endothelium with reduction in capillary luminal caliber has been thought to precipitate trapping of red blood cells and impede further flow of blood (Flores *et al.*, 1972; Summers and Jamison, 1971). Current evidence suggests that cell swelling may be a factor in the causation of "no reflow phenomenon" in the renal medulla immediately following the release of renal artery occlusion. However, the functional failure of the kidney subjected to an ischemic insult appears to result from tubular obstruction and backleak of glomerular filtrate through damaged, leaky tubular epithelia (Arendshorst *et al.*, 1975, 1976; Donohoe *et al.*, 1978; Tanner and Sophasan, 1976).

3.2. The Glycerol Model

The glycerol model of the ischemic type of ATN is induced by intramuscular or subcutaneous injection of 50% glycerol in rats and rabbits (Carroll *et al.*, 1965; Dach and Kurtzman, 1976; Finckh, 1957; Reineck *et al.*, 1980; Solez *et al.*, 1976; Suzuki and Mostofi, 1970a; Thiel *et al.*, 1967; Wardle and Wright, 1973). Vasospasm of the kidneys is accompanied by myoglobinuria and hemoglobinuria; these factors are thought to be important in the genesis of the lesions that subsequently develop. Early changes include the effacement of proximal tubule brush borders and deposition of fibrin in glomeruli. At 24 hr, many proximal tubules are necrotic, and distal nephron segments are obstructed by casts containing heme pigment. By this time, the glomeruli appear to be normal, but renal failure is well developed. When one examines the renal vasculature after rapidly fixing the kidney by rapid freezing or perfusion fixation, vasoconstriction of the interlobular arteries and afferent and efferent arterioles can be demonstrated (Venkatachalam *et al.*, 1976). Resolution of the disease occurs by regeneration of the epithelium and disappearance of casts from the tubules.

3.3. The Norephinephrine Model

Infusions of norepinephrine (0.75–1.0 μg/kg per min) directly into the renal arteries of dogs or rats cause severe renal ischemia and subsequent ATN (Cox *et al.*, 1974; Cronin *et al.*, 1978). There is severe necrosis of the proximal tubule and obstruction of the distal nephron by casts. The lesions are more severe than those encountered following ischemia caused by renal artery occlusion for the same duration. Two-hour infusions of the drug causes marked flattening of the glomerular epithelium with a corresponding reduction of the area of glomerular basement membrane covered by filtration slits. However, shorter duration of infusion (40 min) causes ARF with tubular necrosis and casts but no glomerular abnormalities (Cronin *et al.*, 1978).

3.4. The Mercuric Chloride Model

The nephrotoxic compound mercuric chloride, when injected either subcutaneously or intravenously, causes necrosis of proximal tubular epithelium; with low doses of this poison the necrosis is confined to the pars recta portion of the proximal tubule; higher doses damage other segments of the proximal tubule also (Biber *et al.*, 1968; Cuppage *et al.*, 1972; Cuppage and Tate, 1967; Ganote *et al.*, 1974; Gritzka and Trump, 1968; Oliver *et al.*, 1951; Solez *et al.*, 1976; Zimmerman *et al.*, 1977). Shortly after administration of mercuric chloride, early changes indicative of cell injury are observed. These include the following: vacuolation, dilated cisternae of the endoplasmic reticulum, dispersion of ribosomes, and focal loss of brush border microvilli. Subsequently, membrane systems appear to break down and mitochondria exhibit flocculent, electron-dense material in the matrix. Although by microdissection the mercuric chloride lesion does not usually exhibit tubulorrhexis (Oliver *et al.*, 1951), abnormal permeability of necrotic proximal tubular epithelium has been shown to occur in this model of ATN (Kriz, 1962). Following intravenous injection of trypan blue, kidneys were frozen and freeze substituted in alcohol. Necrotic cells showed permeation by the dye. Others have confirmed this observation by direct intratubular injections of dyes (Bank *et al.*, 1967; Steinhausen *et al.*, 1969).

3.5. The Uranyl Nitrate Model

Uranyl nitrate in doses of 5–20 mg/kg body weight injected intravenously produces consistent ARF (Blantz and Konnen, 1975; Flamenbaum *et al.*, 1972, 1976; Oliver, 1915; Stein *et al.*, 1975). Early after injection, there are morphologic changes in the proximal tubule cells similar to those seen following mercuric chloride injection, including cell swelling, vacuolation, and mitochondrial degeneration. Within 48 hr, there is extensive necrosis of the straight portion of the proximal tubules, often extending into the convoluted portions. Many casts are found in the distal nephron. Sloughing of necrotic tubular epithelium results in exposure of denuded tubular basement membrane. Between 24 and 48 hr following uranyl nitrate administration fusion of glomerular epithelial cell foot processes can be demonstrated, which may correlate with decrease in the glomerular ultrafiltration (Stein *et al.*, 1975; Flamenbaum *et al.*, 1976). However, glomerular podocytes are normal 2–6 hr following uranyl nitrate despite the manifestations of overt renal failure, implying that the glomerular lesion is not necessary for diminution of glomerular filtration at this stage (Blantz and Konnen, 1975; Flamenbaum *et al.*, 1976).

Recently, it has been reported that a marked reduction in the number and diameter of capillary endothelial fenestrae occurs in this model of ATN. The suggestion was made that this lesion might correspond to the ultrafiltration anomaly that can also be detected at this time (Avasthi *et al.*, 1979).

4. *In Vitro* Studies of Anoxia Cell Injury

Little is known about the cellular mechanisms that render cells unable to survive a period of ischemia, anoxia, or substrate deprivation. In fact, the precise cellular event that leads to irreversible ischemic death in any tissue of any species is unknown. In order for us to be able to understand the effects of ischemia or nephrotoxins on individual cell types, the environment of cells has to be strictly controlled. Studies such as these can be carried out *in vitro* with either isolated whole tubules or individual cell types.

In order to determine the effect of ischemia on water and electrolyte transport in various segments of the nephron, isolated tubules from different nephron segments of the rabbit were microperfused (Hanley, 1980). Ischemia reduced proximal convoluted tubule fluid reabsorption by 77%, and cortical proximal straight tubule fluid reabsorption by 88%. Ischemia also reduced the ability of the thick ascending limb of Henle's loop to lower perfusate chloride ion concentration by 60% and its diluting ability by 49%. In addition, the antidiuretic hormone-dependent osmotic water permeability of the cortical collecting tubule was similarly reduced by 59%. Morphologic alterations were noted only in the proximal segments. The conclusion drawn from these studies is that 60 min of ischemia impairs the transport capability of the proximal and distal nephron segments studied.

Studies performed on kidneys flushed with the oncotic agent 8% polyethylene glycol (PEG) and exposed to 45 min of renal artery clamping determined that renal function was protected and proximal tubule cells were prevented from swelling and blebbing (Frega *et al.*, 1979). The addition of 8% PEG to the extracellular fluids during ischemia balances the osmotic activity of the intracellular nondiffusable macromolecules and prevents cells from swelling (Wiggins, 1964; Robinson, 1971; Frega, 1979). The relationship between cell swelling and cell death remains unclear. Deprivation of cell energy supplies prevents extrusion of sodium ions, which accumulate with chloride and water in the cell, resulting in osmotic swelling of the cell (Leaf, 1956). Extracellular sodium normally balances the oncotic swelling pressure of obligatory intracellular colloids, but cell membranes are permeable to sodium ions. Thus the extracellular position of sodium is maintained by the continuous, energy-requiring active transport of sodium out of the cell by membrane-bound Na–K-ATPase. This establishes a steady state and stabilizes cell volume. When energy metabolism is inhibited by anoxia or poisons, this regulatory mechanism fails and cells swell.

Although the results obtained with PEG in the kidney mentioned above confirm the importance of cell swelling to ischemic renal injury, they do not clarify the nature of the relationship. Namely, cell swelling may obstruct renal tubules, increase vascular resistance, render cell membranes permeable to noxious extracellular and essential intracellular ingredients, or disrupt the spatial relationship of critical membrane-bound enzymes. For these reasons, we isolated and cultured renal proximal tubule cells *in vitro*, where the cell environment can be

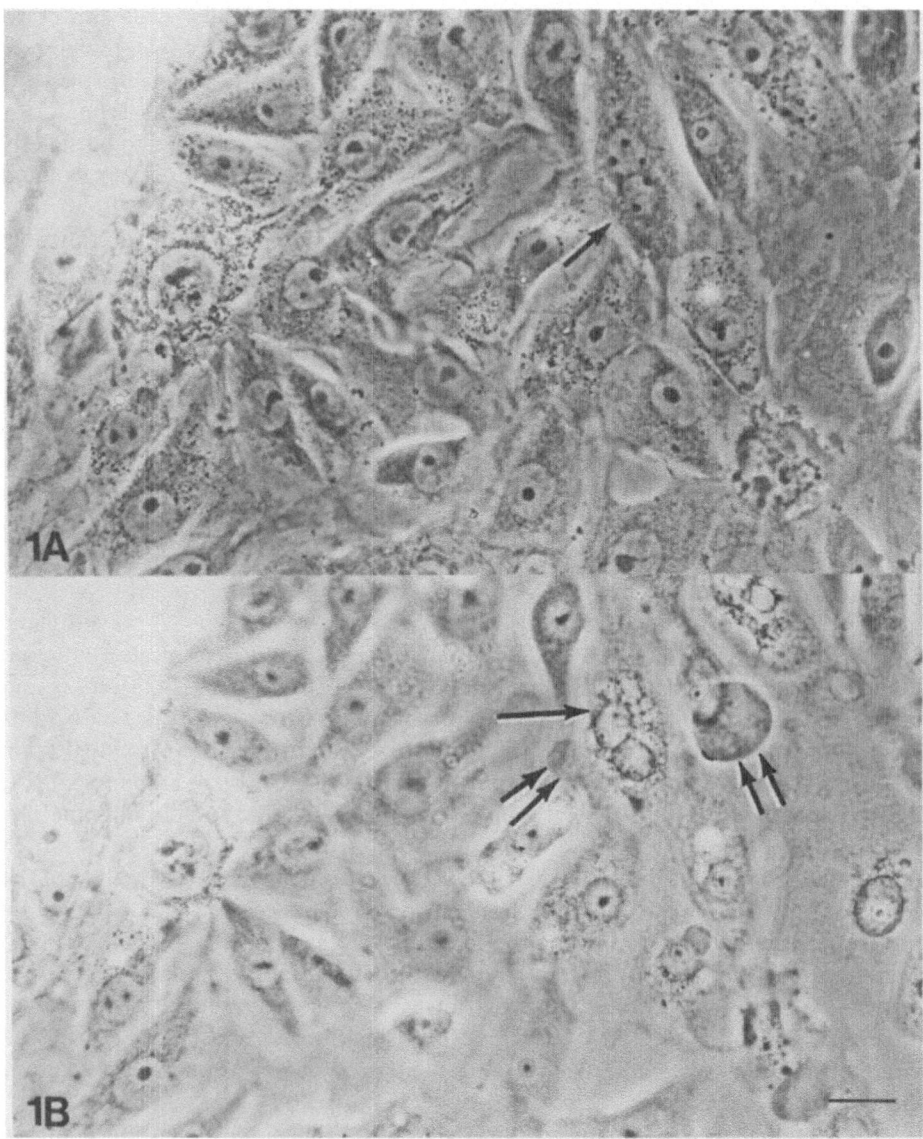

FIGURE 1. (A) Phase-contrast photomicrograph of proximal tubule epithelial cells grown in tissue culture prior to anoxic perfusion. (B) Same field as (A) after 2 hr of continuous perfusion of anoxic medium. Note the swollen appearance of the cells (single arrow points to the same cell in both pictures) and the numerous plasma membrane blebs (double arrows). Bar = 20 μm.

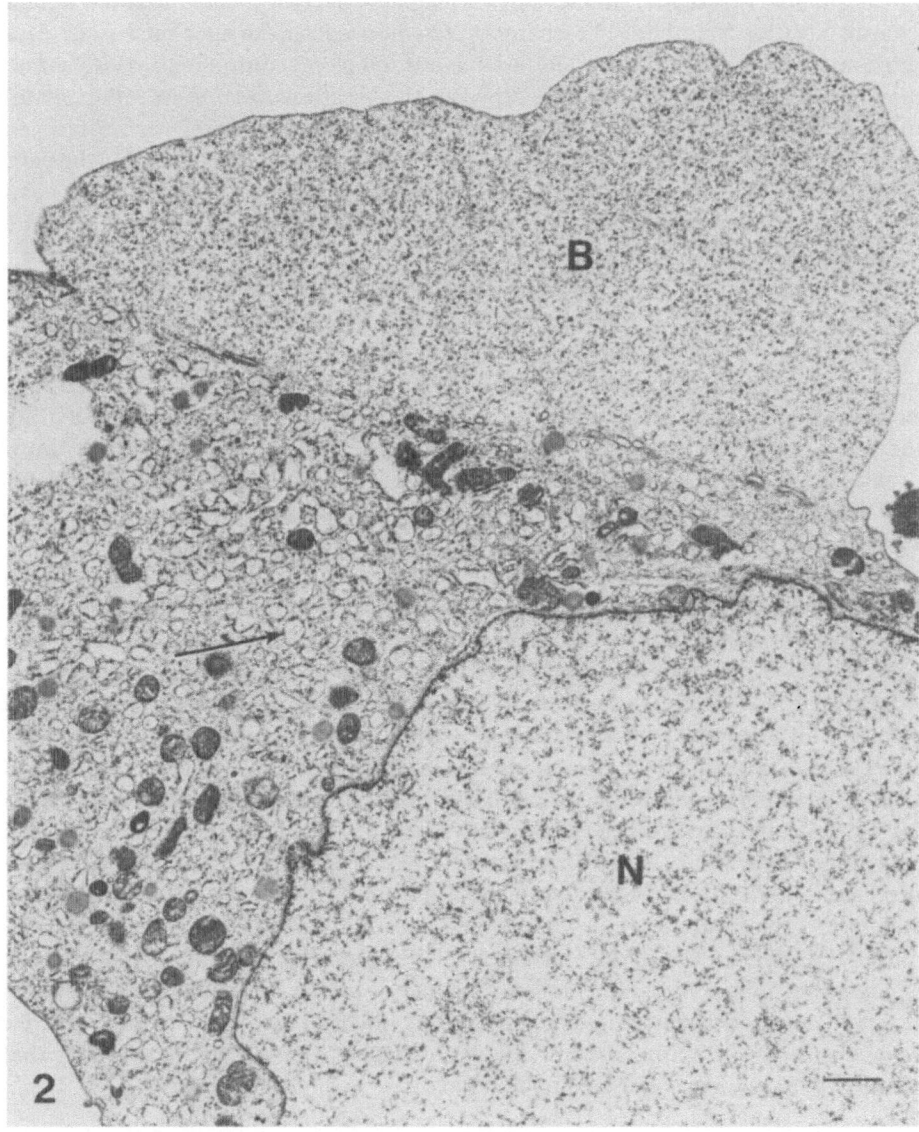

FIGURE 2. Transmission electron micrograph of a cultured proximal tubule epithelial cell after 2 hr of anoxia. Note the swollen appearance of this cell as evidenced by the swollen profiles of rough endoplasmic reticulum (arrow), and the presence of a plasma membrane bleb (B). N, Nucleus. Bar = 1 μm.

strictly controlled, in order to study the pathogenesis of anoxic injury (Kreisberg *et al.*, 1980).

Explants of proximal tubules grown on glass coverslips were mounted onto a Dvorak-Stottler chamber. The chamber was housed on the specimen stage of a photomicroscope. With this model, we have studied two components of ischemia, namely anoxia along with substrate deprivation. Anoxic, substrate-free phosphate-buffered saline was fed by gravity, and the effluent connected to an oxygen electrode. The advantages of this system are as follows: (1) The shape of the cells can be observed continuously and recorded by photography; (2) the perfusate can be readily modified for studying the pathogenesis of anoxia; and (3) the perfusate can be passed over the cells at low flow rates.

During 2 hr of anoxia in the absence of substrate, the cultured cells swelled and blebbed (Figs. 1 and 2). Cells similarly treated in the presence of the oncotic agent 8% PEG did not swell and bleb, and when they were counted 18 hr later (coverslips were returned to normal culture conditions), similar cell counts were obtained as in the untreated cultures. The tubule cells exposed to anoxia in the absence of PEG had 50% fewer cells 18 hr later. Therefore, if cell swelling is prevented during 2 hr of anoxia, cell viability is improved.

We have also observed a similar protection of cultured proximal tubule cells from anoxic cell death when the cells were exposed to anoxia in a low calcium medium (50 μM) (Kreisberg, unpublished observation). Cells treated under such conditions for 2 hr did not demonstrate plasma membrane blebbing, and cell viability approached controls. Additionally, cultured cells exposed to the calcium ionophore A23187 in the presence of calcium demonstrated cell swelling, blebbing, and cell death. The deleterious effects of increased cytosolic calcium have been described in other cell types as well as in other cell injury models (Schanne *et al.*, 1979). Calcium can activate plasma membrane phospholipases, which may result in membrane damage. Additionally, increased cytosolic calcium is associated with the destabilization of certain cytoskeletal structures such as microtubules and microfilaments (Borisy and Olmsted, 1972; Condeelis and Taylor, 1977; Keihart and Inoue, 1976). In this regard, cultured proximal tubule cells exposed to agents that disrupt microtubules (e.g., colchicine) or microfilaments (e.g., cytochalasin B) caused the cells to swell, bleb, and die (Kreisberg, unpublished data). Therefore, it appears that ionized calcium may play a role in proximal tubule cell injury and death *in vitro*. Although the exact nature of this role has yet to be defined, we believe that with this *in vitro* model of cell injury, the sequence of events that culminate in cell death will be elucidated.

References

Anderson, R. J., and Schrier, R. W., 1980, Clinical spectrum of oliguric and nonoliguric renal failure, in: *Contemporary Issues in Nephrology, Acute Renal Failure,* (B. M. Brenner and J. H. Stein, eds.), vol. 6, pp. 1–17, Churchill Livingston, New York.

Anderson, R. J., Linas, S. L., Berns, A. S., Henrich, W. L., Miller, T. R., Gabow, P. A., and Schrier, R. W., 1977, Nonoliguric acute renal failure, *N. Engl. J. Med.* **296**:1134.

Ansari, R. M., and Baldwin, D. S., 1976, Acute renal failure due to radiocontrast agents, *Nephron* **17**:28.

Arendshorst, W. J., Finn, W. F., and Gottschalk, C. W., 1975, Pathogenesis of acute renal failure following temporary renal ischemia in the rat, *Circ. Res.* **37**:558.

Arendshorst, W. J., Finn, W. F., and Gottschalk, C. W., 1976, Micropuncture study of acute renal failure following temporary renal ischemia in the rat, *Kidney Int.* **10**:S100.

Avasthi, P. S., Huser, J., and Evan, A. P., 1979, Glomerular endothelial cells in gentamicin induced acute renal failure in rats, *Kidney Int.* **16**:771.

Bank, N., Mutz, B. R., and Aynedjian, H. S., 1967, The role of "leakage" of tubular fluid in anuria due to mercury poisoning, *J. Clin. Invest.* **46**:695.

Barnes, J. L., Osgood, R. W., Reineck, H. J., and Stein, J. H., 1981, Glomerular alterations in an ischemic model of acute renal failure, *Lab. Invest.* **45**:378.

Biber, T. U. L., Mylle, M., Baines, A. D., Gottschalk, C. W., Oliver, J. R., and MacDowell, M. C., 1968, A study by micropuncture and microdissection of acute renal damage in rats, *Am. J. Med.* **44**:684.

Blantz, R. C., and Konnen, K., 1975, The mechanisms of acute renal failure after uranyl nitrate, *J. Clin. Invest.* **55**:621.

Bohle, A., Jahnecke, J., Meyer, D., and Schubert, G. E., 1976, Morphology of acute renal failure: Comparative data from biopsy and autopsy, *Kidney Int.* **10**:S9.

Borisy, G. G., and Olmsted, J. B., 1972, Nucleated assembly of microtubules in porcine brain extracts, *Science* **177**:1196.

Bushinsky, D. A., Wish, J. B., Hou, S. H., Cohen, J. I., and Harrington, J. T., 1979, Hospital acquired renal insufficiency (1978–1979), *Proc. Am. Soc. Nephrol.* **12**:105A.

Carroll, R., Kavacs, K. and Tapp, E., 1965, The pathogenesis of glycerol induced renal tubular necrosis, *J. Pathol. Bacteriol.* **89**:573.

Carvallo, A., Rakowsky, T. A., Argy, W. P., Jr., and Schreiner, G. E., 1978, Acute renal failure following drug infusion pyelography, *Am. J. Med.* **65**:38.

Condeelis, J. S., and Taylor, D. L., 1977, The contractile basis of ameboid movement. V. The control of gelation, solation, and contraction in extracts from *Dictyostelium discoideum, J. Cell Biol.* **74**:901.

Cox, J. W., Baehler, R. W., Sharma, H., O'Dorisio, T., Osgood, R. W., Stein, J. H., and Ferris, T. F., 1974, Studies on the mechanism of oliguria in a model of acute renal failure: Functional and histological correlates of protection, *Kidney Int.* **14**:115.

Cronin, R. E., 1978, Aminoglycoside nephrotoxicity: Pathogenesis and prevention, *Clin. Nephrol.* **11**:251.

Cronin, R. E., DeTorrente, A., Miller, P. D., Bulger, R. E., Burke, T. J., and Schrier, R. W., 1978, Pathogenetic mechanisms in early norepinephrine-induced acute renal failure: Functional and histological correlates of protection, *Kidney Int.* **14**:115.

Cuppage, F. E., and Tate, A., 1967, Repair of the nephron following injury with mercuric chloride, *Lab. Invest.* **26**:122.

Cuppage, F. E., Chiga, M., and Tate, A., 1972, Cell cycle studies in the regenerating rat nephron following injury with mercuric chloride, *Lab. Invest.* **16**:122.

Dach, J. L., and Kurtzman, N. A., 1976, A scanning electron microscopic study of the glycerol model of acute renal failure, *Lab. Invest.* **34**:409.

Dobyan, D. C., Nagle, R. B., and Bulger, R. E., 1977, Acute tubular necrosis in the rat kidney following sustained hypotension, *Lab. Invest.* **37**:411.

Donohoe, J. F., Venkatachalam, M. A., Bernard, D. B., and Levinsky, N. G., 1978, Tubular leakage and obstruction in acute ischemic renal failure, *Kidney Int.* **13**:208.

Finckh, E. S., 1957, Experimental acute tubular necrosis following subcutaneous injection of glycerol, *J. Pathol. Bacteriol.* **73**:69.

Flamenbaum, W., McNeil, J. S., Kotchen, T. A., and Saladino, A. J., 1972, Experimental acute renal failure induced by uranyl nitrate in the dog, *Circ. Res.* **31**:682.

Flamenbaum, W., Hamburger, R. J., Huddleston, M. L., Kaufman, J., McNeil, J. S., Schwartz, J. H., and Nagle, R., 1976, The initiation phase of experimental acute renal failure: An evaluation of uranyl nitrate-induced acute renal failure in the rat, *Kidney Int.* **10**:S115.

Flores, J., DiBona, D. R., Beck, C. H., and Leaf, A., 1972, The role of cell swelling in ischemic renal damage and the protective effect of hypertonic solute, *J. Clin. Invest.* **51**:118.

Frega, N. S., DiBona, D. R., and Leaf, A., 1979, The protection of renal function from ischemic injury in the rat, *Pfluegers Arch.* **381**:159.

Ganote, C. E., Reimer, K. A., and Jennings, R. B., 1974, Acute mercuric chloride nephrotoxicity, *Lab. Invest.* **31**:633.

Gritzka, T. L., and Trump, B. F., 1968, Renal tubular lesions caused by mercuric chloride. Electron microscopic observations: Degeneration of the pars recta, *Am. J. Pathol.* **52**:1225.

Hanley, M. J., 1980, Isolation of nephron segments in a rabbit model of ischemic acute renal failure, *Am. J. Physiol.* **239**:F17.

Heptinstall, R. H., 1974, *Pathology of the Kidney,* 2nd ed., Little, Brown and Co., Boston, Massachusetts.

Hoyer, J. R., and Seiler, M. W., 1979, Pathophysiology of Tamm-Horsfall protein, *Kidney Int.* **16**:479.

Kant, K. S., Pesce, A. J., Clyne, D. H., and Pollak, V. E., 1976, Urinary cast formation: pH dependence and interaction of Tamm-Horsfall mucoprotein with myoglobin, hemoglobin, Bence-Jones protein and albumin, *Kidney Int.* **10**:559.

Keihart, D. P., and Inoue, S., 1976, Local depolymerization of spindle microtubules by microinjection of calcium ions, *J. Cell Biol.* **70**:230a.

Kreisberg, J. I., Bulger, R. E., Trump, G. F., and Nagle, R. B., 1976, Effects of transient hypotension on the structure and function of rat kidney, *Virchows Arch. Pathol.* **22**:121.

Kreisberg, J. I., Mills, J. W., Jarrell, J. A., Rabito, C. A., and Leaf, A., 1980, Protection of cultured renal tubular epithelial cells from anoxic swelling and cell death, *Proc. Natl. Acad. Sci. USA* **77**:5445.

Kriz, W., 1962, Histophysiologische Untersunchungen en der Rattenniere bei Sublimatvergiftung, *Z. Zellforsch.* **57**:914.

Leaf, A., 1956, On the mechanism of fluid exchange of tissue *in vitro, Biochem. J.* **62**:241.

McQueen, E. G., 1962, The nature of urinary casts, *J. Clin. Path.* **15**:367.

Miller, T. R., Anderson, R. J., Linas, S. L., Henrich, W. L., Berns, A. S., Gabow, P. A., and Schrier, R. W., 1978, Urinary diagnostic indices in acute renal failure. A prospective study, *Ann. Intern. Med.* **89**:47.

Oliver, J., 1915, The histogenesis of chronic uranium nephritis with special reference to epithelial regeneration, *J. Exp. Med.* **21**:425.

Oliver, J., MacDowell, M., and Tracy, A., 1951, The pathogenesis of acute renal failure associated with traumatic and toxic injury. Renal ischemia, nephrotoxic damage and the ischemuric episode, *J. Clin. Invest.* **30**:1305.

Olsen, S., 1976, Renal histopathology in various forms of acute anuria in man, *Kidney Int.* **10**:S2.

Porter, G. A., and Bennett, W. M., 1980, Nephrotoxin-induced acute renal failure, in: *Contemporary Issues in Nephrology,* Volume 6, *Acute Renal Failure* (B. M. Brenner and J. H. Stein, eds.), Churchill Livingston, New York, p. 123.

Reineck, H. S., O'Connor, G. I., Lifschitz, M., and Stein, J. H., 1980, Sequential studies on the pathophysiology of glycerol-induced renal failure, *J. Lab. Clin. Med.* **96**:356.

Robbins, S. L., and Cotran, R. S., 1979, *Pathologic Basis of Disease,* 2nd ed., Saunders, Philadelphia, Pennsylvania, p. 1149.

Robinson, J. R., 1971, Control of water content of non-metabolizing kidney slices of sodium chloride and polyethylene glycol (PEG 6000), *J. Physiol. (Lond.)* **213**:227.

Schanne, F. A. X., Kane, A. B., Young, E. E., and Farber, J. L., 1979, Calcium dependence of toxic cell death: A final common pathway, *Science* **206**:700.

Schreiner, G. E., and Maher, J. F., 1965, Toxic nephropathy, *Am. J. Med.* **38**:409.

Smolens, P., and Stein, J. H., 1980, Hemodynamic factors in acute renal failure: Pathophysiologic and therapeutic implications, in: *Contemporary Issues in Nephrology,* Volume 6, *Acute Renal Failure* (B. M. Brenner and J. H. Stein, eds.), Churchill Livingston, New York, p. 180.

Smolens, P., and Stein, J. H., 1981, Pathophysiology of acute renal failure, *Am. J. Med.* **70**:479.

Smolens, P., Reineck, H. J., Venkatachalam, M., and Stein, J. H., 1981, Studies on the pathogenesis of renal failure in a rat model of multiple myeloma, *Kidney Int.* **19**:214A.

Solez, K., Altman, J., Reinhoff, H. Y., Riela, A. F., Finer, P. M., and Heptinstall, R. H., 1976, Early angiographic and renal blood flow changes after HgCl$_2$ or glycerol administration, *Kidney Int.* **10**:S153.

Solez, K., Morel-Maroger, L., and Sraer, J. D., 1979, The morphology of "acute tubular necrosis" in man: Analysis of 57 renal biopsies and a comparison with the glycerol model, *Medicine* **58**:362.

Stein, J. H., Gottschalk, J., Osgood, R. W., and Ferris, T. F., 1975, Pathophysiology of a nephrotoxic model of acute renal failure, *Kidney Int.* **8**:27.

Steinhausen, M., Eisenbach, G.-M., and Helmstadter, V., 1969, Concentration of lissamine green in proximal tubules of antidiuretic and mercury poisoned rats and the permeability of these tubules, *Pfluegers Arch.* **311**:1.

Summers, W. K., and Jamison, R. L., 1971, The no reflow phenomenon in renal ischemia, *Lab. Invest.* **25**:635.

Suzuki, T., and Mostofi, F. K., 1970a, Electron microscopic studies of acute tubular necrosis. Early changes in the glomeruli of rat kidney after subcutaneous injection of glycerin, *Lab. Invest.* **23**:8.

Suzuki, T., and Mostofi, F. K., 1970b, Electron microscopic studies of acute renal necrosis. Vascular changes in the rat kidney after subcutaneous injection of glycerin, *Lab. Invest.* **23**:29.

Swartz, R. D., Rubin, J. E., Leeming, B. W., and Silva, P., 1978, Amphotericin B nephrotoxicity with irreversible renal failure, *Ann. Intern. Med.* **59**:716.

Tanner, G. A., and Sophasan, S., 1976, Kidney pressures after temporary renal artery occlusion in the rat, *Am. J. Physiol.* **230**:1173.

Thiel, G., Wilson, D. R., Arce, M. L., and Oken, D. E., 1967, Glycerol-induced hemoglobinuric acid renal failure in the rat. II. The experimental model, predisposing factors, and pathophysiologic features, *Nephron* **4**:175.

Venkatachalam, M. A., Rennke, H. G., and Sandstrom, D. J., 1976, The vascular basis for acute renal failure in the rat: Preglomerular and postglomerular vasoconstriction. *Circ. Res.* **38**:267.

Venkatachalam, M. A., Bernard, D. B., Donohoe, J. F., and Levinsky, N. G., 1978, Ischemic damage and repair in the rat proximal tubule. Differences among the S_1, S_2, and S_3 segments. *Kidney Int.* **14**:31.

Wardle, E. N., and Wright, N. A., 1973, Intravascular coagulation and glycerin hemoglobinuric acute renal failure, *Arch. Pathol. Lab. Med.* **95**:271.

Wiggins, P. M., 1964, Selective accumulation of potassium ion gel and kidney slices, *Biochim. Biophys. Acta* **88**:593.

Zimmerman, H. D., Schmidt, E., Weller, E., Becker, C., and Dicken, P., 1977, Intra and extrarenal vascular changes in the acute renal failure of the rat caused by mercuric chloride, *Virchows Arch. Pathol. Anat.* **372**:259.

3

Effect of Heavy Metals on Sodium Transport *in Vitro*

JOHN H. SCHWARTZ

1. Introduction

Heavy metal salts have been extensively used to study the pathophysiology of acute renal failure. There have been numerous studies which have carefully characterized the alterations in renal functions, renal hemodynamics, and vasoactive substances in heavy metal-induced acute renal failure. Based on these studies (Flamenbaum *et al.,* 1974), it was proposed that activation of a tubuloglomerular feedback mechanism may be an important mediator for the initiation of acute renal failure. In this proposal it is assumed that heavy metal salts (uranyl nitrate or mercuric chloride) initially interact with epithelial cells of the renal tubule so as to reduce fluid and electrolyte transport primarily in early nephron segments. As a consequence of the altered transport function in the early segments, the composition of the fluid delivered to the distal nephron and macula densa is changed. The changed fluid composition results in increased renin release and thereby activation of tubuloglomerular feedback (Schnermann *et al.,* 1970). This latter step results in the observed decrease in glomerular filtration rate and changes in renal hemodynamics. The renal hemodynamic changes in turn will tend to

JOHN H. SCHWARTZ • Renal Section, Department of Medicine, Thorndike Memorial Laboratory, Boston City Hospital and Boston University School of Medicine, Boston, Massachusetts 02118.

augment and perpetuate the initial tubular dysfunction so that the usually negative feedback mechanisms for tubuloglomerular feedback do not become operative.

Although abnormalities in tubular function are observed early in acute renal failure, it is uncertain if these functional changes are primary events as described above or occur as a consequence of altered renal hemodynamics or other hormonal perturbations. The studies to be presented were designed to characterize the effects of uranyl nitrate and mercuric chloride on the transport function of urinary epithelial cells. In order to avoid the potential changes in epithelial function which could be associated with perturbations in systemic factors or renal hemodynamics, the effect of these heavy metal salts was examined in an *in vitro* preparation, the isolated urinary bladder of the freshwater turtle.

2. The *in Vitro* Preparation

The urinary bladder of the freshwater turtle *(Pseudemys scripta)*, a mesonephric derivative, resembles in functional terms the mammalian distal nephron. This urinary epithelium *in vitro* has the capacity to reabsorb sodium, secrete hydrogen ions, and respond to aldosterone and vasopressin in an appropriate manner (Steinmetz, 1974). The transport system for sodium depends on passive entry of sodium across the luminal membrane at specific sites within this membrane. This process may be carrier-mediated. This step is inhibitable by amiloride. Extrusion of sodium out of the cell at the basolateral surface is an active process, dependent on a Na–K-ATPase, and is inhibitable by ouabain.

2.1. Methodology

Urinary bladders were mounted in plastic chambers and both sides were bathed in an identical Ringer's solution (Schwartz and Flamenbaum, 1976). Because of the presence of electrogenic active transport processes, the bladder generates a spontaneous electrical potential difference (PD) that is normally oriented such that the serosal (body fluid) side is positive with respect to the mucosal (urinary) side. This PD was continuously nullified by the application of an external short-circuit current (SCC) by means of an automatic voltage clamp. The SCC was recorded on a potentiometric recorder except for brief intervals when the PD and transepithelial resistance were measured. The net rate of sodium transport was either estimated from the SCC or measured with isotopic unidirectional fluxes using ^{24}NaCl and ^{22}NaCl. Rates were expressed as microequivalents per hour per 8 cm^2 (μeq/hr/8 cm^2) or converted to microamperes (μA) in accordance with Faraday's law.

3. Effect of Heavy Metals on Membrane Function

3.1. Electrophysiologic Effects

Before we examined in detail the effects of uranyl nitrate and mercuric chloride on sodium transport, a dose–response curve was determined for these salts by examining their effect on SCC. The SCC can be used as an approximate measure of net sodium transport. With as little as 1×10^{-7} M uranyl nitrate in the mucosal solution, the SCC was inhibited by $12 \pm 2\%$. The degree of inhibition of SCC progressively increased as the concentration of uranyl nitrate was increased, to a maximum value of $69 \pm 6\%$ with 1×10^{-4} M uranyl nitrate. In contrast to the marked effect of uranyl nitrate on SCC, there was no significant alteration in transepithelial resistance. The constancy of transepithelial resistance with inhibition of SCC by uranyl nitrate would indicate that the epithelium remained intact as a barrier to the passive movement of small ions despite the effect of this agent on active transport. Serosal addition of uranyl nitrate had no measurable effects on SCC or PD even at concentrations as high as 1×10^{-2} M.

Based on the dose–response relationship, the minimal concentration of uranyl nitrate (1×10^{-4} M) that reduced SCC to a maximal degree (70%) was studied in further detail. The time course of SCC inhibition after mucosal addition of uranyl nitrate to a final concentration of 1×10^{-4} M is shown in Fig. 1. After the addition of uranyl nitrate the SCC declined rapidly from a stable value of 250 μA, achieving a minimum value of 65 μA within 20 min. Thereafter, there was no further change in SCC whether or not uranyl ions were washed out of the bulk solution.

Additional studies were also performed to examine the effect of a second heavy metal salt, mercuric chloride. Concentrations of mercuric chloride ranging from 1 $\times 10^{-8}$ to 1×10^{-3} M were added to the mucosal solution bathing freshly prepared turtle bladders. The effect of this metal was similar to uranyl nitrate. The inhibition

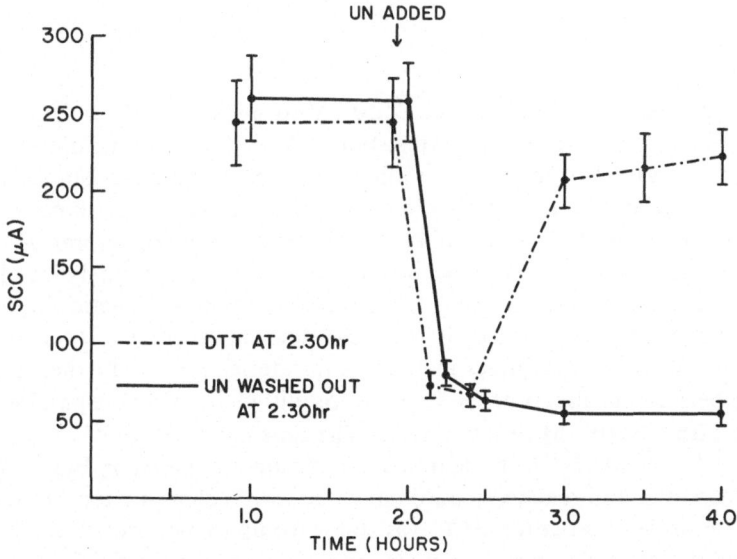

FIGURE 1. Time course of uranyl nitrate (UN) inhibition of SCC and reversal by dithiothreitol (DTT). The UN, 0.1 mM, was added to the mucosal solution at 2.0 hr. After SCC achieved at stable value, UN was removed by washing and 2 mM DTT was added to one group. Values are means ± SEM. [From Schwartz and Flamenbaum (1976).]

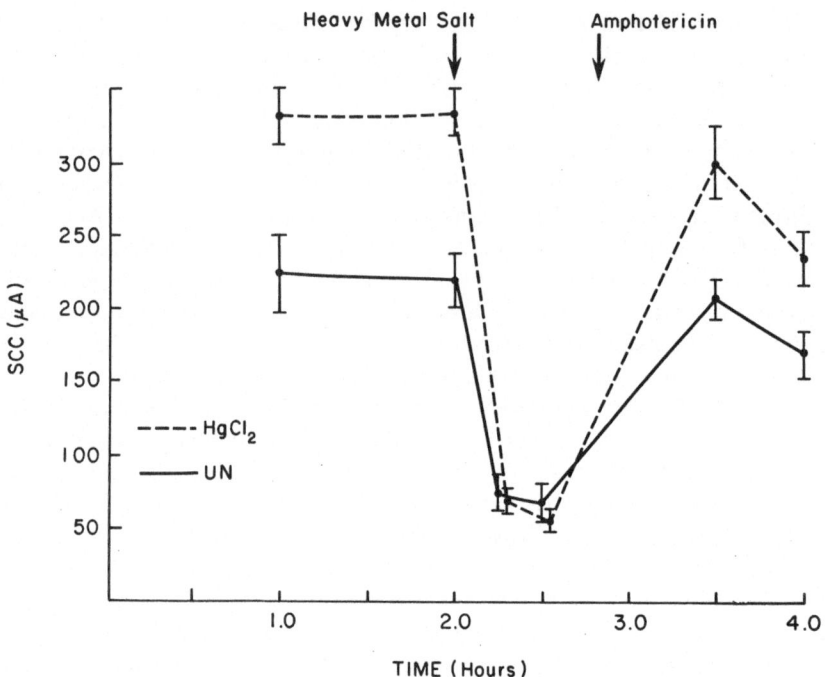

FIGURE 2. Time course of uranyl nitrate, 0.1 mM, and $HgCl_2$, 10 μ M, inhibition of SCC and reversal by amphotericin B, 20 μg/ml. These reagents were added to the mucosal solution. Values are means $\pm$ SEM. [From Schwartz and Flamenbaum (1976).]

of SCC increased from about 10% at 1×10^{-8} M to 80% at 1×10^{-5} M $HgCl_2$ without significant change in transepithelial resistance. At higher concentrations of $HgCl_2$, although there was a further degree of inhibition of SCC, tissue resistance markedly declined. For example, at 1×10^{-3} M $HgCl_2$ transepithelial resistance decreased from 1.4×10^3 Ω cm^{-2} to less than 0.4×10^3 Ω cm^{-2}. At this concentration of $HgCl_2$ the bladder was also observed to become opaque and was extremely friable. These results are consistent with direct and irreversible tissue destruction by mercuric chloride at concentrations of 1×10^{-3} M or greater that preclude further meaningful evaluation. The concentration of mercuric chloride selected for further study was 1×10^{-5} M because its effects on SCC and transepithelial resistance paralleled those observed with 1×10^{-4} M uranyl nitrate. The time course of SCC inhibition produced by 1×10^{15} M $HgCl_2$ in the mucosal solution is depicted in Fig. 2. The SCC rapidly decreased and reached a minimal value within 20 min after the addition of mercuric chloride. Thereafter, SCC was stable for up to 3 hr (not shown in Fig. 2).

Mercuric chloride, unlike uranyl nitrate, also affects SCC after serosal exposure. After the addition of 1×10^{-5} M mercuric chloride to the serosal solution, there was a progressive decline in SCC (Fig. 3). The time course of this decline was significantly slower, -1.0 A/min, than after mucosal exposure, -7 A/min.

Furthermore, a stable value for SCC was never achieved after serosal addition of mercuric chloride.

It would appear from these initial studies that the inhibition of SCC, which probably represents active sodium transport, results from an interaction of these two heavy metal salts with specific elements in the luminal membrane. This proposal is based on the observations that (a) the onset of action occurs rapidly only after mucosal addition, (b) serosal addition has either no effect (uranyl nitrate) or a diminished effect with a prolonged and slow time constant (mercuric chloride), and (c) there is no significant change in the transepithelial resistance, an observation consistent with the notion that the passive ionic conductance of the tissue remains unchanged.

3.2. Alteration in Ion Transport

The SCC in turtle bladder is determined by the sum of two opposing electrogenic transport systems, hydrogen ion secretion and sodium absorption. In addition, this tissue has an electrically silent transport system for chloride, active chloride–bicarbonate exchange (Leslie *et al.,* 1973). To fully characterize the changes in ion transport induced by these heavy metal salts, we also determined their effect on unidirectional fluxes of $^{22}Na^+$ and $^{24}Na^+$ (Table I). Prior to the addition of uranyl nitrate the net rate of sodium transport was 7.5 μeq/hr. After the

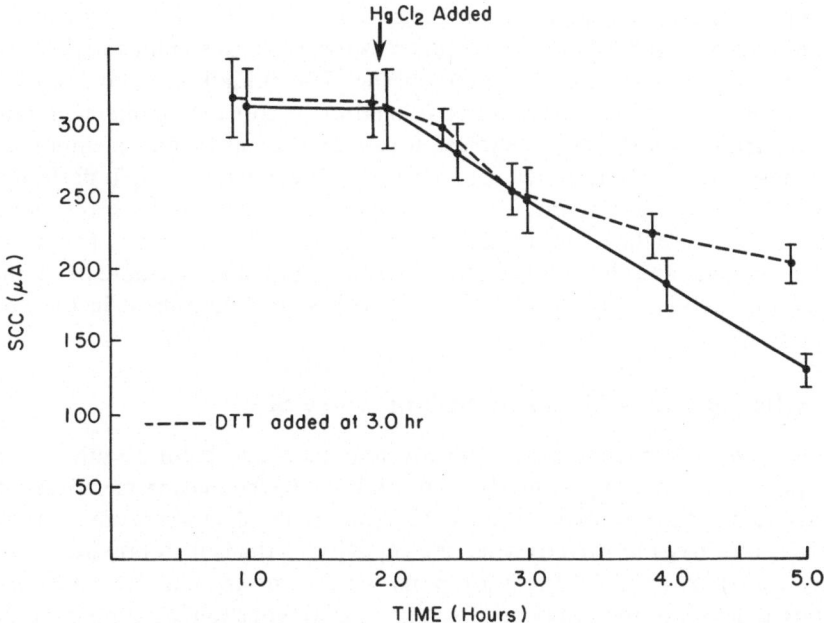

FIGURE 3. Effect of serosal addition of HgCl₂, 10 μM, on SCC. After 1 hr dithiothreitol, 2 mM, was added to the serosal and mucosal solution of one group of bladders. Values are means ± SEM. [From Schwartz and Flamenbaum (1976).]

TABLE I. Effect of Heavy Metal Salts on Na⁺ Transport across Turtle Urinary Bladder[a]

| | Sodium fluxes | | | |
	$J(m \rightarrow s)$	$J(s \rightarrow m)$	J(net)	SCC
Control	9.2 ± 2.1	1.7 ± 1.1	7.5 ± 2.1	6.4 ± 3.1
Uranyl nitrate 1×10^{-4} M	3.9 ± 1.8^{b}	1.6 ± 1.1	2.3 ± 1.0^{b}	1.6 ± 1.1^{b}
Control	10.7 ± 3.1	1.5 ± 1.6	9.2 ± 1.9	7.8 ± 0.9
HgCl₂ 1×10^{-5} M	3.8 ± 1.9^{b}	1.6 ± 1.4	2.2 ± 0.7^{b}	1.2 ± 0.5^{b}

[a] Uranyl nitrate and HgCl₂ were added to the mucosal solution to the final concentration indicated. Rates for sodium fluxes and SCC are means $\pm$ SEM expressed as μeq/hr/8 cm².
[b] $P < 0.05$ ($n = 6$).

addition of uranyl nitrate this net flux declined by approximately 5 μeq/hr, a change not different from the change in the SCC. Thus, the change in SCC with uranyl nitrate addition is an accurate reflection of the change in sodium transport. The decrease in net sodium transport resulted from a decrease in the mucosa to serosa flux (the active component) without a significant alteration in the passive, serosa to mucosa flux of this ion. Similar observations were also made after tissues were exposed to mercuric chloride (Table I). The decline in SCC in these mercuric chloride-treated tissues was caused by a decrease in the active component of sodium transport without measurable change in the passive flux.

These observations confirm our initial suggestion that both uranyl nitrate and mercuric chloride inhibit active sodium transport without significantly altering the passive ionic conductance of the membrane. This inhibitory action could result either from an effect on the entry step for sodium across the luminal membrane or by interfering with the active extrusion step at the basolateral membrane. This latter possibility could result from an alteration in cellular metabolic pathways that provide ATP for the sodium pump or with the pump itself (Na–K-ATPase). However, because the inhibitory effect occurs primarily after mucosal exposure to the heavy metal and the onset of action is quite rapid, it seems more likely that the site of interaction of the heavy metal is with some component in the mucosal membrane.

3.3. Effect of Heavy Metals on Sodium Entry Site

It is proposed that uranyl nitrate and mercuric chloride inhibit sodium transport by interacting primarily with some site in the plasma membrane on the mucosal side of these polarized cells. This interaction decreases sodium entry into the cell, but may not directly alter the capacity of the cell to provide energy for active transport or affect the antiluminal sites involved in active transport. The proposed action of the heavy metals may be analogous to the action of amiloride. Thus, after mucosal exposure to heavy metal, the rate of sodium entry is decreased, but because the active exit step is not directly affected, transport will continue until it is limited by a diminished intracellular sodium pool. If this proposal for the action of heavy metals is correct, a maneuver that would increase the luminal sodium conductance should reverse the inhibition of sodium transport by heavy metals.

Amphotericin B has been demonstrated to increase the ionic conductance of the mucosal barrier of amphibian and reptilian urinary bladders to cations (Finn, 1968; Lichtenstein and Leaf, 1969; Steinmetz and Lawson, 1970). Amphotericin B may create pores in a membrane by its molecular orientation in the lipid phase of the membrane (Andreoli, 1973). In biologic membranes the increased permeability is often cation-selective, as has been demonstrated in turtle bladder (Steinmetz and Lawson, 1970). A reversal of the inhibition of sodium transport by heavy metals after the mucosal addition of amphotericin B would lend support to the hypothesis that these heavy metals decrease transport by their effect on apical sodium conductance.

The effect of 20 μg/ml amphotericin B added to the mucosal solution 20 min after mucosal exposure to 10^{-4} M uranyl nitrate or 10^{-5} M mercuric chloride resulted in a reversal in the inhibition of SCC. These results are depicted in Fig. 2. The SCC increased to a value not different from the initial control values within 1 hr after the addition of amphotericin B. To demonstrate that there was equivalence between the alterations in SCC and net sodium transport after exposure to both heavy metals and amphotericin B, bidirectional sodium fluxes were also determined. The results of these studies are presented in Table II. The bidirectional fluxes under these experimental conditions demonstrated that the increment in SCC is matched by an increment in net sodium transport. The stimulation of sodium transport by amphotericin B in bladders previously exposed to heavy metals demonstrates that the cellular capacity to transport sodium was not directly altered by heavy metals. The action of amphotericin B is consistent with the hypothesis that heavy metals reduce transport by inducing a change in the conductance of the apical membrane to sodium.

3.4. Heavy Metal–Apical Membrane Chemical Interaction

One important consideration is the nature of the interaction of the heavy metal with some constituent of the apical membrane. One possibility is that this heavy metal ion interacts with the sodium entry site in a manner analogous to that of amiloride or guanidinium. If this were the case, removal of the ion from the bulk solution should rapidly reverse the observed inhibition. After ½ hr exposure of the mucosal side of the tissue to uranyl nitrate, this heavy metal salt was removed from the bulk mucosal solution by repeatedly washing the tissue with fresh, uranyl nitrate-free Ringer's. As illustrated in Fig. 1, this maneuver did not reverse the inhibition of sodium transport. Similar results were also obtained with mercuric chloride. Therefore, the interaction of the heavy metal ion with the membrane binding site must involve some chemical interaction other than a concentration-dependent affinity.

There is considerable evidence that mercuric chloride is preferentially bound by sulfhydryl group of intact tissues (Rothstein, 1959; Webb, 1966). It is reasonable to consider the possibility that specific membrane sulfhydryl groups of the turtle

TABLE II. Effect of Amphotericin B on Na Transport across Turtle Urinary Bladder Exposed to Heavy Metal Salts[a]

	Sodium fluxes			
	$J(m \rightarrow s)$	$J(s \rightarrow m)$	$J(net)$	SCC
Uranyl nitrate 0.1 mM in M[b]	4.11 ± 0.11	1.50 ± 0.14	2.67 ± 0.21	2.57 ± 0.30
Uranyl nitrate plus amphotericin B 20 μg/ml in M[b]	9.12 ± 0.21	3.04 ± 0.17	6.08 ± 0.22	5.72 ± 0.46
HgCl$_2$ 10^2 μM in M[b]	4.10 ± 0.37	1.60 ± 0.11	2.54 ± 0.27	2.68 ± 0.10
HgCl$_2$ plus amphotericin B[b] 20 μg/ml in M	13.71 ± 1.23	3.21 ± 0.38	10.50 ± 0.80	11.00 ± 1.00

[a] Mucosal to serosal fluxes $J(m \rightarrow s)$ and serosal to mucosal fluxes $J(s \rightarrow m)$ were measured simultaneously. Rates are means ± SEM expressed as μeq/hr/8 cm^2.
[b] Mucosal solution.

bladder epithelium are the binding sites for both uranyl nitrate and mercuric chloride. To evaluate this possibility, the effects of dithiothreitol on SCC of bladder exposed to heavy metals were examined. Dithiothreitol is a dithiol sugar which is capable of complexing heavy metals, maintaining monothials in a reduced state, and maintaining certain radicals in a reduced state (Cleland, 1964). The effect of uranyl nitrate on SCC was rapidly and completely reversed by the addition of 2.0 mM dithiothreitol to the mucosal solution (Fig. 1). The addition of dithiothreitol to the serosal solution of bladder exposed to mucosal uranyl nitrate or dithiothreitol to the mucosal solution of control bladders had no effect on SCC.

Dithiothreitol had similar effects on the SCC in bladders treated with mucosal exposure to mercuric chloride. Since serosal addition of mercuric chloride also altered SCC, we examined the subsequent effect of this dithiol sugar in this circumstance (Fig. 3). Unlike the rapid reversal in SCC observed after mucosal exposure, dithiothreitol had only minimal effects. Presumably this difference is caused by entry of mercuric ions into the cells with inhibition of multiple enzyme systems (Schwartz and Flamenbaum, 1976).

The rapid reversibility of the effects of mucosal mercuric chloride and uranyl nitrate on SCC by dithiothreitol strongly suggests that there are heavy metal–sulfhydryl interactions in the apical membrane. Similar conclusions were made by other investigators in different biologic systems (Rothstein, 1959; Barron et al., 1948). However, uranyl ions can also form complexes with hydroxyls and phosphates. To demonstrate that uranyl ions can also form complexes with sulfhydryl groups, the ability of dithiol sugars or cysteine to solubilize uranyl-hydroxyl precipitates was examined. Both dithiothreitol and dithioerythritol, as well as cysteine, solubilize these precipitates. However, threitol, erythritol, and cystine did not solubilize these precipitates. These studies demonstrate that the sulfhydryl group can react with uranyl ions. Thus, the reversal of uranyl nitrate-induced inhibition of SCC by dithiothreitol is consistent with the suggestion that uranyl ions may also react with sulfhydryl groups in the apical membrane.

4. Analysis of the Action of Heavy Metals on Transport

It is difficult to predict the specific mode of action of heavy metals on the transport function of intact cells, even though these metals are known potent inhibitors of numerous enzymes (Webb, 1966). Several studies, however, suggest that the initial effect of heavy metals on transport function does not result from an effect on intracellular enzyme systems, but rather results from heavy metal interaction with cellular plasma membrane proteins (Barron *et al.,* 1948; Demis *et al.,* 1954; Rothstein, 1959). The major question that needs to be explored in the series of studies presented in this chapter is whether or not uranyl nitrate and mercuric chloride exert their inhibiting action at a membrane site, at an intracellular site, or at both sites.

The interpretation of the inhibitory action of these heavy metals on net sodium transport must account for the observations that: (a) the onset of action begins rapidly after mucosal, but not after serosal, addition of heavy metal; (b) dithiothreitol rapidly reverses heavy metal inhibition of SCC only after mucosal addition; and (c) agents such as amphotericin B which increase apical membrane cation conductance also reverse the inhibitor effect. An interpretation which I think best accounts for these observations would be that uranyl and/or mercuric ions interact initially with specific sodium entry sites in the apical membrane. These sites function as if they are in series with the antiluminal transport sites (Reuss and Finn, 1975). A decrease in conductance of the apical entry site leads to rapid changes in active transport primarily through a decrease in the transportable pool of sodium in the cells. Amphotericin B alters this apparent coupling by allowing the entry of sodium into the pool through alternative sites in the apical membrane. The action of amphotericin B in heavy metal-treated bladder further demonstrates the capacity of the cell to transport. Once sodium is made available to the cell by the permeability effect of amphotericin B, transport is restored. Finally, it would appear that the reactive chemical groups in the membrane with which these heavy metals complex are sulfhydryl groups. The action of dithiothreitol is consistent with this proposal and dependent on the restoration of free sulfhydryl in the membrane and the formation of soluable, more stable cyclic metal complexes by these dithiol sugars.

The data available are also inconsistent with the proposal that a major interaction occurs between heavy metals and intracellular constituents. If significant intracellular accumulation of uranyl or mercuric ions did occur, most metabolic pathways that provide energy for transport and cell maintenance would be disrupted. Epithelial integrity would be lost and membrane ionic conductance would increase. The fact that passive sodium permeability remained unchanged, as did the electrical resistance, is important evidence against such a cellular effect. The rapid reversal of the heavy metal inhibition of transport by dithiothreitol is also inconsistent with this proposal. In addition, the action of amphotericin B demonstrates that these cells are capable of generating sufficient metabolic energy to maintain high levels of transport. Thus, a major intracellular inhibitory effect could not have occurred after limited exposure to these heavy metals. However, after prolonged exposure mercuric chloride may gain access to the interior of cells. For example, in muscle cells obtained from rat diaphragm, mercuric chloride may have an effect on plasma membranes as well as an effect on cellular enzymes after prolonged exposure to this metal ion (Rothstein, 1959). The initial rapid accumulation of mercuric chloride in these muscle cells, probably at plasma membrane sites, results in the inhibition of glucose transport into the cell. This initial phase of mercuric chloride accumulation, and its physiologic consequences, can be completely reversed by cysteine. A second, slower component of mercuric chloride accumulation, which occurs after prolonged exposure, eventually leads to an inhibition of oxidative respiration. Cysteine did not reverse these latter effects. In turtle bladders similar observations were made. After brief exposure (up to 1 hr) to mercuric chloride, the physiologic effects could be reversed by dithiothreitol, but after prolonged mucosal or serosal exposure, they could not be reversed with dithiothreitol. These results demonstrate that mercuric chloride exerts its effects on ion transport initially by an apical membrane interaction, but it does eventually gain access into the cells and alters intracellular enzymes.

The most likely site of initial heavy metal–membrane interaction is at the luminal or apical membrane of the epithelial cells. I would also postulate that complex formation with the heavy metal and sulfhydryl groups of the macromolecules associated with sodium transport occurs. Although we have not yet characterized the effect of this interaction on these macromolecules, some analogies with other systems may help define the effect of such an interaction. For example, mercuric chloride lowers the O_2 affinity of hemoglobin, whereas organic mercurials, such as para-chloromercuribenzoate (PCMB), do not alter O_2 affinity (Riggs and Walbach, 1956). The differential effect of inorganic versus organic mercurials may result from the capacity of inorganic mercurials to form polymercaptides; organic mercurials can only form monomercaptides with the target protein. In the case of the heme molecule, the spatial arrangements of the sulfhydryl group are such that bridge formation can occur in the presence of inorganic mercurials, but not with organic mercurials (Riggs, 1959). The bridge formation may alter the tertiary structure of the heme molecule and thereby alter its O_2 affinity. In toad and turtle bladders, PCMB does not inhibit sodium transport (Frenkel et al., 1975). This is in contrast to the demonstrated effect of mercuric chloride on sodium transport. It is suggested that mercuric chloride and perhaps uranyl nitrate alter the affinity of the sodium carrier or channel in the apical membrane by forming multiple heavy metal (captide) bridges with sulfhydryl groups that change the tertiary structure of the reactive site. The consequence of the interaction is inhibition of transepithelial ion transport by decreasing the conductance of the apical membrane to sodium. These changes in membrane function could also be initiated by other events which could alter the redox state of the cell, such as anoxia. Thus, the specific functional changes induced by heavy metals may be representative of functional alterations that could occur with a diverse set of pathophysiologic conditions which might lead ultimately to acute renal failure.

Acknowledgments

The author greatly appreciates the administrative and secretarial assistance of Mary Shea.

This work was supported in part by the Walter Reed Army Institute of Research and National Institutes of Health, Grant AM 20611. During the tenure of part of this work, the author was an Established Investigator of the American Heart Association with funds contributed in part by American Heart Association, Massachusetts Affiliate.

References

Andreoli, T. E., 1973, On the anatomy of amphotericin B-cholesterol pores in lipid bilayer membrane, *Kidney Int.* **4**:337.

Barron, E. S. G., Muntz, J. A., and Gasvoda, B., 1948, Regulatory mechanisms of cellular respiration. I.: The role of cell membranes; uranium inhibition of cellular respiration, *J. Gen. Physiol.* **32**:163.

Cleland, W. W., 1964, Dithiothreitol, a new protective agent for SH groups, *Biochemistry* **3**:480.

Demis, D. J., Rothstein, A., and Meier, R., 1954, The relationship of the cell surface to metabolism, X: The location and function of envertase in the yeast cell, *Arch. Biochem.* **48**:55.

Finn, A. L., 1968, Separate effects of sodium and vasopressin on the sodium pump in toad bladder, *Am. J. Physiol.* **215**:849.

Flamenbaum, W., Huddleston, M. L., McNeil, J. S., and Hamburger, R. J., 1974, Uranyl nitrate induced acute renal failure in rat: micropuncture and renal hemodynamic studies, *Kidney Int.* **6**:408.

Frenkel, A., Ekbald, E. B. M., and Edelman, I. S., 1975, Effect of sulfhydryl reagents on basal and vasopressin-stimulated Na^+ transport in toad bladder, in: *Biomembranes 7* (H. Eisenberg, E. Katachelski-Katzir, L. A. Manson, eds.), Plenum Press, New York, p. 167.

Leslie, B. R., Schwartz, J. H., and Steinmetz, P. R., 1973, Coupling between Cl^- absorption and HCO_3^- secretion in turtle bladder, *Am. J. Physiol.* **225**:610.

Lichtenstein, N. F., and Leaf, A., 1969, Effect of amphotericin B on the permeability of toad bladder, *J. Clin. Invest.* **44**:1328.

Reuss, L., and Finn, A. L., 1975, Dependance of serosal membrane potential on mucosal membrane potential in toad urinary bladder, *Biophys. J.* **15**:71.

Riggs, A., 1959, Hemoglobin structure, in: *Sulfur Protein* (R. Benesch, ed.), Academic Press, New York, p. 173.

Riggs, A., and Walbach, R. A., 1956, Sulfhydryl group and the structure of hemoglobin, *J. Gen. Physiol.* **39**:585.

Rothstein, A., 1959, Cell membrane as site of action of heavy metals, *Fed. Proc.* **18**:1026.

Schnermann, J., Wright, F. S., Davis, J. M., Stockelberg, W. V., and Grill, G., 1970, Regulation of superficial nephron filtration rate by tubuloglomerular feedback, *Pfluegers Arch.* **318**:147.

Schwartz, J. H., and Flamenbaum, W., 1976, Heavy metal-induced alteration in ion transport by turtle urinary bladder, *Am. J. Physiol.* **230**:1582.

Steinmetz, P. R., 1974, Cellular mechanism of urinary acidification, *Physiol. Rev.* **54**:890.

Steinmetz, P. R., and Lawson, L. R., 1970, Defect in urinary acidification induced by amphotericin B, *J. Clin. Invest.* **49**:596.

Webb, J. L., 1966, *Enzymes and Metabolic Inhibitors,* Volume II, Academic Press, New York, p. 1237.

4

Renal Hemodynamics in Nephrotoxic Acute Renal Failure

N. LAMEIRE, R. VANHOLDER, L. VAKAET, P. PATTYN, S. RINGOIR, and J. QUATACKER

1. Introduction

Although a large number of clinical and laboratory studies have been conducted to determine the pathogenesis of acute renal failure (ARF), no unifying concept has emerged to account for all the pathophysiologic manifestations of this disease.

Renal ischemia, tubular obstruction, tubular backleak of filtrate, and a decrease in glomerular ultrafiltration coefficient K_f have each individually been considered as the main pathogenetic factor. It becomes more and more clear, however, that in a given clinical or experimental setting, different combinations of these alterations may be present.

Markedly reduced renal cortical blood flow is the key mechanism in all vascular theories that have been proposed. This renal cortical ischemia, in the presence of a normal systemic blood pressure, reflects an increase in intrarenal vascular resistance. If this increased resistance is exclusively due to preglomerular

N. LAMEIRE, R. VANHOLDER, L. VAKAET, P. PATTYN, and S. RINGOIR • Renal Division of the Department of Medicine, State University of Gent, University Hospital, B-9000 Gent, Belgium. J. QUATACKER • Division of Anatomo-Pathology, State University of Gent, University Hospital, B-9000 Gent, Belgium.

vascular constriction, this must result in a lower capillary pressure. The glomerular capillary pressure can even be lower if this afferent arteriolar constriction is coupled with efferent arteriolar dilation. When this occurs, the glomerular filtration may be decreased even in the presence of a normal or supranormal renal blood flow.

Exact localization of the alterations in renal vascular resistance is of course only possible in renal micropuncture studies of superficial glomeruli. The intrarenal hemodynamics at the glomerular level has been directed measured in three studies.

Daugharty *et al.* (1974) have shown that afferent and efferent resistances rise proportionately in a variant of the postischemic model of ARF.

After a low dose of gentamicin, 4 mg/kg for 10 days, no change in nephron plasma flow was noted, but with a higher dose of 40 mg/kg of gentamicin, only a slight fall in glomerular plasma flow was observed, probably a result of an increased vascular tone at the afferent arteriolar site (Baylis *et al.,* 1977).

Blantz (1975) studied glomerular dynamics in rats 2 hr after a challenge with uranyl nitrate; no changes in glomerular vascular resistance were observed.

Finally, attention has recently been given to changes in the glomerular ultrafiltration coefficient K_f as mediator of the fall in glomular filtration rate (GFR). The K_f is the product of the glomerular capillary surface area and the hydraulic conductivity across the capillary membrane.

It is easy to visualize how glomerular vasoconstriction may cause a decrease in surface area of the glomerular capillaries. Two studies have shown that nephrotoxic agents such as uranyl nitrate and gentamicin produce a decrease in K_f, which accounts to a substantial degree for the reduction in GFR (Blantz, 1975; Baylis *et al.,* 1977).

Thus, many mechanisms can be invoked to explain how alterations in renal vascular resistance may lead to a fall in glomerular filtration rate. In this review, we will focus on the role of the changes in renal hemodynamics in the initiation and maintenance of nephrotoxic ARF. For a discussion of other mechanisms, the reader is referred to recent excellent reviews on the general pathophysiology of ARF (Flamenbaum, 1973; Oken, 1975; Levinsky, 1977, Stein *et al.,* 1978; Thurau *et al.,* 1979; Conger and Schrier, 1980).

The remaining discussion will be divided into four sections. In the first two, renal hemodynamics in human ARF and in animal nephrotoxic ARF will be briefly summarized. Emphasis will be placed on the differences which may exist between the initiation and maintenance phases of the disease. Then, possible mediators of renal vasoconstriction will be considered. Finally, mechanisms of renal vasoconstriction, such as tubuloglomerular feedback and tubular obstruction, will be discussed.

2. Renal Hemodynamics in Human ARF

The role of changes in renal hemodynamics in the pathogenesis of ARF was recognized more than 60 years ago. The term *"vasomotorische nephrose"* was probably introduced by Hackradt (1917), a German physician.

Goormaghtigh (1937) in Gent described his histologic findings of the juxtaglomerular apparatus in kidneys from patients who died from ARF. He wrote that in no other pathologic condition yet examined were the hypertrophy and granularity of the juxtaglomerular apparatus so marked. He later proposed that in ARF a vasopressive substance is liberated in excess and causes a persistent spasm at the vascular pole of the glomerular tufts (Goormaghtigh, 1945).

The first attempts to estimate renal blood flow (RBF) in patients with ARF were made by Bull *et al.* (1950), using PAH (*p*-aminohippuric acid) clearance and the renal extraction ratio for PAH in two patients in the late oliguric phase and one in the diuretic phase. They concluded that RBF was reduced to about one-tenth of the normal value.

More accurate techniques for measurement of RBF were developed later. In patients with established ARF, total RBF is reduced by 50–75% as estimated by various methods, particularly dye dilution (Reubi *et al.,* 1962, 1966, 1973) and inert gas washout techniques (Brun *et al.,* 1955; Munck, 1958; Hollenberg *et al.,* 1968, 1970).

On renal angiography, circulation time is prolonged, the arcuate and interlobar vessels are markedly attenuated, and poor filling of the terminal cortical vasculature is demonstrated, probably indicating intense preglomerular vasoconstriction (Hollenberg *et al.,* 1968, 1970). Since total RBF continues in anuric ARF patients with the same magnitude as in chronic renal failure, with normal or increased urine output, attention was given to the possibility of a redistribution of intrarenal bloodflow. Largely based on the analysis of xenon washout curves, it was inferred that total or outer cortical ischemia due to preglomerular vasoconstriction was the main pathogenetic event in established ARF, of either ischemic or nephrotoxic origin (Hollenberg *et al.,* 1968, 1970). These observations in patients were in fact compatible with an older concept formulated by Trueta *et al.* (1947), who proposed that the fundamental defect in the crush kidney was a reduction in glomerular filtration rate as a result of the diversion of blood away from the cortical glomeruli through the medullary, aglomerular regions of the kidney.

It is clear that studies in ARF patients always have been performed several days after the initiating renal insult and it is obvious that clinical investigations, however valuable and carefully performed, are limited by ethical and practical considerations. Therefore, the use of experimental models of ARF in animals is necessary.

In these models, aggressive techniques such as micropuncture and renal hemodynamic measurements can be used at various times during the course of ARF. As recently stated by Olbricht (1980), three different categories of experimental models are needed, in accord with the three most common causes of human ARF, i.e., (1) nephrotoxic models, (2) heme-pigment models, and (3) ischemic models.

Although the model produced by a 40-min intrarenal infusion of norepinephrine seems to correspond best with the human form of ischemic ARF (Cronin *et al.,* 1978), some nephrotoxic animal models, such as the $HgCl_2$ model, have the advantage of being produced by the same causative factor as in human post-mercury ARF.

Investigation of the role of renal hemodynamics in the initiation phase of ARF requires study of nephrotoxic models since most ischemic models, by definition, completely interrupt renal blood flow. Eventual dissociations between changes in RBF and renal function can therefore not be realized in the first hours after the ischemic insult.

3. Renal Hemodynamics in Experimental Nephrotoxic ARF

The evidence supporting a pathogenetic role for hemodynamic mechanisms in the pathophysiology of nephrotoxic ARF is considerable, although conflicting. The reasons for the conflicting results are manifold:

1. Use of widely divergent methods for measurement of total or regional renal blood flow, including clearances, dye dilution, microspheres, and inert-gas washout methods.

2. Use of different renal toxins, sharing a common nephrotoxic action but not necessarily a similar pathogenetic mechanism.

3. Failure to recognize that changes in renal hemodynamics may be either a cause or a consequence of the insult, or even be a completely unrelated event, without any contribution to the decrease in measured GFR.

4. Study of the influence of nephrotoxins on renal function at quite different time intervals after the toxic challenge, and failure to recognize that different pathogenetic factors may play a role at the different stages of initiation, maintenance, and recovery of ARF.

5. Application of different doses and of varying routes of administration of the nephrotoxin.

Nephrotoxic renal damage is produced mainly with the salts of inorganic ions and recently with aminoglycoside antibiotics and is considered to resemble clinical ARF occurring after exposure to toxins. The most commonly used toxins are mercuric chloride, uranyl nitrate, potassium dichromate, and gentamycin in rats, rabbits, and dogs.

In the analysis of the hemodynamic data, the physiologic background of the animal and experimental setting should be taken into account, since it becomes more and more clear that the preinsult extracellular volume status, solute, and water excretion must be standardized before conclusions can be made.

Tables I and II summarize the data obtained on renal hemodynamics in nephrotoxic ARF in rat and rabbits (Table I) and dogs (Table II). Also included are the different methods used for measurement of RBF, the postinsult time interval of the study, the dose of nephrotoxin used, and, where possible, the background experimental setting.

It is clear that nephrotoxic ARF has been induced by a variety of agents, doses, and ways of administration. Furthermore, responses to nephrotoxins may differ in various animal species, particularly the dog and the rat. Even if we take into account all these variables, some conclusions from the data listed in Tables I and II can be made. Let us first consider the data concerning the initiation phase, i.e., measurements of renal hemodynamics up to 6 hr after the insult.

TABLE 1. Renal Blood Flow Values in Nephrotoxic ARF in the Rat and Rabbit

Reference	Method	Model	Time	RBF	Redistribution	Comments
Biber et al. (1968)	PAH clearance + extraction, rat	HgCl₂ 5–10 mg/kg	3 days	Normal	?	—
Foulkes (1971)	Vein effluent, rabbit	Uranyl nitrate 0.2 mg/kg	2 days	Normal	?	—
Solez et al. (1974)	Albumin, rat	HgCl₂ 5 mg/kg	40–48 hr	—	PPF ↑↑	—
Blantz (1975)	Micropuncture, rat	Uranyl nitrate 15–25 mg/kg	2 hr	—	SNPF normal	—
Russell (1975)	Isolated perfused kidney, rat	HgCl₂ 100 µg (bolus)	0–5 min	↑ in vascular resistance	—	Attenuated by mannitol; no effect of alpha-blocking agents
Flamenbaum et al. (1976a,b)	Xenon, rat	Uranyl nitrate 10 mg/kg	6 hr	↓↓	+	—
			48 hr	↓↓	+	Reversed by chronic salt loading
Lameire et al. (1976)	Microspheres, rat	Mercaptomerin 15 mg/kg	24 hr	↓↓	+	—
Solez et al. (1976)	Flowmeter angiography, rabbit	HgCl₂ 1.5 mg/kg	1–3 hr	Normal	—	—
Hsu et al. (1977)	Microspheres, rat	HgCl₂ 4.7 mg/kg	3 hr	↓↓	+	—
			12 hr	↓↓	+	
			24 hr	Normal	↑Cortical flow	Reversed by chronic salt loading
Churchill et al. (1977)	Hydrogen washout Venous cannulation, rat	HgCl₂ 4.7 mg/kg	24 hr	↑	Normal	—
		HgCl₂ 12 mg/kg	24 hr	Normal	Normal	—
			48 hr	Normal	+	—
Siegel et al. (1977)	Microspheres, rat	Dichromate 15 mg/kg	1 day	↓↓	Redistribution outer cortex	—
			4–7 days	↓↓	+	Recovery associated with outer cortical redistribution
Sudo et al. (1977)	Microspheres, rat	Uranyl nitrate 2 mg/kg	14 days	Normal	Normal	Slightly volume-expanded animals
Baylis et al. (1977)	Micropuncture, rat	Gentamycin 10 days, 40 mg/kg	1 day	↑	No redistribution	
			3–21 days	Normal or ↑	↓ Single NPF	↓Kf
Lameire et al. (1979)	Microspheres, rat	HgCl₂ 4.7 mg/kg	—	↓↓	+	
			3 hr	↓↓	+	Salt loading, normal distribution, but ↓RBF
			24 hr	↓	+	Salt loading, normal distribution, normal RBF
Schor et al. (1979)	Micropuncture, rat	Gentamycin 40 mg/kg IP	—	—	↓SNPF	—
Sudo et al. (1980)	Microspheres, rabbit	Uranyl nitrate 0.9–2 mg/kg	7 hr	Normal	—	—
			15 hr	Normal	—	—
			5 days	↑	—	—

TABLE II. Renal Blood Flow Values in Nephrotoxic ARF in the Dog

Reference	Method	Model	Time	RBF	Redistribution	Comments
Bobey et al. (1943)	Diodrast	Uranyl nitrate	2 days	Normal	—	—
Conn et al. (1954)	Nitrous oxide diffusion	HgCl₂ 15–30 mg/kg	2–5 days	↗ or normal	?	—
Eisner et al. (1968)	Venous effluent	Uranyl nitrate 5 mg/kg	48 hr	Normal	—	—
Balint (1969)	Flowmeter Rb uptake	HgCl₂ 2 mg/kg IV	3 days	↗ or normal	?	—
Flamenbaum (1973)	Xenon; microspheres	Uranyl nitrate 10 mg/kg	6 hr	↘	+	—
			24 hr	↘	+	—
			72 hr	↘	+	—
			96 hr	↘	+	—
Sherwood et al. (1974)	Flowmeter arteriography	HgCl₂ 0.5 mg/kg	1–6 hr	↗	?	Restoration of normal perfusion with mannitol
Stein et al. (1975)	Microspheres	Uranyl nitrate 10 mg/kg	6 hr	↘	—	—
		5 mg/kg	48 hr	Normal	+	—
Kleinman et al. (1975)	Flowmeter, microspheres	Uranyl nitrate 10 mg/kg	48 hr	Normal	?	—
			0–3 hr	↘	+	—
Solomon and Hollenberg (1975)	Flowmeter	HgCl₂ intrarenal	—	Dose-related ↘	?	Cardiac output unchanged
Mauk et al. (1977)	Flowmeter, microspheres	Uranyl nitrate 5 mg/kg	3 hr	Normal	—	Prior PGE₂, no protection
			48 hr	Normal	?	
Baehler et al. (1977)	Microspheres	HgCl₂ 2 mg/kg	48 hr	↘	+	Volume expansion; restoration of RBF without effect on GFR
Lindner et al. (1979)	Microspheres	Uranyl nitrate 10 mg/kg	1 hr	↘	+	Water-loaded animals, protection by furosemide + dopamine
			3 hr	↘	+	
			6 hr	↘	+	

3.1. Renal Hemodynamics in the Initiation Phase

In all models, an early reduction in RBF has been observed, and where intrarenal blood flow distribution has been examined, a preferential outer cortical ischemia has been found in the majority of the studies. In some studies, cardiac output and renal fraction of cardiac output have been measured and it appears that the reduction in RBF is due to a rather selective renal vasoconstriction.

Thus, initial renal vasoconstriction is consistently observed and since in most nephrotoxic models a parallel fall in GFR is noted, it was concluded that they were causally related. However, one study by Mauk *et al.* (1977) questions the pathogenetic importance of this initial renal vasoconstriction in the uranyl nitrate model of the dog. These authors increased RBF to 50% above control levels by administration of large doses of prostaglandin E (PGE) previous to and during a 3 hr period following the nephrotoxic insult. Although RBF never decreased below the control level, the fall in inulin clearance was similar to that in the nonvasodilated kidney.

Some comments regarding this important study are warranted. Resetting the initial RBF at a higher level before a nephrotoxic insult of course does not exclude the occurrence of a renal vasoconstriction after the insult, even if the total RBF is not allowed to decrease under the control value.

In addition, in this study, a dose of 5 mg/kg of uranyl nitrate was administered, which does not cause a renal vasoconstriction in the dog kidney, but decreases the GFR by 40%. In a previous study by the same authors (Mauk *et al.*, 1976), a higher dose of 10 mg/kg was utilized and was associated with a diminution in RBF of 40–50%. Intrarenal PGE infusion did not overcome the vasocontrictor effect of this larger dose of uranyl nitrate and a fall in GFR was noted. Finally, the possibility that the PGE infusion could have influenced more or less selectively the efferent arteriolar resistance, while uranyl nitrate could have caused an afferent arteriolar vasoconstriction, is not excluded. This combination of oppositely directed changes in renal resistance could also explain the finding of a normal or even enhanced RBF with a fall in GFR, the latter due to reduction in glomerular capillary pressure. In any case, this study suggests that, at least in this low-dose uranyl nitrate model, the fall in glomerular filtration rate is not primarily due to a general renal vasoconstriction.

This work stimulated a recent study in our laboratory to evaluate the role of systemic, renal, and intrarenal hemodynamics in the initiation of $HgCl_2$-induced ARF in the dog. As can be noted from Table III (left half), a progressive and almost parallel fall in RBF and GFR was observed the first 3 hr after the intravenous injection of 3 mg/kg of $HgCl_2$. A marked increase in urine flow and fractional excretion of sodium and a decrease in urinary osmolality were noted.

That the fall in inulin clearance was not primarily due to tubular obstruction was derived from additional measurements of proximal tubular pressures, which decreased from 19.6 ± 0.34 to 13.9 ± 0.38 mm Hg at 3 hr. Tubular backleak of filtrate in this initial stage was unlikely, for several reasons: the presence of an intact tubular epithelium at transmission electron microscopy at 3 hr (Fig. 1), the finding of an unchanged renal venous PAH extraction, and the fact that the ratio of the postmercurial creatinine-to-inulin clearances remained near 1.0.

TABLE III. Summary of Renal Hemodynamics and Function in HgCl$_2$-Induced Acute Renal Failure in the Dog

	Hydropenic animals				Volume expansion (VE) + phentolamine-treated animals				
	Control	1 hr	2 hr	3 hr	Control hydropenia	Control VE + Ph	1 hr	2 hr	3 hr
Urine flow, ml/min	0.34 ±0.20	1.64^a ±0.45	2.63^b ±0.42	2.49^b ±0.25	0.29 ±0.04	2.29^a ±0.95	6.81^{b,c} ±0.85	8.79^{b,c} ±1.27	8.21^{b,d} ±0.96
GFR, ml/min	45 ±4	40^a ±5	31^b ±3	25^b ±2	64 ±6	64 ±9	51^b ±8	42^b ±5	36^{b,d} ±6
RBF, ml/min	268 ±22	224^b ±20	184^b ±19	161^b ±19	470 ±28	375 ±37	483 ±104	481 ±65	473 ±99
FE$_{Na}$,^{e} %	0.45 ±0.18	3.77^a ±1.17	8.61^b ±1.71	9.32^b ±1.39	0.28 ±0.09	1.48^a ±0.45	10.39^{b,d} ±1.67	20.25^{b,d} ±2.64	21.45^{b,d} ±1.97
U$_{Osm}$, mOsm/liter	1104 ±193	612^a ±134	389^b ±56	369^b ±51	1255 ±293	477 ±203	328^a ±48	342^a ±28	329^a ±23

^{a} $p < 0.05$ (vs. control).
^{b} $p < 0.01$ (vs. control).
^{c} $p < 0.05$ (vs. control VE + phentolamine, Ph.).
^{d} $p < 0.01$ (vs. control VE + phentolamine, Ph.).
^{e} Fractional urinary excretion of sodium.

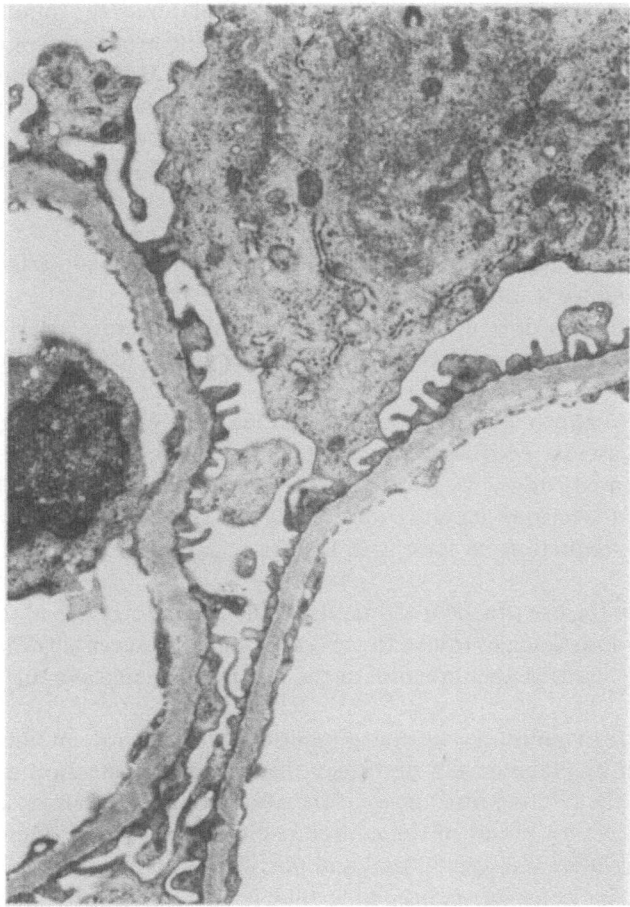

FIGURE 1. Transmission electron micrograph of the tubular epithelium 3 hr after $HgCl_2$. No remarkable alterations are noted.

A further analysis of the systemic and intrarenal hemodynamics with radioactive microspheres revealed a significantly lower renal fraction of the cardiac output and an unchanged cortical blood flow distribution 3 hr after mercury. The latter finding does not agree with previous results obtained in rats 3 hr after a challenge with a low dose of $HgCl_2$ (Lameire *et al.*, 1979).

Thus, a quite selective and progressive renal vasoconstriction, apparently involving all cortical layers, occurs in this model. In an attempt to dissociate the changes in RBF from the fall in GFR, another series of animals was pretreated with a combination of Haemaccel®, a gelatin-like plasma expander, and phentolamine in an alpha-adrenergic-blocking dose. The reasons for the use of the alpha blocker were twofold:

1. The intriguing observation of Solomon and Hollenberg (1975), who showed a considerable reduction in vascular smooth muscle contraction in response to $HgCl_2$ administration by alpha-adrenergic blockade.

2. By administering a systemic dose of phentolamine, the amount of volume expansion by Haemaccel® could be so titrated that the pretreatment per se did not influence the premercurial RBF or GFR. After the $HgCl_2$ administration, the degree of volume expansion was so monitored that no decrease in RBF ever occurred, as measured by an electromagnetic flowmeter.

The results of these studies are also summarized in Table III (right half). Despite a continuous rise in RBF up to 2 hr and a stabilization at 3 hr after $HgCl_2$, a similar fall in GFR occurred as in the control group.

It was concluded that at least a general renal vasoconstriction was not necessary to induce a fall in GFR in this model.

This conclusion does not imply that preferential afferent arteriolar vasoconstriction is not the cause of the impairment of glomerular filtration in $HgCl_2$-induced ARF.

In order to approach this problem in an indirect way, the changes in afferent and efferent vascular resistances were calculated, according to Hall *et al.* (1977). The unpretreated dogs, a parallel increase in resistance at the pre- and postglomerular level after mercury was calculated. This combination should lead to a substantial reduction in effective filtration pressure, provided K_f remains unchanged.

In view of the use of whole-kidney data, the absolute values of these variables in glomerular resistance obtained in this study must be accepted with caution. At least they may indicate the direction of the resistance changes in this experimental setting.

The failure to maintain a normal glomerular filtration rate by continuous renal vasodilatation by Haemaccel® and phentolamine administration may be due to alterations in the relative resistances at the afferent and efferent arterioles. Due to the lack of direct estimates of the colloid osmotic pressure in Haemaccel-loaded animals, the relative changes in pre- and postglomerular resistances could not be calculated. It is, however, conceivable that a marked fall in efferent arteriolar resistance was induced by the volume expansion.

A change in K_f directly induced by $HgCl_2$ could also account for the fall in GFR, despite an augmented RBF. Cachia *et al.* (1981) recently found no change in K_f in isolated glomeruli from rats challenged with a low dose of 4 mg/kg of $HgCl_2$ at 24 and 48 hr prior to the experiment. However, a substantial fall in K_f after a dose of 10 mg/kg of $HgCl_2$ was noted. Indirect evidence for a change in K_f has been found in structural alterations in the glomerular epithelium, as observed with scanning electron microscopy of the dog kidney after uranyl nitrate, $HgCl_2$, and intrarenal norepinephrine administration. In our hands, transmission electron microscopy revealed no detectable changes in glomerular epithelium or basement membrane 3 hr after $HgCl_2$, despite a fall in GFR of 45%. In addition, staining of the glomerular capillary wall with colloidal iron demonstrated a normal distribution of charged anionic groups in the glomerular barrier.

On the other hand, scanning electron microscopy revealed minor swelling of the microvilli on the surface of the cell body of the visceral epithelium 3 hr after $HgCl_2$ (Fig. 2). These morphologic findings do not exclude a subtle alteration in K_f in this model.

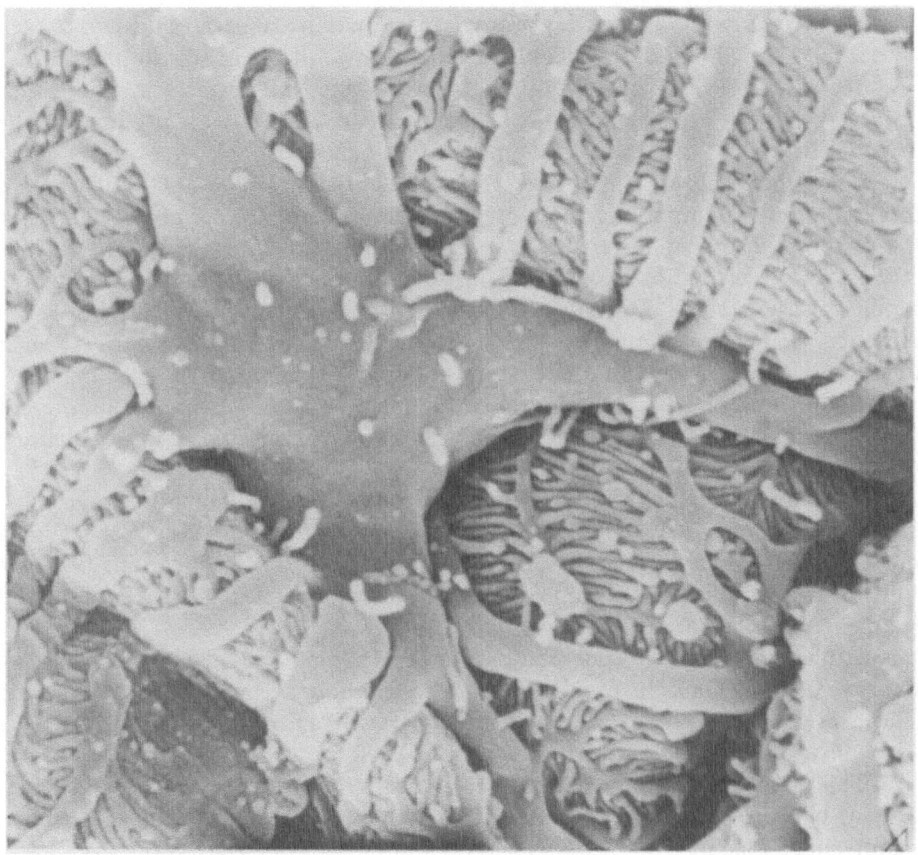

FIGURE 2. Scanning electron micrograph of the visceral epithelium 3 hr after $HgCl_2$. Minor swelling of the microvilli on the cell body surface is noted. ×5000.

Changes in K_f may be involved in this stage, since at least in the uranyl nitrate model, they are present already 2 hr after the insult (Blantz, 1975). Recent interesting morphologic studies by Avasthi *et al.* (1979, 1980) demonstrated an important reduction in diameter and density of the glomerular endothelial fenestrae at different time intervals after uranyl nitrate and gentamicin administration in rats. It is suggestive that these morphologic alterations can explain the reduction in K_f by reducing the filtration area.

3.2. Renal Hemodynamics in the Maintenance Phase

It is obvious from the numerous studies collected from the literature that discrepancies exist concerning whether total renal ischemia persists in the maintenance phase of ARF when the animal is anuric or severely oliguric. A persistent reduction in renal blood flow has been found in some studies after $HgCl_2$

in rats and dogs, while after uranyl nitrate, most investigators find a normal to even an enhanced renal blood flow (Tables I and II). In addition, restoration to normal of a decreased RBF by acute volume expansion does not reverse the oliguria or improve the GFR. The role of a persistent renal vasoconstriction in the maintenance phase of ARF is thus more questionable.

Using the 2 mg/kg $HgCl_2$ model in the dog, Baehler et al. (1977) observed a total and preferential outer cortical ischemia in the maintenance phase 48 hr after the mercury. The primary evidence for a profound fall in nephron GFR at this moment was the failure to visualize intrarenally injected lissamin green in the lumina of the tubules. This suggested that the oliguria was primarily due to an impaired GFR. Restoration of a supranormal renal perfusion by acute Ringer loading was associated with a persistant oliguria in this phase.

In our opinion it is not definitely excluded that opposite changes in afferent and efferent arteriolar resistances are induced by the volume expansion, explaining the lack of improvement of the GFR.

In studies where chronic saline loading has been shown to improve renal function after mercury and dichromate, the protection was associated with a normal RBF (Lameire et al., 1976, 1979; Hsu et al., 1977; Kashgarian, 1979). In some of these studies prevention of preferential outer cortical ischemia was found (Lameire et al., 1976, 1979; Hsu et al., 1977).

Studies of the recovery phase after dichromate-induced ARF show an improvement in GFR in parallel with a restoration of outer cortical ischemia (Siegel et al., 1977). These results are compatible with an additional role for cortical vasoconstriction in the maintenance phase of ARF.

This conclusion should be taken to indicate that other mechanisms, such as tubular backleak and tubular obstruction, are not playing a dominant role in the maintenance phase.

4. Mediators of Renal Vasoconstriction

Several causes for altered renal hemodynamics in the pathophysiology of nephrotoxic ARF have been proposed.

4.1. Increased Adrenergic Activity

There is in vitro evidence that the direct vascular smooth muscle contraction by $HgCl_2$ is adrenergically mediated (Solomon and Hollenberg, 1975). The constricting vascular response of the renal vasculature and of aortic strips to increasing doses of $HgCl_2$ is considerably reduced by alpha-adrenergic blockade with phentolamine and phenoxybenzamine. However, the marked increase in perfusion pressure obtained with $HgCl_2$ in isolated, constant-flow perfused rat kidneys could not be altered by a large standard dose of phentolamine (Russell, 1975). Although not definitely conclusive, the latter study argues against the possibility that the mercury-induced vasoconstriction is adrenergically mediated.

4.2. Intrarenal Renin–Angiotensin Stimulation

There is more evidence for a major role of the intrarenal renin–angiotensin system as mediator of the vasoconstriction in the initiation of nephrotoxic-induced ARF. In support of this concept, an early increase in the renin activity of the juxtaglomerular apparatus in uranyl nitrate- and $HgCl_2$-induced acute renal damage has been reported (Flamenbaum *et al.*, 1976a,b; Kleinman *et al.*, 1977). The highest concentration of renin is found in the outer cortex (Flamenbaum and Hamburger, 1974) and either total renal or outer cortical ischemia has been found early after $HgCl_2$ in rats and dogs (Hsu *et al.*, 1977; Lameire *et al.*, 1979; Vanholder and Lameire, unpublished results) and after uranyl nitrate in the dog (Stein *et al.*, 1975; Kleinman *et al.*, 1975).

The involvement of the renin–angiotensin system in this early vasoconstriction may, however, be questioned. Chronic salt loading with or without DOCA pretreatment, maneuvers which substantially lower both plasma and renal renin concentrations (Flamenbaum *et al.*, 1974; Baehler *et al.*, 1978), do not prevent a marked total renal vasoconstriction up to 3 hr after $HgCl_2$, and in this situation a fall in glomerular filtration rate is also not prevented (Lameire *et al.*, unpublished results; Hsu *et al.*, 1977). Furthermore, Mason *et al.* (1979) explored the effect of a chronic suppression in juxtaglomerular renin activity and tubuloglomerular feedback response on the evolution of glomerular filtration rate up to 3 hr after injection of uranyl nitrate. Despite reduction in juxtaglomerular renin activities as low as 4% of control, no significant attenuation in fall of GFR was detected. These results lead to the suggestion that if the early fall in GFR is a direct or indirect consequence of early renal vasoconstriction, the latter is probably not mediated by intrarenal angiotensin II.

A remarkable correlation between intrarenal renin depletion and functional protection of GFR has frequently been noted in the late or maintenance phase of nephrotoxic ARF (Baehler *et al.*, 1978; Flamenbaum *et al.*, 1976b; Hsu *et al.*, 1977). The persistence of total renal and/or outer cortical vasoconstriction has therefore been attributed to a continued activation of the intrarenal renin–angiotensin system. In the studies where renal hemodynamics have been measured, the protection was associated with a better or even normal perfusion of either the whole cortex or of at least the outer cortical zones. This has been the case after $HgCl_2$ (Lameire *et al.*, 1976; Hsu *et al.*, 1977) and after dichromate (Kashgarian, 1979).

However, recent studies have even questioned the importance of the renin–angiotensin system as mediator of the vasoconstriction in the maintenance phase. Thiel *et al.* (1976) have shown that protection in the mercuric rat model could be achieved without suppression of intrarenal renin, while Bidani *et al.* (1979) obtained a complete dissociation between intrarenal renin content and the degree of functional impairment after both $HgCl_2$ and uranyl nitrate in the rat. The same investigators had previously shown that equal degrees of ARF following $HgCl_2$ developed in the two kidneys of Goldblatt rats, despite large differences in renal renin content (Churchill *et al.*, 1978).

All these studies are certainly not compatible with the view that the renin–

angiotensin system is involved in the total renal and/or outer cortical vasoconstriction observed in the maintenance phase of toxic ARF. It must be added, however, that determination of the cortical renin content is only an index of intracellular renin and angiotensin II production. In fact, recent studies have shown a dissociation between intrarenal angiotensin II and renal cortical renin levels (Mendelsohn and Smith, 1980). Although chronic saline loading suppresses the cortical renin, it has no influence on the intrarenal angiotensin II concentration (Mendelsohn, 1979).

Thus, although there is evidence that an activated renin–angiotensin system can participate in the vasoconstriction observed in the maintenance phase of most nephrotoxic models, there exists as much evidence against this concept and this inevitably leads to the conclusion that other mechanisms must be involved.

Another possible role for either circulating or intrarenal angiotensin is its influence on the glomerular ultrafiltration coefficient K_f. Direct evidence for a fall in K_f has been obtained after 2 hr in the uranyl nitrate model in the rat and after gentamycin administration in the same animal (Blantz, 1975; Baylis *et al.,* 1977; Schor *et al.,* 1979). Indirect, largely morphologic evidence for a change in K_f has been noted 48 hr after uranyl nitrate and mercury chloride in the dog (Stein *et al.,* 1975; Baehler *et al.,* 1978). It has been shown by Blantz *et al.* (1976) that angiotensin II infusion may lead to a reduction in K_f. Furthermore, Hornych and Richet (1977) have shown that angiotensin causes marked capillary constriction and a decrease in surface area of the glomerular capillaries. The existence of a specific angiotensin receptor in the glomerulus has been shown (Sraer *et al.,* 1974).

Finally, the glomerular mesangial cell has many of the characteristics of smooth muscle cells. It is thus possible that angiotensin II has a major direct effect on the glomerular capillary circulation and causes profound alterations in glomerular dynamics.

4.3. Renal Prostaglandins

Studies cited above have mainly focused on a possible role for vasoconstricting substances in the pathogenesis of ARF. On the other hand, deficiency of potent vasodilators could also be important in the hemodynamic alterations which occur in ARF.

Such a role has recently been proposed for the renal prostaglandins and/or kinins. Prostaglandins may function as physiologic antagonists to vasoconstrictor stimuli (Anderson *et al.,* 1976). Failure to release prostaglandins in response to these stimuli could be important in the initiation of ARF. Of interest in this regard is a study by Torres *et al.* (1974), who demonstrated that indomethacin, an inhibitor of prostaglandin synthesis, enhanced the severity of glycerol-induced but not of $HgCl_2$-induced ARF. These findings are surprising, since an increase in medullary prostaglandin concentration has been found in the glycerol-treated rabbit. In addition, these results have not been confirmed by Oken (1976).

Recently, indomethacin has been shown to interfere directly with the metabolism of glycerol (Craig *et al.,* 1980). It is thus likely that if there is increased severity of glycerol-induced ARF by indomethacin, it is simply due to a longer exposure of the kidneys to the toxin. Thus, further studies are needed to evaluate the role of renal prostaglandins in the pathogenesis of ARF.

Attention will certainly be given in the near future to a possible role in the renal vasoconstriction of ARF of thromboxane (TxA_2), a very potent vasoconstrictor. There is some evidence that small amounts of TxA_2 may be generated by isolated glomeruli (Folkert et al., 1979). Exaggerated synthesis of TxA_2 has been found in the isolated hydronephrotic kidney of the rabbit (Morrison et al., 1978). Recently, a markedly enhanced renal release of TxA_2 induced by vasoactive peptides was observed in the rabbit with renal vein constriction (Zipser et al., 1980). This model is associated with increased renal vascular resistance (Haddy et al., 1958; Yoshitoshi et al., 1966). Furthermore, an increased renal capacity to produce TxA_2 has been demonstrated by kidneys in glycerol-induced ARF in the rabbit (Benabe et al., 1980). The role of thromboxanes as possible mediators of vascular resistance in nephrotoxic ARF is certainly worthwhile to investigate.

The influence of other vasoconstrictive material (such as adenosine), released by injured tubular epithelial cells, on the genesis or maintenance of ARF is totally unknown.

5. Mechanisms of Renal Vasoconstriction

5.1. Tubuloglomerular Feedback

In this hypothesis, nephrotoxins such as uranyl nitrate or $HgCl_2$ decrease the sodium reabsorption in proximal nephron segments, so that tubular fluid salt concentration is increased at the macula densa. Because of a tubuloglomerular feedback mechanism, this increased salt delivery activates renin, and causes local angiotensin production and vasoconstriction of the glomerular afferent arterioles. Evidence has been obtained in nephrotoxic models of ARF that transport proximal to the macula densa is impaired (Flamenbaum et al., 1976b), that distal electrolyte load is increased, and that the feedback mechanism remains functional under these circumstances (Flamenbaum et al., 1976b; Mason, 1976; Mason et al., 1978).

Recent results obtained by Mason et al. (1979) are incompatible not only with the participation of the renin–angiotensin system, but also with the mechanism of tubuloglomerular feedback. In this study, as in others (Thiel et al., 1976), the prophylactic effect of chronic saline loading on the early evolution of the GFR was largely attributed to the protective effect of volume expansion.

If previous chronic or acute volume expansion plays a protecting role in nephrotoxic ARF, several factors, such as increase in urine output, high rates of solute excretion, or even enhancement of the excretion rate of the nephrotoxic substance, may be involved.

In order to distinguish between these factors, we applied several chronic prophylactic pretreatment procedures to rats subsequently challenged with a low dose of $HgCl_2$ (Lameire et al., 1979). The procedures included a 4-week pretreatment period with tapwater, glucose 5%, isotonic saline, with or without DOCA injections and isotonic $NaHCO_3$. An additional group of animals was given isotonic NaCl according to a paired feeding schedule with the $NaHCO_3$ group, in order to assure a similar dietary intake in sodium.

In all groups, the renal mercury content was measured by atomic absorption. No consistent correlation between the 24 hr post-mercurial serum creatinine and the pre-mercurial 24-hr urinary volume, urinary sodium excretion, urinary osmolar excretion, or renal tissue mercurial concentration could be found, but there was significant protection in the DOCA-saline-loaded animals.

There are of course many other compositional and hemodynamic alterations which occur with chronic salt loading, but any obvious explanation of the protection is still lacking.

It is highly probable that in the maintenance phase of nephrotoxic ARF, other factors that lower the kidney GFR, such as tubular backleak and tubular obstruction, become more and more important. There are indications that in the late phase of nephrotoxic ARF some degree of tubular backleak exists (Stein *et al.,* 1975).

One interesting theory tries to correlate the finding of tubular obstruction with persistent renal vasoconstriction. Arendshorst *et al.* (1974) observed a markedly decreased glomerular capillary pressure, probably by afferent arteriolar constriction, after a complete ureteral ligation of 24 hr duration. A similar response occurred in tubules from normal kidneys in which an oil block had been placed. The mechanism and mediators of this afferent vasoconstriction are, however, completely unknown. Although there is some evidence for tubular obstruction in the low-dose $HgCl_2$ model in the rat after 24 hr (Flamenbaum *et al.,* 1971) or after a high dose of $HgCl_2$ in the first hours (Mason *et al.,* 1977), this has not been found after uranyl nitrate (Stein *et al.,* 1975; Blantz, 1975; Mason *et al.,* 1977).

6. Conclusions

The participation of renal vasoconstriction in the initiation of nephrotoxic acute renal failure in man and animals cannot be denied. On the other hand, there are serious doubts about the importance of altered renal hemodynamics in the maintenance phase of ARF. The failure of the renal excretory function to improve when renal blood flow is raised to normal or supranormal levels in established ARF is certainly a very important counterargument.

However, changes in overall renal vascular resistance may not correctly reflect changes at individual vascular segments and opposite changes in afferent and efferent arteriolar resistances may have profound effects on the glomerular capillary pressure.

It is also clear that other mechanisms, such as tubular backleak or tubular obstruction, are equally important in the maintenance phase.

The etiology of renal vasoconstriction remains unknown. Evidence exists for a role of a stimulated renin–angiotensin axis, either by a direct action on the glomerular ultrafiltration coefficient and/or on the afferent arteriolar diameter, or indirectly by a tubuloglomerular feedback system. However, recent experiments have seriously questioned this mechanism.

A role for prostaglandins, particularly thromboxane, should be explored in the near future.

Finally, although further efforts in animal research should certainly be encouraged, it should always be kept in mind that no experimental model exactly parallels human ARF.

Acknowledgments

We appreciate the secretarial help of I. Verslycken, L. De Torck, and C. Van Den Berghe.

References

Anderson, R. J., Berl, T., McDonald, K. M., and Schrier, R. W., 1976, Prostaglandins. Effects on blood pressure, renal blood flow, sodium and water excretion, *Kidney Int.* **10**:205.

Arendshorst, W. J., Finn, W. F., and Gottschalk, C. W., 1974, Stop-flow pressure response to obstruction for 24 hours in the rat kidney, *J. Clin. Invest.* **53**:1497.

Arendshorst, W. J., Finn, W. F., and Gottschalk, C. W., 1975, Pathogenesis of acute renal failure following temporary renal ischemia in the rat, *Circ. Res.* **37**:558.

Avasthi, P. S., Huser, J., and Evan, A. P., 1979, Glomerular endothelial cells in gentamicin-induced acute renal failure, *Kidney Int.* **16**:771 (abstract).

Avasthi, P. S., Evan, A. P., and Hay, D., 1980, Glomerular endothelial cells in uranyl nitrate-induced acute renal failure in rats, *J. Clin. Invest.* **65**:121.

Baehler, R. W., Kotchen, T. A., Burke, J. A., Galla, J. H., and Bedathena, D., 1977, Considerations on the pathophysiology of mercuric chloride-induced acute renal failure, *J. Lab. Clin. Med.* **90**:330.

Baehler, R. W., Kotcher, T. A., and Ott, C. E., 1978, Failure of chronic sodium chloride loading to protect against norepinephrine-induced renal failure in dogs, *Circ. Res* **42**:23.

Balint, P., Fekete, A., and Harla, T., 1969, Intrarenal circulation in mercuric chloride-induced renal failure, *Experientia* **25**:722.

Baylis, C., Rennke, H. R., and Brenner, B. M., 1977, Mechanisms of the defect in glomerular ultrafiltration associated with gentamicin administration, *Kidney Int.* **12**:344.

Benabe, J. E., Klahr, S., Hoffman, M. K., and Morrison, A. R., 1980, Production of thromboxane A_2 by the kidney in glycerol-induced acute renal failure in the rabbit, *Prostaglandins* **19**:333.

Biber, T. V. L., Mylle, M., Baines, A. D., Gottschalk, C. W., Oliver, J. R., and McDowell, M. C., 1968, A study by micropuncture and microdissection of acute renal damage in rats, *Am. J. Med.* **44**:664.

Bidani, A., Churchill, P., Fleischmann, L., 1979, Sodium-chloride-induced protection in nephrotoxic acute renal failure: independence from renin, *Kidney Int.* **16**:481.

Blantz, R. C., 1975, The mechanism of acute renal failure after uranylnitrate, *J. Clin. Invest.* **55**:621.

Blantz, R. C., Konner, K. S., and Tucker, B. J., 1976, Angiotensin II effects upon the glomerular microcirculation and ultrafiltration coefficient of the rat, *J. Clin. Invest.* **57**:419.

Bobey, M. E., Longley, L. P., Dickes, R., Price, J. W., and Hayman, J. M., 1943, The effect of uranium poisoning on plasma diodrast and renal plasma flow in the dog, *Am. J. Physiol.* **139**:155.

Brun, C., Crone, C., Davidson, H. G., Fabricius, J., Tybjaerg-Hansen, A., Lassen, N. A., and Munck, O., 1955, Renal blood flow in anuric human subject determined by use of radioactive krypton 85, *Proc. Soc. Exp. Biol. Med.* **89**:687.

Bull, G. M., Jockes, A. M., and Lowe, K. G., 1950, Renal function studies in acute tubular necrosis, *Clin. Sci.* **9**:379.

Cachia, R., Savin, V. J., Patak, R. V., and Ridge, S. M., 1981, Effect of mercuric chloride and uranyl-nitrate on ultrafiltration coefficient in isolated glomeruli, *Kidney Int.* **19**:236 (abstract).

Churchill, S., Zarlengo, M. D., Carvalho, J. S., Gottlieb, M. N., and Oken, D. E., 1977, Normal renocortical blood flow in experimental acute renal failure, *Kidney Int.* **11**:246.

Churchill, P. C., Bidani, A., Fleischman, L., and Becker-McKenna, B., 1978, HgCl$_2$-induced acute renal failure in the Goldblatt rat, *J. Lab. Clin. Med.* **91**:660.

Conger, J. D., and Schrier, R. W., 1980, Renal hemodynamics in acute renal failure, *Ann. Rev. Physiol.* **42**:603.

Conn, H. L., Wilds, L., Helwig, J., and Ibach, P., 1954, A study of the renal circulation, tubular function and morphology and urinary volume and composition in dogs following mercury poisoning and transfusion of human blood, *J. Clin. Invest.* **35**:732.

Craig, E., Cooney, G. J., and Dawson, A. G., 1980, Effects of indomethacin on the metabolism of glycerol by rat kidney tubules; an alternative explanation for the enhancement of glycerol induced acute renal failure by indomethacin, *Clin. Sci.* **58**:337.

Cronin, R. E., Erickson, A. M., De Torrente, A., McDonald, K. M., and Schrier, R. W., 1978, Norepinephrine-induced acute renal failure: A reversible ischemic model of acute renal failure, *Kidney Int.* **14**:187.

Daugharty, T. M., Ueki, I. F., Mercer, P. F., and Brenner, B. M., 1974, Dynamics of glomerular ultra-filtration in the rat. V. Response to ischemic injury, *J. Clin. Invest.* **53**:105.

Eisner, G. M., Slotkoff, L. M., and Lilienfield, L. S., 1968, Distribution volumes in the dog kidney during anuria produced by uraniumnitrate, *Am. J. Physiol.* **214**:929.

Flamenbaum, W., 1973, Pathophysiology of acute renal failure, *Arch. Intern. Med.* **131**:911.

Flamenbaum, W., and Hamburger, R. J., 1974, Superficial and deep juxtaglomerular apparatus renin activity of the rat kidney, *J. Clin. Invest.* **54**:1373.

Flamenbaum, W., McDonald, F. D., DiBona, G. F., and Oken, D. E., 1971, Micropuncture study of renal tubular factors in low dose mercury poisoning, *Nephron* **8**:221.

Flamenbaum, W., McNeil, J. S., Kotchen, T. A., and Saladino, A. J., 1972, Experimental acute renal failure induced by uranylnitrate in the dog, *Circ. Res.* **31**:682.

Flamenbaum, W., Huddleston, M. L., NcNeil, J. S., and Hamburger, R. J., 1974, Uranylnitrate induced acute renal failure in the rat: micropuncture and renal hemodynamic studies, *Kidney Int.* **6**:408.

Flamenbaum, W., Hamburger, R. J., Huddleston, M. L., Kaufman, J., McNeil, J. S., Schwartz, J. H., Nagle, R., 1976a, The initiation of experimental acute renal failure: An evaluation of uranyl-nitrate-induced acute renal failure in the rat, *Kidney Int.* **10**:S115.

Flamenbaum, W., Hamburger, R. J., and Kaufman, J., 1976b, Distal tubule (Na$^+$) and juxtaglomerular apparatus renin activity in uranyl nitrate induced acute renal failure in the rat, *Pfluegers Arch.* **364**:209.

Folkert, V. W., and Schlondorff, D., 1979, Prostaglandin synthesis in isolated glomeruli, *Prostaglandins* **17**:79.

Foulkes, E. C., 1971, Glomerular filtration and renal plasma flow in uranium-poisoned rabbits, *Toxicol. Appl. Pharmacol.* **20**:380.

Goormaghtigh, N., 1937, L'appareil neuro-myo-artériel juxtaglomérulaire du rein; ses réactions en pathologie et ses rapports en pathologie et ses rapports avec le tube urinifère, *C. R. Soc. Biol.* **124**:293.

Goormaghtigh, N., 1945, Vascular and circulatory changes in renal cortex in the anuric crush-syndrome, *Proc. Soc. Exp. Biol. Med.* **59**:303.

Hackradt, A., 1917, Ueber akute, tödliche vasomotorische nephrosen nach Verschüttung, Inaugural Dissertation, Munich.

Haddy, F. J., Scott, J., Fleischman, M., and Emanuel, D., 1958, Effect of change in renal venous pressure upon renal vascular resistance, urine and lymph flow rates, *Am. J. Physiol.* **195**:97.

Hall, J. E., Guyton, A. C., and Cowley, A. W., 1977, Dissociation of renal blood flow and filtration rate autoregulation by renin depletion, *Am. J. Physiol.* **1**:F215.

Haymann, J. M., Jr., Schumway, N. P., Dumke, P., and Miller, M., 1939, Experimental hyposthenuria, *J. Clin. Invest.* **18**:195.

Hollenberg, N. K., Epstein, M., Rosen, S. M., Basch, R. I., Oken, D. E., and Merrill, J. P., 1968, Acute oliguric renal failure in man: Evidence for preferential renal cortical ischemia, *Medicine* **47**:455.

Hollenberg, N. K., Adams, D. F., Oken, D. E., Abrams, H. L., and Merrill, J. P., 1970, Acute renal failure due to nephrotoxins. Renal hemodynamic and angiographic studies in man, *N. Engl. J. Med.* **282:**1329.

Hornych, H., and Richet, G., 1977, Dissociated effect of sodium intake on glomerular and pressor responses to angiotensin, *Kidney Int.* **11:**28.

Hsu, C. H., Kurtz, T. W., Rosenzweig, J., and Weller, J. M., 1977, Renal hemodynamics in $HgCl_2$ induced acute renal failure, *Nephron* **18:**326.

Kashgarian, M., 1979, Acute renal failure, in: *Kidney Disease, Present Status* (B. H. Churg, F. K. Spargo, F. K. Mostofi, and M. R. Abell, eds.), Williams & Wilkins, Baltimore, Maryland, p. 239.

Kleinman, J. G., McNeil, J. S., and Flamenbaum, W., 1975, Uranylnitrate acute renal failure in the dog: Early changes in renal function and haemodynamics, *Clin. Sci. Mol. Med.* **48:**9.

Kleinman, J. G., McNeil, J. S., Schwartz, J. H., Hamburger, R. J., and Flamenbaum, W., 1977, Effect of dithiothreitol on mercuric chloride- and uranyl nitrate-induced acute renal failure in the rat, *Kidney Int.* **12:**115.

Lameire, N., Ringoir, S., and Leusen, I., 1976, Effect of variation in dietary NaCl intake on total and fractional renal blood flow in the normal and mercury-intoxicated rat, *Circ. Res.* **39:**506.

Lameire, N., Praet, M., Van Peteghem, C., Waterloos, M. A., and Vanderbiesen, V., 1979, Pathophysiology of mercurychloride induced acute renal failure in the conscious rat, *Kidney Int.* **15:**194 (abstract).

Levinsky, N. G., 1977, Pathophysiology of acute renal failure, *N. Engl. J. Med.* **296:**1453.

Lindner, A., Cutler, R. E., Goodman, W. G., Pansing, P. A., and Kuester, R., 1979, Synergism of dopamine plus furosemide in preventing acute renal failure in the dog, *Kidney Int.* **16:**158.

Mason, J., 1976, Tubuloglomerular feedback in the early stages of experimental acute renal failure, *Kidney Int.* **10:**S106.

Mason, J., Olbricht, C., Takabatake, T., and Thurau, K., 1977, The early phase of experimental acute renal failure. I. Intratubular pressure and obstruction, *Pfluegers Arch.* **370:**155.

Mason, J., Takabatake, T., Olbricht, C., and Thurau, K., 1978, The early phase of experimental acute renal failure. III. Tubuloglomerular feedback, *Pfluegers Arch.* **373:**69.

Mason, J., Kain, H., Shiigai, T., Welsch, J., Schlecker, H., and Steff, M., 1979, The early phase of experimental acute renal failure. V. The influence of suppressing the renin–angiotensin system, *Pfluegers Arch.* **380:**233.

Mauk, R. H., Lifschitz, M. D., and Stein, J. H., 1976, The role of renal ischemia in the initiation of uranylnitrate induced acute renal failure in the dog, *Circ. Res.* **24:**406A (abstract).

Mauk, R. H., Patak, R. V., Fadem, S. Z., Lifschitz, M. D., and Stein, J. H., 1977, Effect of prostaglandin E administration in a nephrotoxic and a vasoconstriction model of acute renal failure, *Kidney Int.* **12:**122.

Mendelsohn, F. A. O., 1979, Evidence for the local occurrence of angiotensin II in rat kidney and its modulation by dietary sodium intake and converting enzyme blockade, *Clin. Sci. Mol. Med.* **57:**173.

Mendelsohn, F. A. O., and Smith, E. A., 1980, Intrarenal renin, angiotensin II, and plasma renin in rats with uranyl-nitrate induced and glycerol induced acute renal failure, *Kidney Int.* **17:**465.

Morrison, A. R., Nishikawa, K., and Needleman, P., 1978, Thromboxane A_2 biosynthesis in the ureter obstructed isolated perfused kidney of the rabbit, *J. Pharmacol. Exp. Ther.* **205:**1.

Munck, O., 1958, *Renal Circulation in Acute Renal Failure,* Blackwell, Oxford.

Oken, D. E., 1975, On the passive back-flow theory of acute renal failure, *Am. J. Med.* **58:**77.

Oken, D. E., 1976, Local mechanisms in the pathogenesis of acute renal failure, *Kidney Int.* **10:**S94.

Olbricht, C. H. J., 1980, Experimental models of acute renal failure, *Contrib. Nephrol.* **19:**110.

Reubi, F. C., Gurtler, R., and Gossweiler, N., 1962, A dye-dilution technique of measuring renal blood flow in man with special reference to the anuric subject, *Proc. Soc. Exp. Biol. Med.* **111:**760.

Reubi, F. C., Grossweiler, N., and Gurtler, R., 1966, Renal circulation in man studied by means of a dye-dilution method, *Circulation* **33:**426.

Reubi, F. C., Vorburger, C., and Tuckman, J., 1973, Renal distribution volumes of indocyanine green, (^{51}Cr)EDTA and ^{24}Na in man during acute renal failure, *J. Clin. Invest.* **52:**223.

Russell, S. B., 1975, The mechanism of action of mercuric chloride on the isolated perfused rat kidney, *Eur. J. Clin. Invest.* **5**:319.

Schor, N., Ichikawa, I., Rennke, H. R., and Brenner, B. M., 1979, Comparative effects of tobramycin vs. gentamicin on glomerular function in the Munich-Wistar rat, *Kidney Int.* **16**:776 (abstract).

Sherwood, T., Lavender, J. P., and Russell, S. B., 1974, Mercury induced renal vascular shut-down: Observations in experimental acute renal failure, *Eur. J. Clin. Invest.* **4**:1.

Siegel, N. J., Gunstream, S. K., Handler, R. I., and Kashgarian, M., 1977, Renal function and cortical blood flow during the recovery phase of acute renal failure, *Kidney Int.* **12**:199.

Solez, K., Kramer, E. C., Fox, J. A., and Heptinstall, R. H., 1974, Medullary plasma flow and intravascular leukocyte accumulation in acute renal failure, *Kidney Int.* **6**:24.

Solez, K., Altman, J., Rienhoff, H. Y., Riela, R. A., Finer, P. M., and Heptinstall, R. H., 1976, Early angiographic and renal blood flow changes after $HgCl_2$ or glycerol-induced administration, *Kidney Int.* **10**:S153.

Solez, K., Idema, T., and Saito, H., 1979, Role of tubular obstruction, thromboxane synthesis and outer medullary microvascular injury in post-ischemic acute renal failure, *Kidney Int.* **16**:777 (abstract).

Solomon, H. S., and Hollenberg, N. K., 1975, Catecholamine release: Mechanism of mercury-induced vascular smooth muscle contraction, *Am. J. Physiol.* **229**:8.

Sraer, J. D., Sraer, J., Ardaillou, R., and Mimoune, O., 1974, Evidence for renal glomerular receptors for angiotensin II, *Kidney Int.* **5**:241.

Stein, J. H., Gottschall, J., Osgood, R. W., and Ferris, T. F., 1975, Pathophysiology of a nephrotoxic model of acute renal failure, *Kidney Int.* **8**:27.

Stein, J. H., Lifschitz, M. D., and Barnes, L. D., 1978, Current concepts on the pathophysiology of acute renal failure, *Am. J. Physiol.: Renal Fluid Electrolyte Physiol.* **3**:F171.

Sudo, M., Honda, N., Hishida, A., and Nagasa, M., 1977, Renal hemodynamics in uranyl acetate-induced acute renal failure of rabbits, *Kidney Int.* **11**:35.

Sudo, M., Honda, N., Hishida, A., and Nagasa, M., 1980, Renal hemodynamics in oliguric and nonoliguric acute renal failure of rabbits, *Nephron* **25**:144.

Thiel, G., Brunner, F., Wunderlich, P., Huguenin, H., Bienko, B., Torhorst, J., Peters-Haefeli, L., Kirchertz, E. J., and Peters, G., 1976, Protection of rat kidneys against $HgCl_2$-induced renal failure by induction of high urine flow without renin suppression, *Kidney Int.* **10**:S191.

Thurau, K., Boylan, J. W., and Mason, J., 1979, Pathophysiology of acute renal failure, in: *Renal Disease* (D. Black and N. F. Jones, eds.), Blackwell, Oxford, p. 64.

Torres, V. E., Carlos Romero, J., Strong, C. G., Huguenin, M., and Wilson, D. M., 1974, Species difference in the effect of indomethacin on the development of acute circulatory renal failure, *Acta Physiol. Lat.-am.* **24**:539.

Trueta, J., Barclay, A. E., Daniel, P. M., Franklin, K. J., and Prichard, M. M. L., 1947, *Studies of the Renal Circulation,* Oxford, Blackwell.

Yoshita, Y., Honda, N., Morikawa, A., and Seki, K., 1966, Alterations in renal hemodynamics induced by increased renal vein pressure in the rabbit kidney, *Japan Heart J.* **7**:289.

Zipser, R., Myers, S., and Needleman, P., 1980, Exaggerated prostaglandin and thromboxane synthesis in the rabbit with renal vein constriction, *Circ. Res.* **47**:231.

5

The Glomerulus in Acute Renal Failure

ROLAND C. BLANTZ

During the past 10–12 years, investigators in medical renal physiology have altered their focus from the examination of normal regulatory mechanisms to focus in part upon a variety of pathophysiologic models of altered renal function. One of the most active and initial arenas of pathophysiologic evaluation has been examination of the mechanism of acute renal failure (ARF). Several of the initial studies, derived from both this country and Europe, provided an initial unitary hypothesis which postulated a pivotal role for renal vasoconstriction in the pathogenesis of acute renal failure (Ayer *et al.*, 1971; Oken *et al.,*1966; Henry *et al.,*1968; Thiel *et al.,* 1967). Studies supported the concept that major reductions in renal blood flow (RBF) and renal plasma flow (RPF) occurred early in the initiating phase of acute renal failure (Ayer *et al.,* 1971; Chedru *et al.,* 1972; Hsu *et al.,* 1977; Flamenbaum *et al.,* 1972b; Munck, 1958; Hollenberg *et al.,* 1968). These findings led to the view commonly and logically held at that time that acute renal failure was a "vasomotor nephropathy."

Questions were raised as to the mechanism of renal vasoconstriction. During this same period, a normal control mechanism which linked the rate of tubular reabsorption to the nephron filtration rate, a tubuloglomerular feedback mechanism, was also becoming better defined (Schnermann *et al.,* 1970; Thurau, 1964). Studies deriving primarily from Germany demonstrated that increase in the

ROLAND C. BLANTZ • Department of Medicine, University of California, San Diego, School of Medicine, and Veterans Administration Medical Center, San Diego, California 92161.

rate of delivery of normal tubular fluid out of the proximal tubule was associated, at the single nephron level, with reduction in nephron filtration rate (Schnermann *et al.,*1970; Thurau, 1964). Certain investigators cleverly attempted to integrate this developing normal control system into a unitary hypothesis which provided a basis for the vasoconstriction which had been postulated as a primary mechanism of acute renal failure. Figure 1 depicts a simplified diagrammatic representation of the mechanisms proposed by Henry *et al.*(1968) linking activation of the normal tubuloglomerular feedback system to a vasoconstrictor mechanism in acute renal failure. Toxic or ischemic damage of tubules could lead to decreased NaCl and water reabsorption, and increases in distal tubular fluid delivery would then activate tubuloglomerular feedback systems acting to reduce nephron filtration rate, presumably by vasoconstriction.

What was the mediator of the postulated feedback-induced vasoconstriction? Single nephron dissection and microperfusion techniques suggested that the renin content of single JGA cells increases with flow-induced activation of tubuloglomerular feedback systems (Thurau, 1974). Angiotensin II (AII), the most potent biologic vasoconstrictor, became the most logical candidate as the mediator of vasoconstriction in acute renal failure. However, firm evidence that angiotensin II was involved in the normal tubuloglomerular feedback mechanism was lacking (Blantz *et al.,* 1981).

There followed a variety of studies which examined the potential role of AII in initiating ARF. Since specific inhibitors of AII formation and AII receptor antagonists were not available at that time, suppression of AII activity in ARF was attempted by prior acute or chronic NaCl loading or antirenin AB administration

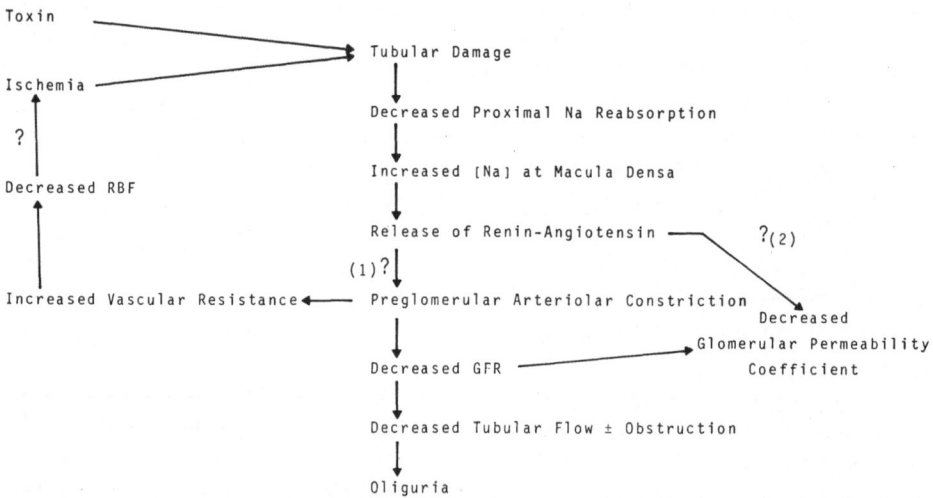

FIGURE 1. Potential mechanisms to explain the oliguria of acute renal failure as originally hypothesized by Henry *et al.* (1968). Pathway 1 leads to vasoconstriction, and an updated hypothesis and mechanism, pathway 2, recognizes the potential action of angiotensin II to decrease the glomerular permeability coefficient.

(Flamenbaum *et al.*, 1972a; McDonald *et al.*, 1969). Studies were not totally conclusive, but the general hypothesis was based upon considerable logic and exerted significant appeal among nephrologists and clinical investigators examining the pathophysiology of acute renal failure.

Several studies later in the past decade raised significant questions regarding "vasomotor nephropathy" as a unitary hypothesis in the pathogenesis of acute renal failure. Studies in established clinical acute renal failure in humans revealed that measured renal blood averaged at least 50% of control values, while glomerular filtration rate (GFR) was concurrently 0–5% of control values (Reubi, 1974). Also, maneuvers which increased RBF to greater than control values exerted no beneficial effect upon GFR or urine volume. Micropuncture studies with certain forms of acute renal failure with heavy metal toxins from Stein *et al.* (1975) and from our laboratory (Blantz, 1975) suggested that major vasomotor events and reductions in renal plasma flow were not a necessary requirement for major reductions in GFR in these forms of ARF.

In studies which appeared in 1975, we examined the specific mechanisms leading to reduction in GFR within a few hours after the administration of uranyl nitrate (UN) (Blantz, 1975). This study examined the earliest stages of a significant form of renal injury, since, in metabolic studies, all rats died within 4 days of uranyl nitrate administration. Within a few hours of UN, the reduction in kidney GFR was always greater than the reduction in superficial nephron filtration rate at both doses of UN. Tubular microinjection studies with inulin and mannitol revealed that the greater reduction in kidney GFR was the result of major transepithelial leak of solutes and water. In addition, single-nephron glomerular filtration rate (SNGFR) at the glomerulus was also decreased. The nephron plasma flow was not reduced below control values at either dose of UN. The major abnormality contributing to initial reductions in SNGFR was a decrease in L_pA, the glomerular permeability coefficient. By transmission electron microscopy, no glomerular membrane abnormalities were observed at this early stage. Stein and co-workers, also in 1975, demonstrated similar physiologic findings in the dog. Scanning electron micrographs of glomeruli several hours after UN in the dog revealed significant alterations in epithelial cell morphology with fusion of foot processes (Stein *et al.*, 1975). Similar morphologic changes were observed in the norepinephrine-infused ischemic model of acute renal failure (Cox *et al.*, 1974).

In 1980, Avashti provided evidence for an interesting morphologic basis for the reduction in glomerular permeability coefficient L_pA which we have observed after uranyl nitrate in the rat (Avasthi *et al.*, 1980). Scanning electron micrographs of the endothelial cell surface of the glomerular membrane at 7 hr after either 15 or 25 mg/kg UN revealed reductions in the radius and the density of endothelial fenestrae which correlated well with the decrease in GFR. Chronic oral saline loading prevented much of the reduction in GFR and this lesser functional alteration correlated well with lesser reductions in fenestral density and radius. Therefore, studies accumulated over the past several years have demonstrated that reductions in L_pA contribute significantly to decreases in GFR in both toxic and ischemic models of ARF. In many cases, there are glomerular morphologic alterations which could serve as a basis for the reduction in L_pA. However, this morphologic basis is less well defined than the morphologic basis for reductions in L_pA documented in acute glomerular immune injury. Accumulation of inflammatory cells within the capillary and alterations in endothelial cells form a reasonable basis for the reduction in L_pA in the renal failure following immune injury (Blantz et al., 1978).

Is there a common mechanism that could explain the reduction of L_pA in ARF and also the morphologic changes? In 1976, we observed that AII infusion produced major reductions in L_pA (Blantz et al., 1976). Baylis and Brenner (1978) have also observed this decrease in L_pA after AII infusion. Since these initial reports, there has been abundant evidence that AII is an important mediator of reduction in L_pA. Ichikawa, Baylis, and Brenner have shown that a variety of hormonal substances can lower L_pA when infused parenterally (Baylis et al., 1976; Ichikawa and Brenner, 1977, 1979; Ichikawa et al., 1978). With the single exception of antidiuretic hormone, these substances appear to be influencing L_pA via the intrarenal release and local action of AII.

In certain physiologic states in which renin–AII activity is increased, such as during reductions in renal artery perfusion pressure and after chronic NaCl depletion in the rat, L_pA is also modestly decreased. Ichikawa and Brenner (1980) have presented data which suggest that blockade of AII activity will prevent this decrease in L_pA during reductions in renal perfusion pressure. After 2 weeks of chronic NaCl depletion in the rat, we have shown that L_pA is decreased to approximately 50% of control values and RPF is also decreased (Steiner et al., 1979). The original study demonstrated that saralasin infusion restored RPF to normal control values but did not effect the reduction in L_pA. However, more recent, unpublished observations from our laboratory demonstrate that L_pA can be restored to normal values by acute volume repletion, and that pretreatment with captopril, an oral converting enzyme inhibitor, totally prevents a reduction in L_pA in rats with chronic NaCl depletion. Therefore, these studies suggest that AII is a potentially important mediator of moderate functional alterations in L_pA in certain altered physiologic states.

How does AII effect the glomerular permeability coefficient? Studies by Hornych et al. (1972) revealed that infusion of AII produced rather dramatic structural alterations in rat glomeruli when evaluated by scanning electron microscope. Contraction of mesangial cells in response to AII could explain much, but possibly not all of the alterations in glomerular architecture observed. Ausiello et al. (1980) have more recently documented contraction of mesangial cells in culture when exposed to physiologic concentrations of AII in the media. Close examination of these electron micrographs suggests that the configuration of epithelial cells and foot processes were also effected by AII. In simplest terms, AII could reduce L_pA through at least three general mechanisms. First, generalized capillary lumen narrowing could result from mesangial contraction with resulting reductions in capillary surface area. Although generally in keeping with findings in the literature, this single explanation is probably not wholly adequate on a hydrodynamic basis when one considers the changes in capillary resistance required. Second, there may be AII-induced structural alterations in the glomerular capillary membrane at the level of either epithelial or endothelial cell which could contribute to decreases in $L_p A$. Third, capillary constriction may be nonuniform, leading to diversion of glomerular plasma flow from conduits with greater effective surface area for ultrafiltration and length to conduits with limited surface area or length, leading again to lower $L_p A$ values (Blantz, 1980).

Is it possible that we have come the full cycle in our attempts to formulate logical hypotheses on the pathogenesis of acute renal failure? Original hypotheses envisioned accelerated AII release after an initial renal insult leading to relentless vasoconstriction. Is it possible that this scenario was correct except that accelerated local AII release may be exerting its major effect upon the glomerulus leading to major decreases in $L_p A$ (Fig. 1) rather than vasoconstriction?

A single report in the recent literature lends some support to such a formulation. Baylis *et al.* (1977) have demonstrated that the modest reduction in nephron filtration rate after gentamicin-induced acute renal failure was due to significant reductions in the glomerular permeability coefficient. A recent publication from this same laboratory by Schor *et al.* (1981) has reported that if captopril, the oral converting enzyme inhibitor, is administered with gentamicin, reductions in the glomerular permeability coefficient and nephron plasma flow were totally prevented and filtration rate remained equal to control rats. Although one must interpret these findings with care to ensure that captopril is not preventing the expression of a normal glomerular response to chronic NaCl depletion (Steiner *et al.*, 1979) as a consequence of gentamicin administration, these results support a role for AII in producing changes in the glomerular permeability coefficient and nephron plasma flow in gentamicin nephrotoxicity. There were no changes in glomerular morphology distinctive for gentamicin nephrotoxicity.

We have recently reexamined the mechanism of the early reduction in $L_p A$ which occurs after 25 mg/kg uranyl nitrate, focusing upon alterations in glomerular function. Rats were pretreated with oral captopril for 4–5 days prior to acute uranyl nitrate administration. These preliminary studies revealed that four of five rats pretreated with captopril died within 4 days after uranyl nitrate administration, findings similar to original observations in untreated rats (Blantz, 1975). However, these fatalities from acute renal failure may have been the consequence of persistent uranium-induced transepithelial backleak of solutes rather than the result of primary alterations in filtration at the glomerulus. Micropuncture studies were also performed within 5 hr of uranyl nitrate administration. These preliminary observations were complicated by lower mean arterial pressure in rats receiving uranyl nitrate and pretreated with captopril, probably reflecting both urinary volume losses in this initially polyuric form of acute renal failure and the inherent effects of captopril on peripheral vascular resistance. As a result of lower blood pressures, nephron plasma flow was reduced sufficiently to result in filtration pressure equilibrium in five of eight rats, a condition in which specific values for $L_p A$ cannot be defined, but only minimum possible values.

Nephron filtration rate was reduced below hydropenic control rats in spite of captopril pretreatment (Table I). The sole reason for reductions in nephron

TABLE I. Effects of Captopril Pretreatment on Uranyl
Nitrate Nephrotoxicity[a]

	Uranyl nitrate 25 mg/kg, SC, + captopril ($n = 8$)	Hydropenia ($n = 9$)	Uranyl nitrate 15 mg/kg, IV ($n = 8$)
Mean arterial pressure, mm Hg	82 ± 2	114 ± 3	119 ± 5
p value	—	<0.01	<0.01
SNGFR,[b] nl min^{-1} gKW^{-1}	27.5 ± 2.6	38.3 ± 2.3	29.1 ± 1.0
p value	—	<0.01	NS
GFR,[b] ml min^{-1} gKW^{-1}	0.61 ± 0.15	1.05 ± 0.04	0.49 ± 0.10
p value	—	<0.02	NS
RPF,[b] nl min^{-1} gKW^{-1}	85.9 ± 9.4	115 ± 12	93.5 ± 5.9
p value	—	NS	NS
ΔP, mm Hg	35.4 ± 1.3	33.8 ± 0.6	27.7 ± 1.8
p value	—	NS	<0.01
$L_p A$,[b] nl sec^{-1} gKW^{-1} mm Hg^{-1}	0.06 ± 0.01	0.08 ± 0.01	0.05 ± 0.01
p value	—	NS	NS

[a] The p values are all comparisons with the uranyl nitrate + captopril group.
[b] gKW = gram of kidney weight.

filtration rate in three rats was the lower value for $L_p A$. However, in five rats, reductions in nephron plasma flow dominated and specific conclusions regarding effects of captopril treatment on $L_p A$ could not be made with confidence.

Based upon these preliminary observations, captopril, a converting enzyme inhibitor, does not prevent a reduction in either GFR or nephron filtration rate after uranyl nitrate administration. Further studies are required to determine if the decrease in $L_p A$ after uranyl nitrate is any way AII-dependent.

Alterations in glomerular function and, specifically, reductions in the glomerular permeability coefficient contribute to the major reductions in glomerular filtration rate in the early stages of many experimental models of acute renal failure. A morphologic basis for these changes in glomerular function has been provided in several experimental models. The question remains whether accelerated generation of local renal AII contributes to these reductions in $L_p A$ in some but not all experimental models of ARF. Further studies which examine the determinants of glomerular ultrafiltration are required to define the specific pathophysiologic models of ARF in which AII contributes significantly to the pathogenesis of the reduction in glomerular filtration rate.

References

Ausiello, D. A., Kreisberg, J. I., Roy, C., and Karnovsky, M. J., 1980, Contraction of cultured rat glomerular mesangial cells after stimulation with angiotensin II and arginine vasopressin, *J. Clin. Invest.* **65**:754.

Avasthi, P. S., Evan, A. P., and Hay, D., 1980, Glomerular endothelial cells in uranyl nitrate-induced acute renal failure in rats, *J. Clin. Invest.* **65**:121.

Ayer, G., Grandchamp, A., Wyler, J., and Truniger, B., 1971, Intrarenal hemodynamics in glycerol-induced myohemoglobinuric acute renal failure in the rat, *Circ. Res.* **29**:128.

Baylis, C., and Brenner, B. M., 1978, Modulation by prostaglandin synthesis inhibitors of the actions of exogenous angiotensin II on glomerular ultrafiltration in the rat, *Circ. Res.* **43**:889.

Baylis, C., Deen, W. M., Myers, B. D., and Brenner, B. M., 1976, Effects of some vasodilator drugs on transcapillary fluid exchange in renal cortex, *Am. J. Physiol.* **230**:1148.

Baylis, C., Rennke, H. G., and Brenner, B. M., 1977, Mechanisms of the defects in glomerular ultrafiltration associated with gentamycin administration, *Kidney Int.* **12**:344.

Blantz, R. C., 1975, The mechanism of acute renal failure after uranyl nitrate, *J. Clin. Invest.* **55**:621.

Blantz, R. C., 1980, The glomerulus, passive filter or regulatory organ?, *Klin. Wochenschr.* **58**:957.

Blantz, R. C., Konnen, K. S., and Tucker, B. J., 1976, Angiotension II effects upon the glomerular microcirculation and ultrafiltration coefficient of the rat, *J. Clin. Invest.* **57**:419.

Blantz, R. C., Tucker, B. J., and Wilson, C. B., 1978, The acute effects of antiglomerular basement membrane antibody on the process of glomerular filtration in the rat. II. Influence of dose and complement depletion, *J. Clin. Invest.* **61**:910.

Blantz, R. C., Steiner, R. W., and Tucker, B. J., 1981, The efferent limb of the tubuloglomerular feedback system, *Fed. Proc.* **40**:104.

Chedru, M. F., Baethke, R., and Oken, D. E., 1972, Renal cortical blood flow and glomerular filtration in myohemolobinuric acute renal failure, *Kidney Int.* **1**:232.

Cox, J. W., Baehler, R. W., Sharma, H., O'Dorisio, T., Osgood, R. W., Stein, J. H., and Ferris, T. F., 1974, Studies on the mechanism of oliguria in a model of unilateral acute renal failure, *J. Clin. Invest.* **53**:1546.

Flamenbaum, W., Kotchen, T. A., and Oken, D. E., 1972a, Effect of renin immunization on mercuric chloride and glycerol-induced renal failure, *Kidney Int.* **1**:406.

Flamenbaum, W., McNeil, J. S., Kotchen, T. A., and Saladino, A. J., 1972b, Experimental acute renal failure induced by uranyl nitrate in the dog, *Circ. Res.* **31**:682.

Henry, L. N., Lane, C. E., and Kashgarian, M., 1968, Micropuncture studies of the pathophysiology of acute renal failure in the rat, *Lab. Invest.* **19**:309.

Hollenberg, N. K., Epstein, M., Rosen, S. M., Basch, R. I., Oken, D., and Merrill, J. P., 1968, Acute oliguric renal failure in man: Evidence for preferential renal cortical ischemia, *Medicine* **47**:455.

Hornych, H., Beaufils, M., and Richet, I., 1972, The effect of exogenous angiotensin on superficial and deep glomeruli in the rat kidney, *Kidney Int.* **2**:336.

Hsu, C. H., Kurtz, T. W., Rosenzweig, J., and Weller, J. M., 1977, Renal hemodynamics in $HgCl_2$-induced acute renal failure, *Nephron* **18**:326.

Ichikawa, I., and Brenner, B. M., 1977, Evidence for glomerular actions of ADH and dibutyryl cyclic AMP in the rat, *Am. J. Physiol.* **233**:F102.

Ichikawa, I., and Brenner, B. M., 1979, Mechanism of action of histamine and histamine antagonists on the glomerular microcirculation of the rat, *Circ. Res.* **45**:737.

Ichikawa, I., and Brenner, B. M., 1980, Importance of efferent arteriolar vascular tone in regulation of proximal tubule fluid reabsorption and glomerulotubular balance in the rat, *J. Clin. Invest.* **65**:1192.

Ichikawa, I., Humes, H. D., Dousa, T. P., and Brenner, B. M., 1978, Influence of parathyroid hormone on glomerular ultrafiltration in the rat, *Am. J. Physiol.* **234**:F393.

McDonald, F. D., Thiel, G., Wilson, D. R., DiBona, G. F., and Oken, D. E., 1969, The prevention of acute renal failure in the rat by long term saline loading. A possible role for the renin–angiotensin axis, *Proc. Soc. Exp. Biol. Med.* **131**:610.

Munck, O., 1958, *Renal Circulation in Acute Renal Failure,* Blackwell, Oxford.

Oken, D. E., Arce, M. L., and Wilson, D. R., 1966, Glycerol-induced hemoglobinuric acute renal failure in the rat I. Micropuncture Study of the Development Oliguria, *J. Clin. Invest.* **45**:724.

Reubi, F. C., 1974, The pathogenesis of anuria following shock, *Kidney Int.* **5**:106.

Schnermann, J., Wright, F. S., Davis, J. M., Stackelberg, V., and Grill, G., 1970, Regulation of superficial nephron filtration rate by tubulo-glomerular feedback, *Pfluegers Arch.* **318**:147.

Schor, N., Ichikawa, I., Rennke, H. G., Troy, J. L., and Brenner, B. M., 1981, Pathophysiology of altered glomerular function in aminoglycoside-treated rats, *Kidney Int.* **19**:288.

Stein, J. S., Gottschall, J., Osgood, R. W., and Ferris, T. F., 1975, Pathophysiology of a nephrotoxic model of acute renal failure, *Kidney Int.* **8**:27.

Steiner, R. W., Tucker, B. J., and Blantz, R. C., 1979, Glomerular hemodynamics in rats with chronic sodium depletion: Effect of saralasin, *J. Clin. Invest.* **64**:503.

Thiel, G., Wilson, D. R., Arce, M. L., and Oken, D. E., 1967, Glycerol induced hemoglobinuric acute renal failure in the rat. II. The experimental model, predisposing factors, and pathophysiologic features, *Nephron* **4**:276.

Thurau, K., 1964, Renal hemodynamics, *Am. J. Med.* **36**:698.

Thurau, K., 1974, JGA renin activity: Constituent of single nephron function and dependence on NaCl at the macula densa, in: *Proc. 5th Int. Congr. Nephrol.,* Volume 2 (H. Zillarreal, ed.), pp. 183–192, S. Karger, Basel, Switzerland.

6

Heavy Metal Models of Experimental Acute Renal Failure

WALTER FLAMENBAUM, JAMES KAUFMAN,
SHREEKANT CHOPRA, MARC GEHR, and
ROBERT HAMBURGER

1. Introduction

A large number of experimental models have been developed to evaluate the pathophysiology of acute renal failure (ARF). The diversity of these models of experimental ARF is a reflection both of the search for the "perfect" model as well as the heterogeneity of clinical ARF. In general, the experimental models of ARF may be subdivided into those induced with a nephrotoxin and those following a period of renal ischemia (Flamenbaum, 1973). A similar subdivision may be made, when an etiologic event is known, for clinical instances of ARF. The lack of a single, universally accepted model used for the study of the pathophysiology of ARF thus reflects both the diversity of clinical ARF as well as a lack of uniform agreement concerning the pathophysiologic mechanism involved in the pathogenesis of ARF.

WALTER FLAMENBAUM • Nephrology Division, Beth Israel Medical Center, New York, New York 10003. JAMES KAUFMAN, SHREEKANT CHOPRA, MARC GEHR, and ROBERT HAMBURGER • Renal Section, Boston Veterans Administration Medical Center, Boston, Massachusetts 02130.

In evaluating experimental models of ARF, one must be extremely cautious. Investigators with a primary interest in this area have developed models for various reasons (Stein *et al.,* 1978). In some instances, they have attempted to replicate a specific or common clinical form of ARF; whereas in other instances they have been more concerned with basic tenets of research methodology. For any model to be of use, it must bear some relationship to clinical ARF, and most also be sufficiently reproducible and homogeneous to allow inferences concerning basic mechanisms. Over the past decade our research efforts in this area have been almost exclusively concerned with heavy metal models of experimental ARF, rather than those induced by renal ischemia. In our laboratory, nephrotoxic ARF has been easy to induce, reproducible, and relatively homogeneous at a given study interval. This has allowed a systematic approach to evaluating pathophysiologic mechanisms and evolving the potential for a rational approach for therapy and the clinical prevention of the processes involved in ARF.

Until recently, both clinicians and clinically oriented experimenters have been dissatisfied with nephrotoxic models of ARF because of an apparent lack of relevance to clinical ARF. It has been pointed out that unless high doses of nephrotoxins are utilized, these models are almost exclusively nonoliguric, a characteristic which was thought to be relatively uncommon in clinical ARF. The recent contributions of Schrier and colleagues have rectified this concern by careful clinical studies indicating that nonoliguric clinical ARF is more common than previously appreciated (Anderson *et al.,* 1977). This observation, along with even more clinically relevant models, has resulted in a renewed interest and enthusiasm concerning nephrotoxic models of ARF.

Before proceeding to a consideration of two specific nephrotoxic models of experimental ARF, a brief review of some general aspects of heavy metal toxicity is in order (Luckey and Venugopal, 1977; Porter and Bennett, 1981). This chapter will deal exclusively with heavy metal nephrotoxins, and the reader is referred to other chapters for more specific details concerning other nephrotoxic models. A number of physiochemical determinants (Table I) influence the toxicity of heavy metals in general. These characteristics, which are inherent in the heavy metal compounds utilized in induced ARF, must be considered when evaluating research in this area and its relevance to clinical medicine by extrapolation. Similarly, a host of other general factors (Table II) influence the physiologic disturbances resulting from the administration of a specific heavy metal compound to any of a host of biologic models, ranging from intact animal studies through and including experiments utilizing isolated membrane preparations. A failure to consider these factors, as well as the biologic suspectibility of the model under study (Table III), may lead to an incomplete understanding of the heavy metal model of ARF under evaluation.

TABLE I. Heavy Metal Toxicity: General Physiochemical Determinants

Solubility, stability, and state of hydration in biologic fluids
Susceptibility to biologic sequestration, metabolism, detoxification, chelation, and excretion
Oxidation state and electrochemical character
Particle size, and form in biologic tissues or fluids

TABLE II. Heavy Metal Toxicity: General Factors Influencing Toxicity

Toxicity of metal
Amount of metal
Absorption and transport of metal to target organ
Binding and combining capacity of metal
Excretion, metabolism (biotransformation), and sequestration of metal

It has also been amply pointed out that there are a host of factors resulting in a relative specificity of a variety of heavy metal toxins for the kidney. As indicated in Table IV, specific renal toxicity may be placed in perspective by considering the characteristics of the kidney in mammals. In general, renal blood flow is usually in the range of 3.5–5.0 ml/min/g of kidney weight, resulting in the delivery of large amounts of circulating nephrotoxins to the kidney. In addition to the amount of blood flowing through the kidney, one must also take into consideration the large endothelial surface area represented by the extensive capillary network within the kidney. Once a nephrotoxin has been delivered to the kidney, it has ample opportunity to interact with an extremely large surface area. In addition, the kidney is both metabolically active and contains certain physiologic mechanisms which promote toxic manifestations. Although clinicians and physiologists may be more concerned with the physical processes of renal blood flow in the formation of a glomerular filtrate, one cannot ignore the many active transport mechanisms, enzyme systems, sites for hormonal synthesis, and the (in general) high metabolic activity of the kidney as an organ. In addition to active transport mechanisms, the ability of the kidney to concentrate tubular fluid using the countercurrent concentration mechanism results in the simultaneous concentration of any toxic compound present within glomerular filtrate.

With these general considerations in mind, one may place in appropriate perspective a variety of nephrotoxic models of ARF. Our laboratories have been concerned with experimental ARF induced by uranyl nitrate and, most recently, the more clinically relevant model resulting from the administration of *cis*-diamminedichlorplatinum (a cancer chemotherapeutic agent).

2. Uranyl Nitrate-Induced Acute Renal Failure

Uranyl nitrate (UN)-induced ARF has been extensively studied over the past decade. A review of some of our observations and the direction they have taken is in order so that our current interest in alternative models of ARF may be better understood. Uranyl nitrate-induced ARF has been studied in intact animals,

TABLE III. Heavy Metal Toxicity: General Biologic Susceptibility

Permeability of cell membranes and subcellular organelles to heavy metal
Ability to alter structure and/or function of proteins, nucleic acids, etc.
Stimulation or inhibition of release/synthesis of modifying substances or hormones

TABLE IV. Heavy Metal Toxicity: General Factors

High blood flow relative to weight
Large endothelial surface area
Active transport mechanisms, high metabolic activity
Countercurrent concentration mechanism
Presence of many enzyme systems and site of multiple hormone synthesis

including dogs, rats, and rabbits, as well as with isolated membrane preparations (Blantz, 1975; Flamenbaum *et al.*, 1972, 1974). After the development of a reproducible UN-induced ARF model, we were able to perform a large number of experiments using a number of different techniques in order to evaluate pathophysiologic mechanisms resulting in renal failure and experimental maneuvers designed to prevent and/or ameloriate the resulting renal insufficiency. The results of studies using an *in vitro* preparation are presented elsewhere (see Chapter 3, this volume) and will not be reviewed within this presentation.

Uranyl nitrate, administered intravenously in the dog and subcutaneously in the rat, results in a reproducible pattern of ARF (Kleinman *et al.*, 1975). In general, UN-induced ARF in our laboratory may best be described as a nonoliguric model characterized by a greater decrease in glomerular filtration rate (GFR) than in renal blood flow, and associated with the loss of urine-concentrating ability and an increase in sodium excretion. Sequential studies carried out in both the rat and the dog allowed us to delineate the temporal course of this model of ARF, resulting in specific descriptions of the following phases: the initiation phase, early after the administration of UN and characterized by parallel decrements in renal blood flow and glomerular filtration rate, in association with diminutions in sodium reabsorption and urine concentration; the maintenance phase, characterized by a more marked and continued decline in glomerular filtration rate than in glomerular blood flow, with continued tubular abnormalities; and the recovery phase, during which renal function tends to return toward normal. These phases, which have been previously described in clinical ARF, were useful experimental struts, permitting the design of specific experiments to evaluate the pathogenetic mechanisms responsible for induced renal abnormalities as a function of the time course of the experimental model under study.

2.1. Tubuloglomerular Feedback Mechanism

The bulk of the studies on UN-induced ARF from our laboratories have been the subject of a recent review (Flamenbaum *et al.*, 1977). There are, however, two sets of experiments of sufficient importance to warrant reviewing them here. The first concerns evidence that the tubuloglomerular feedback mechanism participates in the initiation of this model of experimental ARF (Flamenbaum *et al.*, 1976). Using anatomical and functional evidence, it had been proposed that a tubuloglomerular feedback mechanism related alterations in tubular fluid to changes in glomerular filtration rate. According to this proposed mechanism, an alteration in a sodium-related stimulus within the macula densa segment of the nephron results in feedback regulation of glomerular afferent arteriolar tone, effective filtration pressure, and, therefore, single nephron glomerular filtration rate. It was further suggested that the primary mediator for the vasoactive phenomenon inherent in the tubuloglomerular feedback mechanisms was juxtaglomerular apparatus renin activity. Having described marked alterations in hemodynamics, glomerular filtration rate, and tubular fluid reabsorption after UN-induced ARF in the rat, we further evaluated a possible role for the tubuloglomerular feedback mechanism using micropuncture studies in the rat. Uranyl nitrate was injected subcutaneously, 10 mg/kg of body weight, and the animals were studied using standard micropuncture techniques for up to 6 hr after injection. Additional studies of the renin–angiotensin system were performed, consisting of determinations of plasma renin activity as well as the renin activity within single dissected juxtaglomerular apparatuses. The effect of UN on glomerular filtration rate and fractional excretion of sodium was readily apparent at 6 hr and consisted of a reduction in glomerular filtration rate to 0.6 ± 0.2 (SE) ml/min/100 g body weight (versus control, 1.1 ± 0.1 ml/min/100 g body weight), an increase in the fractional excretion of sodium to $1.8 \pm 0.1\%$ (versus control, $0.7 \pm 0.04\%$); and a concomitant increase in urine volume and decrease in urine concentration. An analysis of single-nephron glomerular filtration rate determined in superficial nephrons revealed a parallel decline at 6 hr to 10.1 ± 0.5 nl/min/100 g body weight, as compared to a control value of 15.9 ± 0.3 nl/min/100 g body weight. Previous studies had indicated a similar 30% decline in total renal blood flow during the initiation phase of UN-induced ARF in the rat.

In order for tubuloglomerular feedback to be entertained as the mechanism responsible for these changes in renal function, the following alterations must be demonstrated: disturbances in tubular fluid handling such that there is an alteration in the sodium chloride-related macula densa stimulus; activation of the renin–angiotensin system, preferably at the level of the individual juxtaglomerular apparatus associated with the involved nephrons; alterations in renal blood flow, and, concomitantly, glomerular filtration rate, consistent with the feedback signal. As noted, there were approximately 30% decrements in renal blood flow and whole-kidney and single-nephron glomerular filtration rates, along with changes indicative of the abnormalities in tubule function (increased fractional excretion of sodium, decreased urinary concentration). Analysis of samples obtained from distal convoluted tubules and analyzed using a helium glow photometer indicated that distal tubule fluid sodium concentration 6 hr after UN had increased to 116.9 ± 2.5 meq/liter, significantly greater than the control value of 53.7 ± 1.2 meq/liter. Furthermore, plasma renin activity increased from a control value of 1.5 ± 0.3 ng/ml/hr to 2.9 ± 0.4 ng/ml/hr, consistent with activation of the renin–angiotensin system. An analysis of the renin activity within individual juxtaglomerular apparatuses indicated marked increases in renin content, from both superficial and deep nephron juxtaglomerular apparatuses. Thus, all of the components necessary to implicate the tubuloglomerular feedback mechanism were clearly demonstrated within this model of experimental ARF. It must be underscored that these studies were carried out early after the administration of UN, and are therefore consistent with a role for the tubuloglomerular feedback mechanism only in the initiation phase of ARF.

2.2. Dithiothreitol Studies

Based on *in vitro* and *in vivo* studies, as well as the physiochemical properties of UN, it was suggested that the initiating event in this model was the binding of heavy metal to sulfhydryl groups present within the epithelial cell membranes of the renal tubules, or binding to sulfhydryl-containing enzymes within the cytoplasm (Kleinman *et al.,* 1977). Subsequently, we used dithiorthreitol, a sulfhydryl reducing agent, in an attempt to prevent and/or reverse this heavy metal–protein interaction. These studies clearly demonstrated that dithiothreitol, when administered prior to the injection of UN, completely prevented the sequence of physiologic abnormalities induced by UN. Furthermore, activation of the renin-angiotensin system, as noted by increased plasma renin activity and juxtamedullary apparatus renin activity, was also prevented, confirming a failure to turn on a mediator of the feedback mechanism.

When viewed retrospectively, our studies suggest a possible sequence of events resulting in the initiation of heavy metal-induced ARF. The initial event, binding the heavy metal to either sulfhydryl groups within the apical membrane of the nephron or the cytoplasmic enzymes, results in an abnormality in tubular fluid handling culminating in an increase in sodium chloride concentration at the macula densa. This abnormality in tubule fluid composition is the signal turning on the tubuloglomerular feedback mechanism, which, through activation of the renin-antiogensin system (with or without the participation of other vasoactive systems), induces renal hemodynamic abnormalities and a reduction in glomerular filtration rate. The failure of feedback inhibition of the tubuloglomerular mechanism is the direct result of heavy metal-induced damage to tubular epithelium. Under normal circumstances the decline in glomerular filtration rate would be expected to result in an enhanced absorption of sodium chloride within proximal nephron segments, thus altering the macula densa signal and turning off or inhibiting feedback. However, continued perturbation of tubular function resulting from heavy metal-protein interaction prevents this normal turnoff. A similar extrapolation may be made with ischemic models of ARF, which could result in a similar pattern of nephron dysfunction. Subsequent to the initiation phase, there is ample opportunity for alternative mechanisms to participate in the pathophysiology of ARF. Depending upon the point in the temporal course of a specific model, the dose of the nephrotoxin, and the experimental techniques utilized, roles for tubule obstruction, tubule fluid leak, and alterations in glomerular hydraulic permeability could also occur, resulting in additional disturbances in renal function.

3. *Cis*-Platinum-Induced Renal Failure

Coordination complexes of platinum have been shown to have antineoplastic properties, although at the present time *cis*-diamminedichlorplatinum (CP) is the only compound currently in clinical use (Connors and Roberts, 1974; Prestayko *et al.*, 1980). The associated nephrotoxicity, demonstrated in experimental and clinical studies, has turned out to be the most significant dose-limiting factor in the clinical use of CP (Krakoss, 1979; Madias and Harrington, 1978). We have recently performed detailed studies specifically designed to evaluate the pathophysiologic basis for CP-induced ARF in the rat. These studies establish a new model of heavy metal-induced ARF, which may be used for detailed investigations of its pathophysiology.

For a model of experimental ARF to be useful for studies designed to evaluate pathophysiology, it must be predictable and reproducible. Our first efforts were therefore designed to establish such a model of ARF. Separate groups of rats, six in each group, were intraperitoneally injected with a single dose of CP at 2.5, 5.0, and 10 mg/kg body weight. Although daily studies of blood urea nitrogen (BUN) concentration were obtained, a significant and reproducible increase in BUN concentration was observed only at 96 hr after CP injection. Animals receiving 2.5 mg/kg body weight of CP had at 96 hr a mean BUN of 20 ± 3 mg/dl, not sufficiently different from the control preinjection value. All animals receiving higher doses of CP had increased BUN concentrations, although the mean peak value at a dose of CP 5.0 mg/kg of body weight, 58 ± 13 mg/dl, was significantly less than the peak BUN concentration of 128 ± 47 mg/dl observed in animals receiving the highest dose of CP. Based on these observations our studies were performed utilizing CP in a dose of 10 mg/kg body weight, administered intraperitoneally.

Additional cage studies were performed using this dose of CP in order to characterize this model of CP-induced ARF. A total of 21 rats were placed in metabolic cages and followed sequentially prior to and after CP injection. Of the 21 rats, one rat died by the fourth day of the study, and four failed to develop renal failure. In the remaining 16 rats, a number of pertinent observations were made. At 96 hr after CP injection, body weight had decreased to 265 ± 11 g (versus a control value of 294 ± 6 g) and packed cell volume had increased to $54 \pm 1\%$, significantly greater than the control value of $47 \pm 0.4\%$. Since the animals ingested less food, and specifically, less water, these observations were felt to be consistent with an associated volume depletion. Further support for this interpretation was apparent upon examining urine volume, the urine to plasma creatinine concentration ratio, and urine osmolality. The CP-induced ARF resulted in a nonoliguric state, with a urine flow at 96 hr of 8.2 ± 7.5 μl/min (versus control 11.4 ± 0.7 μl/min). In addition, there was a significant decline in urine to plasma creatinine concentration ratio to 24 ± 9 (control, 203 ± 12) and in urine osmolality to 834 ± 112 mOsm/kg (versus a control value of 1722 ± 113 mOsm/kg). These changes occurred in association with a marked decrease in renal function, as noted from an increase in serum creatinine concentration to 1.8 ± 0.4 mg/dl (control, 0.4 ± 0.01 mg/dl) and, concomitantly, a decline in endogenous creatinine clearance to 0.3 ± 0.1 ml/min, as compared to a control value of 2.3 ± 0.2 ml/min.

It is to be noted that the urine flow rate after CP was not statistically different from that observed during the control period, although maintenance of urine volume was not a constant finding. Indeed, three of the rats were oliguric on the study day. Additional studies of electrolyte handling by these animals indicated a marked increase in fractional excretion at 96 hr to $2.7 \pm 1.0\%$, as compared to the pre-CP value of $0.6 \pm .02\%$. The sham injections of the dilutent used with CP in six rats failed to cause any change in renal function at any time after the injection.

Based on these observations, additional micropuncture studies were performed 96 hr after the injection of CP (Chopra *et al.*, 1982). Standard micropuncture techniques were utilized to evaluate single-nephron function, with calculation of tubular fluid volume flow rate and the tubular fluid to plasma (TF/P) inulin concentration ratio. During the initiation of these studies it was apparent that intravenously injected lissamine green did not appear in the distal nephron segments, limiting the micropuncture studies to proximal tubule segments. Initial micropuncture studies were performed in control rats and in animals studied 96 hr after CP. Subsequent studies were performed to evaluate the role of volume depletion, as noted in the above cage studies, 96 hr after CP. In this group of animals, the rats were studied using micropuncture techniques 96 hr after CP, before and after volume expansion with Ringer's solution (5% body weight, intravenously, over 30 min). Separate rats were also studied to evaluate the possibility of tubule fluid backleak by the microinjection of radioactive inulin into proximal tubules, using a microinjection technique previously described for this laboratory. In addition, proximal intratubule hydrostatic pressures were measured both in control and CP-injected rats using a servonulling pressure device.

Comparable changes in body weight and packed cell volume were observed in the experimental group of rats subjected to micropuncture as in the initial cage studies noted above. During micropuncture, the kidney was directly observed using a binocular dissecting microscope. The kidney surface appeared homogeneous, with mild and uniform dilatation of the tubules, but without any evidence of intratubular debris or detritus. During the performance of micropuncture, the tubules did not appear to be friable, and were easily subjected to micropuncture. Parallel changes to those noted above were observed in whole-kidney glomerular filtration rate, urine flow rate, and the urine to plasma concentration ratio for inulin. The CP resulted in a decrease in single-nephron GFR (SNGFR) to 6.6 ± 0.4 nl/min/100 g body weight, versus the control value of 10.6 ± 0.5 nl/min/100 g body weight. Proximal tubule fluid volume was also significantly decreased to 2.8 ± 0.2 μl/min/100 g body weight (versus the control of 6.5 ± 0.4 μl/min/100 g body weight). In association with these changes, there was a significant increase in the TF/P inulin concentration ratio to 2.6 ± 0.1, as compared to a control value of 1.8 ± 0.1. Additional micropuncture studies carried out after volume expansion revealed no significant changes in SNGFR, tubular fluid flow rate, and TF/P inulin concentration ratios as compared with values obtained prior to volume expansion. However, volume expansion did result in a statistically significant increase in whole-kidney GFR. Prior to volume expansion at 96 hr after CP, whole–kidney GFR was 0.1 ± 0.01 ml/min/100 g body weight, whereas after volume expansion GFR increased to 0.2 ± 0.02 ml/min/100 g body weight, a value significantly greater than pre-volume expansion, but still significantly less than the control value.

It is evident from consideration of the whole-kidney and micropuncture studies obtained after CP that disproportionate alterations in single-nephron or whole-kidney function were observed. Thus, while whole-kidney glomerular filtration rate decreased to approximately 10% of control values, there was a lesser decrease, of approximately 30%, in superficial SNGFR measured in proximal tubule segments. This discrepancy between whole-kidney and superficial single-nephron GFR may be explained in different ways. First, transepithelial leak of tubule fluid may occur, resulting in a loss of glomerular filtrate at or beyond the site of micropuncture, resulting in a false overestimate of kidney function based on micropuncture results. To evaluate this possibility, SNGFRs measured at early and late puncture sites within the proximal tubule were compared, and microinjection studies were performed. One would anticipate a progressive decline in SNGFR along the length of the tubule if progressively more inulin was being lost due to backleak. The early superficial proximal SNGFR was 7.9 ± 1.3 nl/min/100 g body weight, not significantly different from the value obtained in late superficial proximal segments, 7.0 ± 7.1 nl/min/100 g body weight. This observation suggests that tubule fluid leak was not present within the proximal superficial convoluted tubule. To evaluate the possibility of a transepithelial leak of filtrate at more distal sites, radioactive inulin was microinjected into proximal tubule segments, and urine was collected separately from right and left kidneys. In control animals with intact nephrons, the recovery of microinjected inulin was $98 \pm 1\%$ from the site of injection and $1 \pm 0.2\%$ from the contralateral side, indicating the presence of nephron integrity. In contrast, microinjections performed in rats 96 hr after CP indicated a markedly reduced recovery of radioactive inulin from the injected kidney, $26 \pm 6\%$, and an increase in recovery from the contralateral kidney, $14 \pm 3\%$, as compared to the control group. In addition, total recovery of microinjected radioactivity was significantly decreased in the experimental animals ($40 \pm 9\%$), as compared with $99 \pm 1\%$ in the control group. These data are consistent with transepithelial backdiffusion of glomerular filtrate at a site distal to the puncturable proximal superficial proximal convoluted tubule. This defect was consistent and all nephrons studied indicated evident leak of radiolabeled inulin. Transepithelial leak of this magnitude may explain some but not all of the discrepancy between single-nephron and whole-kidney GFR.

An alternative explanation would be the presence of intratubular obstruction at a site distal to superficial proximal convoluted tubule micropuncture, allowing an overestimate of single-nephron function to occur as a result of "venting." Additional studies of intratubular hydrostatic pressure were therefore performed. In superficial proximal tubules the intratubular hydrostatic pressure in rats 96 hr after CP was 14 ± 1 mm Hg, not significantly different from the value of 14 ± 1 mm Hg obtained in control rats. These results indicate that significant intratubular obstruction was not present. Although it might be predicted that a slight decline in intratubular hydrostatic pressure would occur as the result of the decrement in effective glomerular filtration pressure, this was not observed, suggesting the possibility of mild and relative intratubular obstruction. The last possibility, that of heterogeneity of nephron dysfunction, was not evaluated in our studies. It is possible, but not plausible, that the discrepancies noted in a single-nephron and whole-kidney GFR was due to a greater defect in function among deeper, and not puncturable, nephron segments. This possibility is not considered plausible, since the vast majority of nephron units are located within the outer cortex of the rat kidney, and to achieve this discrepancy, inner cortical nephron units would have to be almost totally functionless.

These studies confirm the dose-related renal toxicity of CP. The major abnormalities in renal function observed 96 hr after the injection of CP consisted of nonoliguric renal failure, diminutions in creatinine clearance associated with increases in serum creatinine concentration, decrease in urine-concentrating ability manifested by diminished urine osmolality and a decline in the urine to plasma creatinine and inulin concentration ratios, and an increase in the fractional excretion of sodium. The disproportionate decline in whole-kidney versus single superficial nephron GFR has been reported in other models of experimental ARF. As noted above, a variety of explanations are available for this discrepancy in single-nephron versus whole-kidney function. The increased backleak of radioactively labeled inulin and the inability to visualize distal tubules after intravenous lissamine green both point to a component of backleak as a pathophysiologic mechanism in this form of experimental ARF. The relatively normal intratubular hydrostatic pressures mitigate against a significant component of tubular obstruction, although venting through necrotic tubular epithelium and normalization of intratubular hydrostatic pressure as a result cannot be excluded. It would not appear, however, that obstruction plays a significant role in CP-induced ARF. A decrease in glomerular filtration rate must therefore be proposed in addition to the decrement in renal function associated with backleak. The exact pathophysiologic mechanism resulting in this decrease in glomerular filtration remain to be elucidated. Certainly one cannot exclude a role for alterations in renal hemodynamics or alterations in glomerular hydraulic conductivity based on the present studies. Preliminary morphologic studies indicate extensive damage to the pars recta and a lack of significant alteration in glomerular structure. While there is not an exact correlation between structural abnormalities and glomerular dysfunction, these would suggest that a major role for alterations in glomerular hydraulic conductivity cannot be postulated in the present model.

Recently, the effect of CP on sodium, chloride, and urea transport by frogskin have been examined by Van denBerg *et al.,* (1981). They observed an increase in the rate of sodium transport, as estimated from short-circuit current, with the mucosal addition of CP (10^{-3}M). This change in short-circuit current could be inhibited by amiloride, but was additive to the effects of vasodepressant on short-circuit current. Since these alterations were specific for CP and did not occur with platinum salts, these investigators concluded that CP induced an increased permeability of the mucosal surface of frogskin to sodium, chloride, and urea. These observations lend additional support for a direct membrane interaction of CP with at least the apical membrane of frogskin. One can infer, therefore, a similar interaction with the luminal membrane of the mammalian nephron. In preliminary studies, we have not, however, been able to reverse the effect of CP on renal function using dithiothreitol, suggesting a different mode of CP–protein/membrane interaction than occurs with either uranyl nitrate or mercuric chloride.

TABLE V. Heavy Metal Toxicity: Renal Mechanisms

Tubular injury
Direct: "ATN," "backleak"
Indirect: ischemia, obstruction, immunologic injury
Vascular response
Ischemic, cortical necrosis
Immunologic (angiitis)
Glomerular damage (permeability, vascular)
Interstitial (nephritis, papillary necrosis)
Coagulation system
Immunologic injury

4. Synthesis

It would appear, based on studies from our laboratory and other groups, that heavy metal models of ARF are clinically significant experimental tools for evaluating the pathophysiologic basis of ARF and therefore for determining potential forms of therapeutic manipulation. As indicated in Table V, a multitude of possible mechanisms are available to explain heavy metal toxicity in general. In contrast to uranyl nitrate or mercuric chloride, in which ample evidence that chelating the heavy metal and/or returning the reduced sulfhydryl groups to their normal state ameliorates the renal failure, there appear to be more complex problems with CP. This may relate to the abilities of CP to combine in a variety of ways with DNA, or to a different form of binding with protein or enzymes than occurs with uranyl nitrate or mercuric chloride. Last, regardless of the type of binding, it may not be amenable to therapy with dithiothreitol, and alternative agents must be sought.

It must be noted that the toxicities associated with heavy metals in general, and platinum compounds in particular, are not limited to the kidney. Platinum compounds have a well-documented carcinogenicity, and in view of their increasing generation within contemporary society, they present additional problems beyond that of the acute renal failure associated with cancer chemotherapy and CP.

References

Anderson, R. J., Linas, S. L., Berns, A. S., Henrich, W. L., Miller, T. R., Gabow, P. A., and Schrier, R. W., 1977, Nonoliguric acute renal failure, *N. Engl. J. Med.* **296:**1134.

Blantz, R. C., 1975, Mechanisms of acute renal failure after uranyl nitrate, *J. Clin. Invest.* **55:**621.

Chopra, S., Kaufman, J. S., Jones, T. W., Hong, W. K., Gehr, M., Hamburger, R. J., Flamenbaum, W., and Trump, B. F., 1982, *Cis*-diamminedichlorplatinum induced acute renal failure in the rat, *Kidney Int.* **21:**54.

Connors, T. A., and Roberts, J. J., 1974, *Platinum Coordination Complexes in Cancer Chemotherapy,* Springer-Verlag, New York, p. 1.

Flamenbaum, W., 1973, Pathophysiology of acute renal failure, *Arch. Intern. Med.* **131**:911.

Flamenbaum, W., McNeil, J. S., Kotchen, T. A., and Saladino, A. J., 1972, Experimental acute renal failure induced by uranyl nitrate in the dog, *Clin. Res.* **31**:682.

Flamenbaum, W., Huddleston, M., McNeil, J. S., and Hamburger, R. J., 1974, Uranyl nitrate induced acute renal failure in the rat: Micropuncture and renal hemodynamic studies, *Kidney Int.* **6**:408.

Flamenbaum, W., Hamburger, R., and Kaufman, J., 1976, Distal tubule sodium concentration and juxtaglomerular apparatus renin activity in uranyl nitrate induced acute renal failure in the rat, *Pflugers Arch.* **364**:209.

Flamenbaum, W., Schwartz, J. H., Hamburger, R. J., and Kaufman, J. S., 1977, The pathogenesis of experimental acute renal failure: The role of membrane dysfunction, in: *Progress in Molecular and Subcellular Biology,* Volume 5 (F. E. Hahn, ed.), Springer Verlag, Berlin, p. 73.

Kleinman, J. G., McNeil, J. S. and Flamenbaum, W., 1975, Uranyl nitrate acute renal failure in the dog: Early changes in renal function and hemodynamics, *Clin. Sci. Mol. Med.* **48**:9.

Kleinman, G. G., McNeil, J. S., Schwartz, J. H., Hamburger, R. J., and Flamenbaum, W., 1977, Effect of dithiothreitol on mercuric chloride and uranyl nitrate induced acute renal failure in the rat, *Kidney Int.* **12**:115.

Krakoss, J. H., 1979, Nephrotoxicity of cis-dichlorodiammineplatinum, *Cancer Treat. Rep.* **63**:1523.

Luckey, T. D., and Venugopal, B., (eds.), 1977, *Metal Toxicity in Mammals: Physiologic and Chemical Basis for Metal Toxicity,* Plenum Press, New York, p. 238.

Madias, N. E., and Harrington, J. T., 1978, Platinum nephrotoxicity, *Am. J. Med.* **65**:307.

Porter, G. A., and Bennett, W. A., 1981, Toxic nephropathies, in: *The Kidney* (Brenner F. Rector, ed.), Saunders, Philadelphia, Pennsylvania, p. 2045.

Prestayko, A. W., Crooke, S. T., and Carter, S. K. (eds.), 1980, *Cisplatinum: Current Status and New Developments,* Academic Press, New York, p. 1.

Stein, J. H., Lifschitz, M. D., and Barnes, L. D., 1978, Current concepts on the pathophysiologies of acute renal failure, *Am. J. Physiol.* **234**(3):F171.

Van denBerg, E., Brazy, P. C., Huang, A. T., and Dennis, V. W., 1981, Cisplatin-induced changes in sodium, chloride and urea transport by the frogskin, *Kidney Int.* **19**:8.

7

Studies on Segmental Transport in Models of Acute Renal Failure

MICHAEL J. HANLEY

1. Introduction

The syndrome of acute renal failure (ARF) has been the subject of intensive investigation for many years (Stein *et al.*, 1978). Much of our understanding of this syndrome is based on animal models using a host of modalities to precipitate the renal failure and an equally impressive list of maneuvers to ameliorate or protect against the development of ARF. Renal artery clamping for 60 min is a model we have recently utilized in order to examine the pathophysiology of this syndrome at the level of the individual nephron segment. The effect of ischemia on segmental nephron function was assessed by the method of isolated tubule microperfusion. The response of nephron segments to ischemic injury provides a rational explanation for the genesis of the syndrome and the laboratory values encountered.

2. Ischemic Acute Renal Failure

Renal artery occlusion is one method commonly used experimentally to induce ARF. For the most part, clearance and morphology methodology was then used to

───────────────────────────────

MICHAEL J. HANLEY • Department of Medicine, Division of Renal Diseases, University of Texas Health Science Center at San Antonio, San Antonio, Texas 78284.

assess the degree and mechanisms of ischemic injury. This was necessary because following a significant insult to the kidney, urine flow was too low for meaningful assessment by surface micropuncture techniques. In addition, subsurface structures are inaccessable with current micropuncture techniques. In the following studies, we developed and characterized a rabbit model of ischemic acute renal failure induced by 60 min of renal artery clamping. The effect of this ischemic insult on the transport capacity of multiple nephron segments was subsequently studied by the method of isolated tubule microperfusion. This technique allowed us to overcome the difficulties of low flow and subsurface inaccessability inherent in a micropuncture examination of this pathologic state. Using this technique, flow rate can be adjusted and subsurface segments can be isolated and examined.

2.1. Methods

Since the female New Zealand white rabbit is the standard laboratory animal used in microperfusion studies, it was first necessary to characterize this animal's response to renal ischemia. Preliminary experiments indicated that a state mimicking acute renal failure could be induced in these animals by 60 min of renal artery occlusion. The animals were anesthetized and through a flank incision the left renal artery was occluded with a snare. Following the ischemic interval the snare was removed and the left flank incision was closed. This procedure was immediately followed by a right nephrectomy. The effect of this procedure on renal function was assessed by serial serum creatinine determinations in these animals. The rabbits used in the microperfusion studies were treated exactly the same as the animals in the clearance studies until the snare was removed. At this point blood flow was reestablished for 10 min before the kidney was removed for dissection and subsequent microperfusion. The 10-min reflow period was used in order to clear clotted blood from the renal parenchyma to facilitate dissection. Following the ischemic period, the medulla remained congested with clotted blood and the kidney was distinctly edematous. Segments of tubules were dissected and perfused by techniques standardly used (Hanley and Kokko, 1978). The tubules were perfused with an artificial ultrafiltrate containing glucose and amino acids. The tubules were bathed with an identical solution containing 5 vol% fetal calf serum. In studies requiring an osmotic gradient raffinose was added to the bathing media to increase the osmolality to 100 mOsm/kg H_2O. We used $[^3H]$-methoxy-inulin was as a volume marker in all studies. Fluid reabsorption J_v in nanoliters per millimeter per minute was calculated by

$$J_v = \frac{V_i - V_0}{L}$$

where the perfusion rate V_i is calculated by dividing the total $[^3H]$inulin counts per minute of the collected fluid by the $[^3H]$inulin counts per minute per nanoliter of perfusion fluid and by the time of the collection period; the collection rate V_0 is obtained directly by a previously calibrated constant-volume pipette and the time of the collection period; and L is the length of the tubule in millimeters. The leakage rate of the [^{3}H]inulin marker into the bath was less than 1% of the perfusion rate. Potential difference (PD) was measured in all tubules by standard methods.

The transepithelial osmotic water permeability coefficient P_f (cm/sec) was computed according to the expression derived by Al-Zahid et al., (1977):

$$P_f = -\frac{V_i C_0}{A \bar{V}_w} \frac{C_0 - C_L}{C_0 C_b C_L} + \frac{1}{(C_b)^2} \ln \frac{(C_L - C_b)C_0}{(C_0 - C_b)C_L}$$

where V_i is the perfusion rate; C_0, C_b, and C_L are the osmolalities of the perfusate, bath, and collected fluids, respectively; A is the luminal surface area; and $\bar{V}_w$ is the partial molar volume of water.

2.2 Results

Renal failure studies. To evaluate the model of acute renal failure used in these studies, the left renal artery of seven rabbits was clamped for 60 min, followed by a right nephrectomy. Sixty minutes of renal artery clamping produced a significant and persistent elevation of serum creatinine (Fig. 1). Four of the rabbits survived for the duration of the serial creatinine determinations. One rabbit died on the third day and two rabbits died on the seventh day following surgery.

Proximal convoluted tubule. Segments of proximal convoluted tubule (PCT) were obtained for the measurement of PD and J_v (Kokko, 1970). The potential difference across this segment was reduced 59% (-2.6 ± 0.4 vs. 0.90 ± 0.30 mV; p <0.05). Corresponding to this reduction in potential difference there was a 77% reduction in J_v (0.72 ± 0.11 vs. 0.14 ± 0.06 nl mm^{-1} min^{-1}, $p < 0.001$).

Tubular perfusion sometimes resulted in prompt formation of membrane-bound vesicles that could be observed in the collecting pipette. Following formation of the vesicles, the apparent cell height of portions of the tubular epithelium was reduced, suggesting that these vesicles represent sloughing of brush border membrane.

Cortical proximal straight tubule. Segments of the cortical proximal straight tubule (CPST) were obtained for the measurement of PD and J_v by methods similar to the methods used in the PCT. Sixty minutes of ischemia reduced the potential difference by 79% (-1.6 ± 0.1 vs. -0.33 ± 0.01 mV, $p < 0.001$). Corresponding to this reduction in potential difference, there was an 88% reduction in J_v (0.54 ± 0.10 vs. 0.06 ± 0.03 nl mm^{-1} min^{-1}, $p < 0.005$) to a value not statistically different from zero. As with the PCT, the CPST also exhibited vesicle formation and cell height reduction. Observations suggested that this process may be more severe in the CPST than in the PCT.

Medullary proximal straight tubule. Medullary segments of the proximal straight tubule (MPST) were found to be difficult to perfuse. Only on rare occasions, by using extremely short segments, was it possible to perfuse an intact segment. As in the PCT and CPST, there was apparent sloughing of the luminal brush border with the formation of membrane-bound vesicles. A second phenomenon occurred almost exclusively in the MPST. Tubular cells exfoliated into the tubule lumen and were carried into the collecting pipette, leaving areas of denuded basement membrane. Unlike the PCT and the CPST, it was not possible to measure fluid reabsorption in the MPST, because of a leakage rate of the [^{3}H]inulin into the bath of greater than 1% of the perfusion rate.

Cortical thick ascending limb of Henle's loop (T-ALH). The functional integrity of this segment was assessed by measurement of its diluting ability and the ability to lower the perfusate chloride ion concentration (Ramsay et al., 1955). The perfusion rates and tubule lengths were not statistically different in any of the groups. Sixty minutes of ischemia reduces the ability of the T-ALH to lower the perfusion fluid chloride ion concentration by 60% (-47 ± 9 vs. -19 ± 3 meq/liter, $p < 0.02$). The diluting ability of the T-ALH was comparably reduced 49% (-87 ± 15 vs. -44 ± 7 mOsm/kg H_2O, $p < 0.01$). The morphologic changes noted in the thick ascending limb of Henle's loop were minimal after 60 min of ischemia.

Cortical collecting tubule (CCT). The functional capacity of this segment was tested by measurement of fluid reabsorption in response to an abrupt increase (100 mOsm/kg H_2O) in bath osmolality with raffinose in the presence of a maximum ADH stimulus (250 μU/ml). A 60-min ischemic episode reduced the ADH-dependent fluid reabsorption and osmotic water permeability P_f of the CCT by 62% (1.06 ± 0.13 vs. 0.40 ± 0.07 nl mm^{-1} min^{-1}, $p < 0.001$) and 59% (0.0203 ± 0.0023 vs. 0.0083 ± 0.0020 cm/sec, $p < 0.01$), respectively.

2.3. Summary

These studies were designed to examine the transport characteristics of multiple nephron segments in a rabbit model of ischemic ARF. The data indicate that the proximal nephron is severely affected both morphologically and functionally by 60 min of ischemia. The distal nephron, while not showing gross morphologic alterations, is also functionally compromised.

The present study suggests explanations of many facets of clinical ischemic ARF. An ischemic insult depresses PCT and CPST fluid reabsorption and also brings about cellular exfoliation in the MPST. Some nephrons are completely blocked by debris and are unable to reabsorb the filtrate that leaks through the basement membrane and reenters the circulation. Those tubules that are not blocked deliver a large NaCl load to the thick ascending limb and collecting tubule. The compromised transport ability of these segments prevents them from coping with the increased sodium delivery and high urinary sodium concentration characteristic of ischemic ARF results. The present study suggests that the usual clinical guides to the differentiation of prerenal azotemia from ischemic ARF (urine Na$^+$, urine-to-plasma creatinine ratio, urine osmolality, and fractional sodium excretion) are, in part, a reflection of the disordered function of the T-ALH and the CCT (Hanley, 1980).

3. Effect of Prior Mannitol and Furosemide Infusion on Ischemic Acute Renal Failure

Furosemide and mannitol are two agents that have been frequently employed in experimental and clinical settings to ameliorate or prevent ARF (DeTorrente et al.,

1978; Burke *et al.*, 1980). The following clearance and microperfusion studies were designed to examine the effects of these agents on glomerular filtration rate (GFR) and segmental nephron function in a rabbit model of ischemic ARF.

3.1. Methods

Initial studies were performed to assess the degree and pattern of renal insufficiency produced by 60 min of total renal ischemia following mannitol or furosemide pretreatment. Mannitol pretreatment consisted of infusion of a 5% mannitol solution in an amount equivalent to 5% body weight over the 60 min preceding the ischemic period. Furosemide pretreatment animals received normal saline in an amount equivalent to 5% body weight in the 60 min preceding the ischemic period. Furosemide was added to this saline infusate providing 20 μg kg^{-1} min^{-1}. After 60 min the snare was removed. Blood samples were obtained at the time of surgery and for up to 2 weeks following the surgical procedure.

The animals used in microperfusion studies were treated in the same manner as described above. After the 60 min of ischemia, the ligature was removed and blood reflow was reestablished before the kidney was removed and sliced for dissection and microperfusion. Segments of renal tubules were perfused as described above.

3.2. Results

Renal function studies. The results of serial serum creatinine determinations performed in these animals is shown in Figure 1. The uppermost curve in the figure depicts the rise in serum creatinine observed in animals that were subjected to renal artery clamping without a protective preinfusion. The middle curve in Fig. 1 (solid squares) shows the rise in serum creatinine observed in eight animals following renal artery clamping with furosemide pretreatment. The lower curve (open circles) in Fig. 1 shows sequential creatinine determinations in nine animals pretreated with mannitol.

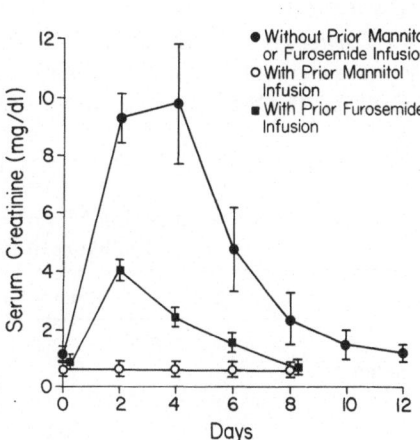

FIGURE 1. Effect of 60-min renal artery clamping (with removal of the opposite kidney) on serum creatinine in the rabbit.

TABLE I. Effect of Prior Mannitol and Furosemide Infusion on PCT Fluid
Reabsorption Following Ischemia

	Mannitol	Mannitol + ischemia	Furosemide	Furosemide + ischemia
J_v, nl mm^{-1} min^{-1}	0.59 ± 0.03	0.52 ± 0.11	0.63 ± 0.05	0.38 ± 0.04
p		NS		<0.01

Proximal convoluted tubule. Segments of proximal convoluted tubule were
obtained for measurement of PD and J_v. Mannitol pretreatment prevented any
significant reduction in PCT fluid reabsorption following ischemia (Table I). The
effect of furosemide pretreatment plus ischemia on PCT function resulted in a
significant reduction in fluid reabsorption in this segment but not to the degree
noted in ischemic tubules without furosemide pretreatment (Table I). No
discernible morphologic changes were noted in either pretreatment group during
tubule microperfusion.

Cortical proximal straight tubule. Segments of cortical proximal straight
tubule were obtained for measurement of J_v and PD. Mannitol pretreatment
prevented any significant reduction in PST fluid reabsorption following the
ischemic episode (Table II). Furosemide and ischemia did result in a significant
reduction in fluid reabsorption in this segment, but not to the degree noted in
ischemic tubules without furosemide pretreatment (Table II).

Cortical thick ascending limb. The functional integrity the T-ALH was
assessed by measurement of its ability to lower the perfusate chloride ion
concentration. Exposure to mannitol with and without ischemia resulted in a
diminution in T-ALH solute transport (Table III). There was no detectable
difference between furosemide-pretreated control and furosemide-pretreated
ischemic tubules (Table III). Exposure to mannitol or furosemide depressed T-
ALH function, but ischemia did not further adversely affect T-ALH solute
reabsorption.

Cortical collecting tubule. The functional capacity of this segment was tested
by measurement of fluid reabsorption in response to an abrupt increase (100
mOsm/kg H_2O) in bath osmolality with raffinose in the presence of a maximum

TABLE II. Effect of Prior Mannitol and Furosemide Infusion on PST Fluid
Reabsorption Following Ischemia

	Mannitol	Mannitol + ischemia	Furosemide	Furosemide + ischemia
J_v, nl mm^{-1} min^{-1}	0.34 ± 0.05	0.39 ± 0.07	0.32 ± 0.04	0.22 ± 0.02
p		NS		<0.05

TABLE III. Effect of Prior Mannitol and Furosemide Infusion on the Ability of the T-ALH to Lower Perfusate Chloride Concentration

	Mannitol	Mannitol + ischemia	Furosemide	Furosemide + ischemia
$\Delta[Cl]$, meq/liter	-11 ± 1	-12 ± 2	-19 ± 3	-17 ± 4
p		NS		NS

TABLE IV. Effect of Prior Mannitol and Furosemide Infusion on ADH-Dependent Water Flow Following Ischemia

	Mannitol	Mannitol + ischemia	Furosemide	Furosemide + ischemia
J_v, nl mm^{-1} min^{-1}	1.32 ± 0.11	0.91 ± 0.11	1.36 ± 0.17	0.67 ± 0.12
p		<0.05		<0.05

ADH stimulus (250 μU/ml). Mannitol pretreatment and furosemide pretreatment did not prevent a significant reduction in ADH-dependent water flow in this segment (Table IV).

3.3. Summary

Mannitol pretreatment in this model of ARF (1) prevented the development of ischemic acute renal failure following 60 min of renal artery clamping, (2) maintained parameters of tubular function at control levels in the PCT and PST, (3) depressed NaCl reabsorption more severely than ischemia alone in the T-ALH, and (4) did not prevent a decrease in ADH-mediated osmotic water flow in the CCT.

Furosemide pretreatment in this model of ARF (1) partially preserved renal function, (2) partially protected proximal nephron function, (3) did not substantially change the absolute transport capacity of the T-ALH or the ADH response of the CCT as compared to ischemia alone (Hanley and Davidson, 1982).

References

Al-Zahid, G., Schafer, J. A., Troutman, S. L., and Andreoli, T. E. 1977, Effect of antidiuretic hormone on water and solute permeation, and the activation energies for these processes in mammalian cortical collecting tubules, *J. Membr. Biol.* **31**:103.

Burke, T. J., Cronin, R. E., Duchin, K. L., Peterson, L. N., and Schrier, R. W., 1980, Eschemia and tubule obstruction during acute renal failure in dogs. Mannitol in protection, *Am. J. Physiol.* **238**:F315.

DeTorrente, A., Miller, P. D., Cronin, R. E., Paulsen, P. E., Erickson, A. L., and Schrier, R. W., 1978, Effects of furosemide and acetylcholine in norepinephrine-induced acute renal failure. *Am. J. Physiol.: Renal Fluid Electrolyte Physiol.* **4:**F131.

Hanley, M. J., 1980, Isolated nephron segments in rabbit model of ischemic renal failure, *Am. J. Physiol.: Renal Fluid Electrolyte Physiol.* **8:**F17.

Hanley, M. J. and Davidson, K., 1981, Effect of prior mannitol and furosemide infusion on a rabbit model of ischemic renal failure, *Am. J. Physiol.,: Renal Fluid Electrolyte Physiol.* **10:**F556.

Hanley, M. J., and Kokko, J. P., 1978, Study of chloride transport across the rabbit cortical collecting tubule, *J. Clin. Invest.* **62:**39.

Kokko, J. P., 1970, Sodium chloride and water transport in the descending limb of Henle, *J. Clin. Invest.* **49:**1838.

Ramsay, J. A., Brown, R. H. J., and Croghan, P. C., 1955, Electrometric titration of chloride in small volumes, *J. Exp. Biol.* **32:**822.

Stein, J. H., Lifschitz, M. D., and Barnes, L. D., 1978, Current concepts on the pathophysiology of acute renal failure, *Am. J. Physiol.: Renal Fluid Electrolyte Physiol.* **3:**F171.

8

Enhancement of Renal Regeneration by Amino Acid Administration

F. GARY TOBACK

1. Introduction

This chapter reviews some clinical and experimental evidence that supports the use of amino acids in the treatment of acute renal failure.

In the predialysis era, death from acute renal failure was usually caused by hyperkalemia and volume overload. Despite effective measures to diagnose and treat these complications, mortality still occurs at a high rate, especially when the syndrome appears after surgery or trauma. Although many of these patients die with, and not of, renal failure, mortality often exceeds 60% (Toback, 1980). Therefore, the conventional dialysis and management strategies used to care for these patients are clearly inadequate.

A major principle that underlies the standard treatment of acute renal failure is that the kidney lesion can heal by itself. Efforts in the past were directed at providing supportive care so that the kidney could fully utilize its regenerative capacity. However, renal regeneration and repair are anabolic processes that must be carried out in a setting of systemic catabolism. That complete restoration of

F. GARY TOBACK • Department of Medicine, Pritzker School of Medicine, The University of Chicago, Chicago, Illinois 60637.

renal function can occur despite reduced caloric intake, retention of toxins and nitrogen, acidosis, electrolyte imbalance, and anemia demonstrates the extraordinary capacity of renal cells to regenerate.

Systemic catabolism during acute renal failure often aggravates nitrogen retention and results in the loss of lean body mass. It is difficult to provide a sufficient supply of protein and nonprotein calories, because anorexia, nausea, and vomiting prevent adequate oral intake, and the loss of renal excretory capacity severely limits the volume of nutrient-containing solutions that can be infused. Protein or amino acids administered without sufficient carbohydrate or fat will be oxidized to provide energy, and not used for the synthesis of protein required for repair of the injured kidney, wound healing, and antibody production. These clinical observations suggest that the rate of renal regeneration may be suboptimal because hypercatabolism increases the demand for calories and nutrients, whereas the adverse effects of the syndrome on renal excretory and gastrointestinal function decrease their supply.

2. Clinical Studies

Some attempts have been made to reverse the catabolic state. The provision of calories as a hypertonic glucose solution has long been used for its protein-sparing effect. The oral administration of essential amino acids by Berlyne et al. (1967) and via the intravenous route by Wilmore and Dudrick (1969) resulted in reduction of the blood urea nitrogen concentration, achievement of positive nitrogen balance, and improvement of uremic symptoms. The provision of α-keto analogs of essential amino acids to patients with chronic renal failure has had similar beneficial effects (Walser et al., 1973). The importance of the so-called nonessential amino acids has been stressed by Pennisi et al. (1978), who reported that mixtures of essential and nonessential amino acids were superior to essential amino acids alone in maintaining the nutritional status of uremic animals.

Abel et al. (1973) assessed the effect of amino acid treatment in 53 adult patients with postsurgical acute renal failure. A double-blind prospective trial was set up to compare the efficacy of a solution containing amino acids and hypertonic glucose with an isocaloric isovolemic glucose infusion (Table I). The most dramatic result of the study was that amino acid treatment increased survival after the acute renal failure episode (Table II). Seventy-five percent of 28 patients treated with amino acids survived as compared to 44% of 25 patients given glucose alone ($p = 0.02$). Amino acid treatment was most effective in patients requiring dialysis and in those who developed complications such as pneumonia, generalized sepsis, or gastrointestinal hemorrhage. Kidney function was studied in nondialyzed patients and in dialyzed subjects up to the time the procedure was initiated. In patients treated with amino acids, the serum creatinine concentration reached its highest value on the second day of treatment and then declined (Fig. 1). In contrast, the creatinine concentration continued to rise until the seventh day in patients treated with the control solution.

TABLE I. Composition of Fluid for Treatment of Acute Renal Failure[a]

Ingredient	Amount
Total nitrogen	1.5 g
Total calories	1452 kcal
L-Amino acids	13.1 g total
L-Isoleucine	1.4 g
L-Leucine	2.2 g
L-Lysine HCl	2.0 g
L-Methionine	2.2 g
L-Phenylalanine	2.2 g
L-Threonine	1.0 g
L-Tryptophan	0.5 g
L-Valine	1.6 g
Glucose	350 g
Water	750 ml
Vitamins	
A	5000 USP U
B_1	25 mg
B_2	5 mg
B_6	7.5 mg
Niacinamide	50 mg
Dexpanthenol	12.5 mg
C	1.5 mg
D	500 USP U
E	2.5 IU
Osmolarity	~2100 mOsm/liter

[a] Reprinted from Abel et al. (1973) with permission.

TABLE II. Amino Acid Treatment of Acute Renal Failure

Patients studied	Percent survivors	
	Glucose	Amino acids and glucose
Acute renal failure episode		
53[a]	44	75
129[a]	30	54
Complications		
Pneumonia 18[a]	13	80
Generalized sepsis[a]	17	100
GI bleeding 17[a]	0	43
Hypotension 57[b]	40	70
Dialyzed patients		
28[a]	18	65
37[b]	22	47
Nondialyzed patients		
25[a]	64	91
92[b]	33	57

[a] Patients studied by Abel et al. (1973).
[b] Patients studied by Baek et al. (1975).

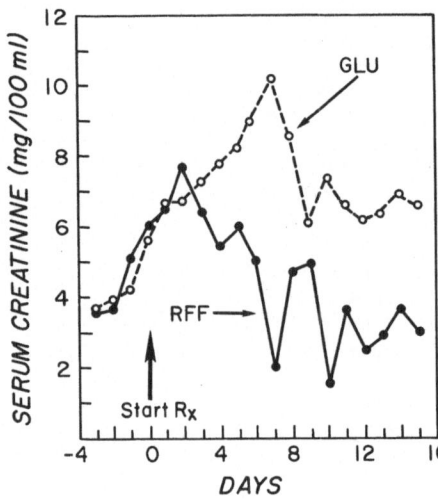

FIGURE 1. Mean daily serum creatinine concentration before and after intravenous administration of glucose (GLU) or amino acids with glucose (RFF) in patients with acute renal failure (see text). [Reprinted from Abel *et al.* (1973) with permission.]

Thus, infusion of essential L-amino acids improved survival from the acute renal failure syndrome, especially in those postoperative patients requiring dialysis and in whom hypercatabolic complications were observed. The recovery of renal function also appeared to be more rapid.

Beneficial effects from amino acid treatment in patients with postoperative acute renal failure were also reported by Baek *et al.* (1975). They administered 13 amino acids in the form of a fibrin hydrolysate with hypertonic glucose, or glucose alone, to 129 consecutive patients with acute renal failure. Some of the results of their study are summarized in Table II.

3. Amino Acid Treatment of Experimental Acute Renal Failure

The mechanism by which amino acids improve survival in patients with acute renal failure is unknown. Studies in human subjects do not permit direct assessment of repair or replacement of injured renal tissue, and leaves the problem unresolved as to whether amino acid infusions act directly on the kidney to speed cellular regeneration. To address this question, the effect of amino acid treatment on the course of acute renal failure was studied in an animal model (Toback, 1977; Toback *et al.,* 1977a).

A reversible syndrome of nonoliguric acute renal failure was induced by injecting mercuric chloride intravenously into male Sprague-Dawley rats at a dose of 1 mg of mercury per kg of body weight (Cuppage and Tate, 1967). On day 1, the serum urea nitrogen concentration was increased, whereas body weight and food intake were decreased. On day 3, the urea nitrogen concentration reached a maximum (13 times normal), and body weight a minimum. By the fifth day, the urea nitrogen concentration returned to normal and body weight and food intake were increasing. Renal histologic examination revealed that necrosis on day 1 was confined to the proximal tubular cells of the inner cortex. By day 3 some necrotic cells were replaced by primitive regenerating epithelial cells.

Membrane metabolism was studied during recovery from acute tubular necrosis, because organelles and surface membranes must be produced by the regenerating cells to restore normal renal structure and function. To assess the effects of amino acid infusion on the kidney, phospholipid synthesis was studied, since cellular membranes are composed largely of phospholipids and proteins. The synthesis of phosphatidylcholine, the major renal phospholipid, occurs primarily via the Kennedy pathway (Toback *et al.,* 1977b; Kennedy and Weiss, 1956). Choline, a normal constituent of plasma, serves as a precursor of phosphatidylcholine. Once inside the cell, choline is phosphorylated by choline kinase to form phosphorylcholine, which then reacts with cytidine triphosphate to yield cytidine diphosphocholine (CDP-choline). This compound reacts with 1,2-diacylglycerol in the cholinephosphotransferase reaction to form phosphatidylcholine. In the kidney, an increased rate of choline incorporation into membrane phosphatidylcholine occurs during the onset of growth, and precedes the formation of cellular membranes (Toback *et al.,* 1974, 1977a,b).

3.1. Renal Phospholipid Metabolism

The effects of an amino acid infusion on renal membrane phospholipid formation were examined in renal cortical slices from rats during the syndrome of mercuric chloride-induced acute renal failure (Toback *et al.,* 1979a; Toback, 1977).

At various times after mercuric chloride administration rats were anesthetized, a jugular vein catheter was placed, and the animals were restrained in individual cages by skin sutures. Each rat received one of two isocaloric solutions by constant infusion for 18 hr. Nine essential and six nonessential L-amino acids were prepared by diluting Freamine II (McGaw Laboratories) with glucose so that 4.4 g of amino acids were delivered. Glucose infusions contained 7.3 g of glucose in 58 ml of water. The survival rate was 82% in both groups of infused rats. At the end of the infusion period rats were killed by decapitation; the kidneys were bisected to identify the necrotic cortical zone; and slices were cut for studies of phospholipid metabolism using [^{14}C]choline as precursor. The rate of choline incorporation into renal phospholipid following amino acid infusion was up to 94% higher than after glucose infusion during the first 4 days of the acute renal failure syndrome. This higher rate observed after amino acid administration was superimposed upon a 71% increase in the rate that occurred during regeneration in noninfused animals (Fig. 2). Separate studies revealed that the breakdown of renal phospholipid was unchanged by amino acid or glucose infusion when the rate of choline incorporation was increased. Therefore, the increased rate of choline incorporation represented net phospholipid biosynthesis.

The mechanism by which amino acids act to enhance renal phospholipid synthesis was examined by studying each step of the Kennedy pathway in regenerating tissue. Net uptake of labeled choline into the intracellular fluid space of renal cortical slices was increased after amino acid infusion compared to glucose

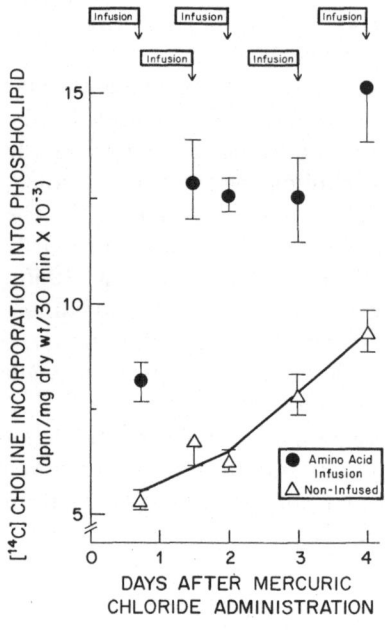

FIGURE 2. [^{14}C]Choline incorporation into renal phospholipid after mercuric chloride-induced acute tubular necrosis in amino acid-infused and noninfused rats. The rate of incorporation in amino acid-treated rats was greater than in noninfused animals at each time. Values are means ± SEM for at least ten determinations. [Reprinted from Toback (1977).]

infusion, which suggests that exogenous amino acids increase choline uptake across the plasma membrane. Furthermore, the V_{max} of both the choline kinase reaction (choline + ATP → phosphorylcholine + ADP) and the cholinephosphotransferase reaction (CDP-choline + 1,2-diacylglycerol → phosphatidycholine + CMP) was increased without a change in the apparent K_m of either. Therefore, amino acid treatment augmented phosphatidylcholine biosynthesis in regenerating renal tissue by increasing the availability of choline for utilization by two stimulated reactions of the biosynthetic pathway.

Additional studies were performed to exclude the possibility that systemic effects induced by infused amino acids mediated the increment in phosphatidylcholine synthesis. The same solution of 15 amino acids was added to slices of regenerating cortical tissue in a glucose-free buffered medium, and phospholipid synthesis and breakdown were then measured. These *in vitro* studies showed that amino acids can act directly on regenerating renal tissue to increase net phosphatidylcholine synthesis even in the absence of glucose.

The amino acid mixture increased [^{14}C]choline incorporation into phospholipid in slices of normal as well as regenerating renal cortex. This effect was observed whether the tissue was exposed to amino acids *in vivo* by infusion or *in vitro* during incubation. Specific amino acids were found that could modify both choline incorporation in intact normal cells and cholinephosphotransferase activity in microsomal preparations, in a concentration-dependent manner (Havener and Toback, 1980). These observations suggested that amino acids act as modifiers of the cholinephosphotransferase reaction which takes place on the endoplasmic reticulum of renal cells. Kinetic analysis indicated that specific amino acids modified the V_{max} but not apparent K_m of the enzyme in microsomal preparations.

These observations suggest that the relative concentrations of certain amino acids in the cell could determine the rate of phosphatidylcholine synthesis by modulating cholinephosphotransferase activity. Furthermore, a mechanism that integrates phosphatidylcholine and protein formation during membrane biogenesis can be proposed. During an amino acid infusion, the amino acid supply within the cell can increase and thereby stimulate protein synthesis. If the balance between different amino acids favors an increase in cholinephosphotransferase activity, phosphatidylcholine synthesis would increase as well so that new membrane formation could be enhanced.

3.2. Renal Protein Metabolism

Renal protein and amino acid metabolism were also examined in the animal model of acute renal failure (Toback et al., 1979b). In regenerating tissue, the free leucine concentration was decreased below normal and was reduced further by the infusion of glucose alone on day 2 of the syndrome. The cellular deficit of this essential amino acid was corrected by amino acid infusion. The reduced leucine concentration in regenerating tissue could restrict protein synthesis for repair and replacement of renal cells to a suboptimal level.

The synthesis of renal protein during regeneration was estimated from measurements of leucine specific radioactivity and cycloheximide-inhibitable [^{14}C]leucine incorporation into protein in cortical slices. Protein synthesis in regenerating tissue was 52% higher than in normal tissue and was not increased by the infusion of glucose. In contrast, the provision of amino acids raised the rate of synthesis 47% above that observed after glucose infusion. As with phospholipid synthesis, amino acid enhancement of renal protein synthesis was superimposed upon the increased rate that occurred during regeneration in untreated animals.

3.3. Renal Function

As in human subjects (Fig. 1), infusion of amino acids in rats with acute renal failure resulted in a reduction of the serum creatinine concentration (Table III). By

TABLE III. Serum Creatinine Concentrations during Amino Acid
Treatment of Experimental Acute Renal Failure[a]

Zero-time injection[b]	Treatment	Serum creatinine,[c] mg/dl	p
Saline	None	0.23 ± 0.02	<0.02
Mercuric chloride	Amino acids	0.49 ± 0.06	<0.01
Mercuric chloride	Glucose	1.03 ± 0.15	<0.001
Mercuric chloride	None	2.90 ± 0.38	

[a] Reprinted from Toback (1977) with permission.
[b] Rats were injected intravenously with mercuric chloride or saline at zero time. Infusions of amino acids or glucose were begun 30 hr later and continued for 18 hr, when blood was obtained.
[c] Values are mean ± SEM for 12–14 rats.

day 2 of the experimental syndrome, the creatinine concentration had increased more than 12-fold to 2.90 mg/dl in non-infused rats. After an 18-hr amino acid infusion the value was only 0.49 mg/dl, which was significantly lower than the concentrations determined in glucose-infused and noninfused animals.

In summary, these results indicate that amino acid infusions act directly on the rat kidney to stimulate phosphatidylcholine biosynthesis for new membrane formation in regenerating cells. The provision of amino acids enhanced cellular uptake of the phospholipid precursor choline, and increased the V_{max} of two reactions of the Kennedy pathway. Renal protein synthesis was also increased, and the cellular deficit in the concentration of at least one essential amino acid was corrected. The enhancement of phosphatidylcholine and protein synthesis after amino acid infusion was superimposed on the increased rates that occur during renal regeneration in untreated animals. Apparently synthesis of these macromolecules is suboptimal in the absence of exogenous amino acids. In addition, there was amelioration of renal functional insufficiency as was observed by Abel *et al.* (1973) in man. Finally, amino acid infusion did not increase mortality.

These results suggest that in patients with acute renal failure, infused amino acids could act directly on the kidney to speed repair of the injury. This therapeutic maneuver may also preserve lean body mass and permit nonrenal tissues to be defended more effectively against infection.

4. Clinical Considerations and Future Perspectives

Not all patients with acute renal failure have improved after amino acid treatment (Leonard *et al.*, 1975; Abel *et al.*, 1974). It is not yet possible to define the patients who are most likely to benefit, but neither the very seriously ill nor those with mild acute renal failure do. Until better clinical criteria for amino acid infusion are defined, several workers have suggested that treatment be given to those patients with acute renal failure who require dialysis (Ng and Suki, 1980; Lee, 1976).

Attention should be given to improving the composition of the solution used by Abel *et al.* (1973). The provision of both essential and nonessential amino acids seems reasonable, because all 20 amino acids are required for optimal protein synthesis (Van Venrooij *et al.*, 1972; Vaughan *et al.*, 1971). The number of calories should be increased above the 1452 kcal/day that they supplied (Table I), because the needs of patients with uncomplicated acute renal failure are probably in the range of 2500 kcal/day, whereas more than 5000 kcal/day may be required when hypercatabolic complications supervene (Ng and Suki, 1980).

Studies will be required to determine if there is a role for hormones such as insulin and glucagon, which potentiate liver growth and repair in mice (Farivar *et al.*, 1976), and thyroxine and testosterone, which enhance renal growth in rats (Katz and Lindheimer, 1973; Korenchevsky *et al.*, 1933). Epidermal growth factor, which is normally found in human blood and urine, stimulates the growth of renal epithelial cells in culture, but its function in man is unknown (Carpenter and Cohen, 1979; Holley *et al.*, 1977). Finally, there may be a place for infusion of magnesium-ATP, which appears to ameliorate some renal functional and structural abnormalities in experimental acute renal failure (Siegel *et al.,* 1980).

It would appear that the design of a solution that can maximally accelerate recovery from acute renal failure ultimately depends on increased understanding of the factors that mediate repair of the injured tissue.

Acknowledgments

Studies performed in my laboratory were supported by USPHS grants AM 18413 and GM 22328 and by the Chicago Heart Association. The collaborative efforts of Leah J. Havener and Dr. Richard C. Dodd are acknowledged with gratitude. This paper was prepared during my tenure as an Established Investigator of the American Heart Association.

References

Abel, R. M., Beck, C. H., Jr., Abbott, W. M., Ryan, J. A., Jr., Barnett, G. O., and Fischer, J. E., 1973, Improved survival from acute renal failure after treatment with intravenous essential L-amino acids and glucose: Results of a prospective, double-blind study, *N. Engl. J. Med.* **288:**695.

Abel, R. M., Abbott, W. M., Beck, C. H., Jr., Ryan, J. A., Jr., and Fischer, J. E., 1974, Essential L-amino acids for hyperalimentation in patients with disordered nitrogen metabolism, *Am. J. Surg.* **128:**317.

Baek, S.-M., Makabali, G. G., Bryan-Brown, C. W., Kusek, J., and Shoemaker, W. C., 1975, The influence of parenteral nutrition on the course of acute renal failure, *Surg. Gynec. Obstet.* **141:**405.

Berlyne, G. M., Bazzard, F. J., Booth, E. M., Janabi, K., and Shaw, A. B., 1967, The dietary treatment of acute renal failure, *Quart. J. Med.* **36:**59.

Carpenter, G., and Cohen, S., 1979, Epidermal growth factor, *Annu. Rev. Biochem.* **48:**193.

Cuppage, F. E., and Tate, A., 1967, Repair of the nephron following injury with mercuric chloride, *Am. J. Pathol.* **51:**405.

Farivar, M., Wands, J. R., Isselbacher, K. J., and Bucher, N. L. R., 1976, Effect of insulin and glucagon on fulminant murine hepatitis, *N. Engl. J. Med.* **295:**1517.

Havener, L. J., and Toback, F. G., 1980, Amino acid modulation of renal phosphatidylcholine biosynthesis in the rat, *J. Clin. Invest.* **65:**741.

Holley, R. W., Armour, R., Baldwin, J. H., Brown, K. D., and Yeh, Y.-C., 1977, Density-dependent regulation of growth of BSC-1 cells in cell culture: Control of growth by serum factors, *Proc. Natl. Acad. Sci. USA* **74:**5046.

Katz, A. I., and Lindheimer, M. D., 1973, Renal sodium- and potassium-activated adenosine triphosphatase and sodium reabsorption in the hypothyroid rat, *J. Clin. Invest.* **52:**796.

Kennedy, E. P., and Weiss, S. B., 1956. The function of cytidine coenzymes in the biosynthesis of phospholipides, *J. Biol. Chem.* **222:**193.

Korenchevsky, V., Dennison, M., and Kohn-Speyer, A., 1933, Changes produced by testicular hormone in normal and in castrated rats, *Biochem. J.* **27:**557.

Lee, H. A., 1976, The role of intravenous nutrition in the management of acute renal failure, *So. Afr. Med. J.* **50:**1703.

Leonard, C. D., Luke, R. G., and Siegel, R. R., 1975, Parenteral essential amino acids in acute renal failure, *Urology* **6:**154.

Ng, R. C. K., and Suki, W. N., 1980, Treatment of acute renal failure, in: *Contemporary Issues in Nephrology,* Volume 6 (B. M. Brenner and J. H. Stein, eds.), Churchill Livingstone, New York, p. 229.

Pennisi, A. J., Wang, M., and Kopple, J. D., 1978, Effects of protein and amino acid diets in chronically uremic and control rats, *Kidney Int.* **13:**472.

Siegel, N. J., Glazier, W. B., Chaudry, I. H., Gaudio, K. M., Lytton, B., Baue, A. E., and Kashgarian, M., 1980, Enhanced recovery from acute renal failure by the postischemic infusion of adenine nucleotides and magnesium chloride in rats, *Kidney Int.* **17**:349.

Toback, F. G., 1977, Amino acid enhancement of renal regeneration after acute tubular necrosis, *Kidney Int.* **12**:193.

Toback, F. G., 1980, Amino acid treatment of acute renal failure, in *Contemporary Issues in Nephrology,* Volume 6 (B. M. Brenner and J. H. Stein, eds.), Churchill Livingstone, New York, p. 202.

Toback, F. G., Smith, P. D., and Lowenstein, L. M., 1974, Phospholipid metabolism in the initiation of renal compensatory growth after acute reduction of renal mass, *J. Clin. Invest.* **54**:91.

Toback, F. G., Havener, L. J., Dodd, R. C., and Spargo, B. H., 1977a, Phospholipid metabolism during renal regeneration after acute tubular necrosis, *Am. J. Physiol.* **232**:E216.

Toback, F. G., Havener, L. J., and Spargo, B. H., 1977b, Stimulation of renal phospholipid formation during potassium depletion, *Am. J. Physiol.* **233**:E212.

Toback, F. G., Teegarden, D. E., and Havener, L. J., 1979a, Amino acid-mediated stimulation of renal phospholipid biosynthesis after acute tubular necrosis, *Kidney Int.* **15**:542.

Toback, F. G., Dodd, R. C., Maier, E. R., and Havener, L. J., 1979b, Amino acid enhancement of renal protein synthesis during regeneration after acute tubular necrosis, *Clin. Res.* **27**:432A.

Van Venrooij, W. J. W., Henshaw, E. C., and Hirsch, C. A., 1972, Effects of deprival of glucose or individual amino acids on polyribosome distribution and rate of protein synthesis in cultured mammalian cells, *Biochim. Biophys. Acta* **259**:127.

Vaughan, M. H., Jr., Pawlowski, P. J., and Forchhammer, J., 1971, Regulation of protein synthesis initiation in HeLa cells deprived of single essential amino acids, *Proc. Natl. Acad. Sci. USA* **68**:2057.

Walser, M., Coulter, A. W., Dighe, S., and Crantz, F. R., 1973, The effect of keto-analogues of essential amino acids in severe chronic uremia, *J. Clin. Invest.* **52**:678.

Wilmore, D. W., and Dudrick, S. J., 1969, Treatment of acute renal failure with intravenous essential L-amino acids, *Arch. Surg.* **99**:669.

9

Amphotericin B Toxicity for Epithelial Cells

PHILIP R. STEINMETZ and RUSSELL F. HUSTED

One of the major goals of the studies of nephrotoxic agents is to determine where exactly a given drug acts. Does it act on the cell membrane and, if so, does it act on the lipid part of the membrane or on some specific transport site or receptor? It has been suggested (see Chapter 3, this volume) that the heavy metal uranyl nitrate may act on the sodium channels of the luminal membrane of the turtle bladder, and that the target structures for the aminoglycosides appear to be intracellular organelles, i.e., lysosomes, in the proximal tubular cells. In most instances, the precise sequence of events leading to injury and functional impairment remains unknown.

In this chapter we would like to focus on the permeability defect induced by the polyene antibiotic amphotericin B. The initial event in the injury is reasonably well defined by studies carried out on lipid bilayers (Finkelstein and Holz, 1973; Andreoli, 1973; Marty and Finkelstein, 1975). Amphotericin B forms aqueous pores in cholesterol-containing lipid bilayers. About eight amphotericin molecules are thought to form a channel, with the chains of hydroxyl groups of each molecule facing the center of the channel (Finkelstein and Holz, 1973). The aqueous pore that results spans half the bilayer. In artificial lipid membranes two such half pores form a complete channel. In biologic membranes the amphotericin must act primarily on the outer leaflet of the lipid bilayer of the cell membrane and form half pores which may link up to existing protein structures within the inner leaflet.

PHILIP R. STEINMETZ and RUSSELL F. HUSTED • Department of Medicine, University of Connecticut School of Medicine, Farmington, Connecticut 06032.

Several studies in biologic membranes have shown that amphotericin B causes a marked increase in their cation permeability (Steinmetz and Lawson, 1970; Finn *et al.*, 1977; Reuss *et al.*, 1978). The evidence that the primary event in the toxicity of amphotericin is such a permeability change is quite convincing. To understand the physiologic consequences of the membrane defect, however, it is necessary to consider the "normal" permeability characteristics of the affected membrane and the impact of the "normal" ion gradients across that membrane.

In terms of the subject of nephrotoxicity it is of special interest that the administration of amphotericin B causes a renal tubular defect which leads to impaired urinary acidification and potassium wasting (McCurdy *et al.*, 1968). The main features of this defect can be reproduced *in vitro* in the urinary bladder of the turtle (Steinmetz and Lawson, 1970; Finn *et al.*, 1977). Some years ago we showed that addition of amphotericin B to the luminal side of the turtle urinary bladder caused a marked increase in the H^+ permeability of the luminal membrane, which normally constitutes a tight barrier for H^+ (Steinmetz and Lawson, 1970; Finn *et al.*, 1977). This permeability change becomes apparent when H^+ secretion is studied during transport against a pH gradient or when the backflux of H^+ is measured at low luminal pH. The permeability for H^+ was increased at least tenfold and the increase was associated with a large increase of the K^+ permeability of the luminal membrane, about 50-fold, and a smaller increase in Na^+ permeability (Steinmetz and Lawson, 1970, 1971; Finn *et al.*, 1977). In the first few hours of amphotericin exposure, there were only small increases in the permeability for Cl^- and HCO_3^- (Finn *et al.*, 1977). These permeability changes must be evaluated in the light of the normal characteristics of the luminal cell membrane, which is extremely tight for H^+ and K^+, but contains Na^+ channels which render it relatively permeable to Na^+.

Dr. Russell Husted in our laboratory has carried out a series of more systematic studies on the transport of K^+ across the two cell membranes of the turtle bladder epithelium (Husted and Steinmetz, 1980, 1981). His studies show how the direction of net K^+, transport is a function of the relative K^+ permeabilities of the luminal and basolateral cell membranes. In the absence of transepithelial electrochemical potential gradients for K^+, there is a small net K^+ absorptive flux which is driven by a K^+ transport system in the luminal cell membrane (Husted and Steinmetz, 1981). We will not discuss the absorptive transport system, but instead focus on the operation of the Na–K pump at the basolateral cell membrane. In turtle bladder as in other epithelial membranes the basolateral Na–K-ATPase is quantitatively the most important transport system (see Fig. 1). The extrusion of Na^+ from the cell across the basolateral cell membrane is coupled to K^+ uptake in a ratio of about 3:2. At the high Na^+ transport rates of the bladder, K^+ would rapidly accumulate within the cell if both cell membranes were impermeable to K^+. The basolateral membrane, however, has a high K^+ conductance and provides exit channels through which the K^+ ions can recycle, as shown in Fig. 1. Under control conditions the accumulated K^+ diffuses out through these basolateral K^+ channels since the luminal membrane is virtually impermeable to K^+ in this epithelium. As shown in the lower part of Fig. 1, luminal addition of amphotericin B introduces aqueous pores into the luminal cell membrane, which greatly increase the K^+ conductance across the membrane. Part of the K^+ accumulated from the basolateral side will now move into

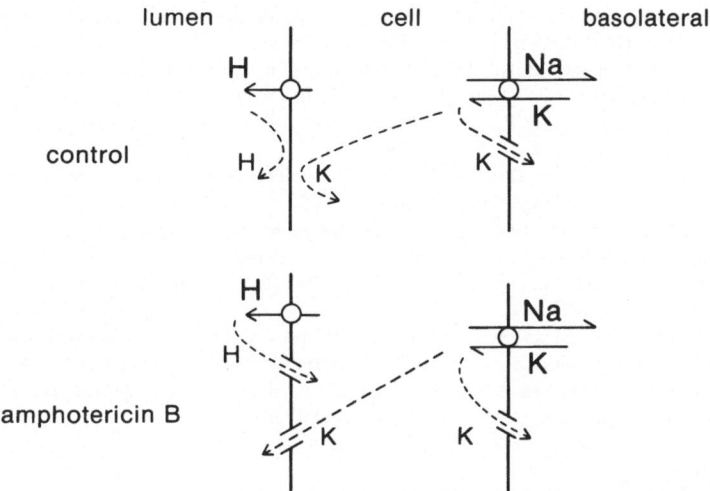

FIGURE 1. Consequences of the amphotericin-induced permeability increase of the luminal membrane of an epithelial cell.

the lumen through the amphotericin channels. The direction of net potassium transport will be a function of the relative permeabilities of the two membranes. Under ordinary conditions, the basolateral membrane remains permeable to K^+ and the rate of K^+ secretion is less than the rate of basolateral K^+ uptake. However, if the basolateral K^+ channel is blocked with barium, the rate of K^+ secretion approaches two-thirds the rate of sodium absorption (Nielsen, 1979; Kirk *et al.,* 1980).

The same amphotericin pores will render the luminal membrane permeable to protons as shown in Fig. 1. In the turtle urinary bladder the backleak of protons will reduce the effective pH gradient that can be generated. This permeability defect is induced within minutes after addition of amphotericin B to the luminal solution at concentrations (5–15 μg/ml) that are close to those observed in the urine of patients treated with amphotericin.

The initial event in amphotericin toxicity is a permeability defect for cations. This defect may lower K^+ activity and increase H^+ activity within the epithelial cell. These intracellular changes may lead to secondary changes in enzyme activities and further cell injury. A better understanding of the initial events of the injury should aid in establishing a safe dose range for this polyene antibiotic and in designing measures that minimize the intracellular changes within the renal tubules.

References

Andreoli, T. E., 1973, On the anatomy of amphotericin B-cholesterol pores in lipid bilayer membranes, *Kidney Int.* **4**:337.

Finkelstein, A., and Holz, R., 1973, Aqueous pores created in thin lipid membranes by the polyene antibiotics nystatin and amphotericin B, in: *Membranes, Lipid Bilayers, and Antibiotics,* Volume 2 (G. Eisenman, ed.), Dekker, New York, p. 377.

Finn, J. T., Cohen, L. H., and Steinmetz, P. R., 1977, Acidifying defect induced by amphotericin B: comparison of HCO_3^- and H^+ permeabilities, *Kidney Int.* **11**:261.

Husted, R. F. and Steinmetz, P. R., 1980, Factors controlling the direction of net K transport in turtle bladder: Two-membrane, two-pump model, *J. Gen. Physiol.* **76**(6):28a.

Husted, R. F., and Steinmetz, P. R., 1981, Potassium absorptive pump at the luminal membrane of the turtle urinary bladder, *Am. J. Physiol.* **241**:F315.

Kirk, K. L., Halm, D. R., and Dawson, D. C., 1980, Active sodium transport by turtle colon via an electrogenic Na–K exchange pump, *Nature* **287**:237.

Marty, A., and Finkelstein, A., 1975, Pores formed in lipid bilayer membranes by nystatin: Differences in its one-sided and two-sided action, *J. Gen. Physiol.* **65**:515.

McCurdy, D. K., Frederic, M., and Elkinton, J. R., 1968, Renal tubular acidosis due to amphotericin B, *N. Engl. J. Med.* **278**:124.

Nielsen, R., 1979, Coupled transepithelial sodium and potassium transport across isolated frog skin: Effect of ouabain, amiloride and the polyene antibiotic filipin, *J. Membr. Biol.* **51**:161.

Reuss, L., Gatzy, J. T., and Finn, A. L., 1978, Dual effects of amphotericin B on ion permeation in toad urinary bladder epithelium, *Am. J. Physiol.* **235**(5):F507.

Steinmetz, P. R., and Lawson, L. R., 1970, Defect in urinary acidification induced in vitro by amphotericin B, *J. Clin. Invest.* **49**:596.

Steinmetz, P. R., and Lawson, L. R., 1971, Effect of luminal pH on ion permeability and flows of Na^+ and H^+ in turtle bladder, *Am. J. Physiol.* **220**:1573.

II

RENAL FAILURE DUE TO ANTIMICROBIAL AGENTS

BRUCE TUNE, Section Editor

II

RENAL FAILURE DUE TO ANTIMICROBIAL AGENTS

10

Antibiotic Nephrotoxicity

An Overview

GEORGE A. PORTER

The percent distribution among recognized causes of acute renal failure has undergone a change during the last decade. Prior to 1970 the frequency of nephrotoxin-induced acute renal failure was 9% (Levinsky and Alexander, 1976), which in the last decade has risen to over 15% in series reporting consecutively occurring cases (McMurray *et al.,* 1978; Anderson *et al.,* 1977; Galpin *et al.,* 1978). Antibiotics continue as the most frequently cited etiology of nephrotoxin-induced acute renal failure, with contrast agents a distant second (Porter and Bennett, 1980). Within the antibiotic category, aminoglycosides are the leader in causing renal injury.

For convenience, antibiotic-induced nephrotoxicity can be subdivided into acute and chronic presentations. In this section we will concentrate on the acute manifestations. The mechanisms proposed to explain antibiotic-induced acute renal failure involve either direct cellular injury or activation of the host immune response. The latter will be reviewed in detail in the section dedicated to tubulointerstitial disease, while the former is the subject of this section. At least two processes may evoke direct cell injury as the result of a toxicant acting in the kidneys. The first involves accumulation of the nephrotoxin within the cell and cytolysis resulting from interference with the normal functions of intracellular organelles. The second is initiated by interaction and/or binding between

GEORGE A. PORTER • Department of Medicine, Oregon Health Sciences University, Portland, Oregon 97201.

nephrotoxic substance and cell membrane, leading to altered permeability; the loss of osmoregulation results in cellular swelling, terminating with rupture. Examples of direct cellular toxicity mediated through interference with intracellular organelle function have been postulated for both aminoglycoside antibiotics and cephalosporins.

Based on experimental observations, principally in rats, a pathophysiologic mechanism of aminoglycoside nephrotoxicity can be outlined. The drug is presented to proximal tubular cells contained in the ultrafiltrate. Low-affinity, high-capacity binding to the brush border of the apical plasma membrane occurs, probably through electrostatic bonds, which is accompanied by measurable brush border enzymuria. Inward migration via pinocytosis ensues, with the resulting vessels fusing with lysosomes. The cell volume occupied by lysosomes rises significantly and myeloid bodies can be identified using transmission electron microscopy. Cellular necrosis, which is patchy and limited to proximal tubular cells, is preceded by mitochondrial dysfunction. The contribution of the proposed angiotensin II-mediated reduction in single-nephron glomerular filtration rate (GFR) of Schor *et al.* (1981) to experimental tubular necrosis and renal failure remains to be clarified. These authors noted only myeloid bodies in proximal tubular cells at a time where whole-kidney GFR had fallen 42% at the tenth day of gentamicin dosing. In the experimental model reported by Parker *et al.* (1982), evidence of proximal tubular necrosis parallels the significant decline in GFR, while myeloid bodies can be seen after only 3 days of gentamicin administration when the only other renal functional abnormality is tubular protenuria. From recent micropuncture studies (Pastoriza-Munoz *et al.,* 1979; Senekjian *et al.,* 1981), evidence of a variable aminoglycoside handling due to nephron heterogenity has been suggested. In particular, aminoglycoside secretion by juxtamedullary nephrons has been postulated to explain the uniform fractional excretion in the urine despite variable fractional delivery to distal nephron segments. Chapters 11–13 will develop this concept in greater detail.

Many similiarities exist between the pathophysiologic mechanism for aminoglycoside nephrotoxicity outlined above and that proposed for cephalosporins, as detailed in Chapter 14. Renal tubular cell entry occurs through the organic acid (catonic) transport system of the basal lateral membrane. Intracellular accumulation of cephalosporins, probably at a key location relative to intracellular organelles, provides a critical concentration which initiates cytolytic activity. Since cephalosporins, either *in vivo* or *in vitro,* cause substantial reduction of mitochondrial respiration, such interference with normal cell energetics could lead to cellular necrosis. An alternative mediator to cephalosporin per se would be a toxic metabolite similar to that proposed by McMurtry and Mitchell (1977).

In contrast to aminoglycosides, cephalosporins, and tetracyclines, amphotericin B represents an example of tubular toxicity in which alteration of the plasma membrane is a prominent feature in the proposed pathophysiologic mechanism of renal injury. There exists a limited amount of animal experimentation which addresses the mechanism of amphotericin B nephrotoxicity. Butler *et al.* (1964)

demonstrated an intense renal vasoconstriction in dogs following infusion of amphotericin B which could not be prevented by either ganglionic or adrenergic blockade. Renal plasma flow, measured as para-aminohippurate (PAH) clearance, was reduced to a greater degree than GFR, suggesting disproportionate afferent arteriolar vasoconstriction. When Fanestil (1968) reported that amphotericin B inhibited the active transport of PAH in renal cortical slices, it was suggested that conventional C_{PAH} may not accurately reflect renal plasma flow under conditions of acute amphotericin B infusion. Gerkens and Branch (1980), using an electromagnetic flowmeter, measured an acute reduction in both renal blood flow and GFR sodium-depleted dogs; however, their finding of a decline in filtration rate three times the fall recorded in blood flow was unexpected. Furthermore, in this same study, the acute renal ischemia of amphotericin B infusion was prevented by sodium loading the dogs, while furosemide pretreatment attenuated the drug-induced deterioration of renal function in sodium-depleted animals. Since measured renal renin production was reduced during amphotericin B-induced renal ischemia, it could not be implicated as a mediator of the renal vasoconstriction.

While most of the animal studies with amphotericin B have been limited to acute infusions, the experiments of Gouge and Andreoli (1971) involved the administration of drug for 25 days. From these studies experimental confirmation was obtained for the clinically derived suggestion that tubular dysfunction heralds the onset of amphotericin B nephrotoxicity. In particular, rats given repeated injections of amphotericin B developed polyuria, hypostheruria, distal tubular renal acidosis, and potassium wasting. In a complementary study Jose et al. (1971) reported a decrease in free water clearance following amphotericin B which was independent of either altered glomerular filtration rate or distal sodium delivery. The later finding is compatible with amphotericin B inducing increased membrane permeability to water at a distal nephron/collecting duct site. Gouge and Andreoli (1971) provided an interesting speculated mechanism of amphotericin B nephrotoxicity. Based on amelioration of amphotericin B distal tubular acidosis during concomitant bicarbonate administration, they suggested that reducing the hydrogen-ion gradient across the more permeable distal tubular lumenal membrane minimized the drug-induced intracellular damage as mediated through a pH effect.

Explanations for the distal tubular toxicity of amphotericin B have usually incorporated Andreoli's (1973) concept of drug–cholesterol pores developing within the membrane to explain changes in both water and electrolyte permeability. A unique feature of the transformed pores is their enhanced anionic perm-selectivity. The increased chloride permeability which is central to the abnormal tubuloglomerular feedback proposed by Gerkens and Branch (1980) was used to link the documented distal tubular injury of amphotericin B with co-existing declines in GFR. Based upon the previously mentioned protection of either sodium loading or furosemide pretreatment in dogs given amphotericin B, these authors constructed the following hypothesis to explain the resulting renal injury. Accepting chloride flux at the macula densa as the modulator of tubuloglomerular feedback regulation of GFR, they reasoned that incorporation of amphotericin B into the luminal membrane of distal tubular cells would increase chloride permeability, and this, in turn, would activate tubuloglomerular feedback at lower than usual chloride delivery rates, leading to renal ischemia and decreased GFR. According to the position advanced by Thurau and Boylan (1976), activation of tubuloglomerular feedback acts as a protective mechanism in acute renal failure to minimize unregulated water and electrolyte losses. However, in the case of amphotericin B-induced renal injury, plasma membranes distal to the macula densa would also have altered electrolyte permeability, thus accounting for the concomitant sodium/potassium loss and failure to maintain a distal hydrogen ion gradient. While their speculation is economical, verification of the altered ion permeability of macula densa/distal nephron is a logical next step.

Since antibiotics are characteristically administered in clinical circumstances for which multiple risk factors for acute renal failure coexist, focus on specific drug effects on the kidney is simplified in animal models. Once a mechanism is identified, then it can be evaluated in man, and, if verified, allows the design of prospective interventions to modify or ameliorate toxic injury.

References

Anderson, R. J., Linas, S. L., Berns, A. S., Henrich, W. L., Miller, T. R., Gabow, P. A., and Schrier, R. W., 1977, Non-oliquric acute renal failure, *N. Engl. J. Med.* **296:**1134.

Andreoli, T., 1973, On the anatomy of amphotericin B-cholesterol pores in lipid bilayer membranes, *Kidney Int.* **4:**337.

Butler, W. T., Bennett, J. E., and Aling, D. W., 1964, Nephrotoxicity of amphotericin B, early and late effect in 81 patients, *Ann. Intern. Med.* **61:**175.

Fanestil, D. D., 1968, Amphotericin B inhibition of active PAH transport, *J. Lab. Clin. Med.* **71:**548.

Galpin, J. E., Shinaberger, J. H., Stanley, T. M., Blumenkrantz, M. J., Bayer, A. S., Friedman, G. S., Montgomerie, J. Z., Guze, L. B., Coburn, J. W., and Glassock, R. J., 1978, Acute interstitial nephritis due to methicillin, *Am. J. Med.* **65:**756.

Gerkens, J. F., and Branch, R. A., 1980, The influence of sodium status and furosemide on canine acute amphotericin B nephrotoxicity, *J. Pharmacol. Exp. Ther.* **214:**306.

Gouge, T. H., and Andriole, V. T., 1971, An experimental model of amphotericin B nephrotoxicity with renal tubular acidosis, *J. Lab. Clin. Med.* **78:**713.

Jose, P. A., Eisner, G. M., Hollerman, C. E., and Calcagno, P. L., 1971, Acute renal effects of amphotericin B, *Proc. Soc. Exp. Biol. Med.* **137:**224.

Levinsky, N. G., and Alexander, E. A., 1976, Acute renal failure, in: *The Kidney* (B. M. Brenner and F. C. Rector, Jr., eds.), Saunders, Philadelphia, Pennsylvania, p. 807.

McMurray, S. D., Luft, F. C., Maxwell, D. R., Hamburger, R. J., Futty, D., Szwed, J. J., Lavelle, K. I., and Kleit, S. A., 1978, Prevailing patterns and predictor variables in patients with acute tubular necrosis, *Arch. Intern. Med.* **138:**950.

McMurtry, R. J., and Mitchell, J. R., 1977, Renal and hepatic necrosis after metabolic activation of 2 substituted furans and thiophenes, including furosemide and cephaloridine, *Toxicol. Appl. Pharmacol.* **42:**285.

Parker, R. A., Bennett, W. M., and Porter, G. A., 1982, Animal models in the study of aminoglycoside nephrotoxicity, in: *The Aminoglycosides: Microbiology, Clinical Use and Toxicity* (A. Whelton and H. Neu, eds.), Dekker, New York, p. 235.

Pastoriza-Munoz, E., Bowman, R. L., and Kaloyanides, G. J., 1979, Renal tubular transport of gentamicin in the rat, *Kidney Int.* **16**:440.

Porter, G. A., and Bennett, W. M., 1980, Nephrotoxin-induced acute renal failure, in: *Contemporary Issues in Nephrology,* Volume 6, *Acute Renal Failure* (B. M. Brenner and J. H. Stein, eds.), Churchill Livingstone, New York, p. 123.

Schor, N., Ichikawa, I., Rennke, H. G., Troy, J. Z., and Brenner, B. M., 1981, Pathophysiology of altered glomerular function in aminoglycoside-treated rats, *Kidney Int.* **19**:288.

Senekjian, H. O., Knight, T. F., and Weinman, E. J., 1981, Micropuncture study of the handling of gentamicin in the rat kidney, *Kidney Int.* **19**:416.

Thurau, K., and Boylan, J. W., 1976, Acute renal success, *Am. J. Med.* **61**:308.

11

Functional Considerations in Aminoglycoside Nephrotoxicity

WILLIAM M. BENNETT, DONALD C. HOUGHTON, and W. CLAYTON ELLIOTT

1. Introduction

Aminoglycoside nephrotoxicity is a relatively common clinical problem. Well-controlled studies have documented up to a 26% incidence of serum creatinine elevations during therapy despite monitoring by frequent aminoglycoside blood level determinations (Smith *et al.*, 1980). Because of the complex clinical settings in which these antibiotics are prescribed, efforts to study the pathophysiology of nephrotoxicity have involved use of animal models (Kosek *et al.*, 1974; Luft *et al.*, 1975; Houghton *et al.*, 1976; Cronin *et al.*, 1980). In this chapter, an attempt will be made to correlate findings in one such model, the rat, with certain variables which may influence aminoglycoside nephrotoxicity. It has become apparent that studies concerning biochemical and subcellular mechanisms of aminoglycoside nephrotoxicity must take into account the considerable regenerative and functional recovery which occurs during continuous treatment with aminoglycosides. It is not

WILLIAM M. BENNETT, DONALD C. HOUGHTON, and W. CLAYTON ELLIOTT • Division of Nephrology, Department of Medicine, Oregon Health Sciences University, Portland, Oregon 97201.

clear at present whether similar caution needs to be exercised in interpreting studies involving other nephrotoxins which cause experimental or clinical acute renal failure. Finally, the consequences of acute nephrotoxic insults on long-term renal functional integrity should be carefully evaluated despite return to normal of such insensitive parameters as serum creatinine after the toxin is withdrawn.

2. The Animal Model

When gentamicin is administered to Fischer 344 male rats, a dose-dependent nonoliguric acute renal failure is produced. In doses of 40 mg/kg of body weight administered in two daily injections, a rise in urine volume and a decrease in urine osmolarity are evident after 1–3 days. At 7 days, maximum urinary concentrating ability and generation of urinary cyclic AMP are impaired in response to water deprivation and exogenous antidiuretic hormone (Plamp et al., 1978). Inulin clearance is depressed by day 3 and reaches a nadir at day 14. Although frank proximal tubular necrosis is seen on light microscopy by day 10–14 of therapy, associated with a rise in serum creatinine and blood urea nitrogen (BUN) concentrations, ultrastructural lesions are noted as early as the first day of treatment. The functional and pathologic features of this model have been previously reported (Houghton et al., 1976) and are summarized in Fig. 1. Gradual accumulation of gentamicin in renal cortical tissue precedes a rise in serum creatinine. Histologic damage of the proximal nephron is most prominent in early segments and is characteristically patchy in distribution. Electron microscopic findings of increased numbers of lysosomes, cytosegresomes, and "myeloid bodies" have been noted with all aminoglycosides, even those that do not produce tubular necrosis, as well as with other pharmacologically unrelated compounds that have a net cationic charge (Lullmann et al., 1978). Associated with the development of overt renal failure, there is depression of the uptake in renal cortical slices of the organic cation N-methylnicotinamide (NMN) and the organic anion para-aminohippurate (PAH). NMN uptake is most sensitive to aminoglycoside treatment and can be depressed as early as day 1. Aminoglycosides such as netilmicin and tobramycin in doses of 40 mg/kg/day produce little tubular necrosis in the rat; however, both are associated with declines in NMN uptake (Ormsby et al., 1979). Toxic aminoglycosides may stimulate PAH uptake for up to 7 days prior to a decline in transport as frank tubular necrosis develops (Bennett et al., 1979b).

Other functional abnormalities have been described with experimental aminoglycoside nephrotoxicity by other laboratories. Cronin et al. (1980) have described renal potassium wasting and hypokalemia in dogs. Although acute studies in the isolated perfused rat kidney have demonstrated increased fractional potassium excretion (Engle et al., 1977), hypokalemia does not develop in rats treated chronically with aminoglycosides (Elliott et al., 1981). Renal glycosuria, with or without Fanconi's syndrome, has been documented in dogs and humans (Cronin et al., 1980; Russo and Adelman, 1980), as has renal magnesium wasting with subsequent hypocalcemia (Cronin et al., 1980; Keating et al., 1977). Tubular proteinuria and enzymuria may also occur, and, indeed, some authors have

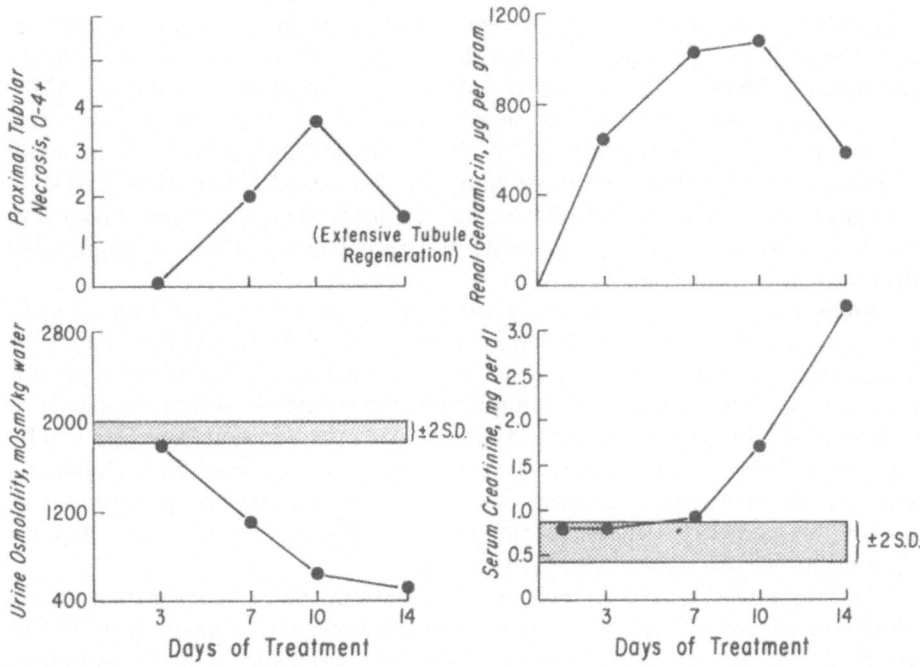

FIGURE 1. Features of experimental gentamicin nephrotoxicity in the Fischer 344 male rat treated with 40 mg/kg subcutaneously BID. A drop in urine osmolality precedes the rise in serum creatinine. Proximal tubular necrosis and accumulation of the drug in renal cortex are maximal on day 10, followed by tubular regeneration and decreases in cortical concentration by day 14.

suggested that serial monitoring of these parameters might assist in early recognition of clinical aminoglycoside nephrotoxicity (Schentag and Plaut, 1980). Prospective studies will be needed to critically examine the role of these tests as useful adjuncts to clinical aminoglycoside administration or as markers of experimental injury, as opposed to a simple consequence of exposure to these compounds.

With continued administration of gentamicin for up to 42 days in the same dose, progressive recovery of renal function occurs along with marked improvement in renal histology. It is difficult to discern any differences between renal tubules in continuously treated animals from those in untreated controls (Gilbert *et al.,* 1979). One of the few distinctive features is the appearance of cortical scars in chronically dosed animals, along with interstitial inflammatory changes. Renal slice uptake of PAH and NMN, inulin clearance, and impairment of renal concentrating ability all return toward normal. Animals in which the drug is discontinued at 14 days cannot be distinguished from continuously treated animals as regards inulin clearance, slice uptake of PAH, and serum creatinine concentrations. NMN uptake in renal cortical slices remains somewhat depressed in animals that are continuously treated as compared to controls or animals in which the drug is stopped after 14 days (Elliott *et al.,* 1981). Likewise, the urinary concentrating ability returns toward normal and partial responsivity to ADH returns. This apparent acquired insensitivity to aminoglycosides has been corroborated by Luft *et al.* (1975) and probably can be inferred from the studies of Cuppage *et al.* (1977). The latter investigators studied male Fischer rats treated with 28 days of continuous gentamicin therapy in doses identical to the model just described. Although no early time points were studied, the late glomerular filtration rate was normal.

Along with functional and structural recovery, tissue concentrations of drug vary cyclically. These relationships are shown in Fig. 2. Thus, any formulation of pathogenesis using drug accumulation in renal tissue as a primary pathogenetic event per se must explain the numerous situations in which tissue drug concentrations, histologic evidence of tubular necrosis, and renal function can be dissociated. For the above data, it seems clear that some caution should be exercised in interpreting studies concerning aminoglycoside nephrotoxicity without the knowledge that experimental manipulations may alter temporal relationships as well as the severity of the insult.

The phenomenon of functional recovery with histologic regeneration during continuous exposure may occur with other nephrotoxins. Suzuki *et al.* (1980) described a similar sequence during long-term cadium chloride injections in rats. This would appear to be a different phenomenon than the resistance to acute renal failure afforded by prior renal failure (Oken *et al.,* 1975). In these latter studies,

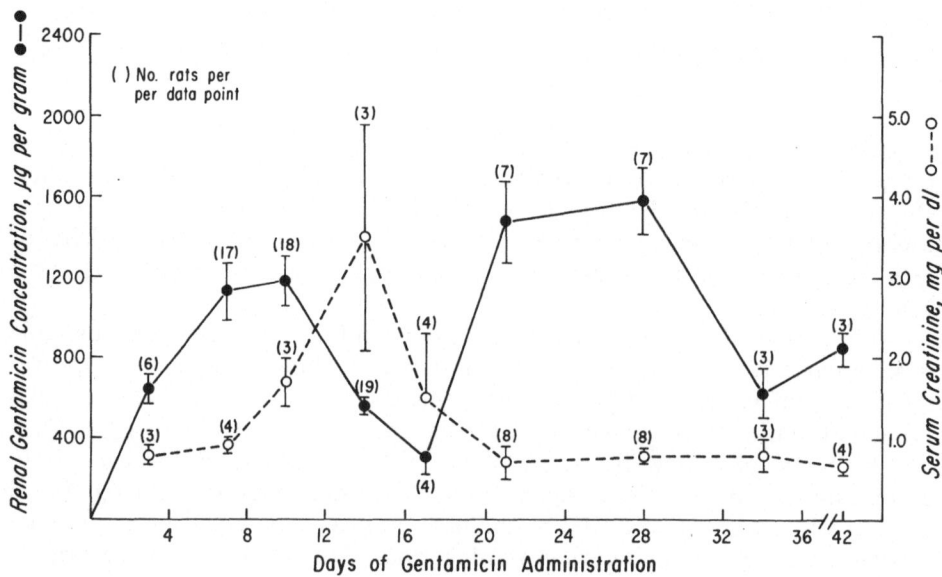

FIGURE 2. Serum creatinine and renal cortical gentamicin concentrations during 42 days of continuous treatment with gentamicin, 40 mg BID, to Fischer 344 male rats. [From Gilbert *et al.* (1979), with permission.]

morphologic evidence of tubular necrosis was produced by a second challenge with a different nephrotoxin despite the failure to decrease glomerular filtration rate. A possible shift in the time course of renal failure was not examined. In addition, female rats were used for some studies (see below). MacNider (1929) described resistance to a second injection of uranyl nitrate in dogs who had made an incomplete recovery from a first injection, as evidenced by immature simplified epithelium. He postulated a resistance to cellular effects on the basis of an inability to transport the poison (MacNider, 1931). Luft *et al.* (1977) administered low-dose gentamicin for 2 weeks following mercuric chloride-induced acute renal failure and observed less severe azotemia but a similar extent of tubular necrosis than when gentamicin was given alone. If gentamicin was given prior to mercuric chloride, more severe functional and morphologic damage occurred. Since mercuric chloride injures the pars recta, the effect of a toxin that primarily injures the early portion of the proximal tubule, given during the recovery phase of aminoglycoside-induced injury, would be of interest. Data from our laboratory show that a single dose of potassium dichromate, 10 mg/kg, modifies the time course and probably the severity of gentamicin nephrotoxicity. The reverse experiment has not been performed. Since aminoglycosides seem to compete with each other for binding sites on proximal tubular cells (Kuo and Hook, 1979), it is possible that pretreatment with less toxic aminoglycosides may provide some protective benefit when a more toxic drug is given later. Preliminary data from our laboratory and the report of Viotte *et al.* (1980) do not support this concept. Further studies are clearly warranted. An additional area for future investigation is the long-term effect on renal structure and function of nephrotoxic acute renal failure. The tubulointerstitial morphologic abnormalities observed in our rat model following continuous dosing, despite return of functional integrity, may be an indication of the relationship between acute clinical or subclinical injury and subsequent chronic renal failure. It is also of interest that renal carcinogens frequently produce acute renal failure in laboratory animals (Elliott *et al.,* 1981; Kluwe and Hook, 1980).

3. Factors Affecting Susceptibility to Experimental Aminoglycoside Nephrotoxicity

3.1. Age

Clinically, increasing age is associated with an increasing susceptibility to aminoglycoside nephrotoxicity. This may be due to a decline in renal function and inadvertent overdosage in the older patient. Pharmacokinetic studies with aminoglycosides in the newborn period reveal a prolonged terminal half-life as compared to adults (Hook and Hewitt, 1977). As the kidney develops, renal excretion increases and terminal half-life decreases. Thus, it appears that delayed excretion is not a satisfactory explanation for the increased toxicity in older patients or animals. Chonko *et al.* (1979) studied the effects of gentamicin on the renal function of mature and immature rabbits. In a dose of 15 mg/kg/day, only mature animals showed significant renal failure and proximal tubular necrosis.

Although the mechanism of this apparent insensitivity is not known, the age of experimental animals needs to be considered in evaluating studies in the literature.

3.2. Sex

In rats, males have a more severe nephrotoxic insult to a given dose of aminoglycoside than do females (Parker *et al.*, 1980). Other toxins, such as chloroform, show a more profound nephrotoxicity in male animals. This is presumably due to sex differences in renal drug metabolism (Ilett *et al.*, 1973). Aminoglycosides, however, have no known metabolism and are recovered quantitatively in the urine in unchanged form. Maunsbach (1966) has reported anatomical differences in the number of lysosomes in proximal tubules between male and female rats. Koenig *et al.* (1980) reported that testosterone augmented the activity of the lysosomal-vacuolar system of mouse proximal tubules. Male mice also had a higher kidney activity of mitochondrial cytochrome oxidase enzymes. Thus, male animals might be more sensitive than females to agents such as gentamicin, which may impair mitochondrial respiratory activity (Simmons *et al.*, 1980). In the experimental model of gentamicin nephrotoxicity described above, female animals had less severe dysfunction as measured by BUN, serum creatinine, and slice uptake of PAH. The peak of renal failure also occurred somewhat earlier. At all time points, cortical gentamicin accumulation in females was less than in males despite similarity in the degree of proximal tubular necrosis. Exogenous testosterone administered to female rats did not enhance the functional disturbance, nor did castration of prepubertal male animals protect against renal failure (Parker *et al.*, 1980). Thus, sex is an important variable to consider in studies of experimental aminoglycoside nephrotoxicity.

3.3. Dosage Regimen, Species, and Animal Strain

Experimental nephrotoxicity due to aminoglycosides is dose-related in all species studied. Doses on a weight basis that correspond to doses administered to patients seldom produce renal failure. The dose of 40 mg/kg, established for our rat model, was intended to reproduce the clinical syndrome of nonoliguric acute renal failure observed in patients. If this dose is correlated to body surface area rather than weight, it more closely simulates those used clinically. Higher doses produce death, although it is not clear whether this is due to renal failure or some other effect of aminoglycosides, such as neuromuscular blockade. Thiel *et al.* (1978), using 200 mg/kg, observed oliguric renal failure in the rat. Thus, the dose of a nephrotoxin, rather than its nature, may explain differences in the character of the insult. It is not clear whether acquired insensitivity develops with doses that produce a more profound acute insult, since many animals do not survive the acute period. Preliminary data from our laboratory reveal that continuous dosing with 80 mg/kg results in the same qualitative effects as lower doses.

The Fischer rat is known to be sensitive to a variety of nephrotoxins (Kosek *et al.*, 1974). Indeed, another well-characterized model in rats, using the Sprague-Dawley strain, requires 60 mg/kg to produce nephrotoxicity equivalent to that seen in the male Fischer rat receiving 40 mg/kg (Luft *et al.*, 1975). However, all the other features of the animal model seem comparable. In one of the few studies comparing rat strains, Cuppage *et al.* (1977) could find no differences in renal function or histology between Sprague-Dawley rats and Fischer rats receiving up to 40 mg/kg of gentamicin. However, only a single 28-day time-point was used for study. It is clear that strain and certainly species differences need to be carefully considered in utilizing information from these animal models.

The actual dosing schedule is of major importance in aminoglycoside nephrotoxicity. In dogs, rats, and rabbits, multiple doses during a 24-hr period are more nephrotoxic than the same dose by weight given as a single daily dose, despite higher peak serum levels in the latter instance (Thompson *et al.*, 1977; Bennett *et al.*, 1979a; Frame *et al.*, 1977). It is of interest in this regard that constant infusions in patients are associated with a high prevalence of nephrotoxicity despite maintenance of so-called therapeutic concentrations (Bodey *et al.*, 1975). This dissociation of the height of peak serum level from renal tissue aminoglycoside concentration and nephrotoxicity is most compatible with a saturable aminoglycoside transport system by which the drug enters cells. Indeed, massive accidental overdose of amikacin did not result in clinical renal dysfunction despite a peak serum concentration many times that achieved during usual therapy (Ho *et al.*, 1979). Experimental studies of aminoglycoside nephrotoxicity cannot be directly compared unless the dosage frequency is known.

3.4. Other Physiologic Variables

As with other models of acute renal failure, preexisting extracellular fluid volume depletion enhances nephrotoxicity. Administration of a low-sodium diet to rats prior to gentamicin treatment produced more severe renal failure, with some early deaths (Bennett *et al.*, 1976). A high sodium intake, although reducing cortical gentamicin concentrations, did not confer protection from acute tubular necrosis (Bennett *et al.*, 1976; Elliott *et al.*, 1980b). Furosemide probably enhances aminoglycoside nephrotoxicity by producing volume depletion rather than a specific drug interaction (Adelman *et al.*, 1979). Metabolic acidosis potentiates aminoglycoside nephrotoxicity (Hsu *et al.*, 1974), possibly by shifting the time course of the injury (Elliott *et al.*, 1980b). Extracellular fluid volume depletion may contribute to the adverse effects of NH_4Cl-induced metabolic acidosis. Sodium bicarbonate pretreatment was associated with intratubular calcifications and similar depression of glomerular filtration rate when compared to that seen in control rats (Elliott *et al.*, 1980b). Thus, although renal concentrations of gentamicin can be reduced by sodium bicarbonate administration (Chiu *et al.*, 1979; Elliott *et al.*, 1980b), no net beneficial effect can be observed. Potassium deficiency is reported to enhance gentamicin nephrotoxicity in dogs with preservation of function by potassium repletion (Brinker *et al.*, 1979). Recently, experimental manipulations such as unilateral nephrectomy (Elliott *et al.*, 1980a) and streptozotocin-induced diabetes have been shown to modify gentamicin nephrotoxicity (Teixeira *et al.*, 1980). Whether some fundamental pathophysiologic mechanism is affected, or whether a change in the temporal course of this toxic nephropathy has occurred, needs further study. Osmotic diuresis produced by isosorbide did not alter pathologic damage or accumulation of gentamicin in rat kidney (Newman *et al.*, 1980), suggesting that this factor alone is insufficient to explain protection in diabetic animals. The experiments in diabetic rats were performed in female animals with a single daily dose which was not increased to adjust for the 40% rise in glomerular filtration rate characteristic of early diabetes (Vaamonde *et al.*, 1980).

References

Adelman, R. D., Spangler, W. L., Beasom, F., Ishizaki, G., and Conzelman, G. M., 1979, Furosemide enhancement of experimental gentamicin nephrotoxicity: Comparison of functional and morphologic changes with activity of urinary enzymes, *J. Infect. Dis.* **140:**342.

Bennett, W. M., Hartnett, M. N., Gilbert, D. N., Houghton, D. C., and Porter, G. A., 1976, Effect of sodium intake on gentamicin nephrotoxicity in the rat, *Proc. Soc. Exp. Biol. Med.* **151:**581.

Bennett, W. M., Plamp, C. E., Gilbert, D. N., Parker, R. A., and Porter, G. A., 1979a, The influence of dosage regimen on experimental gentamicin nephrotoxicity: Dissociation of peak serum levels from renal failure, *J. Infect. Dis.* **140:**576.

Bennett, W. M., Plamp, C. E., Parker, R. A., Gilbert, D. N., Houghton, D. C., and Porter, G. A., 1979b, Renal transport of organic acids and bases in aminoglycoside nephrotoxicity, *Antimicrob. Agents Chemother.* **16:**231.

Bodey, G. P., Chang, H. Y., Rodriquez, V., and Stewart, D., 1975, Feasibility of administering aminoglycoside antibiotics by continuous intravenous infusion, *Antimicrob. Agents Chemother.* **8:**328.

Brinker, K., Cronin, R., Bulger, R., Southern, P., and Henrich, W., 1979, Potassium depletion: Risk factor for and consequence of gentamicin nephrotoxicity, *Proc. Am. Soc. Nephrol.* **12:**79A.

Chiu, P. J. S., Miller, G. J., Long, J. F., and Waity, J. A., 1979, Renal uptake and nephrotoxicity of gentamicin during urinary alkalinization in rats, *Clin. Exp. Pharmacol. Physiol.* **6:**317.

Chonko, A., Savin, V., Stewart, R., Karniski, L., Cuppage, F., and Hodges, G., 1979, The effects of gentamicin on renal function in the mature versus immature rabbit, *Proc. Am. Soc. Nephrol.* **12:**2A.

Cronin, R. E., Bulger, R. E., Southern, P., and Henrich, W. L., 1980, Natural history of aminoglycoside nephrotoxicity in the dog, *J. Lab. Clin. Med.* **95:**463.

Cuppage, F. E., Setter, K., Sullivan, L. P., Reitzes, E. J., and Melnykovych, A. O., 1977, Gentamicin nephrotoxicity. II. Physiological, biochemical and morphological effects of prolonged administration to rats, *Virchows Arch. B Cell Pathol.* **24:**121.

Elliott, W. C., and Bennett, W. M., 1981, Acquired gentamicin insensitivity: Rate of functional recovery with continued drug administration, *Clin. Res.* **29:**72A.

Elliott, W. C., Parker, R. A., Gilbert, D. N., Houghton, D. C., Porter, G. A., DeFehr, J., and Bennett, W. M., 1980a, Effect of compensatory hypertrophy on experimental gentamicin nephrotoxicity, *Clin. Res.* **28:**62A.

Elliott, W. C., Parker, R. A., Houghton, D. C., Gilbert, D. N., Porter, G. A., DeFehr, J., and Bennett, W. M., 1980b, Effect of sodium bicarbonate and ammonium chloride ingestion in experimental gentamicin nephrotoxicity in rats, *Res. Commun. Chem. Pathol. Pharmacol.* **28:**483.

Elliott, W. C., Lynn, R. K., Kennish, J., Houghton, D. C., and Bennett, W. M., 1981, Acute nephrotoxicity of the renal carcinogen TRIS, *Clin. Res.* **29:**72A.

Engle, J. E., Abt, A. B., and Schoolwerth, A. C., 1977, Experimental aminoglycoside nephrotoxicity in the isolated perfused rat kidney, in: *Proc. Internat. Cong. Nephrol.*, Volume 7, p. R-9.

Frame, P. T., Phair, J. P., Watanakunakorn, C., and Bannister, T. W. P., 1977, Pharmacologic factors associated with gentamicin nephrotoxicity in rabbits, *J. Infect. Dis.* **135:**852.

Gilbert, D. N., Houghton, D. C., Bennett, W. M., Plamp, C. E., Reger, K., and Porter, G. A., 1979, Reversibility of gentamicin nephrotoxicity in rats: Recovery during continued drug administration, *Proc. Soc. Exp. Biol. Med.* **160:**99.

Ho, P. W. L., Pien, F. D., and Kominami, N., 1979, Massive amikacin overdose, *Ann. Intern. Med.* **91**:227.

Hook, J. B., and Hewitt, W. R., 1977, Development of mechanisms for drug excretion, *Am. J. Med.* **62**:497.

Houghton, D. C., Hartnett, M. N., Campbell-Boswell, M., Porter, G. A., and Bennett, W. M., 1976, A light and electron microscopic analysis of gentamicin nephrotoxicity in rats, *Am. J. Pathol.* **82**:589.

Hsu, C. H., Kurtz, G., Easterling, R. E., and Weller, J. M., 1974, Potentiation of gentamicin nephrotoxicity by metabolic acidosis, *Proc. Soc. Exp. Biol. Med.* **146**:894.

Ilett, K. F., Watson, D. R., Sipes, I. G., and Krishna, G., 1973, Chloroform toxicity in mice: Correlation of renal and hepatic necrosis with covalent binding of metabolites to tissue macromolecules, *Exp. Mol. Pathol.* **19**:215.

Keating, M. J., Setthi, M. R., Bodey, G. P., and Samaan, N. A., 1977, Hypocalcemia with hypoparathyroidism and renal tubular dysfunction associated with aminoglycoside therapy, *Cancer* **39**:1410.

Kluwe, W. M., and Hook, J. B., 1980, Metabolic activation of nephrotoxic haloalkanes, *Fed. Proc.* **39**:3129.

Koenig, H., Goldstone, A., Blume, G., and Lu, C. Y., 1980, Testosterone-mediated sexual dimorphism of mitochondria and lysosomes in mouse kidney proximal tubules, *Science* **209**:1023.

Kosek J. C., Mazze, R. I., and Cousins, M. J., 1974, Nephrotoxicity of gentamicin, *Lab. Invest.* **30**:48.

Kuo, C., and Hook, J. B., 1979, Specificity of gentamicin accumulation by rat renal cortex, *Life Sci.* **25**:873.

Luft, F. C., Patel, V., Yum, M. N., Patel, B., and Kleit, S. A., 1975, Experimental aminoglycoside nephrotoxicity, *J. Lab. Clin. Med.* **86**:213.

Luft, F. C., Yum, M. N., and Kleit, S., 1977, The effect of concomitant mercuric chloride and gentamicin on kidney function and structure in the rat, *J. Lab. Clin. Med.* **89**:622.

Lullmann, H., Lullmann-Rauch, R. and Wassermann, O., 1978, Lipidosis induced by amphiphilic cationic drugs, *Biochem. Pharmacol.* **27**:1103.

MacNider, W., 1929, The functional and pathological response of the kidney in dogs subjected to a second subcutaneous injection of uranium nitrate, *J. Exp. Med.* **49**:411.

MacNider, W., 1931, The morphological basis for certain tissue resistance, *Science* **73**:103.

Maunsbach, A. B., 1966, Observations on the segmentation of the proximal tubule in the rat kidney, *J. Ultrastruct. Res.* **16**:239.

Newman, R., Weinstock, L. B., Gump, D. W., Hacker, M. P., and Yates, J. W., 1980, Effect of osmotic diuresis on gentamicin-induced nephrotoxicity in rats, *Arch. Toxicol.* **45**:213.

Oken, D. E., Mende, C. W., Taraba, I., and Flamenbaum, W., 1975, Resistance to acute renal failure afforded by prior renal failure, *Nephron* **15**:131.

Ormsby, A., Parker, R. A., Plamp, C., Stevens, P., Houghton, D. C., Gilbert, D. N., and Bennett, W. M., 1979, Comparison of the nephrotoxic potential of gentamicin, tobramycin and netilmicin in the rat, *Curr. Ther. Res.* **25**:335.

Parker, R. A., Bennett, W. M., Plamp, C. E., Houghton, D. C., Gilbert, D. N., and Porter, G. A., 1980, Resistance of female rats to gentamicin nephrotoxicity, *Curr. Chemother.* **1**:601.

Plamp, C. E., Reger, K., Bennett, W. M., McClung, M. R., and Porter, G. A., 1978, Vasopressin resistant polyuria in gentamicin nephrotoxicity, *Clin. Res.* **26**:151A.

Russo, J., and Adelman, R. D., 1980, Gentamicin induced Fanconi syndrome, *J. Pediatr.* **96**:151.

Schentag, J. J., and Plaut, M. E., 1980, Patterns of urinary B-2 microglobulin excretion by patients treated with aminoglycosides, *Kidney Int.* **17**:654.

Simmons, C. F., Bogusky, R. T., and Humes, H. D., 1980, Inhibitory effect of gentamicin on renal cortical mitochondrial oxidative phosphorylation, *J. Pharmacol. Exp. Ther.* **214**:709.

Smith, C. R., Lipsky, J. J., Laskin, O. L., Hellman, D. B., Mellits, E. D., Longstreth, J., and Leitman, P. S., 1980, Double-blind comparison of the nephrotoxicity and auditory toxicity of gentamicin and tobramycin, *N. Engl. J. Med.* **302**:1106.

Suzuki, K. T., Yamamura, M., Yamada, Y. K., and Shimizu, F., 1980, Decreased copper content in rat kidney metallothionein and its relation to acute cadmium nephropathy, *Toxicol. Lett.* **7**:137.

Teixeira, R. B., Morales, J., Kelley, J., Alpert, H., Pardo, V., and Vaamonde, C., 1980, Mechanism of complete protection from gentamicin-induced acute renal failure in the untreated streptozotocin diabetic rat, *Clin. Res.* **28**:463A.

Thiel, G., de Rougemont, D., Konrad, L., Oeschiger, A., Torhorst, J., and Brunner, F., 1978, Genta-
    micin induced acute renal failure in the rat, in *Proc. VII Internat. Cong. Nephrol.,* Volume 7, p.
    D45.
Thompson, L., Reiner, N. E., and Bloxham, D. D., 1977, Gentamicin and tobramycin nephrotoxicity
    in dogs on continuous or once daily intravenous injection, in: *Proc. Internat. Cong. Chemother.,*
    Volume 10, p. 207.
Vaamonde, C. A., Teixeira, R. B., Morales, J., Kelley, J., Alpert, H., and Pardo, V., 1980, A new model
    for studying drug-induced nephrotoxicity: The rat with untreated streptozotocin-induced diabetes
    mellitus, *Proc. Am. Soc. Nephrol.* **13:**108A.
Viotte, G., Morin, J. P., Bendirdjian, J. P., Ducastelle, B., Godin, M., and Fillastre, J. P., 1980, The
    effects of increasing doses of fosfomycin gentamicin and their association on the rat kidney, *Drugs
    Exp. Clin. Res.* **6:**317.

12

Aminoglycoside Nephrotoxicity
Lysosomal and Mitochondrial Alterations in Rat Kidneys after Aminoglycoside Treatment

J. P. MORIN, G. VIOTTE, J. P. BENDIRDJIAN,
B. OLIER, J. P. FILLASTRE, and M. GODIN

1. Introduction

Aminoglycosides are basic organic compounds with pKa varying between 7.5 and 8.0. These molecules have low molecular weights, are highly water-soluble, are lightly bound to plasma proteins, and are not metabolized in the body. In view of their very wide antibacterial spectrum, aminoglycosides are an important class of antimicrobial agents. Although they are oto- and nephrotoxic, these agents are commonly used in clinical treatment.

The renal damage resulting from aminoglycoside therapy is first reflected by polyuria, then by increased urinary enzymatic activity, and by proteinuria, which includes increased $\beta2$ microglobulinuria. Modifications of urinary sediment occur later, involving leukocyturia and cylindruria. Finally, the decrease in the glomerular filtration rate provokes a rise in blood urea nitrogen (BUN) and plasma creatinine levels. Thus, the acute renal failure induced by aminoglycosides attains varying

J. P. MORIN, G. VIOTTE, J. P. BENDIRDJIAN, B. OLIER, J. P. FILLASTRE, and M. GODIN • Tissue Physiology Group, University of Rouen, Rouen, France.

degrees of severity. Most frequently, diuresis is sustained, while in a few cases oligoanuria occurs (Kaloyanides, 1980). This pathophysiologic picture suggests the existence of acute tubular necrosis.

The extent of renal injury induced by aminoglycosides varies among animal species. Acute necrosis of the proximal part of the nephron is seen in the dog after a daily dose of 40 mg/kg for 15 days (Black *et al.,* 1963), in the monkey after a daily dose of 60 mg/kg for 7 days (Vera-Roman *et al.,* 1975), and in both Wistar and Sprague-Dawley rats after 40 mg/kg daily for 14 days. The Fischer 344 rat strain seems to be highly sensitive to aminoglycosides; the first tubular lesions are seen in this strain only 48 hr after the injection of 1 mg/kg of gentamicin (Kosek *et al.,* 1974). At this stage, osmiophilic lamellar material in a concentric deposition is seen by electron microscopy in the lysosomes of the proximal tubule cells. These structures are usually termed myeloid bodies and have been described by numerous authors (Houghton *et al.,* 1978; Wellwood *et al.,* 1976; Cuppage *et al.,* 1977; Vera-Roman *et al.,* 1975; Kosek *et al.,* 1974). Luft *et al.* (1977) and Houghton *et al.* (1978) described identical deposits in the kidneys of patients treated with gentamicin. In addition, Harrison *et al.* (1974), Wellwood *et al.* (1976), and Katz *et al.* (1979) have identified similar material by electron microscopy in the urinary sediment of gentamicin-treated subjects.

Numerous studies (Just *et al.,* 1977; Silverblatt and Kuehn, 1979; Luft *et al.,* 1974) indicate that aminoglycosides accumulate in the kidney tissue, to a larger extent in the cortex than in the medulla. Our own studies in this field have involved the use of autoradiographic techniques:

1. We have shown a preferential localization of ^{14}C-labeled gentamicin on whole-body autoradiograms of rats in cartilage and in the kidneys, to a larger extent in the cortex than in the medulla of the latter (Fresel *et al.,* 1980). Five minutes after the injection of the labeled gentamicin, its distribution is uniform throughout the body. Two hours later, intense labeling remains only in the kidneys and the cartilage.

2. For an autoradiographic study of the uptake of [^{3}H]gentamicin in rabbit renal tubules, isolated tubules were obtained by microdissection following collagenase treatment 4 hr after the administration of [^{3}H]gentamicin. In the proximal tubule, we observed a gradual increase in [^{3}H]gentamicin accumulation from the glomerulus to the end of the pars recta. The silver grain density progressively increases along this structure: five silver grains per 150 μm^2 in the very early proximal tubule and up to 40 silver grains per 150 μm^2 in the terminal pars recta. Almost no gentamicin incorporation ($<$2 silver grains per 150 μm^2) takes place along the distal parts of the nephron, from the beginning of the loop of Henle to the end of the medullary collecting duct. No differences were visible along these parts of the nephron, whether they are cortical or medullary in location.

Fabre *et al.* (1976) confirmed the accumulation of aminoglycosides in the kidney, with a preferential uptake in the cortex. When injected at the dose of 4 mg/kg in the Wistar rat, the serum peak of gentamicin was 6.3 ± 1.8 $\mu g/ml$ at 1 hr after injection. Six hours later, the serum contained only trace amounts of gentamicin. The liver, the heart, and the muscles showed negligible concentrations, while the renal cortex contained 123 ± 20 μg/g of wet tissue. The elimination of the drug from kidney tissue was a very slow process; this tissue contained 7 μg/g at 4 weeks after a single injection (Chauvin *et al.,* 1978). Similar renal accumulation has also been demonstrated in mice, rabbits, and dogs (Reiner *et al.,* 1978; Whelton and Walker, 1974; Frame *et al.,* 1973; Kornguth and Kunin, 1977; Sairio *et al.,* 1978; Alfthan *et al.,* 1973; Edwards *et al.,* 1976; Schentag *et al.,* 1977). The renal accumulation of other aminoglycosides, such as tobramycin, netilmicin, and dibekacin, has also been demonstrated (Whelton, 1980).

Ultrastructural investigations using autoradiographic techniques have permitted determination of the intracellular localization of aminoglycosides (Just *et al.,* 1977); Silverblatt and Kuehn, 1979). Tritiated gentamicin is localized in the apical vesicles of proximal tubule cells 10 min after injection, and almost exclusively in the lysosomes 1 hr after injection. Twenty-four hours later, the silver grains still remain numerous within the lysosomes.

We have shown in parallel, using cell fractionation techniques by density equilibration, that ^{14}C-labeled gentamicin exclusively distributes in the same manner as two lysosomal enzymes 6 hr after the injection. On the other hand, tritiated gentamicin injected 45 min prior to sacrifice shows a slightly bimodal distribution, suggestive of an association partly to lysosomes and, to a lesser extent, to other subcellular structures, such as brush border (as identified by either alanine-aminopeptidase or γ-glutamyltranspeptidase activities). We interpret the more complex distribution of gentamicin early after injection as reflecting its mode of entry into tubular cells, which is most likely absorptive pinocytosis at the tubular lumen. The existence of an absorptive process is supported by the *in vitro* demonstration of binding of gentamicin to brush border membranes (Jerauld and Silverblatt, 1979; Just and Habermann, 1977).

The association constant and the binding capacity revealed by these studies are consistent with a drug uptake rate about 20-fold that of substances that enter cells solely by fluid pinocytosis. Interestingly, a survey of available data shows that the renal uptake of gentamicin is 10- to 25-fold that of inulin or polyvinyl pyrrolidone (PVP) (Morin *et al.,* 1978; Luft and Kleit, 1974; Schiller *et al.,* 1978).

Since aminoglycoside-induced nephrotoxicity involves subcellular targets in the proximal tubule cells, we will limit this chapter to the study of the alterations of subcellular organelles, namely, lysosomes and mitochondria, in the presence of aminoglycosides.

2. *In Vitro* Studies

2.1. Lysosomes

2.1.1. Methods

Lysomes from kidney cortices of Wistar rats were partially purified on linear sucrose gradients after tissue homogenization in 0.25 M sucrose. N-Acetyl-β-D-glucosaminidase assay was used to localize the lysosomal fraction. According to de

Duve *et al.* (1962), the free/total activity ratio of lysosomal enzymes is assumed to reflect the integrity of the lysosomal membrane. The lysosomal fractions were placed in the presence of increasing concentrations of aminoglycosides, and free and total activities of *N*-acetyl-β-D-glucosaminidase were measured before and after a 60-min incubation at 37°C.

2.1.2. Results

Aminoglycosides labilize the lysosomes of rat kidney *in vitro* at concentrations ranging from 5 to 50 μg/ml in standard lysosomal suspension (Fig. 1). Statistical analysis using the Student's *t*-test for mean and variation allowed us to classify these molecules into three groups in terms of the intensity of their labilizing effect. A highly active class includes neomycin, sisomycin, tobramycin, and gentamicin; an intermediately active class includes lividomycin; and a poorly active class includes kanamycin, amikacin, netilmicin, streptomycin, paromomycin, and kasugamycin. Except for neomycin, a rather good correlation is seen between the relative intensity of the aminoglycoside-labilizing effect and the relative nephro-toxicity of these molecules as reported by Appel and Neu (1977).

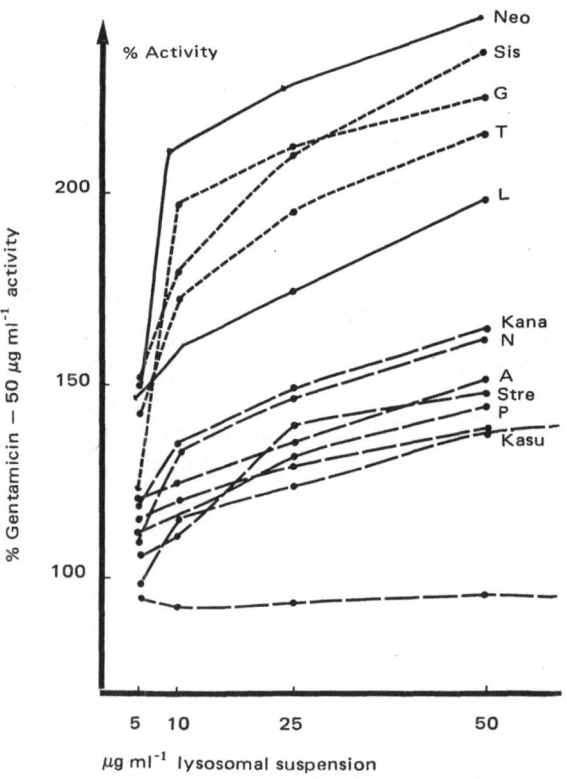

FIGURE 1. Comparative study of the effect of eight aminoglycosides on the structural latency of the lysosomal *N*-acetyl-β-D-glucosaminidase *in vitro*. Results are expressed as the free/total activity ratio. Mean of at least four independent determinations ± SEM.

2.2. Mitochondria

2.2.1. Methods

Kidneys from Sherman rats were homogenized in a medium composed of 0.33 M sucrose, 1 mM EDTA Na_2, and 5 mM Tris-HCl buffer (pH 7.4). The mitochondria were isolated by differential centrifugation at 2°C. The final pellet was suspended in a small quantity of isolating medium to attain 60 mg protein/ml. Oxygen consumption by mitochondria was determined by oxypolarography at 25°C in an incubation medium composed of 80 mM KCl, 10 mM K_2 HPO_4, 20 mM Tris-HCl (pH 7.2). The medium was air-saturated by magnetic agitation, resulting in an oxygen concentration of 240 mM. An aliquot of mitochondrial suspension was added to 1.5 ml of incubation medium to achieve a final concentration of 1 mg/ml of protein. Variations in oxygen concentration were continuously recorded. The effects of various aminoglycosides tested on mitochondria in this study were compared to those of gentamicin C as a reference.

2.2.2. Results

Neomycin B, gentamicin, tobramycin, sisomycin, dibekacin, netilmicin, kanamycin A, and amikacin were studied. These antibiotics share similar activity, i.e., activation of state 4 and inhibition of state 3. The degree of these two effects varies greatly among them; however, the intensity of the activating effect of state 4 is linked to the intensity of the inhibitory effect of state 3. At a concentration of 50 μg/ml, we found it possible to classify aminoglycosides according to the intensity of respiratory modifications which they induced. In decreasing activity we found: neomycin B, sisomicin, tobramycin, dibekacin, gentamicin C, netilmicin, kanamycin A, and amikacin.

A progressive inhibition of the effect of gentamicin on mitochondrial respiratory activity in state 3 or 4 is noted when increasing concentrations of Mg^{++} are added to the incubation medium. This could suggest competition between magnesium and aminoglycosides. We also studied mitochondria isolated in the presence and in the absence of EDTA. In the latter condition, mitochondria are much less affected by aminoglycosides. The presence of EDTA in the isolating medium frees certain sites initially occupied by magnesium, and this effect could sensitize the mitochondrial membrane to gentamicin by chelating part of the membrane cations.

The addition of gentamicin to a mitochondrial suspension in a KCl 80 mM medium induced a very slight swelling of the mitochondria. In contrast, a clear swelling appears when a substrate such as β-hydroxybutyrate or malate-glutamate is added. The swelling is prevented by addition of an uncoupler such as Cl-CCP. When antimycin (0.4 μg/mg protein) is previously added to the incubation medium, gentamicin induces no swelling of the mitochondria in the presence of the respiratory substrate. Thus, the mitochondrial swelling induced by gentamicin requires a source of energy. This energy is provided by respiratory substrate oxidation; since antimycin, which inhibits electron transport, prevents the occurrence of this swelling. The increase in mitochondrial volume implies an active penetration of a substance into the mitochondria. This substance might be either gentamicin or potassium. Evaluations made in media without potassium showed that gentamicin did not induce swelling in these conditions. Swelling appears only when potassium is added to the medium. The phenomenon thus appears to be due to active K^+ penetration into the mitochondria induced by gentamicin. Swelling does not appear in the presence of Mg^{++}, apparently because the cation acts as an antagonist of gentamicin. Gentamicin appears to have no effect on passive transport, and its activity appears linked only to energy.

In conclusion, our opinion on the mechanism of the *in vitro* effect of gentamicin on mitochondria can be summarized as follows. Initially, because of the amine basic characteristics of the amino groups, gentamicin is fixed on the mitochondria membrane sites freed by the effect of EDTA during the isolation process. This fixation would provide access for the antibiotic to the divalent cations Ca^{++} or Mg^{++} situated in a hydrophobic environment and not accessible to EDTA. The properties of the membrane would be altered; in particular, its energy level would be diminished, implying oxidation of the transport chain. This freed energy would activate the H^+ pump of the mitochondrial membrane, which would expel the protons produced by the substrate oxidation. This expulsion of H^+ would be compensated by the penetration of K^+. In the presence of Mg^{++}, the sites that were previously freed by the EDTA would be preferentially occupied by the cation, and gentamicin would no longer be effective.

It seems also that the degree of these alterations increases with the number of the amino groups of the molecules. Neomycin B, the most nephrotoxic of these antibiotics, has the greatest effect on the *in vitro* mitochondrion and has the largest number of free amino groups. Antibiotics having intermediate nephrotoxicity have five free amino groups, and those with lower nephrotoxicity have only four. Sorbistin A_1 has only three free amino groups, and this antibiotic shows very low toxicity in mice and rats (Viotte *et al.,* 1980). This relationship between structure and activity appears to be of value in assessing possible nephrotoxic potential of the various aminoglycosides.

3. *In Vivo* Studies

Because the proximal tubule cells accumulate aminoglycosides within their lysosomes, the aminoglycosides may not reach mitochondria in sufficient amounts to induce alterations of the respiratory control to the same extent as that seen *in vitro*. We have been unable to detect any modification of the respiratory control of mitochondria isolated from the kidney cortices of treated rats (*in vivo* exposure). We shall therefore describe only modifications of the properties of lysosomes after aminoglycoside treatment.

3.1. Methods

3.1.1. Animals

Wistar rats received IP injections of aminoglycoside solutions in saline; control rats of the same strain received daily saline injections. The animals were killed at least 14 hr after the last injection. Renal function was evaluated by blood creatinine level.

3.1.2. Pathology

Three-micrometer slices were prepared after fixation of kidney cortices in Dubosc-Brasyl solution and subsequent paraffin-embedding. Sections were stained with trichrome, PAS, and hematoxylin-eosin.

3.1.3. Electronic Microscopy

Small pieces of superficial renal cortex were first immersed in a 2.5% glutaraldehyde solution, rinsed, and immersed in a 2% osmium tetroxide solution. Ultrafine slices were cut, using a Reichert Om U-2 Ultrotome, and stained with both uranyl acetate (2%) and lead citrate (2%). A Zeiss EM 9-S$_2$ microscope was used.

3.1.4. Stereology

Stereologic study was performed according to the method of Weibel (1969). Analysis is made on micrographs at a final magnification of 20,000.

Fractional cell volumes of lysosomes, mitochondria, brush border, and myeloid bodies were determined in proximal tubule cells.

3.1.5. Antibiotic Assay

Aminoglycoside concentrations were determined using an agar diffusion technique with *Bacillus subtilis* as a test organism. Previous studies showed no requirement for antibiotic extraction prior to the assay.

3.1.6. Biochemical Study

Fragments of kidney cortices were homogenized in distilled water ($1/200 \, w/v$) using a Kontess conical sintered glass tissue grinder. The following enzymes were assayed with chromogenic or fluorogenic substrates according to standard techniques (Van Hoof and Hers, 1968): acid α-L-fucosidase, acid α-D-galactosidase, acid α-D-mannosidase, N-acetyl-β-D-glucosaminidase, cathepsin B, sphingomyelinase, true glucose-6-phosphatase, L-alanine-aminopeptidase, and γ-glutamyltranspeptidase.

Proteins were measured according to the technique of Lowry *et al.* (1951), using bovine serum albumin as a standard.

Lipid phosphorus was determined according to Bligh and Dyer (1959).

3.2. Results

Results will first be discussed with regard to the administered dose and the duration of gentamicin therapy. A comparative study of four molecules currently available for clinical use, gentamicin, tobramycin, netilmicin, and amikacin, given at equi-therapeutic doses, will be presented.

3.2.1. Administration of Increasing Doses of Gentamicin

Wistar rats received daily injections of gentamicin for 8 days at doses of 4, 10, 20, 50, and 100 mg/kg. A significant increase in blood creatinine level was observed as dose of 20 mg/kg. The histopathologic injury to the kidney was evaluated in rats that received doses of 10 and 50 mg/kg. At the lowest dose, no pathologic change was observed. At the dose of 50 mg/kg, however, gentamicin induces extensive alterations in the kidney structure: the brush border of proximal tubule cells is practically absent, the tubular lumens are filled with cellular debris, and severe tubular necrosis is seen.

3.2.1a. Ultrastructural Study. Ultrastructural study shows extensive modifications, even at the dose of 10 mg/kg: numerous myeloid bodies are present within enlarged lysosomes, and the brush border is already much less developed than in the controls. These data will be detailed below.

3.2.1b. Intracortical Levels of Gentamicin. These increased with the administered dose (Fig. 2) up to 50 mg/kg/day for 8 days, and the amount of antibiotic accumulated appeared to be a function of the logarithm of the administered dose. The relatively lower accumulation at a dose of 100 mg/kg is apparently the result of extensive necrosis, leading to a substantial sloughing of proximal tubular cells.

3.2.1c. Lysosomal Structural Latency. The variation in latency of the lysosomal enzyme N-acetyl-β-D-glucosaminidase is shown in Fig. 3. On freshly

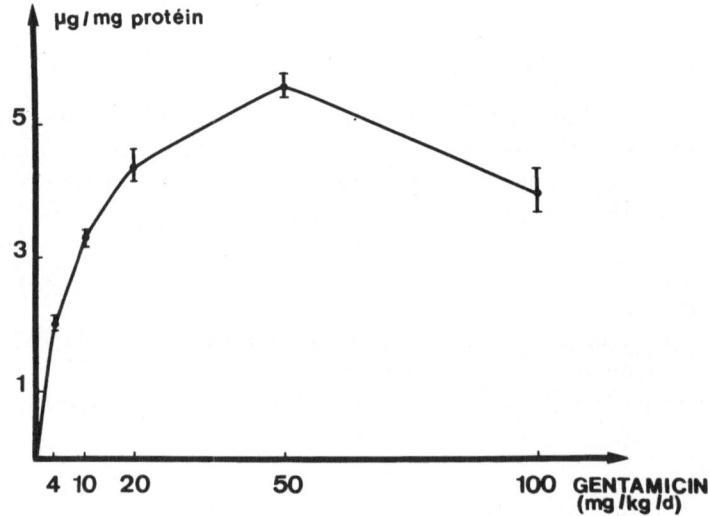

FIGURE 2. Gentamicin content of kidney cortices. Results are expressed as μg/mg of tissue protein. Mean of at least four independent determinations ± SE.

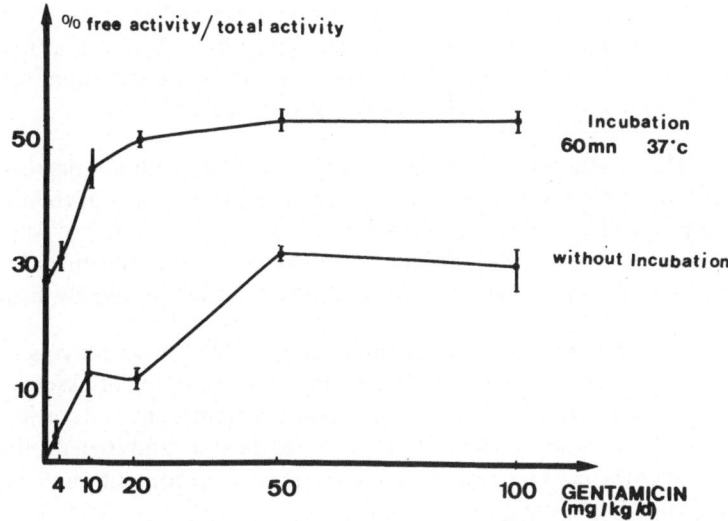

FIGURE 3. Structural latency of the lysosomal N-acetyl-β-D-glucosaminidase following gentamicin treatment. Results are expressed as the free/total activity ratio. Mean of at least four independent determinations ± SEM.

isolated lysosomal fractions, we observed that gentamicin treatment of the animals significantly increases the percentage of free activity, that is, the portion of enzyme activity that is demonstrable in the absence of detergent. The maximum effect is observed at the dose of 50 mg/kg for 8 days. A parallel decrease in lysosomal latency is observed, with or without a 60-min incubation at 37° C. With incubation, however, a higher level of free activity is observed, due to the spontaneous rupture of lysosomes during incubation (Berthet *et al.,* 1951; de Duve, 1962).

3.2.1d. Modification of Enzymatic Activities of the Kidney Cortex. Among the enzymatic activities assayed, alanine-aminopeptidase activity is significantly decreased at 4 mg/kg gentamicin treatment. However, this alteration does not appear to be dose-dependent. The activity of α-D-galactosidase is decreased only after the highest administered dose ($p < 0.05$). The activities of N-acetyl-β-D-glucosaminidase and of true glucose-6-phosphatase remain unaltered, whatever the administered dose. The activity of sphyngomyelinase, an enzyme involved in the catabolism of phospholipids, is greatly impaired by gentamicin treatment. This effect is already significant ($p < 0.001$) after therapy at 4 mg/kg. This impairment appears to be dose-dependent. Cortical DNA content does not vary, whatever the administered dose.

3.2.2. Effect of the Duration of Therapy

The previous study led us to choose the dose of 50 mg/kg/day to record maximal effects. Wistar rats were killed after one, two, four, six, and eight injections and at 2, 4, 6, 7, 15, 24, 31, 46, and 62 days after withdrawal of treatment.

3.2.2a. Blood Creatinine. Blood creatinine increased after the fourth day of treatment and reached maximal value at the eighth day (48 ± 1.8 μmole/liter in controls increased to 87.5 ± 7.5 μmole/liter at 8 days). Blood creatinine returned to control levels 15 days after the withdrawal of the treatment.

3.2.2b. Histopathology. Pathologic alterations of cellular necrosis of the proximal tubule epithelium appeared as early as the fourth day of treatment, and totally disappeared 1 month after the treatment has been withdrawn. At that point the appearance of the brush border, the lysosomes, and the mitochondria is identical to that of controls, and the myeloid bodies have completely disappeared.

3.2.2c. Tissue Concentrations. Intracortical gentamicin levels increased linearly up to day 4 and reached a plateau at about 6 μg/mg of tissue protein, or about 0.9 mg/g of wet tissue. On discontinuation of treatment, the cortical content decreased rapidly to about one-third of the plateau value, and then declined much more slowly over the next 60 days, by which time cortical content was 0.3 μg/mg of tissue protein, or 0.045 mg/g of wet tissue.

3.2.2d. Lysosomal Structural Latency. This parameter was decreased only after one injection at the dose of 50 mg/kg. This alteration is dependent upon the duration of the treatment and is maximum at 8 days. A total and stable recovery is observed 8 days after the withdrawal of treatment.

3.2.2e. Cortical Enzymatic Activities. Sphingomyelinase activity was extensively impaired. This impairment was already significant at day 2 and increased up to the eighth day of administration. At this point, residual activity was about 50% of control activity. Enzyme activity recovered rapidly after the cessation of drug administration.

Alanine-aminopeptidase activity was depressed during the first 2 days of treatment, but this impairment was only transient. Despite continued treatment, this activity recovered to control levels by day 4.

α-D-Galactosidase activity was impaired after the sixth day of treatment. Control levels were recovered 15 days after withdrawal of treatment. The activities of N-acetyl-β-D-glucosaminidase and of true glucose-6-phosphatase remained unaffected during gentamicin treatment.

3.2.2f. Phospholipids. The measurement of lipidic phosphorus failed to reveal any significant accumulation in the whole cell, whatever the dose and the duration of aminoglycoside administration.

3.2.3. Comparison of the Effects of Tobramycin, Netilmicin, Amikacin, and Gentamicin on Rat Kidney Lysosomes

Sprague-Dawley rats were treated daily with intraperitoneal injections of gentamicin, tobramycin, or netilmicin, 10 mg/kg, and amikacin, 30 mg/kg, for 2, 4, 7, and 14 days. These doses in rats, when corrected on the basis of body area, are equivalent to 2 mg/kg/day for gentamicin, tobramycin, and netilmicin and to 6 mg/kg/day for amikacin in humans. These are slightly below therapeutic doses. We will discuss only our results on sterologic and biochemical studies.

3.2.3a. Stereologic. Quantitative data from the stereologic analysis of proximal tubular cells are summarized in Table I.

The fractional cell volume of the lysosomal compartment increases after aminoglycoside therapy. This increase is particularly impressive with gentamicin (from 3.1 to 12.2% of cell volume), while moderate increase was observed with amikacin (3.1 to 4.96%). An intermediate increase was observed with tobramycin and netilmicin.

The lysosomal content was only slightly affected by amikacin and tobramycin administration, while lysosomes are extensively overloaded with myeloid bodies after gentamicin and netilmicin treatment. In these two latter cases about 50% of the lysosomal volume is filled with tightly packed myeloid bodies (Fig. 4).

3.2.3b. Biochemical. Antibiotics concentrations in the kidney cortex of treated rats are summarized in Table II. Netilmicin and gentamicin were accumulated to the same extent and reached concentrations of about 500 μg/g of wet cortex after 14 days of treatment (daily dose 10 mg/kg). Amikacin and tobramycin accumulated less intensively in the kidney cortex, at approximately 185 μg/g (tobramycin) and 555 μg/g (amikacin) for doses of 10 mg/kg daily of tobramycin and 30 mg/kg daily of amikacin, respectively.

3.2.3c. Lysosomal Catabolic Activities. These results are summarized in Table III. Gentamicin, tobramycin, and netilmicin did not affect glycosidase activities except for α-D-galactosidase. This activity was slightly depressed with gentamicin and tobramycin. Amikacin induced markedly increased activity of N-acetyl-β-D-glucosaminidase and of α-mannosidase, while α-D-galactosidase and α-fucosidase activities were increased to a lesser extent.

Activity of cathepsin B (a lysosomal protease) was unaffected whatever the injected aminoglycoside and whatever the duration of therapy.

Sphingomyelinase activity was extensively impaired by gentamicin, slightly affected by tobramycin and netilmicin, and not modified by amikacin.

Glucose-6-phosphatase (a microsomal enzyme) was unaffected by all four aminoglycosides.

TABLE I. Comparative Stereologic Morphometric Study[a]

	Mitochondria	Brush border	Lysosomes	Myeloid bodies
Control	12.1	21.5	3.1	0.12
Gentamicin	25.8	9.35	12.2	5.36
Tobramycin	20.9	4.62	7.05	0.88
Netilmicin	20.3	6.86	7.34	3.86
Amikacin	21.3	20.1	4.96	0.38

[a]Results are expressed as the fractional cell volume (%) occupied by a given subcellular organelle. Results are obtained from at least 16 micrographs at the original magnification of 5000.

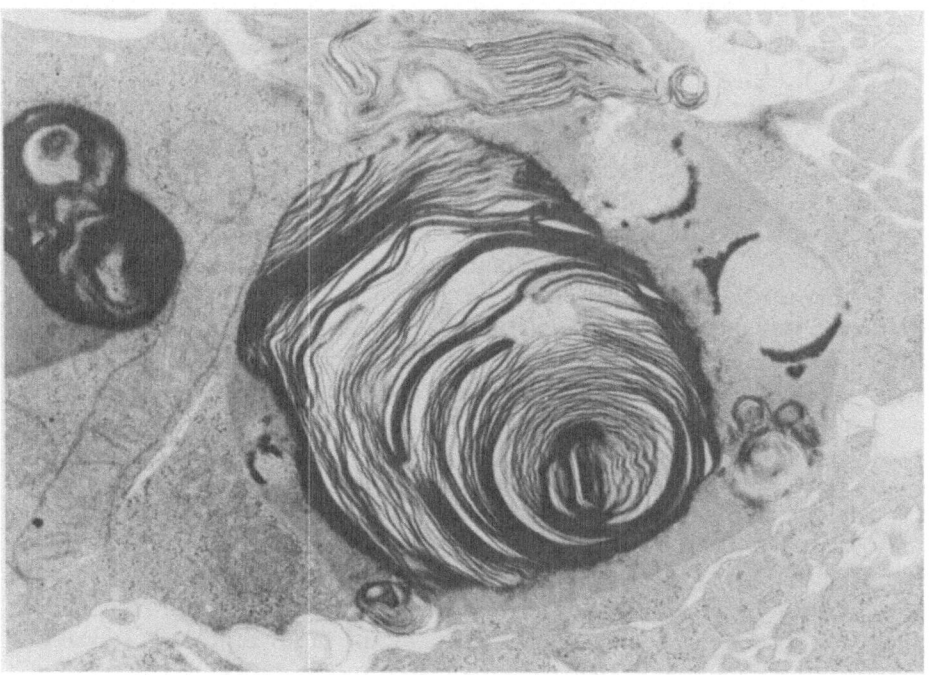

FIGURE 4. A typical myeloid body in a lysosome of rat kidney treated with gentamicin 20 mg/kg daily for 8 days.

Activities of alanine-aminopeptidase and γ-glutamyltranspeptidase (brush border enzymes) were only slightly impaired by aminoglycosides.

Assessment of total cell lipid phosphorus failed to reveal any significant accumulation.

4. Discussion

Aminoglycosides are eliminated from the body exclusively by the kidneys. Aminoglycosides are first filtered by the glomerulus and then penetrate into

TABLE II. Kidney Cortex Concentrations of Aminoglycoside during a 14-Day Therapy in Rats[a]

Days of treatment	Gentamicin	Tobramycin	Netilmicin	Amikacin
2	2.00 ± 0.13	0.92 ± 0.08	1.77 ± 0.07	1.17 ± 0.24
4	3.51 ± 0.22	1.24 ± 0.05	2.68 ± 0.29	2.69 ± 0.27
7	3.00 ± 0.21	1.30 ± 0.09	2.49 ± 0.47	2.62 ± 0.27
14	3.62 ± 0.061	1.27 ± 0.11	3.44 ± 0.18	3.69 ± 0.70

[a]Gentamicin, tobramycin, and netilmicin were given IP at the daily dose of 10 mg/kg. Amikacin was given IP at the daily dose of 30 mg/kg. The results are expressed as μg/mg of cortex protein.

TABLE III. Cortical Levels of Lysosomal Enzyme Activities during Aminoglycoside Treatment in Rats[a]

	SPHINGO	NAG	α-GAL	CAT B	α-FUCO	α-MANNO
Gentamicin[b]						
2 days	94.68	101.96	87.96	88.49	107.09	117.35
4 days	84.14	100.75	90.36	85.12	107.12	107.43
7 days	65.13	99.72	83.74	96.54	107.27	108.26
14 days	55.82	99.46	86.24	97.30	121.17	106.69
Tobramycin[b]						
2 days	97.30	132.13	114.14	111.25	142.12	135.27
4 days	89.96	102.19	85.02	103.72	107.26	117.11
7 days	84.14	110.62	89.22	92.27	113.93	118.67
14 days	87.78	108.41	79.44	101.91	108.09	104.90
Netilmicin[b]						
2 days	99.04	106.75	99.64	100.41	103.80	113.89
4 days	92.36	116.55	95.12	99.03	112.09	136.76
7 days	88.19	88.78	86.96	87.72	102.22	100.65
14 days	81.02	129.46	99.23	107.47	125.05	131.75
Amikacin[c]						
2 days	107.70	121.38	123.17	105.15	123.56	147.57
4 days	102.93	149.38	120.82	94.09	131.81	156.54
7 days	110.65	195.46	137.56	95.55	178.76	205.20
14 days	105.27	191.79	118.83	118.92	157.70	184.01

[a] Values are percentage of control. Results were expressed as mU/mg of tissue proteins $\pm$ SE mean of at least four experiments. SPHINGO, Sphingomyelinase. NAG, N-Acetyl-β-D-glucosaminidase. α-GAL, α-Galactosidase. CAT B, Cathepsin B. α-FUCO, α-Fucosidase. α-MANNO, α-Mannosidase.
[b] 10 mg/kg.
[c] 30 mg/kg.

proximal tubular cells by pinocytosis. Nevertheless, it remains unclear whether pinocytosis is the sole or even the major route by which these drugs enter these cells. Some *in vitro* studies with the renal cortical slice technique suggest the possibility of basolateral membrane transport of aminoglycosides (Hsu, 1977). Apparent lysosomal accumulation of aminoglycosides was demonstrated. Our studies

TABLE IV. Cortical Level of Brush Border Enzyme Activities during Aminoglycoside Treatment in Rats[a]

	Days of treatment	Gentamicin 10 mg/kg	Tobramycin 10 mg/kg	Netilmicin 10 mg/kg	Amikacin 30 mg/kg
Alanine-aminopep-	2	56.64	72.73	72.58	90.70
tidase	4	69.28	76.02	64.13	76.78
	7	60.80	89.35	75.30	72.70
	14	70.31	74.25	78.91	71.45
α-Glutamyl-transpep-	2	79.64	82.68	75.96	89.62
tidase	4	79.31	90.75	97.83	94.37
	7	71.04	108.03	112.6	84.45
	14	91.26	102.76	89.63	94.83

[a] Values and results as in Table III.

demonstrated that intralysosomal aminoglycoside could then impair the intra-lysosomal catabolism as well as alter the stability of the lysosomal membrane, resulting in the appearance of lysosomal hydrolases in the cytosol, and thus leading to cell necrosis.

The cellular uptake and intralysosomal accumulation of gentamicin is not a phenomenon unique to the kidney. Tulkens and Trouet (1978) showed that cultured rat fibroblasts readily take up aminoglycosides and that the accumulation in the cells could reach 100- to 200-fold the concentrations in the culture fluid. Aubert-Tulkens *et al.* (1979) also noted that when intracellular concentrations reached about 4 μg/mg of cell protein (800 μg/g of wet tissue), lysosomes show a conspicuous overloading with myeloid bodies, similar in appearance to those observed in rat kidneys. Other subcellular organelles showed no lesions. Assaying systematically for a large number of acid hydrolases, Aubert-Tulkens *et al.* (1979) showed a constant and highly significant decrease in lysosomal sphingomyelinase activity. They also noted a decrease in acid phospholipase A ($A_1 + A_2$). Concomitantly, the total lipid phosphorus rises to 170% of control value. Thin-layer chromatography indicates that the concentration of all major phospholipids is increased without appreciable change in their relative amounts. We conclude that aminoglycosides may impair the lysosomal catabolism of phospholipids through a decrease of the activities of corresponding hydrolases, and modify the permeability of the lysosomal membranes and thus allow hydrolases to gain access to other cytoplasmic organelles, leading to cell necrosis and cell death.

Aminoglycoside nephrotoxicity should be considered as being the result of at least three factors: (1) the extent to which the drugs accumulate within the lysosomes of the proximal tubule cells, (2) the extent to which the drugs induce phospholipidosis, leading to the overloading of the lysosomal system with myeloid bodies and (3) what we would call the intrinsic toxicity of the molecule as reflected by its effect on either the structural latency of the lysosomal membrane or on the respiratory control of isolated mitochondria.

The degree of intracortical accumulation of the aminoglycoside antibiotics introduces a new variable for consideration by the nephrologist, from whom medullary accumulation of drugs has drawn more attention. Some authors, basing their conclusions on data observed with neomycin and gentamicin (Schentag *et al.,* 1978; Dellinger *et al.,* 1976, Barza *et al.,* 1978) claimed that the potential nephrotoxicity was directly related to the renal accumulation of the molecule. On the other hand, Luft and Kleit (1974) and Trottier *et al.* (1978) showed that the renal accumulation of these molecules does not always allow a good prediction of potential nephrotoxicity. For instance, although netilmicin appears to be less nephrotoxic than gentamicin in humans as well as in animals (Chiu *et al.,* 1977; Soberon *et al.,* 1979; Bowman *et al.,* 1977), the cortical accumulation of netilmicin appears to be greater than or at least equal to that of gentamicin. Kaloyanides and Pastoriza-Munoz (1980) have found similar results in rats treated with neomycin, netilmicin, and gentamicin. They concluded that for a given aminoglycoside, the risk of nephrotoxicity increased as the renal concentration of the drug increases. This accumulation depends on the dose and the frequency of administration until eventually a plateau is reached, reflecting either a saturation of aminoglycoside transport, or the achievement of a balance between uptake and efflux of the drug from the renal cortex; aminoglycoside nephrotoxicity does not depend exclusively on the amount of the drug accumulated in the kidney cortex.

Behavior of aminoglycosides in subcellular organelles. In the course of this chapter, we have described the comparative effect of aminoglycosides on lysosomal structural latency *in vitro* and on the mitochondrial respiratory control *in vitro*, and the effect of gentamicin on the structural latency of kidney cortex lysosomes *in vivo*. A fairly good correlation is found between the degree of subcellular toxicity of these drugs as reflected by *in vitro* studies and the degree of nephrotoxicity as described by Appel and Neu (1977). Thus, aminoglycosides seem capable of affecting the physical properties of intracellular membranes.

In addition, the aminoglycosides also appear capable of affecting the intralysosomal catabolism of phospholipids and of inducing a lysosomal phospholipidosis bearing some similarities to several inborn enzyme defects (Brady, 1978), for instance, Tay-Sachs disease, Niemann-Pick disease, or Fabry disease. Although gentamicin does not seem to induce a defect in a single enzyme, it is perhaps significant that these other storage diseases may provoke renal failure.

Although we have described an elevated accumulation of myeloid bodies, we have been unable to show any significant increase in the total cell content of phospholipids. To our knowledge, the literature contains only one report of a significant rise in phospholipid content in the rat renal cortex (Kaloyanides and Pastoriza-Munoz, 1980). These authors report a 10% increase after 2 days of therapy at the dose of 100 mg/kg, but do not indicate any values for more prolonged administration. We consider that our data reflect a kind of equilibrium attained between the amount of lipidic phosphorus accumulated as myeloid bodies and the amount of lipidic phosphorus lost through disappearance of brush border membranes. This hypothesis is supported by the results of the quantitative stereologic study with gentamicin.

Aminoglycosides can be classified as follows: (1) on the basis of tissue concentrations, GENTA = NETIL >> TOBRA $\simeq$ AMIKA; (2) on the basis of increased lysosomal volume, GENTA > NETIL $\simeq$ TOBRA > AMIKA; (3) on the basis of decreased lysosomal structural latency, GENTA $\simeq$ TOBRA >> NETIL $\simeq$ AMIKA; (4) on the basis of brush border loss, GENTA $\simeq$ TOBRA $\simeq$ NETIL >> AMIKA; (5) on the basis of myeloid body accumulations, GENTA > NETIL >> TOBRA $\simeq$ AMIKA.

In conclusion, relative potential nephrotoxicity of aminoglycosides seems to be determined by the interaction of several factors, including (1) the extent of the lysosomal accumulation of the drug in proximal tubular cells, (2) the extent to which the drug affects the lysosomal catabolism of phospholipids resulting in a phospholipidosis, and (3) the intrinsic toxicity of the molecule as reflected by the effect of the molecule on the structural latency of the lysosomal membrane.

Although animal experiments contribute to the understanding of the basic mechanism of aminoglycoside nephrotoxicity, the extrapolation of these data to the human remains premature. However, it is important to note that similar results are observed in various species, suggesting identical mechanisms of nephrotoxicity. Some discrepancies remain, particularly in comparative studies of aminoglycosides. These discrepancies may be due to the presence or the absence of an associated disease process which may be one more critically important factor for the eventual tolerance of the individual drugs.

Acknowledgments

This work was supported by the International Institute of Cellular and Molecular Pathology and Laboratoire de Chimie Physiologique, Université Catholique de Louvain (Professor C. de Duve), where J. P. M. was a Fellow during 1979–1980. This work was also supported by a grant from Belgium Scientific and Medical Research, No. 3.4516.79 and by the Institut National de la Santé de la Recherche Medicale, Paris (CRL 71.78 103 and CRL 79/2745). We acknowledge Dr. Van Hoof for his help in morphometric studies and we appreciate the secretarial assistance of D. Falize.

References

Alfthan, O., Renkonen, O. V., and Sivonen, A., 1973, Concentration of gentamicin in serum, urine and urogenital tissue in man, *Acta Pathol. Microbiol. Scand. Suppl. B81* **241**:92–94.

Appel, G. B., and Neu, H. C., 1977, The nephrotoxicity of antimicrobial agents, *N. Engl. J. Med.* **269**: 663, 722, 784.

Aubert-Tulkens, G., Van Hoof, F., and Tulkens, P., 1979, Gentamicin-induced lysosomal phospholipidosis in cultured rat fibroblasts — Quantitative ultrastructural and biochemical study, *Lab. Invest.* **40**(4):481.

Barza, M., Pinn, V., Tanguay, P., and Murray, T., 1978, Nephrotoxicity of newer cephalosporins and aminoglycosides alone and in combination in a rat model, *J. Antimicrob. Chemother.* **4**:59.

Berthet, J., Berthet, L., Appelmans, F., and de Duve, C., 1951, Tissue fractionation studies. II. The nature of the linkage between acid phosphatase and mitochondria in rat liver tissue, *Biochem J.* **50**:182–189.

Black, J., Calesnick, B., Williams, D., and Weinstein, M. J., 1963, Pharmacology of gentamicin a new broad-spectrum antibiotic, *Antimicrob. Agents Chemother.* **3**:138.

Bligh, E. G., and Dyer, W. J., 1959, A rapid method of total lipid extraction and purification, *Can. J. Biochem.* **37**:911.

Bowman, R. L. Silverblatt, F. J., and Kaloyanides, G. J., 1977, Comparison of the nephrotoxicity of netilmicin and gentamicin in rats, *Antimicrob. Agents Chemother.* **12**:474.

Brady, R. O., 1978, Elucidation of clinical lysosome deficiencies, in: *Molecular Basis of Biological Processes,* Academic Press, New York, p. 39.

Carbon, C., Contrepois, A., Vigneron, A., and Lamotte-Barillon, S., 1980, Effects of furosemide on extravascular diffusion, protein binding and urinary excretion of cephalosporins and aminoglycosides in rabbits, *J. Pharmacol. Exp. Ther.* **213**:600.

Chauvin, J. M., Rudhardt, M., Blanchard, P., Gaillard, R., and Fabre, J., 1978, Le comportement de la gentamicine dans le parenchyme rénal, *Schweiz. Med. Wochenschr.* **108**:1020.

Chiu, P. J. S., Miller, G. H., Brown, A. D., Long, J. F., and Waitz, J. A., 1977, Renal pharmacology of netilmicin, *Antimicrob. Agents Chemother.* **11**:821.

Cuppage, F. E., Setter, R. K., Sullivan, L. P., Reitzes, E. J., and Melnykovych, A. O., 1977, Gentamicin nephrotoxicity, II: Physiological, biochemical and morphological effects of prolonged administration to rats, *Virchows Arch. B. Cell Pathol.* **24**:121.

De Duve, G., Wattiaux, R., and Wibo, M., 1962, Effects of rat soluble compounds on lysosomes *in vitro, Biochem. Pharmacol.* **9**:97.

Dellinger, P., Murphy, T., Barza, M., Pinn, V., and Weinstein, L., 1976, Effect of cephalothin on renal cortical concentrations of gentamicin in rats, *Antimicrob. Agents Chemother.* **9**:587.

Edwards, C. Q., Smith, C. R., Baughman, K. L., Rogers, J. F., and Lietman, P. S., 1976, Concentrations of gentamicin and amikacin in human kidneys, *Antimicrob. Agents and Chemother.* **9**:925.

Fabre, J., Rudhardt, M., Blanchard, P., and Regamey, C., 1976, Persistence of sisomicin and genta-micin in renal cortex and medulla compared with other organs and serum of rats, *Kidney Int.* **10**:444.

Frame, P., Bannister, T., Tran, J., and Phair, J., 1973, Gentamicin kinetics and nephrotoxicity in rabbits, *Clin. Res.* **2**:842.

Fresel, J., Morin, J. P., Devauchelle, G., and Fillastre, J. P., 1980, Localization viscerale de la ^{14}C gentamicine chez le rat après administration d'une dose unique, *C. R. Acad. Sci. Paris* **29**: 897–899.

Gilbert, D. N., Houghton, D. C., Bennett, W. M., Plamp, C. E, Reger, K., and Porter, G. A., 1979, Reversibility of gentamicin nephrotoxicity in rats: Recovery during continuous drug administra-tion, *Proc. Soc. Exp. Biol. Med.* **160**:99.

Harrison, W. O., Silverblatt, F. J., and Turck, M., 1974, Experimental gentamicin nephrotoxicity, in: *Fourteenth Interscience Conference on Antimicrobial Agents and Chemotherapy,* San Francisco, California, Abstract 70, American Society for Microbiology, Washington, D.C.

Houghton, D. C., Hartnett, M., Campbell-Boswell, M., Porter, G., and Bennett, W., 1976, A light and electron microscopic analysis of gentamicin nephrotoxicity in rats, *J. Pathol.* **82**:589.

Houghton, D. C., Campbell-Boswell, M. V., and Bennett, W. M., 1978, Myeloid bodies in the renal tubules of humans: Relationship to gentamicin therapy, *Clin. Nephrol.* **10**:140.

Hsu, C. H., Kurtz, T. W., and Weller, J. M., 1977, Potentiation of gentamicin nephrotoxicity by meta-bolic acidosis, *Antimicrob. Agents Chemother.* **12**:132.

Jeraud, R. S., and Silverblatt, F. J., 1979, Gentamicin binding to renal proximal tubule brush-border membranes, in: *Proceedings of the 11th International Congress of Chemotheraphy and the 19th Interscience Conference on Antimicrobial Agents and Chemotherapy,* American Society for Microbiology, Washington, D.C., Abstract 935.

Just, M., and Habermann, E., 1977, The renal handling of polybasic drugs, 2. *In vitro* studies with brush border and lysosomal preparations, *Naunyn-Schmiedeberg's Arch. Pharmacol.* **300**:67.

Just, M., Erdmann, G., and Habermann, E., 1977, The renal handling of polybasic drugs. 1. Gentamicin and aprotinin in intact animals, *Nauyn-Schmiedeberg's Arch. Pharmacol.* **300**:57.

Kaloyanides, G., and Pastoriza-Munoz, E., 1980, Aminoglycoside nephrotoxicity, *Kidney Int.* **18**:571.

Katz, S. M., Sufian, S., and Matsumoto, T., 1979, Urinary myelin figures in gentamicin nephrotoxicity, *Am. J. Clin. Pathol.* **72**:621.

Kornguth, M. L., and Kunin, C. M., 1977, Distribution of gentamicin and amikacin in rabbit tissues, *Antimicrob. Agents Chemother.* **11**:974.

Kosek, J. D., Mazze, R. I., and Cousins, M. J., 1974, Nephrotoxicity of gentamicin, *Lab. Invest.* **30**:48.

Lowry, O. H., Rosebrough, N. J., Farr, A. L., and Randall, R. J., 1951, Protein measurement with the Folin phenol reagent, *J. Biol. Chem.* **193**:265.

Luft, F. C., and Kleit, S. A., 1974, Renal parenchymal accumulation of aminoglycosides antibiotics in rats, *J. Infect. Dis.* **130**:656.

Luft, F. C., Yum, M. N., and Kleit, S. A., 1976, Comparative nephrotoxicities of netilmicin and genta-micin in rats, *Antimicrob. Agents Chemother.* **10**:845.

Luft, F. C., Yum, M. N., Walker, P. D., and Kleit, S. A., 1977, Gentamicin gradient patterns and mor-phological changes in human kidneys, *Nephron* **18**:167.

Morin, J. P., Fresel, J., Fillastre, J. P., and Vaillant, R., 1978, Aminoglycosides actions on rat kidney lysosomes *"in vivo"* and "in vitro", in: *Nephrotoxicity, Interaction of Drugs with Membrane Systems: Mitochondria — Lysosomes* (J. P. Fillastre, ed.), Masson, Paris, p. 253.

Morin, J. P., Viotte, G., Vandewalle, A., Van Hoof, F, Tulkens, P., and Fillastre, J. P., 1980, Gentami-cin induced nephrotoxicity. A cell biology approach, *Kidney Int.* **18**:583.

Reiner, N. E., Bloxham, D. D., and Thompson, W. L., 1978, Nephrotoxicity of gentamicin and tobra-mycin given once daily or continuously in dogs, *J. Antimicrob. Chemother.* **4**:85.

Sairio, E., Kasanen, A., Kangas, L., Nieminen, A. L., and Nieminen, L., 1978, The nephrotoxicity and renal accumulation of amikacin, tobramycin and gentamicin in rats, rabbits and guinea pigs, *Exp. Pathol.* **15**:S370.

Schentag, J. J., Jusko, W. J., Plaut, M. E., Cumbo, T. J., Wance, J. W., and Abrutyn, E., 1977, Tissue persistence of gentamicin in man, *J. Am. Med. Assoc.* **238**:327.

Schentag, J. J., Lasezkay, G., Plaut, M. E., Jusko, W. J., and Cumbo, T. J., 1978, Comparative tissue accumulation of gentamicin and tobramycin in patients, *J. Antimicrob. Chemother.* **4**:23.

Schiller, A., Rer, G., and Taughnner, R., 1978, Excretion and intrarenal distribution of low molecular polyvinyl pyrrolidone and inulin in rats, *Arzneim. Forsch,* **28:**2064.

Silverblatt, F. J., and Kuehn, C., 1979, Autoradiography of gentamicin uptake by the rat proximal tubule cell, *Kidney Int.* **15:**335.

Soberon, L., Bowman, R. L., Pastoriza-Munoz, E., and Kaloyanides, G. J., 1979, Comparative nephrotoxicities of gentamicin, netilmicin and tobramycin in the rat, *J. Pharmacol. Exp. Ther.* **210:**334.

Thompson, W. L., Reiner, N. E., and Bloxham, D. D., 1971, Gentamicin and tobramycin nephrotoxicity in dogs on continuous or once daily intravenous injection, in: *Proceedings 10th International Congress of Chemotherapy,* Zurich, Switzerland, Abstract 208, American Society for Microbiology, Washington, D.C.

Trottier, S., Bergeron, M. G., and Gauvreau, L., 1978, Intrarenal concentration of netilmicin and gentamicin, in: *Current Chemotherapy, Proceedings of the Xth Int. Congr. Chemotherapy,* American Society for Microbiology, Washington, D.C., Volume II, p. 953.

Tulkens, P., and Trouet, A., 1978, The uptake and intracellular accumulation of aminoglycosides antibiotics in lysosomes of cultured rat fibroblasts, *Biochem. Pharmacol.* **27:**415.

Van Hoof, F., and Hers, H. G., 1968, The abnormalities of lysosomal enzymes in mucopolysaccharidoses, *Eur. J. Biochem.* **7:**34.

Vera-Roman, J., Krishnakantha, T. P., and Cuppage, F. E., 1975, Gentamicin nephrotoxicity in rats. I. Acute biochemical and ultrastructural effects, *Lab. Invest.* **33:**412.

Viotte, G., Morin, J. P., Bendirdjian, J. P., Godin, M., and Fillastre, J. P., 1980, Sorbistin A_1: A new non nephrotoxic aminoglycoside, in: *Proceedings of the 11th International Congress of Chemotherapy and the 19th Interscience Conference on Antimicrobial Agents and Chemotherapy,* American Society for Microbiology, Washington, D.C., p. 437.

Weibel, E. R., 1969, Stereological principles for morphometry in electron microscopic cytology, *Int. Rev. Cytol.* **26:**235.

Wellwood, J. M., Lovell, D., Thompson, A. E., and Tigne, J. R., 1976, Renal damage caused by gentamicin: A study of the effects on renal morphology and urinary enzyme excretion, *J. Pathol.* **118:**171.

Whelton, A., and Walker, W. G., 1974, Intrarenal antibiotic distribution in health and disease, *Kidney Int.* **6:**131.

Whelton, A., Carter, G. G., Stout, R. L., Herbst, D. V., and Bryant, H. H., 1980, Dibekacin intrarenal distribution characteristics and renal cortical elution kinetics: Comparison with gentamicin and tobramycin, *J. Clin. Pharmacol.* **20:**8.

13

Aminoglycoside Interactions with Other Drugs
Clinical and Toxicologic Implications

FRIEDRICH C. LUFT

1. Introduction

Aminoglycoside antibiotics account for a large percentage of nephrotoxin-induced acute renal failure in humans. Clinically, other drugs are also frequently implicated. Such drugs include chemotherapeutic agents, anesthetics, and several antimicrobial agents. Not infrequently, patients who are critically ill receive one or more potentially nephrotoxic drugs during the course of aminoglycoside therapy. Thus attention has been directed to the possibility that interactions exist between the aminoglycosides and other pharmacologic agents which may modulate aminoglycoside nephrotoxicity. In addition, aminoglycoside nephrotoxicity is influenced by several nonpharmacologic factors. These include the presence of acidemia, chronic osmotic diuresis, and the state of sodium and potassium balance. When considering the risks of aminoglycoside therapy, these features must all be taken into account (Bennett *et al.*, 1980).

Pharmacologic agents that have been implicated in influencing aminoglycoside nephrotoxicity include methoxyflurane, *cis*-diamminedichlorplatinum, furosemide,

FRIEDRICH C. LUFT • Department of Medicine, Nephrology Section, Indiana University School of Medicine, Indianapolis, Indiana 46223.

amphotericin B, clindamycin, and β-lactam antibiotics, particularly cephalothin and cephaloridine. Several of these purportedly toxic combination have been studied in detail in experimental animals. The animal studies have not uniformly verified the human experience and have occasionally produced conflicting results. In this chapter the current clinical experience and the results of animal experimentation will be reviewed. Unfortunately, a conciliation of all the reported findings cannot, as yet, be made.

2. Methoxyflurane

Methoxyflurane (2,2-dichloro-1,1-difluoroethyl methyl ether) is implicated in the production of acute polyuric vasopressin-resistant, renal failure in humans and experimental animals (Barr *et al.,* 1973). Renal injury is related to the dose-dependent accumulation of the toxic metabolites, inorganic fluoride and oxalic acid. Mazze and Cousins (1973) reported a patient with methoxyflurane nephrotoxicity whose renal dysfunction worsened considerably following the administration of gentamicin. In addition, they were able to cite several other similar instances involving methoxyflurane and aminoglycosides. These same investigators then conducted studies in rats confirming the additive tox- icity of methoxyflurane and gentamicin (Barr *et al.*, 1973). Fisher 344 rats receiving gentamicin 20 mg/kg per day prior to and following 0.5% methoxyflurane anesthesia for 3.5 hr developed greater nonoliguric renal failure than rats treated with either drug alone. Characteristic morphologic changes associated with each of these potential nephrotoxins were observed. Based on their findings, the authors recommend that methoxyflurane not be chosen as an anesthetic agent in patients receiving aminoglycosides.

3. Heavy Metals

Cis-platinum is a nephrotoxic chemotherapeutic agent of considerable value in the treatment of non-seminomatous testicular neoplasms. The mechanism of nephrotoxicity is unknown; however, it has been postulated that the nephrotoxic effect of *cis*-platinum is similar to that of other heavy metals. *Cis*-platinum decreases protein-bound sulfhydryl groups, but to a lesser degree than mercuric chloride. In addition, the resultant renal failure is slower in onset (Kaiser *et al.,* 1980).

Dentino *et al.* (1978) studied the effects of long-term *cis*-platinum therapy on renal function in patients with non-seminomatous testicular tumors. Three of 15 patients followed for more than 18 months received gentamicin in conjunction with a course of *cis*-platinum. Each of these patients exhibited a permanent decrease in creatinine clearance of greater than 50 ml/min. Decreases of such magnitude were not observed in patients receiving no aminoglycosides. Whether or not the aminoglycosides or their primary clinical indication were responsible for this decrease in renal function is not known. Both drugs accumulate in renal cortex and may exert influences longer than their respective half-lives in plasma would suggest.

The results of *cis*-platinum–aminoglycoside combination toxicity studies have not been reported; however, the combination of gentamicin–mercuric chloride has been studied in detail (Luft *et al.,* 1977). In that investigation, animals pretreated with gentamicin were significantly more susceptible to mercuric chloride administration than were control animals. The interaction of these drugs was complex. Animals recently recovered from mercuric chloride-induced acute renal failure were resistant to the nephrotoxic effects of gentamicin.

4. Furosemide

The concomitant use of furosemide and gentamicin has been implicated in the development of acute renal failure in humans (Noël and Levy, 1978). This combination has been examined in detail by Adelman *et al.* (1979). They found that gentamicin nephrotoxicity occurred earlier and was more marked in dogs that received furosemide than dogs that did not. Similar results have been reported in rats. Furosemide-induced volume depletion may be responsible for the enhancement of aminoglycoside nephrotoxicity.

5. Amphotericin B

Amphotericin B is a polyene macrolide antibiotic used in the treatment of systemic fungal diseases. This drug may cause decreased glomerular filtration rate, renal tubular acidosis, hypokalemia, and decreased renal concentrating ability. Churchill and Seely (1977) observed that four patients receiving relatively low-dose amphotericin B therapy developed prompt deterioration in renal function when gentamicin was added to the regimen. They postulated that amphotericin B altered the sterol content of cell membranes, thus facilitating the entry of gentamicin. However, both hypokalemia and acidosis resulting from amphotericin B therapy would also increase the chances for aminoglycoside nephrotoxicity.

6. Clindamycin

Butkus *et al.* (1976) described three patients who developed acute renal failure following treatment with clindamycin and gentamicin. The evidence for combined nephrotoxicity consisted of the temporal relationship between administration of the antibiotics and the development of acute renal failure, and the prompt improvement in renal function upon cessation of therapy. In the rat, large doses of clindamycin had no effect on the nephrotoxicity of another aminoglycoside, netilmicin (Hagstrom *et al.,* 1978).

7. Cephalosporins

The effect of the cephalosporin antibiotics on renal function and structure has been the topic of excellent recent reviews (Barza, 1978; Tune and Fravert, 1980). With one notable exception, this class of β-lactam antibiotics causes renal damage only rarely in humans. The cephalosporins resemble the penicillins. As organic anions, they are actively secreted into the lumen of the proximal tubule by the proximal tubular cells. Cephalosporins are actively transported from the capillaries into the cells of the renal cortex via an anionic transport system. The drugs then diffuse down their respective concentration gradients into the tubular lumen.

Cephaloridine is a consistently nephrotoxic cephalosporin. Its effects upon the kidney have been studied in great detail in experimental animals (Tune, 1975). The diffusion of this drug down its concentration gradient into the tubular lumen is relatively impaired, perhaps because of precipitation or polymerization of the drug within the tubular cells (Boyd *et al.,* 1973). Cephaloridine accumulates within the proximal tubular cells and causes tubular damage proportional to the degree of accumulation within the cells (Perkins *et al.* 1968; Tune, 1975). The role of the anionic transport system in the transport of cehaloridine is supported by the observation that cephaloridine nephrotoxicity can be ameliorated by the prior administration of probenecid (Tune, 1975). In addition, Wold *et al.* (1977) have shown that in newborn rabbits, which are ordinarily resistant to the toxic effects of cephaloridine, the susceptibility to nephrotoxicity parallels the maturation of the renal anionic transport system. Substrate stimulation of the anionic transport system by para-aminohippurate or penicillin increases the nephrotoxicity of cephaloridine in newborn rabbits. The nephrotoxicity of cephaloridine in humans and in experimental animals is a dose-related phenomenon. Cefazolin resembles cephaloridine in its effect upon the nephron; however, it is only one-third to one-fourth as toxic (Birkhead *et al.,* 1973).

Recent studies with cephaloglycin, a cephalosporin released only for oral administration, have shown it to be as least as toxic as cephaloridine in rabbits and guinea pigs (Tune and Fravert, 1980). However cephaloglycin does not exhibit renal cortical concentrations as high as those of cephaloridine. Prolonged intracellular trapping does not occur.

In contrast to cephaloridine, cephaloglycin, and cefazolin, the drugs cephalothin, cephalexin, cefamandole, and cephapirin are only minimally nephrotoxic (Barza, 1978). A dose-related nephrotoxic response relationship for these drugs is difficult to establish. Isolated reports of cephalothin toxicity in humans suggest that at least some of the cases are the result of an allergic interstitial nephritis, as observed with methicillin, rather than a direct toxic effect on renal tubular epithelium.

Barza (1978) has emphasized that since the combination of cephalosporins and aminoglycosides provides excellent antibacterial activity *in vivo*, and may provide synergism for certain infections, the question of whether or not such combinations involve increased risk of renal injury assumes significant clinical importance. In addition to numerous case reports and brief clinical series, several prospective randomized trials, one of which was also conducted in a double-blind fashion, suggest that in humans the combination of aminoglycoside plus cephalothin is more nephrotoxic than aminoglycoside plus a penicillin derivative. Studies in experimental animals have produced conflicting results and in general have failed to verify the human experience.

8. The Cephalosporin–Aminoglycoside Combination in Humans

The widespread use of the combination of cephalothin and gentamicin has been followed by numerous reports suggesting an unexpectedly high rate of nephrotoxicity in patients receiving this combination. Opitz *et al.* (1971) reported that ten of 14 patients with bronchopneumonia receiving the combination of cephalothin and gentamicin developed renal insufficiency as compared to none of 30 patients receiving either agent alone. Subsequently, numerous isolated case reports, some but not all of which were documented histologically, have also implicated the combination of cephalothin and gentamicin as being particularly nephrotoxic. Moreover, this combination has recently been implicated in the development of the Fanconi syndrome (Schwartz and Schein, 1978).

Plager (1976) found a high incidence of renal tubular damage among 119 patients with malignant disease, 69 of whom died following recent treatment with the combination of cephalothin and gentamicin. All had normal blood urea nitrogen (BUN) values prior to therapy. In his series, tubular abnormalities were found in 33 (48%) patients who received the combination prior to death, as opposed to only 3% in a control population. Seven of the 33 patients, who showed either hydropic degeneration or acute tubular necrosis, had normal BUN values within the last 3 days of life. Seventeen patients had increasing BUN values (52–98 mg/dl). Death in these 17 patients was attributed to complications of their underlying disease. Sixteen patients showed initial evidence of renal insufficiency during the period of cephalothin–gentamicin administration and died in severe renal failure. Changes of tubular injury at postmortem examination correlated with the clinical evidence of renal failure. In addition, Plager observed that patients receiving the combination appeared particularly susceptible to other nephrotoxic insults, such as blood loss or ongoing bacterial infection.

Hansen and Kaaber (1977) treated 26 patients with 32 courses of cephalothin–gentamicin combination therapy and observed an increase in serum creatinine during six courses of treatment. A transient decrease in serum potassium occurred in 11 patients. Burck and Sörgel (1975) examined the records of 12,000 patients, 355 of whom received 418 courses of gentamicin and cephalothin. The combined application of cephalothin and gentamicin was implicated in the development of acute renal failure in 14% of cases. The acute renal failure was generally reversible. The authors concluded that the combination should be administered only when clearly superior to other regimens and that close monitoring of renal function was mandatory. The studies previously discussed are retrospective surveys. Interpretation must be limited because of the nature of the controls provided. Some of the surveys failed to provide control groups.

Three prospective randomized trials have been performed addressing the question of possible enhanced risk from nephrotoxicity when cephalothin and an aminoglycoside are given in combination. All three yielded affirmative answers. Klastersky *et al.* (1975) randomly administered the combinations of cephalothin–tobramycin, cephalothin–ticarcillin, and ticarcillin–tobramycin to 186 patients with cancer and suspected life-threatening bacterial infection. Although the three combinations were similarly effective, the administration of the cephalothin–tobramycin combination was associated with a significantly higher frequency of nephrotoxicity than the other two regimens.

The European Organization for Research on Treatment of Cancer (EORTC, 1978) reported the results of a prospective randomized trial of three antibiotic combinations: namely, carbenicillin plus cephalothin, carbenicillin plus gentamicin, and cephalothin plus gentamicin. These regimens were employed in 625 trials with granulocytopenic cancer patients suspected of having gram-negative bacteremia. Although the combinations were equally effective in the treatment of infection, the combination of cephalothin and gentamicin was found to be substantially more nephrotoxic than the other regimens. Severe renal dysfunction was observed in 12% of patients receiving that combination, whereas renal dysfunction occurred in only 4% of patients receiving carbenicillin plus cephalothin and 2% of patients receiving carbinicillin plus gentamicin. Older patients, whose creatinine levels were initially in the high normal range, exhibited a 26% incidence of renal dysfunction related to treatment with cephalothin plus gentamicin.

Wade *et al.* (1978) performed a prospective randomized, double-blind trial to determine if cephalothin plus an aminoglycoside is more nephrotoxic than methicillin plus an aminoglycoside. Patients were assigned to one of four treatment groups: cephalothin and gentamicin, cephalothin and tobramycin, methicillin and gentamicin, or methicillin and tobramycin. Nephrotoxicity was defined as a rise in serum creatinine of over 0.4 mg/dl if the initial level was less than 3.0 mg/dl, or a rise of over 0.9 mg/dl if the initial creatinine was 3.0 mg/dl or above. Definite nephrotoxicity was assumed when no other cause for decreasing renal function could be identified in the 72 hr before the rise in serum creatinine. There was no significant difference in nephrotoxicity between the combined gentamicin and tobramycin groups. However, definite nephrotoxicity devloped in 12 of 47 (26%) of the combined cephalothin groups and in only three of 43 (7%) of the combined methicillin groups ($p < 0.05$).

Although the above studies support a strong indictment against the combination of cephalothin and an aminoglycoside, other studies have reached opposite conclusions. When Fanning *et al.* (1976) reported the findings of the Boston Collaborative Drug Surveillance Program, they were unable to document an increase in nephrotoxicity when gentamicin and cephalothin were given in combination, as compared to either drug alone. They reviewed the records of 1073 patients, 334 of whom received gentamicin, 492 of whom received cephalothin, and 247 of whom received the combination. The unadjusted frequency in rise of BUN was 6.6% with gentamicin, 2.0% with cephalothin, and 9.3% with the combination. Increases in the BUN values were evaluated only if the attending physician felt that antibiotics were responsible for their elevation. The mean increase in BUN was 59 mg/dl with gentamicin, 76 mg/dl with cephalothin, and 46 mg/dl with the combination. The increase was not correlated with age, sex, first discharge diagnosis, diuretic administration, or drug dosage, either total or daily; however, it was correlated with mortality. Patients who died had a frequency of rising BUN values of 13.6%, as opposed to a frequency of 4.9% in survivors. When the statistical analysis was adjusted for mortality, no significant difference in the frequency of rising BUN between groups receiving gentamicin alone or the combination could be shown.

Stille and Arndt (1972) reviewed their results obtained from 74 immuno-suppressed patients who received 85 courses of the combination cephalothin plus gentamicin. Renal insufficiency implicating the combination was identified in six patients, in all of whom septic shock may have also played a role. They concluded that, in contrast to the finding of Opitz et al. (1971), an added nephrotoxicity engendered by the combination could not be documented. Wellwood et al. (1975) assessed renal damage by measuring urinary enzyme excretion in 36 patients who received 53 courses of antimicrobial treatment. Only gentamicin caused enzymuria. The enzymuria observed in patients receiving cephalothin and gentamicin was similar to those receiving only gentamicin. Mondorf et al. (1977) observed the effect of aminoglycosides, cephalosporins, and their combination on the urinary excretion of alanine-aminopeptidase in healthy adults. They found that the aminoglycosides all caused increased excretion of this enzyme. Cephalothin and cefazolin given immediately prior to gentamicin prevented enzymuria.

The interpretation of studies examining the possibility of an enhanced nephrotoxicity with the combination cephalothin plus aminoglycoside is complicated by the recent observation that cephalosporin antibiotics interfere with the Jaffé reaction, which is universally used in the determination of creatinine in both manual and automated techniques. Swain and Briggs (1977) found that cephaloglycin, cephalothin, cephaloridine, cefoxitin, and cephacetrile all reacted with alkaline picrate solution to give a creatinine-like response. Carbenicillin and methicillin specifically do not interfere with the Jaffé reaction (R. Swain, personal communication). Rankin et al. (1979) examined the potential implication of these findings in human volunteers with normal renal function. Creatinine concentrations in urine and plasma were measured by two automated techniques, namely the Programachem P1040 method (Indianapolis, Indiana) and the AutoAnalyzer technique (Tarrytown, New York). Glomerular filtration rate was monitored by the clearance of ^{99m}Tc-DTPA. Following a 4-g intravenous dose of cephalothin, the plasma creatinine concentration was significantly elevated as performed by the Programachem technique. Plasma creatinine increased from 1.0 ± 0.1 to 1.26 ± 0.15 mg/dl. Urine creatinine excretion was significantly elevated by either method. The clearance of ^{99m}Tc-DTPA was unchanged. By either method, plasma and urine creatinine concentrations were directly correlated with plasma and urinary cephalothin values.

Whether or not these *in vitro* and *in vivo* findings have particular relevance to studies indicating that the combination of cephalothin and aminoglycoside is nephrotoxic is unclear. The studies by Klastersky *et al.* (1975) and the EORTC (1978) each included groups receiving cephalothin combined with another β-lactam antibiotic. This combination did not lead to an increased incidence of diminished renal function as determined by plasma creatinine concentrations in either study. In addition, Plager (1976) based his conclusions on histologic evidence rather than one renal function studies. Nevertheless, the fact that cephalothin can interfere with the Jaffé reaction underscores the care that must be taken in interpreting the studies that have appeared in the literature.

9. The Cephalosporin–Aminoglycoside Combination in Experimental Animals

To clarify the implication that cephalosporin–aminoglycoside combinations are particularly nephrotoxic, investigators have turned to animal models. Such models present a number of obstacles in scientific investigations. First, it is by no means certain that human kidneys behave similarly to those of any particular animal species with respect to susceptibility to nephrotoxic agents. For instance, it is well established that the rat, rabbit, and dog, all commonly employed as animal models, differ in their susceptibility to aminoglycoside nephrotoxicity. Even within animal species, differences may exist in susceptibility to renal injury. The Wistar rat, studies by Flandre and Damon (1967), appeared considerably more resistant to the nephrotoxic effects of gentamicin than the Fischer rat employed by other investigators (Kosek, 1974). Even within the same strain, differences may be encountered. We have observed considerable variability in susceptibility of the Sprague-Dawley rat to the nephrotoxic effects of gentamicin when switching from the colony of one commercial supplier to another.

Another problem is the question of dose. Drugs are conveniently administered on the basis of weight; however, glomerular filtration rate correlates more realistically with total body surface area than with weight. A 200-g rat has only 1/350 the mass of its prototype 70-kg human counterpart, yet it has approximately 1/70 the surface area. Its glomerular filtration rate also approximates 1/70 that of man. Dosage interval provides an additional obstacle. Although animals are conveniently dosed once or twice daily, patients are given drugs to assure adequate concentrations, which generally requires more frequent drug administration. It is therefore apparent that dosage interval is quite important in the study of antibiotic nephrotoxicity. Whereas aminoglycosides are more toxic in experimental animals when given in smaller divided daily doses or when infused continuously (Thompson *et al.,* 1978), cephaloridine appears more toxic when given as a single daily injection (Barza, 1978). Yet another problem arises because of possible interactions between certain agents. It is well established that β-lactam antibiotics may inactivate aminoglycosides under certain circumstances *in vivo* and *in vitro* (Noone and Pattison, 1971). An additional consideration is the complexity of the acute renal failure model provided by the aminoglycosides. In many models of acute renal failure, those involving ischemia or those in which specific tubular toxins such as mercuric chloride or uranyl nitrate are given, the renal failure occurs abruptly. The administration of aminoglycosides, on the other hand, causes a relatively gradual acute renal failure, which becomes established over a period of days to weeks. As tubular necrosis progresses, active regeneration occurs, and the animal may recover normal renal function even with continued administration of the drug. The course of acute renal failure may be modulated by the state of sodium balance, as well as by the administration of agents that promote an osmotic diuresis. Since such influences may be imposed by the administration of a β-lactam antibiotic concurrently with an aminoglycoside, the interpretation of results obtained under such circumstances becomes highly complicated.

The results of animal studies examining nephrotoxicity following the concurrent administration of β-lactam antibiotics and aminoglycosides are outlined in Table I. Hautmann *et al.* (1977) studied the effects of the cephalothin–gentamicin combination on rabbits. The doses outlined on the table were given as two equally divided IM injections daily over 14 days. Urinary output, plasma urea, and plasma creatinine concentrations were monitored and renal tissue was obtained prior to and at the end of the experimental period. The authors' data indicate that the addition of cephalothin in increasing doses resulted in increased nephrotoxicity as compared to animals that received only gentamicin. Interpretation of their data is complicated by the fact that conventional statistical analysis was not applied, and by the fact that the authors do not provide a discrete grading scale of the pathologic changes observed.

Harrison *et al.* (1975) studied the effects of increasing doses of cephaloridine, cephalothin, cefazolin, and gentamicin, either alone or in combination, on the renal morphology of Sprague-Dawley rats. None of the combination regimens produced more renal injury than did gentamicin alone. Plasma constituents reflecting glomerular filtration rate were not examined by these investigators.

Dolislager *et al.* (1979) examined the interaction of the aminoglycosides gentamicin and neomycin and of cephaloridine in the rabbit kidney. The aminoglycosides had no effect on renal cortical cephaloridine uptake and there was no additive nephrotoxicity. The authors concluded that the mechanisms of nephrotoxicity of the cephalosporins and aminoglycosides are not closely related.

Hagstrom *et al.* (1978) gave Sprague-Dawley rats netilmicin 30 or 60 mg/kg per day for 14 days either separately or combined with cefamandole 400 mg/kg per day, a variety of penicillins, or clindamycin. None of the combinations resulted in increased nephrotoxicity.

Dellinger *et al.* (1976b) studied the effects of 6, 12, 25, or 50 mg/kg per day gentamicin either alone or combined with 200, 400, or 800 mg/kg per day cephalothin on the renal function and structure of the Fischer 344 rat. They found that the administration of the two drugs simultaneously resulted in a significant protective effect of cephalothin against gentamicin-related nephrotoxicity. They also observed that an equiosmolar amount of sodium sulfate provided a similar degree of protection and suggested that the phenomenon might be related to the presence of a nonresorbable anion in the urine. In additional studies (Dellinger *et al.* 1976a), these same investigators noted that renal cortical concentrations of gentamicin were significantly lower in rats given gentamicin and cephalothin

TABLE I. Nephrotoxicity of Cephalosporins Administration with Aminoglycosides to Animals

Animal	Cephalosporin (dose)	Aminoglycoside (dose)	Result	Reference
Rabbit	Cephalothin (100, 500, 1000 mg/kg/day)	Gentamicin (5, 25 mg/kg/day)	Synergistic nephrotoxicity (function, histology)	Hautman et al. (1977)
Rat	Cephaloridine (500, 1100 mg/kg/day) Cephalothin (1100 mg/kg/day) Cefazolin (1100 mg/kg/day)	Gentamicin (15, 20, 60, 120 mg/kg/day)	Cephalosporins failed to potentiate renal injury	Harrison et al. (1975)
Rabbit	Cephaloridine (75–150 mg/kg/day)	Gentamicin, neomycin (100, 150 mg/kg/day)	No additive nephrotoxic effect	Dolislager et al. (1979)
Rat	Cefamandole (500 mg/kg/day)	Netilmicin (30, 60 mg/kg/day)	Cefamandole failed to potentiate renal injury	Hagstrom et al. (1973)
Rat	Cephalothin (200, 400, 800 mg/kg/day)	Gentamicin (6, 12, 25, 50 mg/kg/day)	Simultaneous administration had protective effect; separation of doses: no interaction.	Dellinger et al. (1976b)
Rat	Cephalothin (400 mg/kg)	Gentamicin (12 mg/kg)	Concomitant cephalothin reduced renal cortical concentration of gentamicin	Dellinger et al. (1976a)
Rat	Cephalothin (100 mg/kg every 4 hr) Cefazolin (20, 50 mg/kg every 4 hr) Cephaloridine (50 mg/kg every 4 hr)	Gentamicin (5 mg/kg every 4 hr)	Cefazolin 50 mg/kg and cephaloridine had protective effect	Luft et al. (1976)
Rat	Cephalothin (500 mg/kg/day)	Gentamicin (10, 100 mg/kg/day)	Simultaneous administration had protective effect	Sugarman et al. (1976)
Rat	Cephalothin (400 mg/kg/day) Cefamandole (400 mg/kg/day) Cefazolin (200 mg/kg/day)	Gentamicin (25 mg/kg/day) Tobramycin (25 mg/kg/day) Netilmicin (25 mg/kg/day) Amikacin (25 mg/kg/day)	Protective effect; greatest reduction in creatinine observed with gentamicin and tobramycin	Barza et al. (1978)
Rat	Cephalothin (200, 400, 800 mg/kg/day)	Gentamicin (12 mg/kg/day)	Dose-dependent protective effect; greatest when cephalothin was given before or concomitant with gentamicin.	Roos and Jackson (1978)
Rat	Cephalothin (100, 500, 1000 mg/kg/day)	Gentamicin (60 mg/kg/day)	Dose-dependent protective effect; also observed with carbenicillin at a lower dose	Bloch et al. (1979)

simultaneously than in animals given gentamicin alone. They suggested that the reduction in nephrotoxicity observed in their initial studies may have been related to a lower cortical concentration of gentamicin in rats receiving cephalothin.

That cephalosporins given to rats receiving gentamicin could ameliorate gentamicin-induced nephrotoxicity was confirmed by observations from our laboratory (Luft *et al.,* 1976) and by Sugarman and associates (1976), who also found that an equivalent improvement was provided by either sodium chloride or para-aminohippurate. Subsequently, Barza and associates (1978) have studied the comparative nephrotoxicity of newer cephalosporins and aminoglycosides alone and in combination in the Fischer 344 rat. Cephalothin, cefamandole, and cefazolin were given concomitantly with gentamicin, tobramycin, netilmicin, or amikacin. Their results confirmed their previous observations with cephalothin and gentamicin (Dellinger *et al.,* 1976b), and extended their findings to other cephalosporin and aminoglycoside congeners. In addition, they observed that cortical concentration of aminoglycosides was generally decreased when cephalosporins were given concomitantly.

Roos and Jackson (1978) gave rats gentamicin 12 mg/kg alone, concomitantly with cephalothin 200, 400, or 800 mg/kg per day, or with sodium chloride 32, 64, or 129 mg/kg per day. Cephalothin 800 mg/kg per day was also given either 4 hr before or 4 hr after gentamicin 12 mg/kg per day. The experiments were conducted for either 6 or 15 days. The investigators monitored the renal concentrations of gentamicin at sacrifice as well as the total excretion of gentamicin in the urine. Histologic injury was computed via a discrete scoring system. Sodium chloride provided no protection. The protective effect of cephalothin was dose-dependent and required a molar ratio of cephalothin to gentamicin of at least 500:1 when the drugs were given concomitantly. The renal cortical concentration of gentamicin was reduced by concomitant cephalothin administration, but not by sodium chloride. The protective effect of cephalothin was enhanced when it was given 4 hr prior to gentamicin, and the urinary excretion of the drug was significantly increased as well. However, under these circumstances the cortical concentration of gentamicin was not reduced. Cephalothin given after gentamicin was also protective; however, neither the cortical concentration of gentamicin nor its urinary excretion was significantly altered. The authors suggested that the cephalothin protective effect may in part be related to a cephalothin-induced increase in gentamicin excretion.

An additional study from our laboratory (Bloch *et al.,* 1979) examined the possibility that a non-cephalosporin β-lactam antibiotic, namely carbenicillin, may also ameliorate gentamicin-induced nephrotoxicity in the Sprague-Dawley rat. Studies in humans have compared the cephalothin–gentamicin combination to gentamicin in combination with a penicillin derivative rather than to gentamicin alone. The possibility thus existed that cephalothin was not necessarily contributing an additional nephrotoxic insult, but rather than β-lactam antibiotics provided variable degrees of protection from gentamicin nephrotoxicity. Cephalothin was given at 100, 500, or 1000 mg/kg per day. Carbenicillin was given at 50, 100, 250, 500, or 1000 mg/kg per day. The gentamicin dose was 60 mg/kg per day.

Five percent glucose solution was used as a diluent. The drugs were given concomitantly, but were injected at separate sites for 14 days. According to creatinine clearance measurements, cephalothin provided protection at 500 mg/kg per day, whereas carbenicillin provided protection at 100 mg/kg per day. Increasing the doses of either drug failed to provide additional protection. Amelioration of the histologic damage did not occur until cephalothin 1000 mg/kg per day or carbenicillin 500 mg/kg per day was given. Although the doses of cephalothin and carbenicillin employed in the study were extremely large when compared to man on a weight basis, the findings may have clinical relevance. Maximum doses of carbenicillin and cephalothin in humans approach 30 and 16 g/day, respectively. A 30 g/day carbenicillin dose in humans provides 167 mg carbenicillin per liter of glomerular filtrate per day, assuming a normal glomerular filtration rate of 125 ml/min. A 16 g/day cephalothin dose provides 88 mg cephalothin per liter of glomerular filtrate. In the rat, protection was observed at a dose of 8.3 mg carbenicillin per liter of filtrate per day and 42 mg cephalothin per liter of filtrate per day. When compared on the basis of glomerular filtration rate, doses of carbenicillin and cephalothin employed in humans assume a magnitude that could conceivably modulate the effects of gentamicin upon the kidney. Since we could not demonstrate that the β-lactam antibiotics employed in our study lowered the renal parenchymal concentration of gentamicin, we were unable to confirm the observations of previous investigators suggesting that the protective effect is in part related to an enhanced excretion of gentamicin. It is possible that the bioassay employed in our study failed to provide the necessary sensitivity to show subtle changes in renal parenchymal gentamicin concentrations. The use of radiolabeled aminoglycoside would circumvent such problems and would provide a direct assessment of aminoglycoside concentrations and excretion.

Recently we have examined the effects of cephalothin, methicillin, carbenicillin, sodium chloride, and sodium sulfate on gentamicin nephrotoxicity. Rats received gentamicin 100 mg/kg per day alone or combined with the following regimens: cephalothin 1000 mg/kg per day, methicillin 1000 mg/kg per day, carbenicillin 1000 mg/kg per day, sodium chloride 52 mg/kg per day, or sodium sulfate 83.7 mg/day. The doses of sodium chloride and sodium sulfate were selected so as to provide an osmolar load equivalent to 1000 mg/day carbenicillin. The effects of these regimens on plasma creatinine and tubular necrosis score appear in Table II. Necrosis severity was graded as follows: grade 1, 25% of cortex; grade 2, 50% of cortex; grade 3, 75% of cortex; grade 4, 100% of cortex.

Gentamicin alone caused severe renal failure and marked tubular necrosis. Carbenicillin provided almost complete protection from the functional and structural damage. Methicillin provided somewhat more protection than cephalothin. Both sodium chloride and sodium sulfate also afforded substantial protection. Sodium combined with a nonreabsorbable anion provided a modest advantage over sodium chloride. In these studies the effect of the various combinations on the intrarenal pharmacokinetics of gentamicin was not addressed.

Whelton et al. (1978) performed a systematic series of studies to identify active secretory or reabsorptive transport pathways that might play a role in the production of high intracellular aminoglycoside concentrations in the proximal

TABLE II. Effect of Regimen on Plasma Creatinine[a] and Necrosis Score

Regimen	Day 0	Day 7	Day 14	Necrosis score (0–4)	
				Median	Range
Diluent (5% glucose)	0.44 ± 0.09	0.49 ± 0.07	0.47 ± 0.06	0	0
Gentamicin	0.46 ± 0.04	4.82 ± 0.45	6.37 ± 0.68	3	2–4
Gentamicin–cephalothin	0.49 ± 0.04	2.03 ± 0.42	2.77 ± 0.79	2	0–3
Gentamicin–methicillin	0.41 ± 0.04	1.92 ± 0.35	1.76 ± 0.66	1	0–2
Gentamicin–carbenicillin	0.39 ± 0.04	0.61 ± 0.25	0.66 ± 0.03	0	0–2
Gentamicin–NaCl	0.45 ± 0.06	2.06 ± 0.47	2.35 ± 0.83	2	1–4
Gentamicin–Na$_2$SO$_4$	0.39 ± 0.04	1.82 ± 0.36	2.01 ± 0.59	1	0–3

[a] In mg/dl (mean ± SEM).

tubular epithelium of the canine renal cortex. Quinine, a potent competitor for organic cation transport, failed to significantly influence renal cortical tobramycin accumulation. Cephalothin, which competes for the organic anion secretory pathway of the renal proximal tubule, also failed to influence the renal cortical concentration of tobramycin in 12 acute experiments. A solution of mixed amino acids was used to inhibit the active proximal tubular absorption of amino acids. At a relatively high rate of infusion, the mixed amino acid solution caused a significant reduction in renal cortical tobramycin uptake.

An additional study pertinent to the question of a possible influence by cephalosporins on the renal cortical accumulation of aminoglycosides was performed by Morin *et al.* (1978). These investigators observed that aminoglycoside antibiotics affect rat renal-cell lysosomes *in vitro*. The toxic effect of the aminoglycosides was assessed by incubating each drug at concentrations from 5 to 50 μg/ml with lysosomal suspensions and measuring the release of n-acetyl-β-D-glucosaminidase after 60 min of incubation at 37° C. The addition of cefoxitin, cephaloridine, or cefazolin inhibited n-acetyl-β-D-glucosaminidase release to a significant degree in lysosomal suspensions containing 50 μg/ml gentamicin, sisomicin, netilmicin, or amikacin. No protective effect was observed for kanamycin or streptomycin. In additional studies, the investigators gave rats ^{14}C-labeled gentamicin 50 mg/kg alone, or in combination with cefazolin 500 mg/kg. The animals were killed at 6 and 24 hr after injection. Both regimens resulted in high renal cortical concentrations of gentamicin at 6 and 24 hr; however, the concentrations in animals receiving cefazolin were not significantly different from those receiving gentamicin alone. The authors concluded that cephalosporins probably do not interfere with the renal concentration and distribution of gentamicin.

10. Conclusion

Clinical experience suggests that the risk of aminoglycoside-induced nephrotoxicity is significantly enhanced in patients receiving methoxyflurane anesthesia.

Studies in experimental animals support this view. The implication that heavy metals, notably *cis*-platinum, may increase the risk of aminoglycoside nephrotoxicity is not yet conclusively proved; however, studies in experimental animals, utilizing a different heavy metal, suggest that such enhancement may well exist. Furosemide increases the severity of aminoglycoside nephrotoxicity in experimental animals, as does dietary sodium restriction. It is prudent to assume that dehydration in humans would do likewise.

The issue concerning deleterious interactions between aminoglycosides and beta-lactam antibiotics is not entirely clarified. Whereas a single study provides evidence that the combination of cephalothin and gentamicin is more toxic than either drug alone in the rabbit, other experiments in rats and rabbits suggest either no added toxicity or demonstrate an ameliorative effect. The relevance of these findings to the human situation remains unclear. Species difference may account for the discrepancy between the results obtained in rabbits and rats. Additional studies in rabbits and other experimental animal models are necessary to clarify the issue. Although the animal studies have not verified the results reported in many of the human studies, they have offered additional insight into the mechanism of aminoglycoside nephrotoxicity and drug interaction within the renal cortex. Particularly intriguing are findings indicating that cephalosporins after the kinetics of aminoglycosides within the renal cortex and may facilitate urinary aminoglycoside excretion. This area warrants particular additional investigation.

Although the information obtained in humans remains controversial, three prospective studies, one of which was conducted in a double-blind fashion, indicate that the combination of cephalothin plus aminoglycoside is more nephrotoxic than penicillin derivative plus aminoglycoside. The data from human studies thus far address only cephalothin. Whether or not other cephalosporins should engender similar concern is presently unknown. The prudent physician should consider the data at hand. It would appear reasonable to reserve the combination cephalothin plus aminoglycoside to those circumstances in which it provides a particular therapeutic advantage.

References

Adelman, R. D., Spangler, W. L., Beasom, F., Ishizaki, G., and Conzelman, G. M., 1979, Furosemide enhancement of experimental gentamicin nephrotoxicity, *J. Infect. Dis.* **140:**342.

Barr, G. A., Mazze, R. I., Cousins, J. J., and Kosek, J. C., 1973, An animal model for combined methoxyflurance and gentamicin nephrotoxicity, *Br. J. Anaesth.* **45:**306.

Barza, M., 1978, The nephrotoxicity of cephalosporins: An overview, *J. Infect. Dis.* **137:**S60.

Barza, M., Pinn, V., Tanguay, P., and Murray, T., 1978, Comparative nephrotoxicity of newer cephalosporins and aminoglycosides alone and in combination in a rat model, in: *Current Chemotherapy,* American Society for Microbiology, Washington, D.C., 964.

Bennett, W. M., Luft, F. C., and Porter, G. A., 1980, Pathogenesis of renal failure due to aminoglycosides and contrast media used in roentgenography, *Am. J. Med.* **69:**767.

Birkhead, H. A., Briggs, G. B., and Saunders, L. Q., 1973, Toxicology of cefazolin in animals, *J. Infect. Dis.* **128:**S379.

Bloch, R., Luft, F. C., Rankin, L. I., Sloan, R. S., Yum, M. N., and Maxwell, D. R., 1979, Protection from gentamicin nephrotoxicity by cephalothin and carbenicillin, *Antimicrob. Agents Chemother.* **15:**46.

Boyd, J., Burcher, B. T., and Stewart, G. T., 1973, The nephrotoxic effect of cephaloridine and its polymers, *Int. J. Clin. Pharmacol.* **7**:307.

Burck, H. C., and Sörgel, G., 1975, Nephrotoxicity of the combined application of cephalothin and gentamicin, in: *The Proceedings of the Sixth International Congress of Nephrology,* International Congree of Nephrology, Florence, Italy, Abstract 700.

Butkus, D. E., de Torrente, A., and Terman, D. S., 1976, Renal failure following gentamicin in combination with clindamycin, *Nephron* **17**:307.

Churchill, D. N., and Seely, J, 1977, Nephrotoxicity associated with combined gentamicin–amphotericin B therapy, *Nephron* **19**:176.

Dellinger, P., Murphy, T., Barza, M., Pinn, V., and Weinstein, L., 1976a, Effects of cephalothin on renal cortical concentrations of gentamicin in rats, *Antimicrob. Agents Chemother.* **9**:587.

Dellinger, P., Murphy, T., Pinn, V., Barza, M., and Weinstein, L., 1976b, The protective effect of cephalothin against gentamicin-induced nephrotoxicity in rats, *Antimicrob. Agents Chemother.* **9**:172.

Dentino, M. E., Luft, F. C., Yum, M. N., and Einhorn, L. H., 1978, Long term effect of *cis*-diamminedichloride platinum on renal function and structure in man, *Cancer* **41**:1274.

Dolislager D., Fravert, D., Tune, B. M., 1979, Interaction of aminoglycosides and cephaloridine in the rabbit kidney, *Res. Comm. Chem. Pathol. Pharmacol.* **26**:13.

EORTC International Antimicrobial Therapy Project Group, 1978, Three antibiotic regimens in the treatment of infection in febrile granulocytopenic patients with cancer, *J. Infect. Dis.* **137**:14.

Fanning, W. L., Gump, D., and Jick, H., 1976, Gentamicin- and cephalothin-associated rises in blood urea nitrogen, *Antimicrob. Agents Chemother.* **10**:80.

Flandre, O., and Damon, M., 1967, Experimental study of the nephrotoxicity of gentamicin in rats, in: *Gentamicin, First International Symposium,* Schwabe and Co., Basle, Switzerland, p. 47.

Frame, P. T., Phair, J. P., Watanakunakorn, C., and Bannister, T. W. P., 1977, Pharmacologic factors associated with gentamicin nephrotoxicity in rabbits, *J. Infect. Dis.* **135**:952.

Hagstrom, G. L., Luft, F. C., Yum, M. N., Sloan, R. S., and Maxwell, D. R., 1978, Nephrotoxicity of netilimicin in combination with nonaminoglycoside antibiotics, *Antimicrob. Agents Chemother.* **13**:490.

Hansen, M. M., and Kaaber, K., 1977, Nephrotoxicity in combined cephalothin and gentamicin therapy, *Acta Med. Scand.* **201**:463.

Harrison, W. O., Silverblatt, F. J., and Turck, M., 1975, Gentamicin nephrotoxicity failure of three cephalosporins to potentiate injury in rats, *Antimicrob. Agents Chemother.* **8**:209.

Hautmann, R., Kurth, M., Buss, H., and Lutzeyer, W., 1977, Zur Nephrotoxizitat der Gentamycin–Cephalotin Kombinationstherapie, *Med. Welt* **28**:1617.

Kaiser, B. A., Hendrik, J. V., and Weiner, M. W., 1980, Effects of *cis*-platinum, mercuric chloride, and glycerol on protein bound sulfhydryl groups during acute renal failure, *Proc. Am. Soc. Nephrol.* **13**:97A.

Klastersky, J., Hensgens, C., and Debusscher, L., 1975, Empiric therapy for cancer patients: Comparative study of ticarcillin–tobramycin, ticarcillin–cephalothin, and cephalothin–tobramycin, *Antimicrob. Agents Chemother.* **7**:640.

Kosek, J. C., Mazze, R. I., and Cousins, M. J., 1974, Nephrotoxicity of gentamicin, *Lab. Invest.* **30**:48.

Luft, F. C., Patel, V., Yum., M. N., and Kleit, S. A., 1976, The nephrotoxicity of cephalosporin–gemtamicin combinations in rats, *Antimicrob. Agents Chemother.* **9**:831.

Luft, F. C., Yum, M. N., and Kleit, S. A., 1977, The effect of concomitant mercuric chloride and gentamicin on kidney function and structure in the rat, *J. Lab. Clin. Med.* **89**:622.

Mazze, R. I., and Cousins, M. J., 1973, Combined nephrotoxicity of gentamicin and methoxyflurance anaesthesia in man, *Br. J. Anaesth.* **45**:394.

Mondorf, A. W., Klose, J., Breier, J., Hendus, J., and Schoeppe, W., 1977, The effect of cephalosporins and aminoglycosides on proximal tubular membrane of human kidney, in: *Program and Abstracts of the Seventeenth Interscience Conference on Antimicrobial Agents and Chemotherapy* (New York), American Society for Microbiology, Washington, D. C., Abstract 201.

Morin, J. P., Fillastre, J. P., and Vallant, R., 1978, Prediction of aminoglycoside–cephalosporin nephrotoxicity, in: *Current Chemotherapy* Volume 2, American Society for Microbiology, Washington, D. C., p. 960.

Nöel, P., and Levy, V. G., 1978, Toxicité renale de l'association gentamicin–furosemide, *Nouv. Presse Med.* **7**:351.

Noone, P., and Pattison, J. R., 1971, Therapeutic implications of interaction of gentamicin and penicillins, *Lancet* **2**:578.

Opitz, A., Herrman, I., v. Herrath, D., and Schaefer, K., 1971, Akute Niereninsuffizienz nach Gentamycin-cephalosporin-Bombinationstherapie, *Med. Welt* **22**:434.

Perkins, R. L., Apicella, M. A., Lee, I. S., Cuppage, F. E., and Saslaw, S., 1968, Cephaloridine and cephalothin: Comparative studies of potential nephrotoxicity, *J. Lab. Clin. Med.* **71**:75.

Plager, J. E., 1976, Association of renal injury with combined cephalothin–gentamicin therapy among patients severely ill with malignant disease, *Cancer* **37**:1937.

Rankin, L. I., Swain, R. R., and Luft, F. C., 1979, The effect of cephalothin on creatinine measurements in man, *Antimicrob. Agents Chemother.* **15**:666.

Roos, R., and Jackson, G. G., 1978, Protective effect of cephalothin on gentamicin nephrotoxicity: Effect of cephalothin anion, not sodium cation, in: *Current Chemotherapy,* American Society for Microbiology, Washington, D. C., p. 962.

Schwartz, J. H., and Schein, P., 1978, Fanconi syndrome associated with cephalothin and gentamicin nephrotoxicity, *Cancer* **41**:769.

Stille, W., and Arndt, I., 1972, Argumente gegen eine Nephrotoxizitat von Cephalothin und Gentamycin, *Med. Welt* **23**:1603.

Sugarman, A., Brown, R. S., and Rosen, S., 1976, Gentamicin nephrotoxicity and the beneficial effect of simultaneous administration of cephalothin or other sodium salts, *Proc. Am. Soc. Nephrol.* **10**:80A.

Swain, R. R., and Briggs, S. L., 1977, Positive interference with the Jaffe reaction by cephalosporin antibiotics, *Clin. Chem.* **23**:1340.

Thompson, W. L., Reiner, N. R., and Bloxham, D. D., 1978, Gentamicin and tobramycin nephrotoxicity in dogs given continuous or once daily intravenous injections, in: *Current Chemotherapy,* Volume 2, American Society for Microbiology, Washington, D. C., p. 941.

Tune, B. M., 1975, Relationship between the transport and toxicity of cephalosporins in the kidney, *J. Infect. Dis.* **132**:189.

Tune, B. M., and Fravert, D., 1980, Mechanisms of cephalosporin nephrotoxicity: A comparison of cephaloridine and cephaloglycin, *Kidney Int.* **18**:591.

Tune, B. M., and Kempson, R. L., 1973, Nephrotoxic drugs: Letter to the editor, *Br. Med. J.* **3**:635.

Wade, J. C., Smith, C. R., Petty, B. G., Lipsky, J. J., Conrad, G., Ellner, J., and Lietman, P. S., 1978, Cephalothin plus an aminoglycoside is more nephrotoxic than methicillin plus an aminoglycoside, *Lancet* **3**:604.

Wellwood, J. M., Simpson, P. M., Tighe, J. R., and Thompson, A. E., 1975, Evidence of gentamicin nephrotoxicity in patients with renal allografts, *Br. Med. J.* **3**:278.

Whelton, A., Carter, G. G., Bryant, H. H., Cody, T. S., Craig, T. J., and Walker, W. G., 1978, Tobramycin and gentamicin intrarenal kinetic comparisons: Therapeutic and toxicologic answers, in: *Current Chemotherapy,* Volume 2, American Society for Microbiology, Washington, D. C., p. 951.

Wold, J. S., Joost, R. R., and Owen, N. V., 1977, Nephrotoxicity of cephaloridine in newborn rabbits: Rate of the renal anionic transport system, *J. Pharmacol. Exp. Ther.* **201**:788.

14

Nephrotoxicity of Cephalosporin Antibiotics
Mechanisms and Modifying Factors

BRUCE M. TUNE

1. Introduction

Several of the cephalosporins produce acute proximal tubular necrosis when given in large single doses (Tune and Fravert, 1980b). The degree of nephrotoxicity varies considerably among the individual cephalosporins. As recently reviewed in a clinical context (Prime and Tune, 1981), toxicity is severe enough with some to restrict their use significantly, but ranges from relatively mild to absent with others (Tune and Fravert, 1980b). The reasons for these individual differences have not been fully elucidated.

Even though highly effective nontoxic newer cephalosporins are being developed, there remain several reasons to study the nephrotoxicity of this group of antibiotics. First, an individual drug may be essentially nontoxic in animal studies and in carefully controlled human testing, but may later prove to have some nephrotoxic risk under certain conditions, such as inappropriately large dosage (in prerenal or renal azotemia) or where there are associated potentiating factors (as in combined use with aminoglycosides) (Foord, 1975; Barza, 1978). The study of the

BRUCE M. TUNE • Division of Nephrology, Department of Pediatrics, Stanford University School of Medicine, Stanford, California 94305.

mechanisms of cephalosporin nephrotoxicity has led to an understanding of certain potentiating factors, to be discussed below, which could influence decisions regarding the selection or dosage of the individual antibiotics in clinical use. Second, other highly effective beta-lactam antibiotics are currently being developed (Kahan *et al.*, 1978; Butterworth *et al.*, 1979; Sakamoto *et al.*, 1979; Kropp *et al.*, 1980), certain of which are also nephrotoxic. The inhibition of the toxicity of these newer, broad-spectrum agents by methods recognized to be effective in cephalosporin toxicity should permit their use with little or no concern about this complication. Finally, the development of an understanding of the mechanisms of cephalosporin nephrotoxicity may provide important insights into mechanisms of drug-induced cytotoxicity in general.

2. Renal Transport — Relationship to Toxicity of Cephaloridine

An understanding of the mechanisms of cephalosporin toxicity should begin with a brief review of the cellular mechanism of transport of the secreted organic anions. Para-aminohippurate (PAH) is secreted across the proximal tubular cell after a primary active transport step at the antiluminal (blood) side (Tune *et al.*, 1969). The resulting intracellular concentrations are considerably greater than those in the extracellular fluids, and favor subsequent movement of PAH down a concentration gradient into the tubular fluid, and thus into the urine.

Probenecid inhibits the secretion of PAH by a reduction of active transport at the antiluminal side, with a resulting lowering of both intracellular and tubular fluid concentrations (Tune *et al.*, 1969). Although there is certainly more than one organic anion transport system in the proximal tubule (Barany, 1973), the active transport of the various penicillin and cephalosporin antibiotics appears to be related to that of PAH. Inhibition of secretion of the penicillins by PAH (Beyer *et al.*, 1944) and that of the cephalosporins by PAH and the penicillins (Tune, 1972; Tune and Fernholt, 1973) are suggestive of a common transport system for both the hippurates and the beta-lactam antibiotics.

Studies of the toxicity of the cephalosporins in this laboratory began with a concern over an apparent contradiction in the transport–toxicity relationships of cephaloridine. This cephalosporin undergoes little or no net secretion across the proximal tubule (Welles *et al.*, 1966; Child and Dodds, 1967; Tune *et al.*, 1974), but has a probenecid-inhibitable selective toxicity to that segment of the nephron (Child and Dodds, 1967; Tune, 1972). Early studies were therefore performed, which demonstrated an unusual transport process: the active transport of cephaloridine into the cell at the antiluminal side, but a severely limited movement from cell to tubular fluid across the luminal membrane (Tune, 1972; Tune and Fernholt, 1973; Tune *et al.*, 1974). The resulting intracellular concentrations of cephaloridine are higher and more sustained than those of any other cephalosporin studied (Whelton and Walker, 1974; Luft *et al.*, 1976; Tune and Fravert, 1980a,b) and may be partly pathogenic in the toxicity of this antibiotic.

3. Cephaloglycin — Transport and Toxicity

Cephaloglycin is normally secreted across the tubular cell, with limited and comparatively transient intracellular concentrations (Tune and Fravert, 1980a,b). The discovery of nephrotoxicity of cephaloglycin comparable to that of cephaloridine led to the development of the concept that there may be two potential mechanisms whereby a cephalosporin can be significantly toxic (Tune and Fravert, 1980a,b). The first, as seen with cephaloridine, is that of prolonged intracellular trapping, but with apparently limited or reversible affinity for or damage to its molecular target. The second, as seen with cephaloglycin, is that of a toxin with a relatively irreversible binding or injury to the target receptor, with comparable cytotoxicity resulting after a lower and less sustained intracellular concentration. It is not yet clear whether the molecular target(s) of cephaloridine and cephaloglycin is (or are) the same.

Several observations lend support to the two-model concept (Tune and Fravert, 1980a,b). These findings suggest, in each case, that cephaloridine does have the relatively low and cephaloglycin the relatively high receptor-affinity properties described above. First, the nephrotoxicity of cephaloridine can be prevented by the intravenous infusion of a bolus of probenecid as late as 20 min after its administration, while that of cephaloglycin cannot. Second, cephaloridine shows little or no cumulative toxicity when given in a series of marginally toxic doses, whereas cephaloglycin shows striking cumulative toxicity when given in a series of injections of even lower single-dose toxicity. Finally, although the respiratory toxicity of a dose of cephaloridine appears to be significantly reversed during the process of mitochondrial isolation and washing (Tune *et al.,* 1979), that of a comparably nephrotoxic dose of cephaloglycin is not.

4. Metabolite Hypothesis

It has been suggested that the cytochrome P-450-dependent mixed function oxidase (MFO) system has a pathogenic role in the nephrotoxicity of cephaloridine (McMurtry and Mitchell, 1977). As evidence for this role, two inhibitors of the MFO system, cobaltous chloride and piperonyl butoxide, were shown to reduce the toxicity of cephaloridine in mice and rats. It was proposed that the MFO system produces a highly reactive metabolite of cephaloridine by epoxidation of its thiophene ring. The fact that cephalothin, which is minimally toxic (Perkins *et al.,* 1968), also has a thiopene side-ring need not represent a contradiction of this hypothesis, because the renal cortical concentrations of cephalothin are considerably lower than those of cephaloridine (Luft *et al.,* 1976).

The question of cephalosporin metabolism remains controversial and somewhat confusing. There has been no demonstration of a reactive metabolite, or of a fraction of cephaloridine bound firmly to microsomes, as has been seen with other MFO-activated toxins (Boyd, 1976; Mitchell *et al.,* 1977). Further, the other toxic cephalosporins lack a thiophene side ring (Tune and Fravert, 1980b). Both cephaloglycin (toxic) and cephalexin (nontoxic) share a phenylglycyl radical group and reach nearly equal cortical concentrations, yet differ greatly in their nephrotoxic potential. Whatever the chemical basis for the large differences in the nephrotoxicity of the various cephalosporins, the basic mechanism of this insult must involve the core structure to a great extent.

Studies of the effects of the MFO inhibitors on cephalosporin toxicity in this laboratory (Tune *et al.,* 1983) have not fully supported the earlier observations with cephaloridine in the mouse and rat. The rabbit is not protected against cephaloridine by cobalt. Piperonyl butoxide is protective, but produces, first, a 50% reduction in very early (15 min) and later (peak) cortical cephaloridine concentrations and, second, a doubling of the rate of disappearance of the antibiotic when the MFO inhibitor is administered after maximal cortical accumulation has occurred.

One possible explanation of this observation would be the elimination by piperonyl butoxide of a large pool of metabolized cephaloridine, one which is measured along with the parent compound by our fluorometric assay for the pyridyl sidering (Tune, 1972). However, studies of cephaloridine with high-performance liquid chromatography have failed to define separate peaks of drug and metabolite in extracts of renal cortex (Wold and Turnipseed, 1977; Kuo and Hook, unpublished). A second possible explanation could be that piperonyl butoxide reduces the net cortical uptake of cephaloridine by an effect on either cortical blood flow or organic anion transport. Preliminary studies have failed to document such an effect on the net cortical uptake of PAH *in vitro* or *in vivo,* but have shown a significant increase by piperonyl butoxide of the rate of runout of PAH from renal cortex *in vitro* (Tune *et al.,* 1983).

Whatever the explanation, piperonyl butoxide has no effect on the nephrotoxicity of cephaloglycin (Tune *et al.,* 1982b). The question of a possible role of the MFO system will therefore require further study before an understanding of the variable effects of certain MFO inhibitors on cephaloridine toxicity can be achieved.

5. Mildly Toxic Cephalosporins

The fundamental differences between the toxicity of cephaloridine and that of cephaloglycin lead to certain concerns beyond the question of whether the two antibiotics act upon the same molecular target(s). Central to the problem of cephalosporin nephrotoxicity is whether the less toxic members of this class more closely resemble cephaloglycin or cephaloridine in their action.

We have had two areas of study pertinent to this question under recent investigation. The first is whether the irreversible or cumulative properties of cellular injury are seen with the mildly toxic cephalosporins, which lack the prolonged intracellular trapping seen with cephaloridine (Table I). The second is whether the various mildly toxic secreted cephalosporins, because their intracellular concentrations are determined by the balance of active transport into the cell at the antiluminal side and rapid movement out of the cell into the tubular fluid, might

TABLE I. Cortical Concentrations of Cefaclor in the Rabbit Kidney[a]

	Cortical concentration,[b] $\mu g/g$	Serum concentration,[b] $\mu g/ml$	Cortex-to-serum ratio[b]
300 mg/kg			
Steady state	1127 ± 111	339 ± 24	3.5 ± 0.5
40 min	358 ± 72	61 ± 14	6.1 ± 0.3
100 mg/kg, steady state	499 ± 48	77 ± 9	6.8 ± 1.0

[a] Cefaclor, 300 or 100 mg/kg, infused over 30 min; sacrifice at time 0 or 40 min after termination of the infusion.
[b] Mean ± SEM; five animals in each group.

have properties in common with cephaloglycin that would make it more suitable than cephaloridine for the study of the effects of changes in tubular fluid flow rate on the risk of toxicity.

The study of two potentiating factors, that of combined use with aminoglycosides and that of acute obstruction of the urinary tract, has produced results which lead us to suspect that there is a greater similarity of the less toxic cephalosporins cefaclor and cefazolin to cephaloglycin than to cephaloridine.

6. Additive Aminoglycoside–Cephalosporin Toxicity

Because the aminoglycosides carry some risk of nephrotoxicity, there has long been the concern of a potential toxic synergy with the cephalosporins. Although one early clinical survey failed to demonstrate statistical significance of a trend toward additive toxicity (Fanning et al., 1976), there is a growing body of evidence that such an interaction does exist in human use (Klastersky et al., 1975; Wade et al., 1978; EORTC, 1978).

The results of early studies of this interaction in animals were surprisingly negative. No additive toxicity was seen in studies from several different laboratories with the combined use of several aminoglycosides and cephalosporins in the rat (Harrison et al., 1975; Dellinger et al., 1976; Luft et al., 1976; Barza et al., 1978). In fact, a protective effect of very large doses of the cephalosporins against the aminoglycosides was demonstrated (Dellinger et al., 1976; Luft et al., 1976), possibly as a result of the delivery of a large load of nonreabsorbable solute to the renal tubule (Dellinger et al., 1976; Roos and Jackson, 1978).

The rat is relatively resistant to cephalosporin toxicity, however (Atkinson et al., 1966; Welles et al., 1966), and no toxicity was demonstrated in single-drug cephalosporin control groups in any of the above-mentioned studies. Further work was therefore done in this laboratory with the rabbit, using cephaloridine in combination with either neomycin or gentamicin (Dolislager et al., 1979). Although some cytotoxicity was seen in both aminoglycoside and cephalosporin single-drug control groups, there was no additive toxicity in several regimens of combined administration. At the time of those studies, however, we had not yet become familiar with the distinctive properties of cephaloglycin. Because of the development of the high-affinity model of cephalosporin toxicity, later studies were designed to examine the interaction of three aminoglycosides (neomycin, gentamicin, and tobramycin) and three normally secreted toxic cephalosporins (cephaloglycin, cefazolin, and cefaclor) in the rabbit.

The results of combined treatment in rabbits showed almost consistently additive toxicity of these particular aminoglycosides and cephalosporins (Bendirdjian *et al.,* 1981). The degree of augmentation in the various combinations appears to occur in proportion to the nephrotoxicities of the individual antibiotics used. This additive toxicity is seen with short series of combined daily antibiotic administration, in doses which produce little or no tubular necrosis or elevation of serum creatinine concentration in single-drug control groups.

In animals given only a three-day course of neomycin pretreatment (nontoxic), followed by exposure to a single, mildly toxic dose of cephaloglycin, additive toxicity is seen in the absence of an effect of the aminoglycoside on either the peak or the rate of decline of cortical or serum concentrations of the cephalosporin. Further, correlated studies of the mitochondrial toxicity of cephaloglycin reveal a potentiation of this respiratory toxicity by the aminoglycoside within 1 hr of administration of the cephalosporin. Thus, we have evidence that potentiation occurs very early, at the subcellular level, and is unrelated to any effect of the aminoglycoside on the renal transport or elimination of the cephalosporin.

It is concluded from the results of these studies that the tendency toward additive aminoglycoside–cephalosporin toxicity seen in humans can be demonstrated in healthy rabbits and that the less toxic cephalosporins resemble cephaloglycin, rather than cephaloridine, in having this potential.

7. Effects of Ureteral Obstruction on Cephalosporin Toxicity

Because of the unusual transport of cephaloridine, intracellular concentrations are not significantly influenced by ureteral ligation (Tune *et al.,* 1974). However, ligation doubles the cortical concentrations of PAH in the same protocols. Concentrations of PAH are increased in both tubular fluid and cell water after obstruction, because the exit of PAH from cell water to tubular fluid to urine is restricted, while active transport into the cell continues (Tune *et al.,* 1969). To the extent that the secreted cephalosporins undergo a similar elevation of intracellular concentrations during ureteral obstruction, their toxicity should be proportionately increased.

In studies of unilateral obstruction applied for 1 hr shortly after the intravenous administration of a mildly toxic dose of cephaloglycin (Wang *et al.,* 1982), toxicity is significantly greater in the obstructed than in the contralateral control kidney. No similar effect is seen with cephaloridine, or when the same dose of cephaloglycin is given several hours before application of or shortly after release of obstruction. In correlated studies of uptake, the early (6-min) cortical concentrations and cortex-to-serum concentration ratios of cephaloglycin in the obstructed kidneys are slightly (15%) but significantly higher than those in the contralateral control kidneys. Although the early effects of obstruction are not nearly as great as are seen with PAH, cortical concentrations and cortex-to-serum ratios of cephaloglycin are approximately half again as high in the obstructed as in the control kidneys by the time of the release of obstruction. It is somewhat more difficult to demonstrate a potentiating effect of transient unilateral obstruction on the toxicity of a single dose of cefaclor, possibly as a result of its rapid elimination by the nonobstructed kidney. However, an increase in toxicity is produced with this cephalosporin if the obstruction is maintained for 2 hr.

Further studies were done with the less toxic cephalosporins using 2 hr of bilateral ligation (Wang *et al.,* 1982). In these studies one must take into account the effects of obstruction on both peak cortical concentrations and the rates of decline of concentrations of the cephalosporins in renal cortex and serum. Bilateral obstruction results in an augmentation of the toxicity of cefaclor, which is rapidly secreted (Table I), but not of cefazolin, which is not (Brogard *et al.,* 1978). It is of interest that pretreatment of animals with neomycin for 3 days (nontoxic by itself) further augments the effect of 2 hr of bilateral obstruction on cefaclor toxicity.

It is concluded that unilateral ureteral ligation significantly increases the toxicity of cephaloglycin, but not of cephaloridine, in keeping with the effects of this procedure on the tubular cell concentrations of these two antibiotics. Results with the less toxic cephalosporin cefaclor show a similar effect of unilateral obstruction. Bilateral obstruction augments the damage produced by the less toxic cephalosporin more significantly, and this augmentation is additive with the potentiating effect of the aminoglycoside neomycin.

8. Exceptions to the Transport–Toxicity Relationship

The fundamental relationship between the transport and the cytotoxicity of the cephalosporins in the proximal tubule has been mentioned above. As studied in great detail with cephaloridine, there are several lines of evidence that this relationship is quantitative. First, the nephrotoxicity of cephaloridine in rabbits, guinea pigs, rats, and mice, in order of decreasing toxicity, is proportional to degree of carrier-mediated uptake by the kidney cortex in each of these four species (Tune, 1975). Second, protection is not seen with the prior infusion of a single dose of the rapidly secreted organic anion PAH, which significantly but only transiently inhibits cortical cephaloridine uptake (Tune *et al.,* 1977b). Protection by PAH is seen only when the inhibiting dose is followed by a constant infusion sufficient to maintain inhibition of cephaloridine uptake until its serum levels have declined substantially. Finally, in both rabbits and guinea pigs the doses of probenecid necessary to prevent toxicity are the same as those that result in a significant sustained reduction of active transport (Tune *et al.,* 1977a).

We have recently found, however, some important exceptions to the rule of prevention of toxicity of the cephalosporins by transport inhibition.

Because of the relative impermeability of the luminal cell membrane to cephaloridine, it can be expected that entry of filtered antibiotic from the tubular fluid into the cell will be minimal. In support of this concept, cortex-to-serum concentration ratios of cephaloridine are reduced by probenecid from eight times those of inulin to values slightly lower than those of inulin (Tune, 1972), in contrast to inhibited cortex-to-serum ratios of PAH (Tune *et al.,* 1974), an anion which can move freely from luminal fluid to cell water (Tune *et al.,* 1969).

To the extent that cephaloglycin movement across the luminal cell membrane may be bidirectional, it could produce toxicity in the presence of transport inhibition if large enough antibiotic doses were used. In a set of studies possibly demonstrating such luminal-side entry as a cause of toxicity (Table II), a very large dose of cephaloglycin produces significant damage in half of the treated animals after aggressive treatment with probenecid, while the equivalent dose of cephaloridine in the same protocol produces no toxicity.

Thus, there are two circumstances under which transport inhibition may be ineffective in preventing cephalosporin toxicity. The first, discussed earlier, is where the inhibiting agent has a shorter effective half-life than that of the cephalosporin. The second is when the serum (and proximal tubular fluid) concentrations of the cephalosporin are high enough to result in toxic intracellular concentrations following entry across the luminal cell membrane. This latter circumstance may account for our difficulty in eliminating multiple-dose additive aminoglycoside–cephalosporin toxicity with probenecid (Table III).

A more important exception to the rule of a close relationship between transport inhibition and prevention of toxicity is seen in the effects of less toxic or nontoxic cephalosporins and penicillins on cephaloridine and cephaloglycin toxicity. Two observations caused us to pursue this area of investigation. First, in earlier studies (Tune and Kempson, 1973), protection against cephaloridine was possible after a single limited dose of cephalothin, even though the serum half-life of cephalothin is considerably shorter than that of cephaloridine (Brogard *et al.,*

TABLE II. Protection by Probenecid against Cephaloridine and
Cephaloglycin Nephrotoxicity

	Proximal tubular necrosis[c]			
	None	Mild	Moderate	Severe
Cephaloridine				
150 mg/kg				
No probenecid	0	1	3	2
Probenecid 100 mg/kg[a]	6	0	0	0
1200 mg/kg				
Probenecid 150 mg/kg[b]	6	0	0	0
Cephaloglycin				
100 mg/kg				
No probenecid	0	2	4	0
Probenecid 100 mg/kg[a]	6	0	0	0
800 mg/kg				
Probenecid 150 mg/kg[b]	3	1	0	2

[a] Single SC dose 30 min before cephalosporin.
[b] 100 mg/kg SC 30 min before and 50 mg/kg 3 hr after cephalosporin.
[c] Numbers of animals; scored with randomly assorted, coded slides.

TABLE III. Protective Action of Probenecid against Cefaclor–Gentamicin Toxicity in the Rabbit Kidney

Probenecid,[a] 100 mg/kg	Cefaclor,[a] 300 mg/kg	Gentamicin,[a] 100 mg/kg	Proximal tubular necrosis[b]				Serum creatinine,[c] mg/dl
			None	Mild	Moderate	Severe	
0	0	0	6	0	0	0	0.74 ± 0.05
0	+	0	10	2	0	0	0.95 ± 0.07
0	0	+	7	1	0	0	0.79 ± 0.04
0	+	+	0	4	2	2	7.30 ± 2.24
+	+	+	0	5	2	1	8.22 ± 3.10

[a] Single daily SC doses for five consecutive mornings; sacrifice on day 8.
[b] Numbers of animals; scored with randomly assorted, coded slides.
[c] Mean ± SEM.

1978). This observation was in striking contrast to subsequent findings using PAH as an inhibitor (Tune *et al.*, 1977b). Second, a very mild inhibition by cephalexin of respiration in mitochondria recovered from drug-treated animals led to the speculation that even nontoxic cephalosporins may have some affinity for the toxic receptor(s), without necessarily resulting in cell death (Tune and Fravert, 1980a).

Studies were therefore done with other cephalosporins and subsequently with penicillins to determine, first, the minimal doses that substantially reduce cephaloridine nephrotoxicity, and second, the effects of the same regimens on the cortical concentrations of cephaloridine (Tune *et al.*, 1982). Four protecting antibiotics were used, cefazolin, cephalexin, benzylpenicillin, and ampicillin. It was found that single limited doses of each of these antibiotics, which produce little or no reduction of peak or declining cortical concentrations of cephaloridine (a specific fluorometric assay was used), result in a significant reduction or elimination of its nephrotoxicity. Pretreatment with equal doses of benzylpenicillin before mildly to moderately toxic doses of cephaloglycin also results in significant reduction of toxicity. As with cephaloridine, in identical-dose protocols, pretreatment with the penicillin produces no significant effect on the cortical concentrations or cortex-to-serum ratios of cephaloglycin (^{14}C isotope).

It is concluded that the less toxic or nontoxic cephalosporins and the penicillins can protect against the nephrotoxicity of both cephaloridine and cephaloglycin by a mechanism other than the overall inhibition of cortical accumulation of either toxic antibiotic. It is possible that this effect is mediated by competition of the nontoxic beta-lactam for the toxic receptor site, but an inhibiting effect on the formation of an activated toxin must also be considered. In spite of the previously noted differences between cephaloridine and cephaloglycin, it is perhaps significant that this process of selective protection as seen with both drugs.

It is speculated that the penicillins might provide some protective effect against cephalosporin nephrotoxicity in therapeutic-range dosage, without the significant reduction of excretion and elevation of blood levels that can result from protective doses of probenecid (Brogard *et al.*, 1978). As preliminary confirmatory evidence,

TABLE IV. Protective Action of Benzylpenicillin against Cefazolin–Neomycin Toxicity in the Rabbit Kidney

Penicillin,[a] 500 mg/kg	Cefazolin,[a] 500 mg/kg	Neomycin,[a] 100 mg/kg	Proximal tubular necrosis[b]				Serum creatinine,[c] mg/dl
			None	Mild	Moderate	Severe	
0	0	0	6	0	0	0	0.74 ± 0.05
0	+	0	6	0	0	0	0.73 ± 0.08
0	0	+	7	5	0	0	0.82 ± 0.05
0	+	+	3	1	0	4	4.19 ± 1.41[d]
+	+	+	4	4	0	0	1.11 ± 0.09

[a] Single daily SC doses for five consecutive mornings; sacrifice on day 8.
[b] Numbers of animals; scored with randomly assorted, coded slides.
[c] Mean ± SEM.
[d] Serum creatinine: none–mild 0.76 ± 0.04; severe 7.12 ± 1.00.

equal doses of benzylpenicillin given immediately before cefazolin (Table IV) prevent the cephalosporin-induced increases in toxicity of neomycin (Bendirdjian *et al.,* 1981).

9. Summary (Table V)

The cephalosporin antibiotics are variably nephrotoxic. Certain of them can produce acute renal failure by causing selective proximal tubular necrosis. This toxicity is preventable by the prior administration of inhibitors of organic anion transport, and this protection is complete if: (1) transport inhibition is of sufficient duration, and (2) the serum concentrations of the toxic antibiotic are within the reasonable pharmacologic range.

Two important cellular mechanisms of nephrotoxicity of the individual cephalosporins have been discussed. The first, as seen with cephaloridine, is the occurrence of very high and trapped intracellular concentrations, resulting from active transport into the cell at the antiluminal side, with very limited subsequent

TABLE V. Nephrotoxic Properties of the Mildly Toxic Cephalosporins Compared to Those of Cephaloridine and Cephaloglycin[a]

	C_L	C_G	C_{Cl}–C_Z
Proximal tubular necrosis	+	+	+
Protection by probenecid	+	+	+
Secretion like PAH	−	+	+
Obstruction increases toxicity	−	+	+
Limits to probenecid protection	−	+	+[b]
Additively toxic with aminoglycosides	−	+	+
In vivo mitochondrial toxicity	+	+	?
Protection by low doses of beta-lactams	+	+	+[b]

[a] C_L, Cephaloridine; C_G, cephaloglycin; C_{Cl}, cefaclor; C_Z, cefazolin.
[b] Studied in aminoglycoside–cephalosporin interaction.

movement into the tubular fluid. This unusual transport phenomenon may be the result of the fixed cationic charge on the pyridyl side ring of cephaloridine (Tune and Fravert, 1980a), and might be expected to occur if other cephalosporins are developed with this fixed charge (if they are also actively transported). The second mechanism, as seen with cephaloglycin, is a high and relatively irreversible affinity of the cephalosporin for the target receptor (or the prompt production of an irreversible insult to that target), resulting in toxicity comparable to that of cephaloridine after a more limited and transient accumulation within the tubular cell.

To the extent that the less toxic cephalosporins, which are normally secreted like cephaloglycin, are toxic because of their relative receptor affinities, they might be expected to share the properties of cumulative toxicity which characterize the cephaloglycin model. As indirect evidence in support of this speculation, although cephaloridine is not additively toxic with the aminoglycosides (which are cumulatively toxic), cephaloglycin, cefaclor, and cefazolin are.

Another feature which might be expected to be shared by the normally secreted cephalosporins would be an increase in intracellular concentrations and nephrotoxicity after unilateral ureteral obstruction. This phenomenon has been demonstrated with cephaloglycin and cefaclor, and to a limited extent with cefazolin. In contrast, because there is no effect of ureteral obstruction on intracellular concentrations of cephaloridine, its toxicity is not significantly affected by unilateral obstruction.

Although these important differences between the cephaloridine and cephaloglycin models have been demonstrated, the damage produced by both is characterized by very early inhibition of tubular cell mitochondrial respiration. In addition, the toxicity of cephaloridine and cephaloglycin can be prevented by the penicillins and less toxic or nontoxic cephalosporins given in amounts which are insufficient to inhibit the cellular uptake of either of the highly toxic cephalosporins (selective protection).

The results of the various studies discussed above may have several implications in the clinical use of cephalosporin and other nephrotoxic beta-lactam antibiotics. Prevention of toxicity may be complete if a satisfactory inhibitor of organic anion transport or a selectively protecting beta-lactam compound is coadministered. However, toxicity might be augmented by combined administration with the aminoglycosides or by obstructive uropathy, especially with bilateral obstruction.

Most interesting is the fact that two of the manipulations described may alter cephalosporin toxicity without changing peak or declining antibiotic concentrations in renal cortex or in serum. The first is the administration of an identical dose of a nontoxic beta-lactam antibiotic, such as benzylpenicillin, immediately before the toxic cephalosporin (toxicity prevented). The second is the administration of an aminoglycoside, such as neomycin, for as little as 3 days before the cephalosporin (toxicity augmented). These manipulations may represent valuable probes in future studies of the molecular basis of cephalosporin and other beta-lactam antibiotic nephrotoxicity.

Acknowledgments

This research was supported by grants from the NIH (AM 17647), the Eli Lilly Company, and the American Heart Association. The author is indebted to the following for their valuable contributions to this work: Kai Yin Wu, Doris Fravert, Chieh-Yin Hsu; Drs. Donna Dolislager, Donald J. Prime, Marc C. Browning, and Peter Wang. The assistance of Juanita Tayabas in organizing and tabulating data and in preparing the manuscript is equally appreciated.

References

Atkinson, R. M., Currie, J. P., Davis, B., Pratt, D. A. H., Sharpe, H. M., and Tomich, E. G., 1966, Acute toxicity of cephaloridine, an antibiotic derived from cephalosporin C, *Toxicol. Appl. Pharmacol.* **8**:398.

Barany, E. H., 1973, The liver-like anion transport system in rabbit kidney, uvea and choroid plexus. I. Selectivity of some inhibitors, direction of transport, possible physiological substrates, *Acta Physiol. Scand.* **88**:412.

Barza, M., 1978, The nephrotoxicity of cephalosporins: An overview, *J. Infect. Dis.* **137**:S60.

Barza, M., Pinn, V., Tanguay, P., and Murray, T., 1978, Comparative nephrotoxicity of newer cephalosporins and aminoglycosides alone and in combination in a rat model, in: *Current Chemotherapy,* Volume II (W. Siegenthaler and R. Luthy, eds.), American Society for Microbiology, Washington, D.C., p. 964.

Bendirdjian, J.-P., Prime, D. J., Browning, M. C., Hsu, C.-Y., and Tune, B. M., 1981, Additive nephrotoxicity of cephalosporins and aminoglycosides in the rabbit, *J. Pharmacol. Exp. Ther.* **218**:681.

Beyer, K. H., Woodward, F., Peters, L., Verwey, W. F., and Mattis, P. A., 1944, Prolongation of penicillin retention in body by means of para-aminohippuric acid, *Science* **100**:107.

Boyd, M. R., 1976, Role of metabolic activation in the pathogenesis of chemically induced pulmonary disease: Mechanism of action of the lung-toxic furan, 4-ipomeanol, *Environ. Health Persp.* **16**:127.

Brogard, J. M., Comte, F., and Pinget, M., 1978, Pharmacokinetics of cephalosporin antibiotics, *Antibiot. Chemother.* **25**:123.

Butterworth, D., Cole, M., Hanscomb, G., and Rolinson, G. N., 1979, Olivanic acids, a family of β-lactam antibiotics with β-lactamase inhibitory properties produced by streptomyces species. I. Detection, properties and fermentation studies, *J. Antibiot.* **32**:287.

Child, K. J., and Dodds, M. G., 1967, Nephron transport and renal tubular effects of cephaloridine in animals, *Br. J. Pharmacol. Chemother.* **30**:354.

Dellinger, P., Murphy, T., Pinn, V., Barza, M., and Weinstein, L., 1976, Protective effect of cephalothin against gentamicin-induced nephrotoxicity in rats, *Antimicrob. Agents Chemother.* **9**:172.

Dolislager, D., Fravert, D., and Tune, B. M., 1979, Interaction of aminoglycosides and cephaloridine in the rabbit kidney, *Res. Comm. Chem. Pathol. Pharmacol.* **26**:13.

EORTC International Antimicrobial Therapy Project Group, 1978, Three antibiotic regimens in the treatment of infection in febrile granulocytopenic patients with cancer, *J. Infect. Dis.* **137**:14.

Fanning, W. L., Gump, D., and Jick, H., 1976, Gentamicin and cephalothin associated rises in blood urea nitrogen, *Antimicrob. Agents Chemother.* **10**:80.

Foord, R. D., 1975, Cephaloridine, cephalothin and the kidney, *J. Antimicrob. Chemother.* **1**:119.

Harrison, W. O., Silverblatt, F. J., and Turck, M., 1975, Gentamicin nephrotoxicity: Failure of three cephalosporins to potentiate injury in rats, *Antimicrob. Agents Chemother.* **8**:209.

Kahan, J. S., Kahan, F. M., Goegelman, R., Currie, S. A., Jackson, M., Stapley, E. O., Miller, T. W., Miller, A. K., Hendlin, D., Mochales, S., Hernandez, S., Woodruff, H. B., and Birnbaum, J., 1978, Thienamycin, a new β-lactam antibiotic. I. Discovery, taxonomy isolation and physical properties, *J. Antibiot.* **32**:1.

Klastersky, J., Hensgens, C., and Debusscher, L., 1975, Empiric therapy for cancer patients: Comparative study of ticarcillin–tobramycin, ticarcillin–cephalothin, and cephalothin–tobramycin, *Antimicrob. Agents Chemother.* **7**:640.

Kropp, H., Sundelof, J. G., Kahan, J. S., Kahan, F. M., and Birnbaum, J., 1980, MK0787 (*N*-formimidoyl thienamycin): Evaluation of *in vitro* and *in vivo* activities, *Antimicrob. Agents Chemother.* **17**:993.

Luft, F. C., Patel, V., Yum, M. N., and Kleit, S. A., 1976, Nephrotoxicity of cephalosporin–gentamicin combinations in rats, *Antimicrob. Agents Chemother.* **9**:831.

McMurtry, R. J., and Mitchell, J. R., 1977, Renal and hepatic necrosis after metabolic activation of 2-substituted furans and thiophenes, including furosemide and cephaloridine, *Toxicol. Appl. Pharmacol.* **42**:285.

Mitchell, J. R., McMurtry, R. J., Statham, C. N., and Nelson, S. D., 1977, Molecular basis for several drug-induced nephropathies, *Am. J. Med.* **62**:518.

Perkins, R. L., Apicella, M. A., Lee, I. S., Cuppage, F. E., and Saslaw, S., 1968, Cephaloridine and cephalothin: Comparative studies of potential nephrotoxicity, *J. Lab. Clin. Med.* **71**:75.

Prime, D. L., and Tune, B. M., 1981, The nephrotoxicity of antimicrobial drugs, in: *Pediatrics Update* (A. J. Moss, ed.), Elsevier/North-Holland, New York.

Roos, R., and Jackson, G. G., 1978, Protective effect of cephalothin on gentamicin nephrotoxicity: Effect of cephalothin anion, not sodium cation, in: *Current Chemotherapy,* Volume II (W. Siegenthaler and R. Luthy, eds.), American Society for Microbiology, Washington, D.C., p. 962.

Sakamoto, M., Iguchi, H., Okamura, K., Hori, S., Fukugawa, Y., and Ishikura, T., 1979 PS-5, a new β-lactam antibiotic. II. Antimicrobial activity, *J. Antibiot.* **32**:272.

Tune, B. M., 1972, Effect of organic acid transport inhibitors on renal cortical uptake and proximal tubular toxicity of cephaloridine, *J. Pharmacol. Exp. Ther.* **181**:250.

Tune, B. M., 1975, Relationship between the transport and toxicity of cephalosporins in the kidney, *J. Infect. Dis.* **132**:189.

Tune, B. M., and Fernholt, M., 1973, Relationship between cephaloridine and *p*-aminohippurate transport in the kidney, *Am. J. Physiol.* **225**:1114.

Tune, B. M., and Fravert, D., 1980a, Cephalosporin nephrotoxicity. Transport, cytotoxicity and mitochondrial toxicity of cephaloglycin, *J. Pharmacol. Exp. Ther.* **215**:186.

Tune, B. M., and Fravert, D., 1980b, Mechanisms of cephalosporin nephrotoxicity. A comparison of cephaloridine and cephaloglycin, *Kidney Int.* **18**:591.

Tune, B. M., and Kempson, R. L., 1973, Nephrotoxic drugs, *Br. Med. J.* **3**:635.

Tune, B. M., Burg, M. B., and Patlak, C. S., 1969, Characteristics of *p*-aminohippurate transport in proximal renal tubules, *Am. J. Physiol.* **217**:1057.

Tune, B. M., Fernholt, M., and Schwartz, A., 1974, Mechanism of cephaloridine transport in the kidney, *J. Pharmacol. Exp. Ther.* **191**:311.

Tune, B. M., Wu, K. Y., and Kempson, R. L., 1977a, Inhibition of transport and prevention of toxicity of cephaloridine in the kidney. Dose-responsiveness of the rabbit and the guinea pig to probenecid, *J. Pharmacol. Exp. Ther.* **202**:466.

Tune, B. M., Wu, K. Y., Longerbeam, D. F., and Kempson, R. L., 1977b, Transport and toxicity of cephaloridine in the kidney. Effect of furosemide, *p*-aminohippurate and saline diuresis, *J. Pharmacol. Exp. Ther.* **202**:472.

Tune, B. M., Wu, K. Y., Fravert, D., and Holtzman, D., 1979, Effect of cephaloridine on respiration by renal cortical mitochondria, *J. Pharmacol. Exp. Ther.* **210**:98.

Tune, B. M., Browning, M. C., Hsu, C.-Y., and Fravert, D., 1982. Prevention of cephalosporin nephrotoxicity in other cephalosporins and by penicillins without significant inhibition of renal cortical uptake. *J. Infect. Dis.* **145**:174–180.

Tune, B. M., Kuo, S., Hook, J. B., Fravert, D., Hsu, C.-Y., 1983, Effects of inhibitors of mixed function oxidase activity on cephalosporin nephrotoxicity in the rabbit. A possible effect on transport, in press.

Wade, J. C., Petty, B. G., Conrad, G., Smith, C. R., Lipsky, J. J., Ellner, J., and Lietman, P. S., 1978, Cephalothin plus an aminoglycoside is more nephrotoxic than methicillin plus an aminoglycoside, *Lancet* **2**:604.

Wang, P. L., Prime, D. J., Hsu, C.-Y., and Tune, B. M., 1982, Effects of ureteral obstruction on the toxicity of the cephalosporin in the rabbit kidney, *J. Infect. Dis.* **145:**574.

Welles, J. S., Gibson, W. R., Harris, P. N., Small, R. M., and Anderson, R. C., 1966, Toxicity, distribution, and excretion of cephaloridine in laboratory animals, *Antimicrob. Agents Chemother.* **1965:** 863.

Whelton, A., and Walker, W. G., 1974, Intrarenal antibiotic distribution in health and disease, *Kidney Int.* **6:**131.

Wold, J. S., and Turnipseed, S. A., 1977, Determination of cephaloridine in serum and tissue by high-performance liquid chromatography, *J. Chromatog.* **136:**170.

15

Tetracycline Nephrotoxicity

MALCOLM COX

1. Introduction

Over three decades have elapsed since the identification of chlortetracycline, the first tetracycline antibiotic. Since that time a large number of other tetracycline congeners have been isolated (from the products of fungal fermentation) and synthesized (by chemical alterations of the basic molecule). Seven congeners are currently available as broad-spectrum antibiotics in the United States: chlortetracycline (CTC), oxytetracycline (OTC), tetracycline (TC), demeclocycline (DMCTC), methacycline (MOTC), doxycycline (DOOTC), and minocycline (DDDTC).

The tetracyclines are closely related derivatives of the polycyclic compound naphthacenecarboximide. Tetracycline is considered to be the parent compound and the structural formulas of the clinically available congeners are illustrated in Fig. 1. The differences among these congeners are in positions 5, 6, and 7 of the basic naphthacene ring system. An intact ring structure and the particular organization of the functional groups at positions 1–4 and 10–12a are prerequisites for antibiotic activity (Mitscher, 1978).

Not surprisingly, as the use of these antibiotics became more widespread, an increasing number of side effects were reported. From the nephrologist's

MALCOLM COX • Renal-Electrolyte Section, Department of Medicine, Philadelphia Veterans Administration Medical Center and University of Pennsylvania School of Medicine, Philadelphia, Pennsylvania 19104.

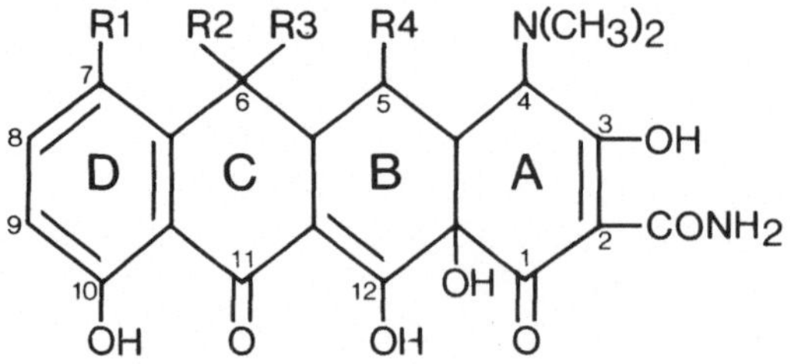

	R1	R2	R3	R4
TC	H	CH₃	OH	H
CTC	Cl	CH₃	OH	H
DMCTC	Cl	H	OH	H
OTC	H	CH₃	OH	OH
MOTC	H	=CH₂		OH
DOOTC	H	CH₃	H	OH
DDDTC	N(CH₃)₂	H	H	H

FIGURE 1. Tetracycline structure. The basic structural formula of the tetracyclines and the substituents (at carbons 5, 6, and 7) that are present in the seven clinically available congeners are provided. [The stereochemistry of the tetracyclines has been excluded; details can be found in Mitscher (1978).] TC, Tetracycline; CTC, chlortetracycline; DMCTC, demeclocycline; OTC, oxytetracycline; MOTC, methacycline; DOOTC, doxycycline; DDDTC, minocycline.

perspective there are three major areas of interest in this regard: (1) selective proximal tubular dysfunction (Fanconi syndrome), (2) selective distal tubular dysfunction (nephrogenic diabetes insipidus), and (3) tetracycline-associated natriuresis and azotemia.

2. Fanconi Syndrome

In the early and mid-1960s a number of case reports linked the development of an acquired, and usually reversible, Fanconi syndrome in both children and adults to the use of outdated tetracycline (Frimpter *et al.,* 1963; Gross, 1963; Fulop and Drapkin, 1965; Cleveland *et al.,* 1965). Proximal tubular dysfunction appears to be due to anhydro-4-epi-tetracycline, one of a number of degradation products of tetracycline formed under the influence of heat, moisture, and low pH. This compound produces proximal tubular necrosis in rats and dogs (Benitz and Diermeier, 1964; Lowe and Tapp, 1966), but the cellular mechanisms underlying its toxicity are unknown. The reformulation of tetracycline preparations and the use of expiration dates has essentially eliminated this as a problem of any clinical importance.

3. Nephrogenic Diabetes Insipidus

Experiments with rats during the early pharmacologic testing of the tetracyclines had revealed a slight diuretic action, the nature of which was apparently not further investigated. In addition, a reduction in renal concentrating ability had been observed in several patients taking therapeutic doses of DMCTC (Goldman and Hall, 1960). However, it was not until 1965 that the first in a series of well-documented reports appeared linking the use of DMCTC to the development of a dose-related, reversible form of nephrogenic diabetes insipidus (Castell and Sparks, 1965; Roth *et al.*, 1967; Wilson *et al.*, 1973; Singer and Rotenberg, 1973). With the exception of TC (which had a very much smaller effect than DMCTC) (Wilson *et al.*, 1973), the effects of other tetracyclines on water homeostasis in humans have not been formally evaluated. There are no reports of polyuria following the use of these agents in the clinical literature, however.

DMCTC impairs the hydroosmotic action of antidiuretic hormone (ADH) in humans, in rats, and in toad urinary bladders, the latter an *in vitro* model of the mammalian distal nephron. Humans with DMCTC-induced polyuria are essentially unresponsive to exogenous ADH and, as one would expect, have reduced solute-free water reabsorptive capacities (Wilson *et al.*, 1973; Singer and Rotenberg, 1973). This defect in renal function is remarkably selective: in otherwise healthy individuals, DMCTC has little or no effect on other tubular transport processes or on glomerular function. In particular, in such individuals, DMCTC-induced nephrogenic diabetes insipidus is not associated with alterations in sodium and potassium balance or with reductions in glomerular filtration rate (Wilson *et al.*, 1973; Singer and Rotenberg, 1973).

Inhibition of the hydroosmotic effect of ADH is the basis for the therapeutic efficacy of DMCTC in the treatment of the hyponatremia associated with the chronic syndrome of inappropriate ADH secretion (SIADH) (DeTroyer, 1977; Forrest *et al.*, 1978). Doses of DMCTC from 600 to 1200 mg/day generally restore the serum sodium concentration to normal within 1–2 weeks of the initiation of therapy. Dehydration and hypernatremia are rarely observed, presumably because of normal thirst sensation in these patients. As in normal subjects, DMCTC-induced nephrogenic diabetes insipidus in patients with SIADH is reversible within several weeks of discontinuing the drug.

Although only limited data are available, DMCTC-induced nephrogenic diabetes insipidus in patients with SIADH does not appear to be associated with other derangements of tubular function. In particular, renal sodium wasting has not been observed (DeTroyer, 1977; Forrest *et al.*, 1978). A small but definite increase in the serum creatinine concentration and a mild to moderate increase in the blood urea nitrogen concentration has been observed in some of these patients. However, this rarely presents a clinically significant problem and has been ascribed to extracellular fluid volume depletion (secondary to drug-induced water loss) and/or to the antianabolic effects of the tetracyclines (see below).

In order to delineate the underlying mechanisms by which DMCTC inhibits the hydroosmotic effect of ADH, animal and *in vitro* model systems were needed. DMCTC has been shown to produce polyuria and ADH unresponsiveness in both normal and Brattleboro rats (Chan, 1979). This animal model may prove useful in the future, but at present most of the information available has been derived from *in vitro* studies using the toad urinary bladder.

In this model system, and in renal epithelia in general, ADH is thought to stimulate osmotic water flow by activating the cyclic AMP second messenger system. For example, ADH activates adenylate cyclase to produce cyclic AMP, cyclic AMP mimics most of the effects of ADH, and cyclic AMP phosphodiesterase inhibitors (e.g., theophylline) enhance the actions of ADH and cyclic AMP. However, the precise cellular and subcellular mechanisms by which ADH mediates epithelial water transport are unknown. On a morphologic level the importance of alterations in mucosal cell membrane structure and in microtubules and microfilaments is being assessed. On a biochemical level, analogies with the second messenger hypothesis have prompted studies of the proteins derived from ADH-responsive epithelia. Since ADH stimulates adenylate cyclase with a resulting action thought to be mediated by cyclic AMP, and since cyclic AMP is known to stimulate protein kinases and protein phosphatases in a variety of tissues (including renal epithelia), one likely biochemical effector of the cellular action of cyclic AMP is an alteration in phosphoprotein metabolism.

Two important modulators of ADH action in renal epithelia should also be mentioned: prostaglandins and calcium. Prostaglandin E inhibits ADH-induced osmotic water flow in renal epithelia by inhibiting ADH-responsive adenylate cyclase. Since ADH has also been shown to enhance prostaglandin E synthesis in renal epithelia, ADH has the ability to internally modulate its own hydroosmotic effect.

The cyclic AMP second messenger system and calcium (and calcium-binding regulatory proteins such as calmodulin) interrelate ubiquitously in many biologic systems. For example, calcium modulates adenylate cyclase and cyclic AMP phosphodiesterase activity in many tissues; conversely, cyclic AMP influences intracellular calcium flux across cellular and subcellular membranes. For example, the hydroosmotic response to ADH or theophylline (but not that to cyclic AMP) in the toad urinary bladder is inhibited by increasing the serosal calcium concentration and ADH-responsive adenylate cyclase in broken-cell preparations of renal epithelia is inhibited both in the absence of calcium and in the presence of 10^{-5}–10^{-3} M calcium. Thus, a low but critical calcium concentration appears to be necessary for optimal adenylate cyclase activation, and deviations from this value in either direction may decrease enzyme activity. Further evidence that calcium has an important role in the hydroosmotic response to ADH derives from studies with a variety of agents (e.g., quinidine, verapimil, A23187, protamine sulfate, lanthanum) which are known to influence the interactions of calcium with membranes. The effects of these agents on the water permeability of renal epithelia has been ascribed to their ability to alter the ionic environment of certain critical intracellular compartments.

Early studies revealed that DMCTC inhibits both ADH- and cyclic AMP-induced water flow in the toad urinary bladder (Singer and Rotenberg, 1973). Consequently, one effect of this agent is to inhibit the cellular action of cyclic AMP. Whether all of the antihydroosmotic effect of DMCTC can be ascribed to this locus of action or whether the drug also inhibits cyclic AMP production is less clear. Since, under certain circumstances, DMCTC inhibits ADH-induced water flow out of proportion to its effects on cAMP-induced water flow (Singer and Rotenberg, 1973; Feldman and Singer, 1974), the latter is certainly a possibility.

Biochemical studies have also indicated both sites of action: DMCTC inhibits both ADH-stimulated adenylate cyclase and cyclic AMP-dependent protein kinase activities in human renal medullary homogenates (Dousa and Wilson, 1974).

When these studies were extended to compare several tetracycline congeners with different properties it was found that only those congeners that exhibited strong binding to human (or canine) serum proteins were effective ADH antagonists (Feldman and Singer, 1974). Other pharmacologic properties, such as lipid solubility or divalent cation chelating ability, did not appear to correlate with ADH inhibition. On the basis of these results it was suggested that tetracycline-induced inhibition of osmotic water flow in renal epithelia involved the binding of these drugs to a cellular protein that has a major role in ADH-mediated water transport (Feldman and Singer, 1974).

In order to further examine this possibility, we have conducted a series of studies designed to correlate the effects of various tetracycline congeners on ADH-induced water flow with their ability to bind to proteins in a crude 10,000g supernatant fraction derived from toad urinary bladder epithelial cells. Initial studies utilizing the intrinsic fluorescence of the tetracyclines and Sephadex gel exclusion chromotography demonstrated that the degree of tetracycline–protein binding varied directly with the inhibitory effect of the particular tetracycline on ADH-induced water flow: DMCTC > TC > OTC (Cox and Singer, 1976a,b). These studies also appeared to demonstrate that the tetracyclines were preferentially bound to high-molecular-weight proteins (peak 1, Fig. 2). However, subsequent

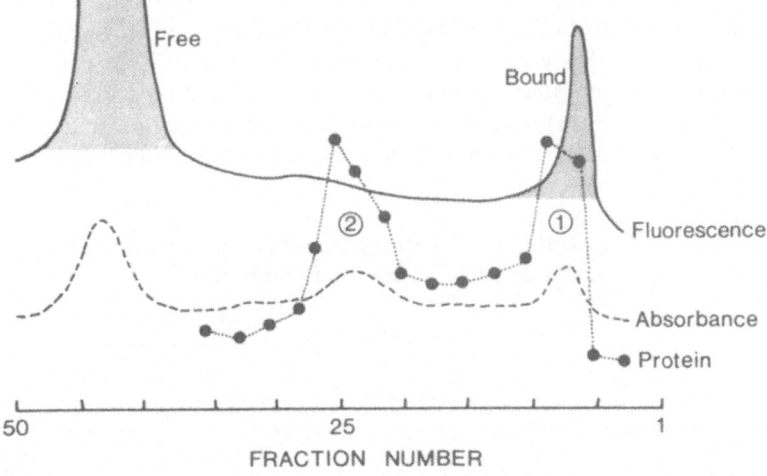

FIGURE 2. Demeclocycline binding study. An aliquot of the 10,000g toad urinary bladder epithelial cell supernatant was incubated with DMCTC (0.5 mg/ml) for 1 hr at pH 7.3 and then separated by Sephadex G-200 gel exclusion chromatography. Two major protein fractions (peaks 1 and 2) are resolved by this technique as indicated by the absorbance curve (280 nm) and the protein curve (Lowry). The fluorescence of the Sephadex column effluent was continuously monitored using a fluorescence spectrometer with a flowthrough cell attachment (excitation and emission wavelengths 380 and 518 nm, respectively). DMCTC fluorescence is associated with peak 1 but not with peak 2. Unbound (free) DMCTC produces a large fluorescence peak between fractions 35 and 50.

TABLE I. Correlation of Tetracycline–Protein Binding with Inhibition of
ADH-Induced Water Flow[a]

| | Drug bound, mg/mg protein | | ADH-induced water flow | |
Drug	Peak 1	Peak 2	Percent inhibition	p
TC	0.06 ± 0.01	0.04 ± 0.01 (5)	−6.7 ± 2.4 (5)	NS
DMCTC	0.13 ± 0.02	0.11 ± 0.02 (7)	−24.5 ± 4.1 (9)	<0.001

[a] Aliquots of the 10,000g toad urinary bladder epithelial cell supernatant were incubated with tritium-labeled TC or DMCTC (0.5 mg/ml) for 1 hr at pH 7.3 and then separated by Sephadex G-200 gel exclusion chromatography. Peaks 1 and 2 refer to the two major chromotographic fractions (see Fig. 2). ADH (100 mU/ml)-induced water flow was determined gravimetrically (2:1 osmotic gradient) in paired hemibladders in the presence and absence of drug (0.5 mg/ml) at a medium pH = 7.3. The numbers represent mean ± SE (n). (Note that these studies were performed under non-equilibrium binding conditions and that the drug concentrations are much greater than those employed in the equilibrium-binding studies reported in Tables II and III.)

studies using tritiated tetracyclines showed that the latter conclusion was incorrect. DMCTC and TC bound with equal affinity to both high and lower molecular weight proteins; however, DMCTC bound with greater affinity than TC (Table I). We presume that the apparent lack of binding to proteins comprising peak 2 (Fig. 2) in the earlier studies was due to quenching of the tetracycline fluorescence by one or more of the components in this fraction. More recent studies, utilizing an equilibrium dialysis technique and reliable fluorescence assay systems, have confirmed that the degree of tetracycline–protein binding varies directly with the inhibitory effect of the particular tetracycline on ADH-mediated water flow: anhydrodemeclocycline (anhydro-DMCTC) > DMCTC > OTC (Table II).

Thus, there is little question that the tetracyclines bind to a variety of toad urinary bladder epithelial cell proteins and that this binding correlates with the effects of these drugs on ADH-induced osmotic water flow. Three major questions are raised by these studies: (1) Can specific proteins be identified which have greater

TABLE II. Correlation of Tetracycline–Protein Binding with
Inhibition of ADH-Induced Water Flow[a]

| | | ADH-induced water flow | |
Drug	Drug bound, μg/mg protein	Percent inhibition	p
OTC	1.8 ± 1.0 (6)	0.6 ± 6.1 (4)	NS
DMCTC	13.8 ± 2.2 (8)	−12.2 ± 4.0 (8)	<0.02
Anhydro-DMCTC	70.4 ± 7.6 (8)	−43.8 ± 3.1 (8)	<0.001

[a] Aliquots of the 10,000g toad urinary bladder epithelial cell supernatant were dialyzed against Tris-buffered Ringer's solution (pH = 7.3, total calcium concentration 0.75 mM) containing OTC, DMCTC, or anhydro-DMCTC (20 μg/ml) for 24 hr. Drug-binding was quantitated using fluorescence assays for OTC and DMCTC (Kohn, 1961a) and anhydro-DMCTC (Kelly et al., 1969). ADH (2 mU/ml)-induced water flow was determined gravimetrically (5:1 osmotic gradient) in paired hemibladders in the presence and absence of drug (20 μg/ml) at a medium pH = 7.3. The numbers represent mean ± SE (n).

TABLE III. Correlation of Tetracycline–Protein
Binding with Medium Calcium Concentration[a]

Total [Ca^{++}], mM	Drug bound, μg/mg protein	
	Anhydro-DMCTC	DMCTC
0	Undetectable	Undetectable
0.10	8.4 ± 1.8 (4)	9.5 ± 1.6 (4)
0.30	39.7 ± 5.4 (4)	6.8 ± 1.1 (4)
0.50	81.9 ± 8.2 (4)	14.5 ± 3.3 (4)
0.75	70.4 ± 7.6 (8)	13.8 ± 2.2 (8)

[a]These experiments were carried out as described in Table II, except that the medium calcium concentration was varied as indicated. The numbers represent mean ± SE (n).

affinities for the tetracyclines than the remaining epithelial cell proteins? (2) Do the affinities of the tetracyclines for such proteins correlate with their inhibitory effects on ADH-induced water flow? (3) What is the cellular distribution and function of these specific tetracycline-binding proteins? Further studies will be needed to address these questions, but techniques are available to answer all of them.

Although these studies indicate new directions for research into the cellular mechanisms underlying the inhibitory effects of tetracyclines on renal epithelial water transport, it is well to keep in mind that other potential clues may also warrant investigation. Three examples (related to chelation, prostaglandins, and structural alterations of the tetracyclines) should suffice to emphasize this point. Since calcium modulates ADH-induced water flow and since tetracyclines chelate a variety of divalent cations, including calcium and magnesium (Albert and Rees, 1956; Doluisio and Martin, 1963a; Martin, 1979), the role of chelation in the effects of tetracyclines on ADH-induced water flow warrants much more careful investigation. It is certainly conceivable that tetracycline-related chelation alters the divalent cation microenvironment in one or more critical subcellular compartments in renal epithelia. In addition, divalent cations may be involved in the binding of tetracyclines to proteins and other macromolecules (Kohn, 1961b; Doluisio and Martin, 1963b). In this regard, preliminary studies in our laboratory have shown that calcium has a marked effect on the binding of tetracyclines to toad urinary bladder epithelial cell proteins (Table III).

The fact that the inhibitory activity of the tetracyclines does not correlate with their divalent ion stability constants (Feldman and Singer, 1974) should not deter further investigation into tetracycline–calcium or tetracycline–magnesium interactions as a basis for their effects on water transport. The quoted constants refer to cupric, nickel, and zinc ions, not to calcium or magnesium. Indeed, the exact nature of the interaction of calcium and magnesium with tetracyclines is still a matter of some controversy (Martin, 1979). Even in the event that the calcium- or magnesium-chelating ability of the various tetracycline congeners did not correlate with their inhibitory activity, this would still not exclude a role for chelation in the action of these agents. For example, properties other than chelating ability (e.g., protein binding or lipid solubility) could account for translocation of the active congeners to their site of action; once at this site, chelation could still form the basis for the inhibition of water flow. Thus, the inhibitory potential of a particular congener could be a complex function of several of its physicochemical properties rather than being a simple function of only one of them.

Prostaglandin metabolism and function also need to be considered as potential loci of tetracycline action. Since DMCTC has no effect on prostaglandin E synthesis in the toad urinary bladder at a time when ADH-induced water flow is inhibited (Zusman *et al.,* 1978), tetracyclines probably do not inhibit adenylate cyclase activity by enhancing prostaglandin E biosynthesis. However, the possibility that tetracyclines enhance the inhibitory effect of prostaglandin E on the ADH-responsive adenylate cyclase complex needs to be evaluated.

Just as structure–activity relationships have played a central role in the development of the semisynthetic tetracycline antibiotics (Blackwood and English, 1970), it is likely that similar correlative studies will prove helpful in the elucidation of the mechanisms underlying the effects of these agents on water transport. Such studies have been initiated in our laboratory, with a number of interesting findings. For example, 4-dedimethylamino-tetracycline has a small, but definite, effect on ADH-induced water flow and 5a,6-anhydro-DMCTC is 4–5 times more potent than equimolar concentrations of its parent compound DMCTC (Cox *et al.,* 1977a). Since neither of these congeners has significant tetracycline antibiotic activity (Blackwood and English, 1970), the latter is not a prerequisite for the effects of the tetracyclines on water transport.

The 5a,6-anhydro derivative of DMCTC is of considerable interest. For example, the greater potency of this congener as compared to DMCTC itself correlates with the binding of these two congeners to toad urinary bladder epithelial cells proteins (Table II). 5a,6-Anhydro derivatives can be produced from any tetracycline with a hydroxyl group at carbon-6. In acidic media 6-hydroxylated tetracyclines undergo a dehydration reaction which results in aromatization of the C ring and a change in the phenolic diketone group at carbons 10, 11, and 12 (Fig. 3). Since the pka of this group is ~7.5 (Mitscher, 1978), it is likely to contribute significantly to the physicochemical properties of the tetracyclines at physiologic pH. Consequently, alterations in this portion of the molecule might be expected to have important physiologic implications. In addition, the phenolic diketone group may be involved in the calcium-chelating activity of the tetracyclines (Martin, 1979). Whether such alterations account for the enhanced activity of anhydro-DMCTC relative to DMCTC is conjectural. However, it is clear that 5a,6-dehydration is not an absolute prerequisite for ADH-inhibitory activity, since DOOTC and DDDTC (which lack hydroxyl groups at carbon-6) both inhibit ADH-induced water flow (Feldman and Singer, 1974).

4. Tetracycline-Associated Natriuresis and Azotemia

Early studies (Shils, 1962, 1963) had shown that under certain circumstances (most notably in patients with some degree of preexisting renal insufficiency) the

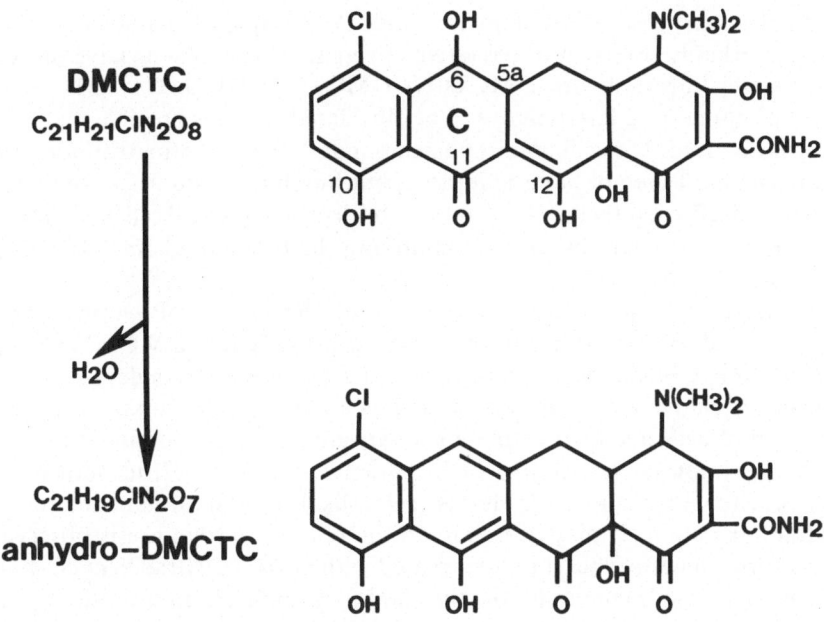

FIGURE 3. Relationship of DMCTC to anhydro-DMCTC (see text for details).

administration of tetracycline could be associated with a natriuresis. However, this phenomenon was not further investigated, since a natriuresis did not appear to occur in individuals with normal renal function who were treated with tetracyclines (Singer and Rotenberg, 1973; Wilson *et al.*, 1973) or in patients with SIADH treated with DMCTC (DeTroyer, 1977; Forrest *et al.*, 1978).

Tetracycline-associated natriuresis was "rediscovered" when DMCTC was employed to treat the chronic hyponatremia often associated with the pathologic edema-forming states, in which high circulating levels of ADH may contribute to the water retention characteristic of the severe forms of these disorders. In some patients with cirrhosis or congestive heart failure, DMCTC causes a water diuresis and amelioration of the hyponatremia (Oster *et al.*, 1976; DeTroyer *et al.*, 1976; Carrilho *et al.*, 1977; Cox *et al.*, 1977b; Zegers de Beyl *et al.*, 1978; Miller *et al.*, 1980; Geheb and Cox, 1980; Braden *et al.*, 1980a). The development of nephrogenic diabetes insipidus was predictable; what was not expected was the concomitant development of rather striking renal sodium wasting (and azotemia) in a large proportion of these patients.

The etiology of DMCTC-induced natriuresis in edematous patients is not clear, but studies performed in the toad urinary bladder model system have provided some insight. In this tissue, DMCTC inhibits both basal and aldosterone-mediated Na^+ transport but has little or no effect on ADH- or insulin-mediated Na^+ transport; other tetracycline congeners have similar, although quantitatively different, effects (Guzzo *et al.*, 1978; Cox *et al.*, 1979; Braden *et al.*, 1980b). As an example, the effects of several tetracyclines on aldosterone-mediated Na^+ transport in the toad urinary bladder are shown in Table IV. A sequence of potency is evident that is remarkably similar to the relative effects that these agents have on ADH-induced water flow (anhydro-DMCTC > DMCTC > DDDTC > OTC, in both cases). The connecting link remains to be elucidated, but it is possible that protein binding is involved. The cellular mechanisms underlying the antinatriferic effects of the tetracyclines have not been delineated, but biophysical studies have indicated that they may affect epithelial Na^+ transport by decreasing apical cell membrane Na^+ conductance rather than by directly inhibiting the basolateral Na^+-K^+-ATPase pump (Guzzo et al., 1978).

We have been particularly interested in the effects of tetracyclines on aldosterone-induced Na^+ transport because it is possible that DMCTC exerts part of its natriuretic effect in edematous patients (in whom secondary hyperaldosteronism is relatively common) by acting as an aldosterone antagonist. Such a mechanism of action may explain why a natreuresis is not prominent in normal subjects on a normal Na^+ intake or in patients with SIADH; in both of these situations, circulating aldosterone levels are less likely to be important determinants of sodium excretion. It may also explain the relative antikaliuresis observed in patients with congestive heart failure treated with DMCTC (Braden et al., 1980a).

However, the magnitude of the DMCTC-induced natriuresis in some edematous patients is so great that it is unlikely to be completely accounted for by inhibition of the tubular action of aldosterone. Several other possibilities obviously exist. For example, the natriuresis may simply be a reflection of a nonspecific toxic effect of the drug (Obek et al., 1974), especially since renal insufficiency appears to be a common concomitant of the natriuresis.

The azotemia associated with the use of tetracyclines has generally been ascribed to the antianabolic effects of these drugs (Shils, 1962, 1963). The antibiotic effect of these agents is related to inhibition of ribosomal function (Pestka, 1971). Furtunately, eukaryotic ribosomes are more resistant, but tetracyclines do inhibit protein synthesis in mammals, although to a lesser extent than in bacteria

TABLE IV. Effects of Tetracyclines on Aldosterone (Aldo)-Induced Na^+ Transport[a]

Drug	Aldo-induced Na^+ transport	
	Percent inhibition	p
OTC	-18.1 ± 3.9 (6)	<0.01
DDDTC	-62.2 ± 5.5 (6)	<0.001
DMCTC	-78.4 ± 1.3 (6)	<0.001
Anhydro-DMCTC	-83.7 ± 3.6 (6)	<0.001

[a] Toad urinary bladders were mounted in modified Ussing chambers and treated with aldosterone (1.4×10^{-7} M in the serosal medium) for 16 hr. Thereafter, one bladder (of a matched pair) received a tetracycline (0.5 mg/ml) and the other bladder served as a control. Short-circuit current (which was used to measure Na^+ transport) was then monitored for 4 hr. The numbers represent mean $\pm$ SE (n).

(Franklin, 1963). This antianabolic effect is particularly evident in patients with pre-existing renal insufficiency, because many tetracyclines (including DMCTC) are excreted by the kidneys, and consequently they may accumulate in the presence of renal disease (Anderson *et al.,* 1976).

In at least some cirrhotic patients with DMCTC-induced natriuresis and azotemia serum DMCTC levels are strikingly elevated (Oster *et al.,* 1976; Miller *et al.,* 1980) despite the ingestion of a normal therapeutic dose of the drug. The cause of the increased serum DMCTC levels in such patients is unknown. However, since the glomerular filtration rate is often reduced in patients with severe cirrhosis, it is possible that preexisting functional renal insufficiency contributes to the elevated serum DMCTC levels in such patients. Hepatic disease, by reducing the amount of the drug excreted in the bile, may also have a role. Whether an analogous situation occurs in patients with congestive heart failure is less clear. However, recent studies in our laboratory in two patients with congestive cardiomyopathies have indicated that natriuresis and azotemia can occur in the absence, as well as in the presence, of markedly elevated serum DMCTC levels.

While an antianabolic effect may contribute to DMCTC-induced azotemia in edematous individuals, a true reduction in glomerular filtration rate also occurs, at least in some patients (Miller *et al.,* 1980; Braden *et al.,* 1980a). The etiology of the renal insufficiency is unknown, but it cannot be entirely accounted for oñ the basis of sodium depletion per se. For example, the glomerular filtration rate fell in two cirrhotic patients despite replacement of the sodium deficit and improved in yet another cirrhotic patient after DMCTC was discontinued despite ongoing natriuresis (Miller *et al.,* 1980). Similar observations have been made in patients with congestive heart failure treated with DMCTC (Braden *et al.,* 1980a). Presumably, therefore, at least in some edematous patients, DMCTC can be directly nephrotoxic.

The problem of nephrotoxicity clearly needs more extensive evaluation. For example, if the natriuresis and azotemia could be avoided by decreasing the dose of DMCTC employed, then this drug may yet prove to be a useful adjunct to the therapy of hyponatremia associated with far-advanced cardiac and hepatic disease. Until such time as this problem is resolved, however, the clinical use of DMCTC in such situations cannot be condoned.

Acknowledgments

Many individuals have contributed to the formulation of the ideas expressed above. In particular, I would like to express my appreciation to Irwin Singer, M.D., Michael Geheb, M.D., Gregory Braden, M.D., and Joseph Guzzo, M.D. The studies would have been impossible without the skilled technical assistance of Gary Huber, Susan Follansbee, and A. Bernard Kelly. James DiStefano provided able secretarial assistance. These studies were supported in part by grants from the NIH (AM-17344 and AM-21454), the Philadelphia Veterans Administration Medical Center (Research Fund 103.14M), and Lederle Laboratories. The author is a Research and Education Associate of the Veterans Administration.

References

Albert, A., and Rees, C. W., 1956, Avidity of the tetracyclines for the cations of metals, *Nature* **177**:433.

Anderson, R. J., Gambertoglio, J. G., and Schrier, R. W., 1976, *Clinical Use of Drugs in Renal Failure,* Charles C. Thomas, Springfield, Illinois, p. 53.

Benitz, K.-F., and Diermeier, H. F., 1964, Renal toxicity of tetracycline degradation products, *Proc. Soc. Exp. Biol. Med.* **115**:930.

Blackwood, R. K., and English, A. R., 1970, Structure–activity relationships in the tetracycline series, in: *Advances in Applied Microbiology,* (D. Perlman and W. Umbreit, eds.), Volume 13, Academic Press, New York, p. 237.

Braden, G., Geheb, M., and Cox, M., 1980a, Renal toxicity of demeclocycline in patients with edematous disorders, *Clin. Res.* **28**:655.

Braden, G., Geheb, M., Singer, I., and Cox, M., 1980b, Lithium- and demeclocycline-induced natriuresis: Studies in toad urinary bladders, *Clin. Res.* **28**:439.

Carrilho, F., Bosch, J., Arroyo, V., Antoni, M., Viver, J., and Rodes, J., 1977, Renal failure associated with demeclocycline in cirrhosis, *Ann. Intern. Med.* **87**:195.

Castell, D. O., and Sparks, H. A., 1965, Nephrogenic diabetes insipidus due to demethylchlortetracycline hydrochloride, *J. Am. Med. Assoc.* **193**:237.

Chan, S., 1979, Water diuretic activity of demeclocycline in rats, *Fed. proc.* **38**:749.

Cleveland, W. W., Adams, W. C., Mann, J. B., and Nyhan, W. L., 1965, Acquired Fanconi syndrome following degraded tetracycline, *J. Pediatr.* **66**:333.

Cox, M., and Singer, I., 1976a, Inhibition of ADH-induced water flow by tetracyclines in toad urinary bladder: Fluorescence studies of drug–protein binding, *Fed. Proc.* **35**:915.

Cox, M., and Singer, I., 1976b, Inhibition of ADH-induced water flow in the toad urinary bladder by tetracyclines: Correlation of physiological effect with epithelial cell protein binding, *Clin. Res.* **24**:397.

Cox, M., Guzzo, J., Huber, G., and Singer, I., 1977a, Inhibition of ADH-stimulated water flow in toad urinary bladders by tetracyclines: pH-dependence and lack of correlation with antibacterial activity, *Kidney Int.* **12**:554.

Cox, M., Guzzo, J., Morrison, G., and Singer, I., 1977b, Demeclocycline and therapy of hyponatremia, *Ann. Intern. Med.* **86**:113.

Cox, M., Guzzo, J., Shook, A., Huber, G., and Singer, I., 1979, Effects of tetracyclines on aldosterone- and insulin-mediated Na^+ transport in the toad urinary bladder, *Biochim. Biophys. Acta* **552**:162.

DeTroyer, A., 1977, Demeclocycline treatment for syndrome of inappropriate antidiuretic hormone secretion, *J. Am. Med. Assoc.* **237**:2723.

DeTroyer, A., Pilloy, W., Broeckaert, I., and Demanet, J.-C., 1976, Demeclocycline treatment of water retention in cirrhosis, *Ann. Intern. Med.* **85**:336.

Doluisio, J. T., and Martin, A. N., 1963a, Metal complexation of the tetracycline hydrochlorides, *J. Med. Chem.* **6**:16.

Doluisio, J. T., and Martin, A. N., 1963b, The binding of tetracycline analogs to conalbumin in the absence and presence of cupric ions, *J. Med. Chem.* **6**:20.

Dousa, T. P., and Wilson, D. M., 1974, Effects of demethylchlortetracycline on cellular action of antidiuretic hormone in vitro, *Kidney Int.* **5**:279.

Feldman, H., and Singer, I., 1974, Comparative effects of tetracyclines on water flow across toad urinary bladders, *J. Pharmacol. Exp. Ther.* **190**:358.

Forrest, J. N., Jr., Cox, M., Hong, C., Morrison, G., Bia, M., and Singer, I., 1978, Superiority of demeclocycline over lithium in the treatment of chronic syndrome of inappropriate secretion of antidiuretic hormone, *N. Engl. J. Med.* **298**:173.

Franklin, T. J., 1963, The inhibition of incorporation of leucine into protein of cell-free systems from rat liver and *Escherichia coli* by chlortetracycline, *Biochem. J.* **87**:449.

Frimpter, G. W., Timpanelli, A. E., Eisenmenger, W. J., Stein, H. S., and Ehrlich, L. I., 1963, Reversible "Fanconi syndrome" caused by degraded tetracycline, *J. Am. Med. Assoc.* **184**:111.

Fulop, M., and Drapkin, A., 1965, Potassium-depletion syndrome secondary to nephropathy apparently caused by "outdated tetracycline," *N. Engl. J. Med.* **272**:986.

Geheb, M., and Cox, M., 1980, Renal effects of demeclocycline, *J. Am. Med. Assoc.* **243**:2519.

Goldman, R., and Hall, L. M., 1960, Studies on renal toxicity of demethylchlortetracycline, Science Information Exchange (final report of research project Z0-10444-C2, May 29, 1961) (listed in Med. Res. Vet. Adm. 1960, p. 130).

Gross, J. M., 1963, Fanconi syndrome (adult type) developing secondary to the ingestion of outdated tetracycline, *Ann. Intern. Med.* **68**:523.

Guzzo, J., Cox, M., Kelly, A. B., and Singer, I., 1978, Tetracycline-induced inhibition of Na^+ transport in the toad urinary bladder, *Am. J. Physiol.* **235**:F359.

Kelly, R. G., Peets, L. M., and Hoyt, K. D., 1969, A fluorometric method of analysis for tetracycline, *Analytical Biochem.* **28**:222.

Kohn, K. W., 1961a, Determination of tetracyclines by extraction of fluorescent complexes. Application to biological materials, *Analytical Chem.* **33**:862.

Kohn, K. W., 1961b, Mediation of divalent metal ions in the binding of tetracycline to macromolecules, *Nature* **191**:1156.

Lowe, M. B., and Tapp, E., 1966, Renal damage caused by anhydro-4-epi-tetracycline, *Arch. Pathol.* **81**:362.

Martin, S. R., 1979, Equilibrium and kinetic studies on the interactions of tetracyclines with calcium and magnesium, *Biophys. Chem.* **10**:319.

Miller, P. D., Linas, S. L., and Schrier, R. W., 1980, Plasma demeclocycline levels and nephrotoxicity—correlation in hyponatremic cirrhotic patients, *J. Am. Med. Assoc.* **243**:2513.

Mitscher, L. A., 1978, *The Chemistry of the Tetracycline Antibiotics. Medicinal Research Series,* Volume 9 (G. L. Grunewald, ed.), Dekker, New York.

Obek, A., Petorak, I., Eroglu, L., and Gurkan, A., 1974, Effects of tetracycline on the dog kidney—a functional and ultrastructural study, *Israel J. Med. Sci.* **10**:765.

Oster, J. R., Epstein, M., and Ulano, H. B., 1976, Deterioration of renal function with demeclocycline administration, *Curr. Ther. Res.* **20**:794.

Pestka, S., 1971, Inhibitors of ribosome functions, *Annu. Rev. Microbiol.* **25**:487.

Roth, H., Becker, K. L., Shalhoub, R. J., and Katz, S., 1967, Nephrotoxicity of demethylchlortetracycline hydrochloride, *Arch. Intern. Med.* **120**:433.

Shils, M. E., 1962, Some metabolic aspects of tetracyclines, *Clin. Pharmacol. Ther.* **3**:321.

Shils, M. E., 1963, Renal disease and the metabolic effects of tetracycline, *Ann. Intern. Med.* **58**:389.

Singer, I., and Rotenberg, D., 1973, Demeclocycline-induced nephrogenic diabetes insipidus: in-vivo and in-vitro studies, *Ann. Intern. Med.* **79**:679.

Wilson, D. M., Perry, H. O., Sams, W. M., Jr., and Dousa, T. P., 1973, Selective inhibition of human distal tubular function by demeclocycline, *Curr. Ther. Res.* **15**:734.

Zegers de Beyl, D., Naeije, R., and DeTroyer, A., 1978, Demeclocycline treatment of water retention in congestive heart failure, *Br. Med. J.* **1**:760.

Zusman, R. M., Keiser, H. R., and Handler, J. S., 1978, Effect of adrenal steroids on vasopressin-stimulated PGE synthesis and water flow, *Am. J. Physiol.* **234**:F532.

III

TUBULOINTERSTITIAL NEPHROPATHY DUE TO DRUGS AND ENVIRONMENTAL TOXICANTS

GEORGE A. PORTER, Section Editor

16

Tubulointerstitial Nephropathy
An Overview

GEORGE A. PORTER

As with many categories of renal disease, tubulointerstitial nephropathy represents an admixture of morphologic and functional terms. Thus, it is a hybrid term used to simultaneously denote a disease entity that possesses unique clinical and histologic features. While the former are easily tabulated and compared with uniform clinically derived criteria, the latter requires sampling of renal tissue, which is less often available. For the purist, combined clinical–histologic criteria are mandatory and yet extenuating circumstances may preclude an invasive procedure. This should not lead to abandonment of the term, which has significant conceptual implications, as detailed by Cotran in Chapter 17, but rather it should be applied with caution in patients without histologic confirmation. Clearly, both clinical and histologic criteria must be accepted to define tubulointerstitial nephropathy, since the most global definition would include all renal parenchymal disease after excluding glomerulopathies and pyelocalyceal disease.

In Chapter 17, Cotran summarizes current thinking regarding both acute and chronic tubulointerstitial nephropathy based upon clinicopathologic interpretations. The importance of preserving a combined morphologic and clinical classification is defended. He emphasizes that the etiologic information is far more abundant for tubulointerstitial disease than for glomerulopathies. Pathologic

GEORGE A. PORTER • Department of Medicine, Oregon Health Sciences University, Portland, Oregon 97201.

discrimination between acute and chronic tubulointerstitial disease is determined by the absence of interstitial fibrosis and tubular atrophy, which characterize the chronic presentation. Since up to one-third of biopsy-confirmed cases of acute tubulointerstitial nephritis cannot be linked to either an infectious or drug-related cause, unrecognized environmental hazards may be responsible. Knowledge concerning chronic tubulointerstitial nephritis is expanding rapidly, especially with the segregation of chronic pyelonephritis. The morphologic contributions to better understanding of chronic tubulointerstitial nephritis include: the microangiopathy associated with analgesic nephropathy, the interstitial deposition of Tamm-Horsfall proteins in obstructed or refluxing kidneys, and the role of glomerulosclerosis in explaining progressive deterioration of glomerular filtration rate. The last contribution may serve to link morphologic changes with a functional counterpart mediated by changes in renal hemodynamics.

Traditionally, clinically measured renal function is confined to an estimate of glomerular filtration rate without coincidental evaluation of tubular function. However, while attempts to correlate glomerular pathology with changes in filtration rate have been disappointing, changes in interstitial volume provide statistically significant relationships (Riemenschneider *et al.,* 1980). A potential explanation for this apparent discrepancy may come from recent observations by Hostetter *et al.* (1981). In essence, they speculate that the eventual glomerulosclerosis is due to changes in blood flow and single-nephron filtration rate that result from disease-induced alterations of glomerular perm-selectivity. Extension of these observations to multiple animal models of glomerulonephritis will assist acceptance of the concept.

Acute renal failure as a consequence of acute tubulointerstitial nephritis, while easily discriminated by histologic criteria from acute tubular necrosis, can be far more confusing when only clinical criteria are available (Van Ypersele de Strihou, 1979). When oliguria is present, then renal failure indices will not differentiate between acute renal failure due to acute tubular necrosis and that due to acute tubulointerstitial nephropathy irrespective of the proposed mechanisms of injury, which are quite different. Experimentally, three different immunologic maneuvers will produce acute tubulointerstitial injury. These include: immune complexes, antibodies to tubular basement membrane, and delayed hypersensitivity reactions. Recently, Husby *et al.* (1981) have demonstrated immunocompetent, T lymphocytes in tissue from patients with interstitial nephritis, consistent with a major role for cell-mediated immunity. However, Roberts *et al.* (1981), by exposing primates to live versus heat-killed bacteria, concluded that while cell-mediated immunity might contribute to the renal inflammatory response, it was not critical. They found that inflammatory response followed by fibrosis was triggered by active bacterial infection plus trauma. A related study by Miller and Phillips (1981) concluded that the number and rapidity with which bacteria reach renal parenchyma are the critical factors in determining the extent of renal damage. Active infections, in contrast to acquiescent infections, stimulate host inflammatory response.

Although less dramatic in clinical presentation, chronic tubulointerstitial disease with resulting progressive renal failure is emerging as a much more significant form of toxic nephropathy. This can be appreciated by comparing the change in distribution of renal diagnostic categories for patients with End Stage Renal Disease (ESRD) as a function of reporting year (Table 1). Accordingly, in Europe the category of pyelonephritis, or interstitial nephritis, increased three percentage points between 1975 and 1978. If it is assumed that the incidence of pyelonephritis is either stable or slowly declining, then the major increase has occurred in tubulointerstitial disease. While the contribution of repetitive low-dose exposure to either drugs or environmental hazards remains speculative, circumstantial evidence concerning analgesics, lead, and cadmium, to mention but a few, will be detailed in subsequent chapters. Finally, the potential exists that multiple, sequential, acute renal insults of a subclinical or clinical character may culminate in progressive tubulointerstitial disease. This may become the most important area for future research in nephrotoxicity.

The prevalence of tubulointerstitial disease varies depending upon the biases of the report. Kleinknecht *et al.* (1978) reported that acute tubulointerstitial disease accounts for 3% of the acute renal failure in unselected patients, while Linton *et al.* (1980), writing about acute interstitial nephritis, placed the percentage between 10 and 14% of unexplained acute renal failures. From ESRD statistics, 20% of patients are considered to have chronic interstitial nephritis.

While the clinical presentation of chronic tubulointerstitial disease is not distinguishable from other causes of chronic uremia, there are historical clues which must be appreciated. Familial incidence characterizes the hereditary chronic nephritities, while Murray reviews the clinical features of chronic analgesic nephropathy in Chapter 18. Presence of coexisting systemic disease, e.g., Sjogren's syndrome or systemic lupus erythematosus, will also point in that direction. The situation for acute tubulointerstitial nephritis is much different, with a more characteristic cluster of signs and symptoms. According to a recent literature survey reported by Linton *et al.* (1980), systemic symptoms include: fever 60–100%, rash 40–50%, and arthralgia <20%. Renal findings include: oliguria 60–70%,

TABLE I. Diagnosis of Primary Renal Disease. 1975 versus 1978[a]

	Percent of patients		
	1975 (8,574)	1978 (10,233)	Δ
Chronic renal failure	7.6	10.5	+2.9
Glomerulonephritis	41.1	32.7	−8.4
Pyelo- or interstitial nephritis	17.9	20.9	+3.0
Drug nephropathy	3.2	3.0	−0.2
Cystic kidney disease	8.8	9.2	+0.4
Heredo-family renal disease	3.1	2.8	−0.3
Renal vascular disease	5.6	8.3	+2.7
Multisystem disease	6.8	8.5	+1.7
Others	7.2	4.2	−3.0

[a] European Dialysis and Transplant Association.

TABLE II. Agents Associated with Tubulointerstitial Nephritis

Acute	
Penicillins	Phenindoine
β-Lactams	Rifampicin
Sulfonamides	Allopurinol
Chronic	
Analgesics	Heavy metals
Lithium	Radiation

hematuria 70–100%, proteinuria 40–100%, and pyuria 70–100%. Elevated blood eosinophilic counts occur in between 60 and 100% of reported cases, while the presence of urinary eosinophils, i.e., more than one-third of all urinary leukocytes, is an even more consistent finding, when evaluated. Finally, an elevated serum IgE has been noted in approximately 50% of patients with renal biopsy-confirmed acute tubulointerstitial nephritis (Ooi et al., 1971).

The potential value of gallium^{-67} scanning to discriminate between acute tubular necrosis and acute tubulointerstitial nephritis in patients presenting with acute oliguric renal failure has been evaluated by both Linton et al. (1980) and ourselves. For 27 patients selected because of acute renal disease, 16 had positive scans according to Linton and co-workers, including all nine patients with the diagnosis of drug-induced interstitial nephritis. None of six patients with suspected acute tubular necrosis yielded positive scans. Our results, which incorporated 17 patients, include consistently positive scans in patients with proliferative glomerulonephritis and minimal change nephrotic syndrome in addition to acute tubulointerstitial nephritis, while those with diabetic renal disease do not have persistent visualizations. The ultimate role of gallium^{-67} in the differential diagnosis of acute oliguric renal failure must await definition of specificity.

While the potential list (Van Ypersele de Strihou, 1979) of agents implicated as causing tubulointerstitial disease is long, a much more succinct tabulation can be achieved based upon frequency and firm literature documentation. My preference is summarized in Table II. The prototype of acute tubulointerstitial disease in man is methacillin nephritis. Originally, it was thought to be induced by antitubular basement membrane antibodies (Baldwin et al., 1968; Border et al., 1974); however, subsequent observations have revealed that such a mechanism exists only in a minority of cases (Galpin et al., 1978). Both natural and semisynthetic penicillins have been implicated from literature reports (Porter et al., 1981a), i.e., penicillin G, ampicillin, oxacillin, nafcillin, and carbenicillin. The beta-lactams, i.e., cephalosporins, contribute significant numbers of cases of acute tubulointerstitial nephritis, distinct from the acute tubular necrosis reported in association with cephalothin (Porter and Bennett, 1981b). While the frequency of sulfonamide hypersensitivity reaction is low, the total amount of parent and derivative drugs, e.g., furosemide and chlorothiazide, that are prescribed accounts for their important contribution to the prevalence of tubulointerstitial disease.

Most of the chapters included in this section deal with various causes of chronic tubulointerstitial disease. Since elimination of the offending agent will either stabilize or reverse functional deterioration of the kidneys while avoidance will allow prevention, the problems of analgesic nephropathy and occupational lead nephropathy deserve special consideration.

In Chapter 18, Murray initiates his review of clinical analgesic nephropathy by reminding us that there is yet to be agreement on the offending agent, amount, duration, or prevalence. With this disclaimer, he provides a critique of current "facts" regarding analgesic nephropathy, the preferred terminology, plus providing an update of current North American registry. While the causal relationship between large doses of analgesics and the development of renal diseases is well supported by both epidemiologic data and animal experimentation, that phenacetin is the culprit is becoming less defensible with each passing year. A recognizable profile of renal disease can be constructed to assist in identifying patients with analgesic nephropathy, although all features are not mandatory. Abnormal intravenous urograms are common, and evidence of papillary necrosis frequent. Sterile pyuria, tubular proteinuria, and hypertension are usual and the risk of ureterovesicle cancer is substantially increased, based on population statistics. Relative risk, based upon limited objective data for (1) amount ingested based on population studies, (2) incidence of renal disease in analgesic users, and (3) incidence of analgesic use in chronic renal failure patients, suggests that cumulative ingestion of 2–3 kg of analgesic drug carries a three to ten times increased risk of chronic renal failure.

According to Mudge, Chapter 19, the localization of experimental analgesic nephropathy to the medullary region of the kidney relates to unique physical-clinical characteristics which allow preferential accumulation. The role of "reactive metabolites" as the potential toxins in analgesic nephropathy is evaluated. There exists both parallel and divergent observations between hepatotoxicity and nephrotoxicity of analgesic metabolites, especially regarding the depletion of intracellular glutathione content. Based upon studies of covalent binding, both liver and kidney can represent sites of metabolite formation. The chemical characteristics of several "reactive metabolites" are constructed and their potential relevance as etiologies in analgesic nephropathy is discussed. Furthermore, the problems of applying classic toxicologic interpretation to substantiate the proposition that the acetaminophen (APAP–p-aminophenol (PAP) conversion is comparable with "reactive metabolite" hypothesis are presented. After reviewing the various metabolic pathways available for PAP formation, he concludes that available data favor PAP as the putative "reactive metabolite" in experimentally induced analgesic renal injury.

In stark contrast to the recent recognition of analgesic nephropathy is the long and circuitous history associated with lead intoxication, which is elegantly detailed by Wedeen in Chapter 22. He emphasizes the close association between plumbism and gout. He indicates that while early authors were cognizant of the coincidental occurrence of the two entities, Ollivier's contention in 1863 that chronic lead nephropathy was simply another form of Bright's disease initiated a separation that is now being rectified. Epidemiologic studies during this century have reestablished both the risk and consequences of chronic lead exposure. While occupational lead nephropathy was a recognized entity in Europe, it was only recently considered to exist in the United States. Data which detail diagnostic criteria, prevalence, and response to chelation therapy are present from the author's extensive experience. After chronicling 3 centuries of the reported association between lead toxicity and gout, he provides data from a prospective study which demonstrates an increased body burden of lead in both gouty patients and hypertensive patients with evidence of renal insufficiency.

Hammond *et al.,* in Chapter 23, pose the question of whether a suitable animal model exists for chronic lead nephropathy. They characterize the temporal sequence of altered renal function encountered in clinical lead nephropathy to define model requirements. They include data from prospective studies of lead-exposed workers to confirm renal profile. After reviewing data available regarding renal function in rats chronically exposed to lead, they conclude that reliable adult rat models must await documentation.

While endemic Balkan nephropathy may appear an unusual subject for a symposium on toxicants, the enlightening review by Hall in Chapter 20 justifies its inclusion. This represents a well-studied example of chronic tubulointerstitial nephropathy which progresses to ESRD. From epidemiologic studies, tubular proteinuria has been confirmed to have increased frequency in endemic regions. Farming the land emerged as an important disease linkage, as did $HLA-B_{18}$ tissue type. While data support an environment toxicant as etiologic after review of the potential agents, none can be verified. The similarities between endemic Balkan nephropathy and the nephropathy of light-chain disease prompted speculation that B2-microglobulins might cause tubular injury. B2-Microglobulinemia has been confirmed in patients from endemic regions. Rational management must await the unraveling of this fascinating mystery. An intriguing clue may be the similarities between experimental ochratoxin A nephrotoxicity and endemic Balkan nephropathy.

The experimental features of the nephrotoxicity of citrinin, a mycotoxin in the rat, are detailed by Berndt in Chapter 21. Fungal infections are not required, but frequent environmental contact by domestic animals represents a principal indirect source for human ingestion. Because significantly increased amounts of citrinin metabolites occur with nephrotoxic doses, the question of active metabolite versus parent compound as toxicant remains. Interestingly, citrinin causes glutathione depletion in both liver and kidneys of treated animals. While the covalent binding of citrinin to kidney tissue is compatible with the presence of "relative metabolite," the enhanced cellular uptake of calcium with *in vitro* additions of either citrinin or ochratoxin A provides support for parent compound as toxicant.

In conclusion, although examples of drug-induced acute tubulointerstitial disease are commonplace, chronic tubulointerstitial disease, because of its potential for dire consequences, is a more potent problem to the clinician.

References

Baldwin, D. S., Levine, B. B., McCluskey, R. T., and Gallo, G. R., 1968, Renal failure and interstitial nephritis due to penicillin and methicillin, *N. Engl. J. Med.* **279**:1245.

Border, W. A., Lehman, D. H., Egan, J. D., Sass, H. J., Glode, J. E., and Wilson, C. B., 1974, Antitubular basement-membrane antibiotics in methicillin-associated interstitial nephritis, *N. Engl. J. Med.* **291**:381.

Galpin, J. E., Shinaberger, J. H., Stanley, T. M., Blumenkrantz, M. J., Bayer, A. S., Friedman, G. S., Montgomerie, J. Z., Guze, L. B., Coburn, J. W., and Glassock, R. J., 1978, Acute interstitial nephritis due to methicillin, *Am. J. Med.* **65**:756.

Hostetter, T. H., Olson, J. L., Rennke, H. G., Venkatachalam, M. A., and Brenner, B. M., 1981, Hyperfiltration in remnant nephrons: A potentially adverse response to renal ablation, *Am. J. Physiol.* **10**:F85.

Husby, G., Tung, K. S. K., and Williams, R. C., Jr., 1981, Characterization of renal tissue lymphocytes in patients with interstitial nephritis, *Am. J. Med.* **70**:31.

Kleinknecht, D., Kanfer, A., Morel-Maroger, L., and Mery, J. Ph., 1978, Immunologically mediated drug-induced acute renal failure, *Contrib. Nephrol.* **10**:42.

Linton, A. L., Clark, W. F., Driedger, A. A., Turnbull, D. I., and Lindsay, R. M., 1980, Acute interstitial nephritis due to drugs, *Ann. Intern. Med.* **93**:735.

Miller, T., and Phillips, S., 1981, Pyelonephritis: The relationship between infection renal scarring and antimicrobial therapy, *Kidney Int.* **19**:654.

Ooi, B. S., Jao, W., First, M. R., Mancilla, R., and Pollak, V. E., 1971, Acute interstitial nephritis: A clinical and pathologic study based on renal biopsies, *Am. J. Med.* **59**:614.

Porter, G. A., and Bennett, W. M., 1981a, Toxic nephropathies, in: *The Kidney* (B. M. Brenner, and F. C. Rector, Jr., eds.), Volume II, Philadelphia, Pennsylvania, p. 2045.

Porter, G. A, and Bennett, W. M., 1981b, Nephrotoxic acute renal failure due to common drugs, *Am. J. Physiol.* **10**:F1.

Roberts, J. A., Dominque, G. J., Martin, L. N., and Kim, J. C. S., 1981, Immunology of pyelonephritis in the primate model: Live versus heat-killed bacteria, *Kidney Int.* **19**:297.

Riemenschneider, T., MacKensen-Haen, S., Christ, H., and Bohle, A., 1980, Correlation between endogenous creatinine clearance and relative interstitial volume of the renal cortex in patients with diffuse membranous glomerulonephritis having a normal serum creatinine concentration, *Lab. Invest.* **43**:145.

Van Ypersele de Strihou, C., 1979, Acute oliquric interstitial nephritis, *Kidney Int.* **16**:751.

17

Current Concepts of Tubulointerstitial Nephritis

RAMZI S. COTRAN

1. Introduction

The purpose of this chapter is to summarize current views on the morphologic and clinical definition of interstitial nephritis, since many of the environmental agents, toxins, and drugs discussed in this book cause renal injury which affects primarily renal tubules and interstitium. It is generally agreed that a purely morphologic definition or classification of interstitial nephritis is of little value, since the histologic alterations are relatively nonspecific, and give little indication of etiology and pathogenesis. Rather, the terms interstitial nephritis or tubulointerstitial nephritis (or nephropathy) are used to describe a diverse group of renal disorders, caused by a variety of etiologic agents, in which the predominant morphologic involvement is in the tubules and the interstitium.

Table I lists the diseases that are ordinarily included under the heading of tubulointerstitial nephropathy. As is evident, these include some of the most common diseases of the kidney, such as chronic pyelonephritis and the drug-induced nephropathies. There are some who would restrict the term interstitial nephritis to those diseases that appear to start in the interstitium, and separate them from tubulointerstitial diseases. However, in the vast majority of instances both

---

RAMZI S. COTRAN • Department of Pathology, Brigham and Women's Hospital and Harvard Medical School, Boston, Massachusetts 02115.

TABLE I. Causes of Tubulointerstitial Renal Diseases

Bacterial infection
  Acute pyelonephritis
  Chronic pyelonephritis

Vesicoureteral reflux (reflux nephropathy)

Drugs
  Acute drug-induced hypersensitivity
    nephropathy
  Analgesic nephropathy
  Lithium nephropathy
  Miscellaneous drugs (methyl-CCNU,
    antibiotics, radiographic
    contrast media)

Metabolic disturbances
  Urate nephropathy
  Hypercalcemic nephropathy
  Hypokalemic nephropathy
  Oxalate nephropathy

Heavy metals
  Lead nephropathy
  Cadmium nephropathy

Physical factors
  Radiation nephritis
  Obstructive uropathy

Immune disorders
  Acute drug-induced hypersensitivity
    nephropathy
  Transplant rejection
    Tubulointerstitial injury associated with
      glomerulonephritis
  Sjogren's syndrome

Vascular disorders

Neoplastic diseases
  Leukemic and lymphomatous infiltration
  Myeloma kidney

Hereditary renal diseases
  Alport's syndrome (predominantly
    glomerular disease)
  Medullary cystic disease
  Familial interstitial nephritis

Miscellaneous disorders
  Balkan nephropathy
  Granulomatous diseases (tuberculosis,
    sarcoidosis, leprosy)
  Idiopathic or nonspecific tubulointerstitial
    nephritis
  Acute interstitial nephritis associated with
    infection
  Acute idiopathic interstitial nephritis
  Chronic idiopathic interstitial nephritis

interstitial and tubular injury coexist, and even in experimental models it is difficult to pinpoint stages in which interstitial injury is isolated. In contrast, particularly in the case of chemical and heavy metal poisoning, it appears more likely that tubular damage prededes the interstitial alterations.

One can criticize grouping of these disorders under one heading. In the first place, the list includes almost all diseases of the kidney excluding glomerulonephritis, and it can be questioned whether a term with such broad definition is of any practical value (Gentile *et al.,* 1978). Second, the lack of glomerular involvement in the categorization of these diseases is inaccurate. Interstitial injury can be an important component of some forms of glomerulonephritis, such as lupus nephritis (Andres and McCluskey, 1975), and it is now well recognized that glomerulosclerosis is an important complication of some chronic interstitial diseases, such as chronic pyelonephritis and reflux nephropathy (Kincaid-Smith, 1975; Torres *et al.,* 1980). Third, as stated by Freedman (1978), serious chronic interstitial disease is often multifactorial, and thus patients with analgesic nephropathy may also have infection and obstruction, and as Wedeen describes in Chapter 22, patients with gout or urate nephropathy have evidence of increased lead exposure as the cause of their renal dysfunction (Batuman *et al.,* 1981). The most important practical benefit of categorizing a renal illness as tubulointerstitial, rather than glomerular, is that whereas the etiologic agents in glomerular injury are largely unknown, a thorough investigation of the patient's clinical and occupational history, as well as detailed radiologic and biochemical studies, can often reveal the specific cause or causes of tubulointerstitial damage.

2. Acute Tubulointerstitial Nephritis (TIN)

It is useful, on the basis of morphology and clinical history, to divide interstitial diseases into acute and chronic forms (Heptinstall, 1976). Morphologically the acute interstitial diseases are characterized by interstitial edema and infiltration with a variety of inflammatory cells — monocytes, lymphocytes, plasma cells, neutrophils, and eosinophils — as well as evidence of tubular injury and regeneration. Unless the disease is superimposed on a chronic disorder, there is little interstitial fibrosis and no tubular atrophy.

The two most common causes of acute TIN are acute bacterial infection and acute drug-induced hypersensitivity (Heptinstall, 1976; Cotran, 1981; Linton *et al.,* 1980) (Table II). However, a proportion of patients with the typical clinical presentation and histologic features of acute drug-induced hypersensitivity give no history of having taken any of the drugs usually associated with this condition. It is tempting, when such a case is seen, to implicate a henceforth unrecognized drug to which the patient has been exposed. Thus the list of drugs allegedly causing acute TIN is long, but a cause and effect relationship in many case reports is not convincing, and histologic confirmation is often equivocal. Before the advent of antibiotics, acute interstitial nephritis was well recognized as a complication of certain infections, particularly streptococcal infection, and was evidence that the renal lesions were not due to intrarenal bacterial proliferation (Kannerstein, 1942; Kimmelsteil, 1938; Mallory and Keefer, 1941). A number of recent case reports implicate streptococci, toxoplasma, Legionnaire's agent, and other infections in this condition (Brass *et al.,* 1974; Guignard and Torrado, 1974; Graber *et al.,* 1978); to

TABLE II. Acute Tubulointerstitial Nephritis (TIN)

Acute bacterial pyelonephritis
Acute drug-induced hypersensitivity reaction (e.g., methicillin)
Nonspecific acute TIN (associated with infection)
    Streptococcal disease
    Diphtheria
    Toxoplasmosis
    Brucellosis
    Staphylococcal disease
    (?) Viral diseases
    (?) Legionnaires' disease
Acute idiopathic TIN

this group of cases I apply the descriptive designation "acute nonspecific interstitial nephritis, associated with infection."

Finally, there are some cases of acute TIN without obvious cause (Graber *et al.,* 1978; Chazan *et al.,* 1972). In a recent study (Laberke and Bohle, 1980), one-third of biopsy-proven cases of acute interstitial nephritis had no obvious etiology. Since the clinical and morphologic manifestations of such cases are similar to those of drug-induced reactions, it is possible that other, heretofore unrecognized environmental toxins may act as haptens in the causation of these "idiopathic" cases.

3. Chronic Tubulointerstitial Nephritis

The most important point to make in regard to the morphologic definition of chronic tubulointerstitial disease is the necessity to distinguish and separate *chronic pyelonephritis* from the list of other conditions causing chronic tubulointerstitial injury. Without reviewing the controversy that continues to surround the term "chronic pyelonephritis," most authors now agree that the term should be restricted to the condition in which corticomedullary scarring is associate with pyelocalceal damage (Heptinstall, 1976; Cotran, 1979). Only a limited number of conditions can lead to this morphologic picture (Cotran and Pennington, 1981). These include: (1) reflux nephropathy; (2) bacterial infection superimposed on focal obstruction; (3) analgesic nephropathy, which can usually be differentiated because of the widespread papillary necrosis; (4) segmental hypoplasia (The Ask-Upmark kidney), which is now thought to be also caused by vesicoureteral reflux; and (5) certain rare cases of papillary necrosis caused by sickle cell disease, dehydration of the newborn, and renal tuberculosis. The diagnosis of chronic pyelonephritis therefore rests on the macroscopic appearance of the kidney, seen either by gross examination or more commonly by radiologic studies. It is frequently impossible to differentiate the chronic tubuloinsterstitial diseases by the histologic alterations in the cortex.

I would like to highlight three rather recently recognized morphologic findings in various forms of chronic tubulointerstitial disease.

Analgesic microangiopathy. In recent years it has become apparent that a consistent finding in the kidneys of patients with analgesic abuse nephropathy is the presence of markedly thickened small blood vessels in the renal medulla and also in the submucosa of the pelvis, ureter, and urinary bladder (Abrahams *et al.,* 1978; Mihatsch *et al.,* 1978). Electron microscopy reveals that this thickening is composed of multiple layers of vascular basement membrane, together with fragments of cellular debris within the basement membrane material. The microangiopathy is thought to represent successive phases of endothelial injury and regeneration, caused presumably by toxic analgesic metabolites, and is somewhat similar to the microangiopathy of diabetics. These changes may be obvious in a deep kidney biopsy, allowing a presumptive diagnosis of analgesic nephropathy (A. Bohle, personal communication).

Interstitial deposits of Tamm-Horsfall protein. Several authors have recently described interstitial deposits of PAS-positive, fibrillar or amorphous material in chronic tubulointerstitial diseases, and particularly in reflux nephropathy and urinary obstruction (Zager et al., 1978; Resnick et al., 1978; Solez and Heptinstall, 1977). These deposits are sometimes surrounded by an inflammatory reaction composed of mononuclear cells, occasional plasma cells and neutrophils, and even giant cells. By immunofluorescence techniques they are shown to contain Tamm-Horsfall protein (THP), the uromucoid that is normally localized to the epithelial cells of the ascending thick segment and distal tubules and to urinary casts. Although small amounts of interstitial THP can be found in a variety of renal disorders, by far the largest amounts are present in patients with reflux nephropathy, urinary obstruction, and medullary cystic disease. These deposits form as a consequence of tubular injury and extravasation of formed urine and also due to forniceal rupture in the presence of severe obstruction. While the pathogenetic significance of extravascular Tamm-Horsfall protein deposits has yet to be established, their presence in large amounts in tissue sections suggests vesicoureteral reflux or urinary obstruction.

Focal glomerulosclerosis in chronic tubulointerstitial disease. Although glomerular alterations have long been known to occur in the kidneys of patients with chronic pyelonephritis (Heptinstall, 1974), it has only been recently recognized that some of these glomerular changes may play a role in the progressive renal failure in this condition. Kincaid-Smith (1975, 1979) first noted the association of chronic pyelonephritic scarring with proteinuria and a lesion in the glomeruli which is best described as focal and segmental sclerosis and hyalinosis (Kincaid-Smith, 1975, 1979). When biopsy specimens were obtained from patients with proteinuria exceeding 0.2 g/day, 78% showed focal glomerulosclerosis, which was more widespread in patients with a serum creatinine of over 2.5 mg %. She found that proteinuria proved to be the best guide to prognosis in these patients, even in the absence of persistent infection, vesicoureteral reflux, or hypertension. Subsequently Bhathena et al. (1980) found focal glomerulosclerosis in all 23 patients with end-stage reflux nephropathy undergoing transplantation, and Torres et al. (1980) found in 54 patients with reflux nephropathy that all patients with progressive renal disease had significant proteinuria and prominent focal and segmental glomerulosclerosis. In a retrospective clinicopathologic study of 51 surgical nephrectomies associated with reflux or calculus and noncalculus obstruction, we found focal glomerulosclerosis in 28% (14 cases). There was a significantly higher incidence of proteinuria, bilateral involvement, and increased serum creatinine in patients with focal sclerosis compared to those without focal sclerosis (Cotran, 1981b).

All these clinicopathologic studies show that focal and segmental glomerulosclerosis develops in some patients with chronic pyelonephritis and reflux nephropathy and that the lesion may be one of the causes of progressive deterioration of renal function. The mechanisms by which proteinuria and glomerulosclerosis develop are subjects of current investigation. The presence of IGM and C3 in the sclerotic areas in over half the cases studied have suggested an immunologic reaction, either to bacterial antigen or to autologous tubular antigen, but the most attractive explanation is that the glomerulosclerosis and proteinuria are the result of the hemodynamic changes which occur in glomeruli as an adaptation to reductions of renal mass by the primary tubulointerstitial injury. Such lesions, associated with proteinuria and progressive azoemia, can be produced readily in rats by ablation of about 75% of renal mass, and recent physiologic studies strongly implicate hemodynamic alterations (increase in single-nephron glomerular filtration rate, capillary plasma flow, and mean glomerular transcapillary pressure difference) in the causation of these lesions (Hostetter *et al.*, 1981; Olson *et al.*, 1979). Such hemodynamic changes cause endothelial and epithelial injury and a defect in both size and charge selective properties of the glomerulus, leading to increased permeability to proteins, trapping of macromolecules in the mesangium, and progressive glomerulosclerosis. It may well be that this mechanism of progressive glomerular damage also occurs in other chronic renal diseases, since proteinuria and glomerulosclerosis have been recognized as complications of analgesic nephropathy, renal cortical necrosis, segmental hypoplasia (the Ask-Upmark kidney), and oligomeganephronia [reviewed in Cotran (1981b)].

References

Abrahams, C., Furman, K. I., and Salant, D., 1978, Analgesic abuse and microvascular changes, *Am. Heart J.* **95**:268.

Andres, G. A., and McCluskey, R. T., 1975, Tubular and interstitial renal disease due to immunologic mechanisms, *Kidney Int.* **7**:271.

Batuman, V., Maesaka, J. K., Haddad, B., Tepper, E., Landy, E., and Wedeen, R. P., 1981, The role of lead in gout nephropathy, *N. Engl. J. Med.* **304**:520.

Bhathena, D. B., Weiss, J. H., Holland, N. H., McMorrow, R. G., Curtis, J. J., Lucas, B. A., and Luke, R. G., 1980, Focal and segmental glomerular sclerosis in reflux nephropathy (chronic pyelonephritis), *Am. J. Med.* **68**:886.

Brass, H., Lapp, H., and Hertz, R., 1974, Akute interstitielle Nephritise, *Dtsch. Med. Wochenschr.* **99**:2335.

Chazan, J. A., Geller, A. S., and Esparza, A., 1972, Acute interstitial nephritis: A distinct clinicopathological entity? *Nephron* **9**:10.

Cotran, R. S., 1979, Interstitial nephritis, in: *Renal Disease—Present Status* (J. Churg, B. Spargo, K. Mostofi, and M. Abel, eds.), Williams & Wilkins, Baltimore, Maryland, p. 254.

Cotran, R. S., 1981, Tubulointerstitial diseases, in: *The Kidney* (B. M. Brenner and F. C. Rector Jr., eds.), Saunders, Philadelphia, Pennsylvania, p. 1633.

Cotran, R. S., 1982, Glomerulosclerosis in reflux nephropathy, *Kidney Int.,* **21**:528.

Cotran, R. S., and Pennington, J. E., 1981, Urinary tract infection, pyelonephritis, and reflux nephropathy, in: *The Kidney* (B. M. Brenner and F. C. Rector, Jr., eds.), Saunders, Philadelphia, Pennsylvania, p. 1571.

Freedman, L. R., 1978, Interstitial renal inflammation (interstitial nephritis), in: *Prevention of Kidney and Urinary Tract Diseases* (C. H. Coggins and N. B. Cummings, eds.), DHEW Publications #78-855 U.S. Government Printing Office, Washington, D.C., p. 243.

Gentile, D. E., Rodiles, H. A., and Berman, L. B., 1978, Systemic disorders and interstitial nephritis, in: *Current Nephrology*, Volume 2, (H. C. Gonick, ed.), Houghton Mifflin, Boston, Massachusetts, p. 263.

Graber, M. L., Cogan, M. G., and Connor, D. G., 1978, Idiopathic acute interstitial nephritis, *West. J. Med.* **129**:72.

Guignard, J. P., and Torrado, A., 1974, Interstitial nephritis with toxoplasmosis in a ten-year old child, *J. Pediatr.* **85**:381.

Heptinstall, R. H., 1974, *Pathology of the Kidney,* 2nd ed., Little, Brown and Co., Boston, Massachusetts.

Heptinstall, R. H., 1976, Interstitial nephritis. A brief review, *Am. J. Pathol.* **83:**214.

Hostetter, T. H., Olson, J. L., Rennke, H. G., Venkatachalam, M. A., and Brenner, B. M., 1981, Increased glomerular pressure and flow: A potentially adverse adaptation to reduced renal mass, *Am. J. Physiol.,* in press.

Kannerstein, M., 1942, Histologic kidney changes in the common acute infectious diseases. *Am. J. Med. Sci.* **203:**65.

Kimmelstiel, P., 1938, Acute hematogenous interstitial nephritis, *Am. J. Pathol.* **14:**737.

Kincaid-Smith, P., 1975, Glomerular lesions in atrophic pyelonephritis and reflux nephropathy, *Kidney Int.* **8**(Suppl.):81–83.

Kincaid-Smith, P., 1979, Glomerular lesions in atropic pyelonephritis (RN), in: *Reflux Nephropathy* (C. J. Hodson and P. Kincaid-Smith, eds.), Masson, New York, p. 268.

Laberke, H. G., and Bohle, A., 1980, Acute interstitial nephritis: Relations between clinical and morphological findings, *Clin. Nephrol.* **14:**263.

Linton, A. L., Clark, W. F., Driedger, A. A., Turnbull, I., and Lindsay, R. M., 1980, Acute interstitial nephritis due to drugs: Review of the literature and a report of 9 cases, *Ann. Intern. Med.* **93:**735.

Mallory, G. K., and Keefer, C. S., 1941, Tissue reactions in fatal cases of *Streptococcus haemolyticus* infection, *Arch. Pathol.* **32:**334.

Mihatsch, M. H., Torhorst, J., Amslet, B., and Zollinger, H., 1978, Cappillosclerosis of the lower urinary tract in analgesic (phenacetin) abuse: An electron microscopic study, *Virchows Arch. Pathol. Anat. Histol.* **381:**41.

Olson, J. L., Hostetter, T. H., Rennke, H. G., Brenner, B. M., and Venkatachalam, M. A., 1979, Mechanisms of altered glomerular permselectivity and progressive glomerulosclerosis following ablation of renal mass, *Kidney Int.* **16:**857.

Resnick, J. S., Sisson, S., and Vernier, R. L., 1978, Tamm-Horsfall protein: Abnormal localization in renal disease, *Lab. Invest.* **38:**550.

Solez, K., and Heptinstall, R. H., 1977, Intrarenal urinary extravasation with formation of venous polyps containing Tamm-Horsfall protein, *J. Urol.* **119:**180.

Torres, V. E., Velosa, J. A., Holley, K. E., Kelalis, P. P., Sticker, G. B., and Kurtz, S. B., 1980, The progression of vesicoureteral reflux nephropathy, *Ann. Intern. Med.* **92:**776.

Zager, R. A., Cotran, R. S., and Hoyer, J. R., 1978, Histological localization of Tamm-Horsfall protein in interstitial deposits in renal disease, *Lab. Invest.* **38:**52.

18

Analgesic Nephropathy
Clinical and Epidemiologic Factors

THOMAS G. MURRAY

1. Introduction

The association between the use of analgesics and the subsequent development of renal disease has been studied for more than 25 years. The causal relationship between the two has now been accepted by most students of this problem and the characteristics of the typical patient with analgesic nephropathy are known to most nephrologists and to many other physicians. There remain, however, many unanswered questions concerning this disease. Neither the precise analgesic(s) nor the daily dose or cumulative amount of the analgesic(s) that causes the renal disease is known. The proportion of individuals exposed to any given amount or type of analgesic who eventually develop analgesic nephropathy is also unknown. Of particular concern to physicians in the United States is the fact that the importance of analgesic-induced renal disease in this country is still undetermined. This chapter begins with a review of a number of widely accepted "facts" concerning analgesic nephropathy. This review will be brief, since most readers are already aware of these data and since many recent reviews of this problem exist (Kincaid-Smith 1978; T. Murray and Goldberg 1975b). The second part of this chapter is a more detailed review of available data, which deals with the magnitude of the risk of developing renal disease after the use of analgesics. The final section briefly outlines

THOMAS G. MURRAY (deceased) • Hospital of the University of Pennsylvania, Philadelphia, Pennsylvania 19104. Inquiries may be sent to Dr. P. D. Stolley at this address.

some preliminary data from an ongoing study of the importance of analgesic nephropathy in the United States.

2. Currently Accepted "Facts" Concerning Analgesic Nephropathy

Throughout this chapter, the renal disease associated with the use of analgesics will be referred to as analgesic nephropathy (AN). Other names (e.g., phenacetin nephropathy) will not be used, because they imply more than is currently known about the exact cause of this disease. The word abuse is also not included in the name of the disease, because it is at a minimum pejorative and may in some cases not be correct.

Spuhler and Zollinger (1953) described 44 patients with chronic interstitial nephritis and noted that fourteen of them had taken combination analgesics for many years. Investigators in Switzerland and Sweden pursued this initial suggestion, and by 1960 had provided evidence of a causal relationship between the use of large doses of combination analgesics and the subsequent development of renal disease as well as evidence that analgesic nephropathy was a common disease in their countries. In the early 1960s in Australia attention was drawn to the correlation between the high incidence of papillary necrosis seen at autopsy in many parts of that country and the frequency of regular analgesic use in the same areas. Studies of the pathology, the pathogenesis, and the epidemiology of analgesic nephropathy have continued since that time. These studies have generated the data upon which the currently accepted "facts" concerning this disease rest.

Fact 1: There is a cause and effect relationship between the regular use of large doses of analgesics and the development of renal disease — specifically, of papillary necrosis and secondary chronic interstitial disease. Evidence supporting this causal relationship comes from four different types of studies. There are studies which demonstrate that both chronic renal disease and papillary necrosis develop more often in those who have ingested substantial quantities of analgesics than in those who have not taken analgesics (Grimlund, 1963; Larsen and Møller, 1959). Other studies show that there is a history of previous analgesic use in a higher proportion of patients who have papillary necrosis than there is in those who do not have this disease (Olafsson *et al.,* 1966; Bengtsson, 1962). Although most of the studies which demonstrate these points suffer from one or more methodologic shortcomings, their conclusion that there is a relationship between analgesic use and renal disease appears irrefutable. There are, in addition, studies in groups of individuals with documented analgesic intake in which the incidence of renal disease has been shown to correlate with the amount of analgesics taken (Lindeneg *et al.,* 1959). Finally, studies in animals fed various analgesics support an association between analgesic use and the development of papillary necrosis (Kincaid-Smith, 1978). Although the animal data have added uncertainty to the issue of which analgesic(s) is responsible for the papillary necrosis, it is clear that papillary necrosis can be induced in animals by some analgesics.

Fact 2: The second commonly held opinion is that despite the lack of definite proof, phenacetin is the major cause of analgesic nephropathy (Kincaid-Smith, 1978; T. Murray and Goldberg, 1975b). This opinion is certainly less well supported than it was previously and may in fact be incorrect. The original basis for the belief that phenacetin was the principal cause of analgesic nephropathy were the findings that phenacetin had been taken by all of the early cases of analgesic nephropathy, that it was the only analgesic common to all of these cases, and that the other drugs taken by these cases were, at that time, thought to be free of renal toxicity. There are currently a number of reasons to question the implication that phenacetin is the principal cause of analgesic nephropathy. Animal studies cast doubt on the role of phenacetin because they consistently show that it, as well as its principal metabolite (acetaminophen), cause papillary necrosis in only a small proportion of animals and then only if they are given very large doses. Recently, cases of what is otherwise typical analgesic nephropathy have been reported in individuals who took analgesics which contained no phenacetin (or acetaminophen). Finally, each of the analgesics taken in combination with phenacetin in the early cases have now been shown to be capable of causing papillary necrosis in at least some settings.

Aspirin is the analgesic taken with phenacetin in most of the recently reported cases of analgesic nephropathy. Originally, aspirin was thought not to be involved in the pathogenesis of analgesic nephropathy, because it had not been taken by all of the cases and because the literature suggested that its use alone, even in large amounts, was not associated with an increased incidence of renal disease or papillary necrosis (mostly in patients with rheumatoid arthritis). Currently, there is an accumulating body of evidence which suggests that aspirin may be involved in the pathogenesis of analgesic nephropathy (Kincaid-Smith, 1978, 1980). The initial suggestion of aspirin's possible involvement came from the results of animal studies. In animals it was demonstrated consistently that the ingestion of aspirin caused papillary necrosis much more commonly than did the ingestion of phenacetin or acetaminophen. A number of recent studies in patients with rheumatoid arthritis who have consumed only aspirin suggest that papillary necrosis is not uncommon in these patients (although other studies continue to find no evidence of a correlation). Finally, cases of what appears to be typical analgesic nephropathy have been reported in patients who had taken no analgesics other than aspirin.

The possibility that acetaminophen would prove to be a cause of renal disease has been raised repeatedly in the literature (Koutsaimanis and de Wardener, 1970). This fear is based on the fact that acetaminophen is structurally identical to the principal metabolite of phenacetin — it is in fact mainly acetaminophen rather than phenacetin that is concentrated in the kidney after the ingestion of phenacetin. Of course, if phenacetin proves not to be the cause of analgesic nephropathy, this worry will be groundless. In any case, there is presently no hint of an association between the use of acetaminophen alone and the development of analgesic nephropathy.

A number of studies demonstrating the potential of various nonsteroidal antiinflammatory drugs for causing papillary necrosis in animals (Arnold et al., 1976) have been reported and a few cases of papillary necrosis in individuals consuming these agents have appeared (Hussert *et al.*, 1979; Lourie *et al.*, 1977). There is as yet, however, no proof of an important association between the use of nonsteroidal antiinflammatory drugs and the development of chronic renal disease.

At the present time, there appears to be no way to resolve the conflicing data concerning the analgesic(s) responsible for analgesic nephropathy. Clearly, the majority of cases have ingested combinations of analgesics, most often ones that contain at least phenacetin and aspirin. This may reflect the renal toxicity of phenacetin and the fact that it is not generally available alone but only in combination with other analgesics. In that case, the studies in animals must be assumed not to apply to humans. On the other hand, aspirin may be the cause of analgesic nephropathy, but may be taken in sufficient quantities only by individuals who take combination analgesics. In that case, phenacetin may contribute to the high levels of intake because of its central nervous system effects. Finally, the combination of analgesics such as phenacetin and aspirin may be more toxic than any single analgesic alone, so that analgesic nephropathy develops much more readily when two or more analgesics are consumed concurrently.

Fact 3: There is a set of clinical factors which are shared by the majority of cases of analgesic nephropathy. This appears to be a correct opinion. The presence of one or more of these clinical factors in a patient with renal disease should, therefore, suggest the possibility that analgesic use is responsible for the patient's renal disease (T. Murray and Goldberg, 1975a; Gault *et al.*, 1968). On the other hand, many patients with analgesic nephropathy do not present with a constellation of features characteristic of analgesic nephropathy. Thus, the absence of these features should not be used to rule out analgesic nephropathy.

Many of the common clinical features of analgesic nephropathy are a reflection of the fact that it is a form of chronic interstitial nephritis. Ninety percent of the cases have abnormal intravenous pyelograms — markedly shrunken kidneys, nonspecific calyceal abnormalitis, and classic pyelonephritic scars have been reported. More than half of the cases have radiologic evidence of papillary necrosis; however, a substantial number do not. Some of the patients in whom papillary necrosis is not demonstrated have advanced azotemia and consequently a poor-quality radiologic study. In those individuals it is possible that papillary necrosis is present but not detected. In some patients, a technically adequate intravenous pyelogram does not show papillary necrosis. (It is, of course, possible to have papillary necrosis histologically but yet not demonstrate it radiologically). Many patients have sterile pyuria ($>$60%). Proteinuria is present in most cases but quantitatively seldom amounts to more than 2 g/day. A few cases of analgesic nephropathy have been reported to develop heavy proteinuria (nephrotic range) as their azotemia advances (Kincaid-Smith, 1980). Despite a clinical history suggestive of urinary tract infections in many patients, infection has been documented in less than 30% of the reported cases. Hypertension is present in more than 50% of patients. Occasionally, the hypertension enters an accelerated phase as the renal failure advances (Kincaid-Smith, 1980). Hyperchloremic metabolic acidosis (supposedly secondary to distal renal tubular acidosis) and hyperkalemia secondary to hyporeninemic hypoaldosteronism have both been reported in cases of analgesic nephropathy (Steele *et al.,* 1969), but it is not known whether they occur more commonly in this disease than in other forms of chronic interstitial nephritis.

There is increasing evidence of an association between analgesic nephropathy (and perhaps even heavy analgesic use without overt nephropathy) and renal pelvic carcinoma (Kincaid-Smith, 1978). The incidence of these rare urinary tract tumors is reported to be more than ten times as great in analgesic users as in nonusers. It is not known whether the carcinomas develop in response to a toxic effect of one of the analgesics (or its metabolites) or whether they result from the chonic inflammation of the papillae and calyces.

Other features commonly seen in patients with analgesic nephropathy are the result of damage by the analgesics to organs other than the kidneys (T. Murray and Goldberg, 1975b). Upper gastrointestinal tract disease is present in more than 40% of patients. Anemia is present in more than 60% of patients with analgesic nephropathy. It is apparently the result of the combined effects of azotemia, gastrointestinal blood loss, and, in some cases, a direct effect of the analgesics on the red blood cells. Preliminary data suggest that there is an increased risk of artherosclerotic cardiovascular disease and of premature aging in patients with analgesic nephropathy, but these suggestions are not yet well substantiated (Kincaid-Smith, 1980; Krishaswamy *et al.,* 1974).

The remainder of the common clinical factors of analgesic nephropathy are actually characteristics of individuals who are heavy users of analgesics whether they have renal disease or not (Kincaid-Smith, 1978; T. Murray and Goldberg, 1975b; R. Murray *et al.,* 1971). The vast majority of regular analgesic users are women. The only studies in which this has not been the case are those done in factories where only males are employed or in Veterans Administration Hospitals. The majority of cases of analgesic nephropathy occur in individuals more than 45 years of age and the onset of this disease before the age of 30 is distinctly uncommon. The habit of regular analgesic use, however, often begins much earlier — occasionally before the age of 20.

The stated reason for the consumption of analgesics is most commonly headaches (R. Murray, 1973). The headaches seldom have an organic basis and do not have the characteristics of migraines. It has been demonstrated that the headaches often abate if the intake of analgesic is curtailed. This suggests that in many cases the headaches are the result rather than the cause of the analgesic intake. Many of the individuals who initially list headaches as the reason for using analgesics and the majority of the remaining patients with analgesic nephropathy actually use analgesics for reasons other than the relief of pain. Analgesics are used by these patients because of the view that they increase productivity at work, or as a relief for tedium, as a general stimulant, or simply as a "habit." Only a small proportion of patients with analgesic nephropathy have taken analgesics for the relief of objectively demonstratable painful conditions.

Patients who use large doses of analgesics often deny or grossly underestimate the use of analgesics (T. Murray and Goldberg, 1975b; R. Murray, 1973). The reason they do this is not known, but it obviously can interfere with attempts to make the correct diagnosis. These patients also commonly continue their use of analgesics, or restart their use after initially stopping, despite a thorough explanation of the consequences of their continued use (Furman *et al.,* 1976)

Fact 4: Analgesic nephropathy is an important medical problem worldwide. Analgesic nephropathy is said to be responsible for a large proportion of all chronic renal failure and of all end-stage renal disease in many countries (Kincaid-Smith, 1978). In reports from Australia, analgesic nephropathy is said to account for more than 25% of all end-stage renal disease. One study from South Africa suggests that 20% of patients on dialysis have analgesic nephropathy. Comparable figures reported from other countries include: Switzerland 15%, the United Kingdom 10%, Canada 6%, and Western Europe 2–3%. Some of these studies originated from parts of the country where analgesic use was particularly common and many do not include sufficient information concerning the frequency of analgesic use in individuals without renal disease from the same area. Despite these shortcomings, they all suggest that from the nephrologist's viewpoint this disease is important.

The magnitude of the problem of analgesic nephropathy in the United States is still uncertain. The first case of analgesic nephropathy in the United States was reported in 1960, but by 1970 only 100 cases had been described. In 1975, a Pennsylvania study suggested that analgesic use might be responsible for 20% of all cases of chronic interstitial nephritis (T. Murray and Goldberg, 1975a). A survey of dialysis patients from California published only in abstract form in 1977 suggested that between 5 and 10% of these patients might have analgesic nephropathy. Recently, a study from North Carolina (Gonwa *et al.,* 1979) suggested that 13% of patients with chronic renal failure and 9% of those with end-stage renal disease had analgesic nephropathy. None of these studies include data from a control group.

3. Strength of the Relationship between Analgesic Use and Renal Disease: Relative and Absolute Risk

Although it appears to be well established that there is a cause-and-effect relationship between the use of analgesics and the development of papillary necrosis and chronic interstitial nephritis, the strength of this relationship is not established. To determine the importance of the problem of analgesic-induced renal disease, it is necessary to know what the risk of developing renal disease is in an individual who takes analgesics relative to the risk in an individual who does not take analgesics (relative risk) and the total number of individuals who develop renal disease as a result of the use of analgesics (absolute risk). There are only a few studies which include sufficient data to shed light on this question.

4. Frequency of Regular Analgesic Use

There are a number of studies which demonstrate that the regular use of large doses of analgesics is fairly common. Thirty percent of consecutive admissions to a

Denmark hospital gave a history of daily use of combination analgesics containing phenacetin prior to admission (Larsen and Møller, 1959). In Sweden, 13% of male factory workers admitted to taking ten or more phenacetin-containing powders daily for more than 1 year (Grimlund, 1963). In a Swiss study, 17% of female factory workers took two or more phenacetin-containing analgesics per week for greater than 1 year (Duback *et al.,* 1968). In a number of Australian community surveys, the frequency of daily analgesic use of more than 1 year duration varied from 14 to 20% (Gillies *et al.,* 1972). In the Australian studies, more than 60% of the analgesic users took phenacetin-containing analgesics and the frequency of regular use increased with age, reaching more than 35% in subjects over 50 years of age. In a Canadian study (Gault *et al.,* 1968a), 37% of males at a Veterans Administration Hospital admitted to the use of three or more analgesics per week for greater than 1 year prior to admission. Seven percent of this group took more than five tablets daily; more than 50% of these took phenacetin-containing analgesics. Only 1.9% of Welsh women in a community survey (Waters *et al.,* 1973) took four or more phenacetin-containing analgesics for more than 1 year.

In the United States, the only direct investigation of the frequency of regular analgesic use was carried out by the Boston Collaborative Drug Surveillance Program (Lawson, 1973). In that study, 7% of hospitalized patients gave a history of daily analgesic use prior to admission. In 3% of the patients the daily use had extended over 1 year or more. However, only one-third of these patients took phenacetin-containing analgesics.

The frequency of regular analgesic use is, perhaps not surprisingly, not uniform throughout the population. The incidence of regular analgesic use is higher in women, in older individuals, and in those from lower socioeconomic classes. There is also geographic variation. In the town in Sweden where analgesic nephropathy was described in the late 1950s, the per capita consumption of analgesics was ten times that in other similar sized cities of Sweden. In the United Kingdom, the frequency of analgesic use appears to be higher in Glasgow than in other cities. There is also thought to be a higher frequency of regular analgesic use in the Southeastern part of the United States.

5. Incidence of Renal Disease in Subjects Who Take Analgesics

One method of studying the strength of the relationship between analgesic use and renal disease is to determine the incidence of renal disease (measured either as decreased renal function, azotemia, radiologic papillary necrosis, or pathologic papillary necrosis) in individuals who are regular analgesic users (cases) and compare it to the incidence of renal disease in control subjects who are matched with the cases for all important variables except analgesic intake. There are four controlled studies in the literature which demonstrate the strength of this cause-and-effect relationship. In a study of workers at a Swedish factory (Grimlund, 1963) where analgesic use was very prevalent, the incidence of renal failure (defined as a creatinine of 1.5 mg % or greater) was found to be greater in analgesic users than in nonusers (Grimlund, 1963). Those who took no analgesics had an incidence of renal failure of 2.4%, while those who took between 1 and 5 kg of phenacetin had an incidence of 19%. The incidence was 50% in those taking 5–10 kg and 80% in those taking more than 10 kg. In a retrospective review of all intravenous pyelograms done at an Australian hospital (Sorensen, 1966), the patients were divided into those who had taken more than 2 kg of phenacetin and those who did not take phenacetin regularly. The incidence of papillary necrosis was 11% in analgesic users and only 2.3% in nonusers. In a Danish study of consecutive hospital admissions, the incidence of renal functional abnormalities (broadly defined) was compared in those who had taken more than 1 kg of phenacetin and those who had not used phenacetin (Larsen and Møller, 1959). Thirty-three percent of the analgesic users but only 7% of the nonusers had evidence of abnormal renal function. In Switzerland, 417 analgesic users who took phenacetin-containing tablets daily and 464 controls were followed for 5 years (Dubach et al., 1975). After 5 years, 5.4% of those taking five or more tablets per day but only 0.5% of those not taking phenacetin had developed a consistently elevated serum creatinine.

There are three controlled studies in the literature which report no difference in the incidence of renal failure in regular analgesic users compared to controls. A study in Australia (Christie et al., 1976) compared the incidence of kidney disease in women who took more than ten analgesic tablets per week with matched controls and found no difference. There were 50 analgesic users and less than half took combination analgesics. In a community study of Welsh women (Waters et al., 1973), the 56 women who had taken four or more phenacetin-containing tablets daily for greater than 1 year had no more renal disease than the controls. Finally, in the study by the Boston Collaborative Drug Surveillance Program (Lawson, 1973), no difference in the incidence of renal diseases between daily users and nonusers was demonstrated.

6. Incidence of Analgesic Use in Patients Who Have Renal Disease

Another method of investigating the strength of the relationship between analgesic use and renal disease is to determine the frequency of past analgesic use in cases with renal disease or a particular form of renal disease (e.g., chronic interstitial nephritis or papillary necrosis) and to compare it to the frequency of use in a matched-control population, i.e., those without renal disease or without that form of renal disease present in the cases. There are only two well-controlled studies of this type in the literature (Olafsson et al., 1966; Bengtsson, 1962). In both of these, the proportion of patients with papillary necrosis who used analgesics regularly is compared to the proportion of patients with chronic nonobstructive interstitial nephritis, but no papillary necrosis, who used analgesics. In one study (Bengtsson, 1962), 50% of patients with papillary necrosis had taken 1 g of phenacetin a day for greater than 10 years, while in those with chronic interstitial disease, but no papillary necrosis, the figure was 17%. In the second study (Olafsson et al., 1966), 85% of the subjects with papillary necrosis had ingested more than 1 kg of phenacetin, while only 30% of those with chronic interstitial nephritis, but no papillary necrosis, had done so.

7. Correlation of Dose of Analgesic with Frequency and Severity of Renal Disease

There are three studies which show a correlation between increasing cumulative intakes of analgesics and the incidence or severity of renal disease (Grimlund, 1963; Larsen and Møller, 1959; Lindeneg *et al.,* 1959). Studies concerning the relationship between decreases in the availability of analgesics (usually brought about by changes in public policy) and changes in the incidence of analgesic nephropathy cannot be interpreted because they did not adequately control for other possibly important variables.

The studies summarized above suggest that the intake of large amounts of analgesics causes renal disease. The risk of developing renal disease after the use of any given amount of analgesic, however, cannot be stated with certainty. If very large amounts are taken, more than half of the subjects may eventually develop renal failure. The risk of renal failure may be 3–10 times higher in those who consume more than 2 or 3 kg of phenacetin or aspirin in combination analgesics. On the other hand, it is clear that some individuals who have consumed enormous quantities of analgesics have not developed renal disease. It is not known how these subjects differ from those who develop renal disease.

8. The Importance of Analgesic Nephropathy in the United States

It is presently impossible to determine how important a problem analgesic nephropathy is in the United States. There is not enough information available to determine how common regular analgesic use is in the general population, and the strength of the relationship between such use and the development of renal failure is not precisely known. The existing studies which address the importance of analgesic nephropathy in the United States are not well controlled, and can, as a consequence, only be used to determine that the use of analgesics may be responsible for a substantial fraction of the chronic renal disease seen in this country.

I am currently involved (in collaboration with Dr. Paul Stolley) in a study which is investigating the epidemiology of analgesic nephropathy in the United States. This study is of the case–control type and is designed to determine the risk of developing end-stage renal disease after the ingestion of various types and amounts of analgesics relative to the risk of developing end-stage renal disease if no analgesics have been used (relative risk). Five hundred and twenty seven dialysis patients served as cases and 1047 matched hospitalized patients served as controls in this study. The past exposure to analgesics in the cases and controls was determined by a structured interview. Data concerning the type of renal disease present in the cases and the presence of etiologic factors other than analgesic use which may have been responsible for the renal disease was obtained by abstracting the medical records of the cases. Analgesic use was considered to have occurred over a period of time if during that period doses of analgesics were taken at least every other day.

The results of the study are not yet fully analyzed, but a number of points can for more than 1 year by 6.0% of cases and 6.2% of controls; but the use for more than 3 years was reported by 4.7% of cases and 3.0% of controls. Cumulative analgesic intake exceeded 1 kg in 3.2% of cases and 3.4% of controls, and exceeded 3 kg in 1.7% and 1.9%, respectively. There was no difference in the proportion of cases and controls in regard to the intake of single analgesic agents, e.g., aspirin, acetaminophen.

The percentage of cases who took large doses of phenacetin and aspirin (e.g., greater than 3 kg) or took these agents for prolonged periods (e.g., greater than 3 years) was greater than the percentage of controls. The total numbers of cases who took large amounts was, however, quite small. When the data from this study are fully evaluated, it appears that the relative risk of developing end-stage renal disease after the use of large doses of combined analgesics will be shown not to be high and possibly much smaller than in many other countries. This conclusion will not eliminate the possibility that in some areas of the United States the absolute risk may be much higher. Of course, since analgesic nephropathy is a preventable disease, efforts to eradicate it will be warranted even if it is responsible for only a small proportion of chronic renal failure.

9. Conclusion

The ingestion of large doses of some analgesics is a cause of renal disease — specifically of papillary necrosis and chronic interstitial nephritis. There are, however, still many unanswered questions concerning this relationship. Some of these include:

1. What analgesic(s) is responsible for the renal disease? Is it phenacetin, and if so, is acetaminophen also a concern in this regard? Is it aspirin, and if so, will other nonsteroidal antiinflammatory drugs be demonstrated to be associated also? Does renal disease only occur (with regularity) if combination analgesics are taken, and if so, is it because the combinations lend themselves to large intakes or because of combined toxicity?

2. What amount of intake puts an individual at significant risk for development of renal disease? Is there a dose level below which there is no risk, or is the risk present at all levels of intake but greater, the greater the intake? Does the presence of other causes of renal disease increase the risk? Does preexisting renal failure increase the risk?

3. What proportion of the population takes the responsible analgesic(s) in sufficient amounts to be at risk? Do those who do so represent "abusers" of the drugs in question or only one end of the spectrum of those who use drugs for valid reasons?

4. What distinguishes those who after the ingestion of large amounts of the responsible analgesics get renal disease from those who do not? Does dehydration play a role in this difference?

5. What is responsible for the geographic difference in the incidence of

analgesic nephropathy? Is it the number of individuals who take analgesics, the type of analgesics used, or factors unrelated to the analgesics?

6. How should the answers to these questions influence public policy issues regarding analgesics? How should these issues be handled until these answers are available?

The answers to these questions await further research concerning this disease, which has been studied for more than 25 years.

References

Arnold, L., Collins, C., and Starmer, G., 1976, Further studies on the acute effects of phenylbutazone, oxyphenbutazone and indomethacin in the rat kidney, *Pathology* **8**:135.

Bengtsson V., 1962, A comparative study of chronic non-obstructive pyelonephritis and renal papillary necrosis, *Acta Med. Scand.* **165**:321.

Christie, D., McPherson, L., and Kincaid-Smith, P., 1976, Analgesics and the kidney: Community based study, *Med. J. Aust.* **2**:527.

Dubach, U., Levy, P., and Minder, F., 1968, Epidemiological study of analgesic intake and its relationship to urinary tract disorders in Switzerland, *Helv. Med. Acta* **34**:297.

Dubach, U., Levy, P., Posner, B., Baumela, H., Muller, A., Peier, A., and Ehrensperger, T., 1975, Relation between regular intake of phenacetin-containing analgesics and laboratory evidence of urorenal disorders in a working female population of Switzerland, *Lancet* **1**:539.

Furman, K., Galaska, G., and Myers, A., 1976, Post-transplantation analgesic dependence in patients who formerly suffered from analgesic nephropathy, *Clin. Nephrol.* **5**:54.

Gault, M., Rudwual, T., and Redman, N., 1968a, Analgesic habits of 500 veterans: Incidence and complications of abuse, *Can. Med. Assoc. J.* **98**:619.

Gault, M., Rudwual, T., Englis, W., and Dossetor, B., 1968b, Syndrome associated with the abuse of analgesics, *Ann. Intern. Med.* **68**:906.

Gillies, M., Skyring, A., and Livingstone, E., 1972, The patterns and prevalance of aspirin ingestion as determined by interview of 2,921 inhabitants of Sydney, *Med. J. Aust.* **1**:974.

Gonwa, T., Hamilton, R., and Bucklew, T., 1979, Importance of analgesic nephropathy (AAN) in the etiology of end-stage renal disease in Northwest North Carolina, *Kidney Int.* **16**:930.

Grimlund, K., 1963, Renal papillary necrosis at a Swedish factory, *Acta Med. Scand.* **174**(S405):1.

Hussert, F., Lange, R., and Kantrow, C., 1979, Renal papillary necrosis and pyelonephritis accompanying fenoprofen therapy, *J. Am. Med. Assoc.* **242**:1896.

Kincaid-Smith, P., 1978, Analgesic nephropathy, *Kidney Int.* **13**:1.

Kincaid-Smith, P., 1980, Analgesic abuse and the kidney, *Kidney Int.* **15**:250.

Koutsaimanis, K., and deWardener, H., 1970, Phenacetin nephropathy with particular reference to the effect of surgery, *Br. Med. J.* **4**:131.

Krishaswamy, S., Wallace, D., and Nanra, R., 1974, Ischemic heart disease in analgesic nephropathy, *Austral. NZ Med. J.* **4**:426.

Larsen, K., and Møller, C., 1959, A renal lesion caused by abuse of phenacetin, *Acta Med. Scand.* **164**:53.

Lawson, D., 1973, Analgesic consumption and impaired renal function. *J. Chronic Dis.* **26**:39.

Lindeneg, O., Fischer, S., Peterson, J., and Nissen, N., 1959, Necrosis of the renal papillae and prolonged abuse of phenacetin, *Acta Med. Scand.* **165**:321.

Lourie, S., Denman, S., and Schroeder, E., 1977, Association of renal papillary necrosis and ankylosing spondylitis, *Arth. Rheum.* **20**:917.

Murray, R., 1973, Dependence on analgesics in analgesic nephropathy, *Br. J. Addict.* **68**:265.

Murray, R., Lawson, D., and Linton, A., 1971, Analgesic nephropathy: The clinical syndrome and prognosis, *Br. Med. J.* **1**:479.

Murray, T., and Goldberg, M., 1975a, Chronic interstitial nephritis: Etiologic factors, *Ann. Intern. Med.* **82**:453.

Murray, T., and Goldberg, M., 1975b, Analgesic abuse and renal disease, *Annu. Rev. Med.* **26**:537.

Olafsson, O., Gudmundsson, K., and Brekkan, A., 1966, Migraine, gastritis and renal papillary necrosis; *Acta Med. Scand.* **179**:121.

Sorensen, A., 1966, Is the relationship between analgesics and renal disease coincidental and not causal, *Nephron* **3**:3366.

Spuhler, O., and Zollinger, H. U., 1953, Die chronischinterstitielle nephritis, *Z. Klin. Med.* **151**:1.

Steele, T., Gyory, A., and Edwards, K., 1969, Renal function in analgesic nephropathy, *Br. Med. J.* **2**:213.

Walters, W., Edward P., and Asschar, P., 1973, Community survey of analgesic consumption and kidney function in women, *Lancet,* **1**:341.

19

Analgesic Nephropathy
Renal Drug Distribution and Metabolism

GILBERT H. MUDGE

1. Introduction

The pharmacologic factors which underlie analgesic nephropathy may best be considered in relation to several features which have been adequately documented by clinical experience: first, the disease results from the ingestion of analgesics over a long period of time and in fairly large daily doses; second, unlike many other drug-induced nephropathies, the primary lesion occurs in the medulla with secondary involvement of the cortex; and third, the disorder may be acutely aggravated by dehydration and oliguria (Duggin, 1980). Taken together, these three features strongly suggest that the disorder results from a high concentration of the offending agent or agents in the urine or, more properly, in the tubular fluid of the distal nephron. Parenthetically, for most of this discussion medulla and papilla will be used somewhat interchangeably.

There are several other features that warrant mention. Some authors have suggested that the lesion results from true sensitivity to the ingested drugs, but the evidence is not convincing and this will not be considered further (Dubach, 1978). The nature and number of the offending drugs is important. Certainly from the clinical experience phenacetin has been incriminated. However, this is promptly metabolized in a first pass through the liver to its deethylated derivative. This is

GILBERT H. MUDGE • Departments of Medicine, Pharmacology, and Toxicology, Dartmouth Medical School, Hanover, New Hampshire 03755.

acetaminophen in the United States and paracetamol elsewhere. Acetaminophen, or APAP (acetyl-*p*-aminophenol), is an over-the-counter preparation for which annual consumption is on the increase. In recent years it has been extensively studied for its toxicity to both kidney and liver. Hepatotoxicity occurs with large acute doses often taken with suicidal intent. Sometimes this may be accompanied by acute renal failure, but this is not the type of nephrotoxicity under present consideration. There is also the important question of the number of different drugs which may be involved with chronic usage. There has also been a major problem in producing the disease experimentally in laboratory animals. Well-known analgesics have been administered either alone or in combination, over variable periods of time, to a variety of species, over an extreme range of dosage, and with variable dietary and fluid intake (Rosner, 1976). One is impressed that the lesion is difficult to produce, that the results are somewhat inconsistent, and that the doses are so large that in some instances the relevance of the findings has been challenged. An excellent summary of the morphologic findings has recently been published (Burry *et al.,* 1977).

2. General Aspects of Drug Metabolism

At the slight risk of oversimplification, drug metabolism may be divided into two parts. First, there are the metabolites that are more polar and less lipophilic than the parent drug. The renal tubule has a low permeability to these metabolites. They are poorly reabsorbed and hence excreted at high rates. Low lipid solubility is responsible for high excretory rates, their status as end products, and probably also for their low order of pharmacologic activity. The second type of drug metabolism involves the generation of compounds which have a high degree of chemical reactivity. In a few instances this reactivity is responsible for the primary pharmacologic action of the drug. In addition, there is an increasing number of examples in which these metabolites appear responsible for drug-induced cytotoxicity. The degree of lipid solubility is not consistent and is of less importance than chemical reactivity. As a general rule, and almost as a truism, metabolites of this type are generated in small amounts in comparison to the inactive polar end products.

3. Mechanisms of Drug Excretion by the Kidney

Drugs and their metabolites appear in the voided urine as the net result of three separate processes: (1) filtration, (2) active secretion or reabsorption, and (3) passive reabsorption. In the case of the weak analgesics there is only a modest degree of binding to plasma protein, so that the amounts filtered are largely determined by the plasma concentration and the filtration rate itself. Many of the drugs exist principally as anions at body pH and are subject to the anionic transport mechanism of the proximal tubule. This occurs dominantly in a secretory direction, although reabsorption may be unmasked under special circumstances. From a quantitative point of view, the most important mechanism regulating urinary excretion is that of passive reabsorption.

There is perhaps no example in pharmacology in which the above principles of drug metabolism and renal excretion are so clearly interrelated as in the case of phenacetin. The drug itself is highly lipid-soluble. On absorption from the gastrointestinal tract it is almost completely converted in the liver to APAP. This involves deethylation and the conversion of $-OC_2H_5$ to the more polar phenolic hydroxyl (see Fig. 1). The phenolic group of APAP then undergoes conjugation to either the glucuronide or sulfate ($\emptyset$–O–R in Fig. 1). Both of these are highly polar and lipid-insoluble. Some of the APAP escapes conjugation and is excreted unchanged in the urine. Thus, following the ingestion of phenacetin, three types of compounds are excreted in the urine, albeit in far different amounts: (1) phenacetin itself, which is highly lipid-soluble, (2) APAP, which is moderately lipid-soluble, and (3) the phenolic conjugates of APAP, which are virtually lipid-insoluble. As the tubular fluid traverses down the nephron its volume is diminished by the

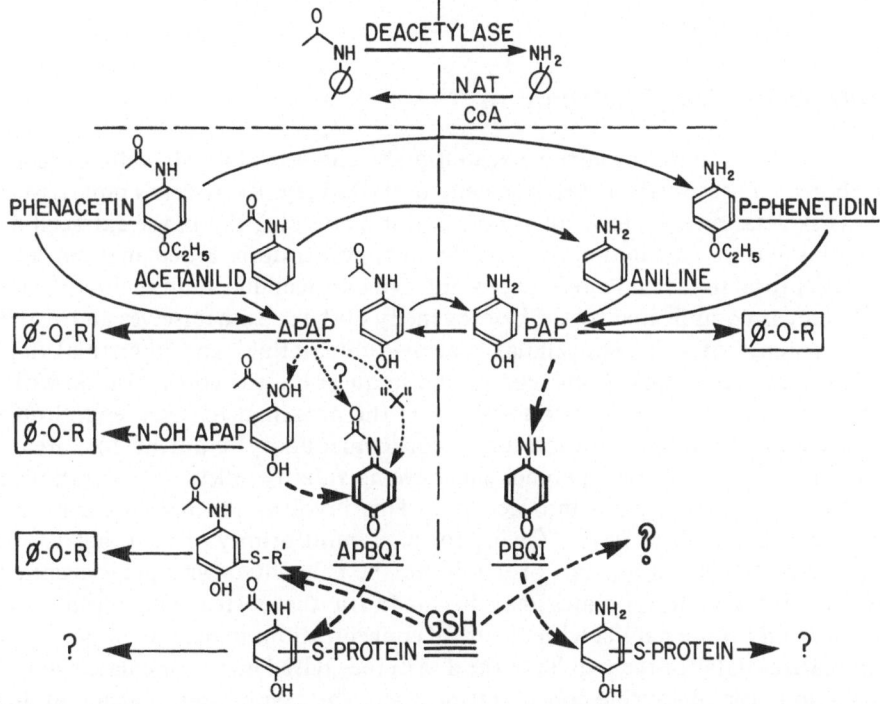

FIGURE 1. Summary of metabolism of phenacetin, APAP, and related compounds. From top to bottom are depicted parent drugs, reactive metabolites, and reactions with tissue GSH and proteins. Vertical dashed line divides compounds (to left) in which amino group is acetylated and those (to right) in which amino group is free. $\emptyset$–O–R indicates phenolic conjugates (glucuronides and sulfates) of parent compound and represents major excretory products. At top, NAT is N-acetyltransferase. Solid arrows indicate enzyme-catalyzed reaction; heavy dashed arrows indicate reaction is nonenzymatic. For discussion see text.

reabsorption of water. This tends to increase the concentration of drugs within the lumen. Phenacetin is so lipid-soluble that it readily backdiffuses across the tubule and thus does not become concentrated within the urine. The urine-to-plasma concentration ratio for phenacetin is unity under all conditions (Duggin and Mudge, 1976b). APAP, which is moderately lipid soluble, backdiffuses to only a limited degree and thus becomes concentrated in the tubular fluid and eventually in the voided urine. However, the concentrations achieved by APAP are less than those of its phenolic conjugates, to which the tubule is impermeable (Duggin and Mudge, 1975).

There are no data available on the renal clearance of the trace or minor metabolites of phenacetin or APAP. Such experiments would require accurate measurements of concentrations in plasma and binding to plasma proteins. Nevertheless, it is reasonable to asume that all glucuronides and sulfates behave in essentially the same manner and that their clearance is high. This may also apply to the 3-mercapturate. In their unconjugated forms, many of the trace metabolites are soluble in organic solvents to about the same degree as APAP. It is probable that renal tubular permeability is likewise similar.

4. Intrarenal Drug Distribution

Direct measurement of drug distribution by analysis of whole-kidney tissue is complicated by the intrinsic heterogeneity of the kidney. Relatively simple tissues, such as skeletal muscle, have an almost uniform composition in the extracellular fluid. This permits calculation of intracellular concentrations based on the primary analysis of total tissue and a reference sample of extracellular fluid such as plasma. The kidney is complicated by the heterogeneity of the cells themselves and also by the coexistence of two extracellular spaces—tubular fluid and interstitial fluid. Furthermore, depending on the compound in question, the concentration within the tubular fluid may vary enormously from the proximal to distal end. Studies which claim the demonstration of a corticomedullary gradient for organic compounds within the renal parenchyma have no validity unless concentrations in the tubular fluid are taken into account. This involves the measurement of a reference solute such as inulin. Owing to the nonuniformity of the tubular fluid, intracellular concentration can probably not be calculated with great accuracy. However, reasonable estimates may be made for the cortex with plasma as a reference and for the papilla with the use of concentrations in bladder urine. Figure 2 summarizes data for the papilla of the dog kidney based on the calculated volume distributions (papilla/urine concentrations) and the assumption that inulin does not penetrate cells. The results are in agreement with the mechanisms of excretion and reabsorption previously described. For compounds that have tubular permeabilities similar to APAP, it should be emphasized that concentrations within the tubular fluid as well as within the cells of the papilla both rise during oliguria and fall during diuresis. The rate of urine flow has no effect on the intracellular concentrations of those compounds to which the tubule is

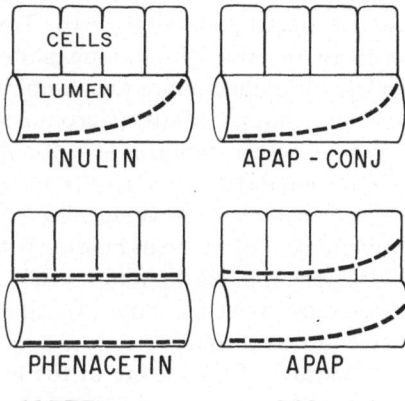

CONCENTRATIONS IN COLLECTING DUCT

FIGURE 2. Schematic summary of drug concentrations in distal nephron. Inulin and APAP conjugates achieve high concentrations but are confined to tubular fluid; phenacetin is at a low concentration in both lumen and cells; APAP becomes concentrated in both tubular fluid and cells.

impermeable (inulin and APAP—conjugates) or to which it is completely permeable (phenacetin) (Duggin and Mudge, 1976a,b).

It should be pointed out that the factors that control the intrarenal distribution of the weak analgesics are different from those that establish a high cortical concentration for certain antibiotics. This subject is covered elsewhere in this book.

5. Reactive Metabolites and Analgesic Nephropathy

5.1. Background

Dubach and Raaflaub (1969) were probably the first to suggest the possibility that analgesic nephropathy might be caused by a metabolite rather than by the parent drug. They observed that with increasing dosage of phenacetin there was a disproportionate increase in the excretion of phenetidine (see Fig. 1) and 2-hydroxyphenetidine-sulfate, but they were unable to show in acute studies that these metabolites were specifically nephrotoxic (Dubach, 1978). Interest was renewed by the series of studies initiated by Brodie, Gillette, Mitchell, and their colleagues on the mechanism of acetaminophen-induced hepatic necrosis (Nery, 1971; Margetts, 1976; Hinson, 1980). Many of the concepts derived from these studies may be applicable to the renal disease, but fundamental differences should be recognized. Of these probably the most important are (1) that hepatotoxicity is acute and directly related to excessive dosage, while analgesic nephropathy results from chronic abuse with daily doses far less than those that produce acute toxicity, and (2) that the kidney, and not the liver, possesses a mechanism for developing local drug concentrations far higher than those of the circulating plasma. Ingested phenacetin and APAP are almost completely recovered in the urine. This implies little biliary excretion. However, no data are available on the concentration of minor metabolites in the bile.

The essential features of the mechanism that has emerged from the studies of hepatotoxicity are as follows. A fraction of the parent drug is metabolized to a reactive electrophilic derivative. This combines with intracellular glutathione to form an inactive glutathione adduct. Once cellular stores of glutathione are depleted, the metabolite then combines with tissue macromolecules and becomes covalently bound. Many macromolecules have vital enzymatic functions which are disrupted by the binding of a foreign compound. Normal cellular functions are thereby impaired. This results in cytotoxicity and cellular necrosis. The above thesis is based on several different types of evidence, which include (1) the measurement of covalent binding with radiolabeled drugs, (2) the urinary excretion of drug metabolites of glutathione, (3) the characterization of putative metabolites in *in vitro* systems, and (4) direct determination of metabolite toxicity by conventional regimens.

Because of the nature of the primary lesion in chronic analgesic abuse, it is essential to distinguish between cortex and medulla. In addition, many experimental studies are of necessity of an acute nature. To our knowledge there are no data on the patterns of drug metabolism with long-term dosage either with or without the coadministration of other drugs known to either stimulate or inhibit drug-metabolizing enzymes. In addition, no matter how remote, the theoretical possibility cannot be completely discarded that various parent drugs might be directly nephrotoxic in the concentrations developed within the distal nephron. There are a number of analgesics and related drugs that display some nephrotoxicity, and reactive minor metabolites have so far been described for only a few.

5.2. Tissue Glutathione

Control levels of glutathione are in the rank order liver > cortex > papilla. With an acute dose of APAP, the greatest depletion occurs in the liver. In the cortex and papilla depletion of glutathione of about 50% is observed in mice and to a lesser extent in the rat. The interpretation of these effects in the kidney is complicated by the fact that renal glutathione may exist in separate cellular pools (Mudge *et al.,* 1978). The effect of chronic dosing is not known.

In isolated cells APAP-*S*-glutathione is formed by liver cells but not by kidney cells (Jones *et al.,* 1979). This apparent discrepancy between *in vivo* and *in vitro* findings may be reconciled by the observation that APAP-*S*-glutathione is rapidly converted to APAP-*S*-cysteine and APAP-*S*-mercapturate by the kidney.

5.3. Covalent Binding

In our initial experiments on the covalent binding of APAP administered *in vivo*, binding to tissue protein was far greater for the papilla than either the cortex or liver. Subsequent experience showed that the injected APAP had contained significant radiochemical impurities. Upon further purification the radiochemical purity, as measured by chromatography, rose to from 99.0 to 99.8% in different batches, while the covalent binding decreased to about 10% of the original value for cortex and papilla and to 50% for liver. The results clearly indicate that a radiochemical impurity binds with greater affinity to the kidney than to the liver. Since the total amount of radioactivity that is covalently bound to the proteins of liver, cortex, and papilla represents approximately 0.1, 0.01, and 0.001%, respectively, of the injected dose, a highly accurate measurement of covalent binding would require radiochemical purity higher than has thus far been achieved (Mudge *et al.,* 1978). Nevertheless, consistent results have been obtained using adequate controls for each batch of radioisotope. After a single dose the half-life for covalent binding is about 150 hr for the papilla. This is approximately 2–3 times longer than for cortex and liver.

Covalent binding is distributed to all subcellular fractions, as shown in Fig. 3. At a low dose the only difference between cortex and papilla is the lesser degree of binding to the nuclei from the papilla. However, at a large dose, binding to both the mitochondrial and miscrosomal fractions is higher for the papilla than cortex. These observations are based on acute experiments and the length of time that these differences persist is not known.

5.4. Site of Formation

From the original studies on hepatic necrosis it was clear that the reactive metabolite of APAP is formed within the liver (Hinson, 1980). Subsequent work

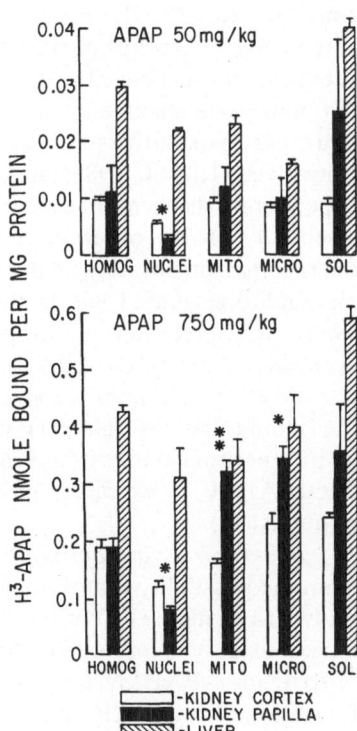

FIGURE 3. Subcellular distribution of tritiated APAP covalently bound to protein. Fischer rats, killed 4 hr after intraperitoneal injection. *, $p < 0.05$; **, $p < 0.01$, in comparison of cortex to papilla.

has shown that the kidney itself has the capacity to generate metabolites that are potentially nephrotoxic. This is supported by several types of evidence. First, when animals are pretreated with compounds such as 3-methylcholanthrene, which induces drug-metabolizing hepatic enzymes, the amount of covalently bound APAP increases about threefold for liver and plasma protein, but remains unchanged for protein from cortex and papilla (Mudge *et al.*, 1978). Second, with the isolated perfused kidney, when APAP is added to the perfusion the urine contains 3-S-adducts derived from glutathione. These must have arisen from a "toxic metabolite" generated within the kidney itself (Ross *et al.*, 1980). Essentially the same findings have been observed with isolated kidney cells (Jones *et al.*, 1979). Third, tritiated APAP is covalently bound to kidney microsomes in a system dependent on NADPH, oxygen, and active enzymes (McMurtry *et al.*, 1978). This is similar to what has been found for liver.

5.5. Identification of Reactive Metabolite(s)

a. N-*Hydroxy acetaminophen* (*N*-O H-A P A P) *and* N-*acetyl-p-benzoquinoneimine* (APBQI). In APAP-induced hepatotoxicity it has been well established that the production of the reactive metabolite requires NADPH, oxygen, and the cytochrome P-450 mixed-function oxidase system (Hinson, 1980). On this basis it was hypothesized that *N*-OH-APAP was the initial compound directly derived from APAP. It was then proposed that *N*-OH-APAP was converted to APBQI, an electrophilic quinone, which spontaneously would react with sulfhydryl groups of either glutathione or protein. This sequence of events was subsequently challenged and, as indicated in Fig. 1, the exact route from APAP to the immediate reactive metabolite is not definitely established at this time. It was clear that an oxidative step is essential and there is substantial evidence that APBQI is involved (Hinson, 1980; Gemborys *et al.*, 1980; Miner and Kissinger, 1979). The experiments that tend to exclude *N*-OH-APAP in the reaction sequence are based essentially on the observation that when synthetic *N*-OH-APAP is added to a microsomal preparation it does not behave in the same manner as the reactive metabolite generated within the microsomes (Hinson *et al.*, 1979; Nelson *et al.*, 1980). Interpretation is complicated by the absence of data on microsomal permeability to *N*-OH-APAP.

Possibly of greater relevance to the pathogenesis of analgesic nephropathy is the finding that *N*-OH-APAP is sufficiently stable to be excreted in the urine after its intraperitoneal injection. However, following the administration of APAP, no *N*-OH-APAP is excreted, either free or conjugated (Gemborys and Mudge, unpublished).

b. 3-Hydroxyacetaminophen (3-OH-APAP). Interest in 3-OH-APAP stems from the possibility that it might be derived from an epoxide, a type of metabolite known to be highly reactive and hence potentially toxic. Although 3-OH-APAP is a trace urinary metabolite of APAP in some species, and although it is formed *in vitro* from APAP by hepatic microsomes, present evidence indicates that the formation of 3-OH-APAP is unrelated to the formation of the reactive metabolite of APAP (Hinson *et al.,* 1980). Although these experiments were limited to the problem of hepatic toxicity, there is no reason at present to suspect that the results would be different if renal toxicity were examined.

    c. p-*Aminophenol* (PAP). In the early studies on the metabolic fate of phenacetin and related drugs (Brodie and Axelrod, 1949; Smith and Williams, 1949) it was found that phenacetin and acetanilid were deacetylated to a slight extent but that no products of deacetylation were excreted after the administration of APAP (see Fig. 1). Deacetylation of APAP would generate PAP. When PAP itself was administered it was excreted to a modest extent as unchanged drug, but principally as the phenolic conjugates (glucuronide and sulfate) and also as the acetylated amino conjugate (Smith and Williams, 1949). Since that time it has been widely held that APAP is not subject to deacetylation and that PAP is not generated from either phenacetin or APAP. However, Welch *et al.* (1966) reported the excretion of a free diazotizable aromatic amine after dosage with APAP. Large amounts were found in the cat and trace amounts in humans. The amine was not identified conclusively, and although it was claimed that it was not PAP, the possibility cannot be excluded that it was a phenolic conjugate of PAP. Smith (1958), in an abstract which was not published at greater length, used APAP labeled in the acetyl group and studied the possibility that APAP might be both deacetylated and then reacetylated. Although the findings were not consistent, definite evidence of this type of reaction was obtained in some subjects.

    PAP is of particular interest in relation to the pathogenesis of analgesic nephropathy. Calder, in a series of papers (Calder *et al.,* 1975, 1979), has shown that PAP is highly nephrotoxic in acute experiments. His group proposed it as a "model" for the study of nephrotoxicity but presumed that PAP itself was not directly involved in the human disease of chronic abuse, apparently for two reasons—first, that after acute dosage lesions were limited to the kidney cortex, and second, that PAP had not been reported as a metabolite of either phenacetin or APAP.

    Carpenter (1978) studied acetylation and deacetylation in the mouse kidney. In tissue homogenates with added CoA, PAP is acetylated to APAP at a high rate. One would anticipate this from previous *in vivo* experiments. However, in the absence of added CoA and with the use of APAP as substrate, there is significant deacetylation of APAP to PAP. The studies were then extended by Gemborys and Mudge (unpublished) with the development of a sensitive HPLC method for the detection of PAP in urine. In the hamster the fractional abundance of the products of the deacetylation of APAP rises with the dosage of APAP from 0.1% at 50 mg/kg to 2.9% at 600 mg/kg. On the basis of enzymatic hydrolysis approximately 1% of the PAP is free; the remainder is conjugated as the phenolic glucuronide or sulfate.

    We have recently examined the deacetylation of APAP in homogenates of rat tissue (Fig. 4). Enzyme activity is in the rank order liver > kidney cortex > kidney medulla. Many metabolic inhibitors are inactive. This is in keeping with the hydrolytic nature of the reaction. It should be noted that 10 mM NaF is inhibitory. Deacetylase activity resides in the soluble fraction, since it is unaffected by the removal of the microsomal or mitochondrial fractions. As to the lesser activity of the medulla, it should be noted that the experiments shown in Fig. 4 were all conducted at the same concentration of APAP. In the tissue homogenates the rate of PAP formation is approximately proportional to the initial concentration of APAP. In the intact animal, at normal or reduced rates of urine flow, the concentration of APAP would be far higher in the medulla than in the cortex. Thus it is possible that the rate of PAP formation *in vivo* might be the same in both portions of the kidney. The hepatic and kidney deacetylases show quantitative differences in response to a number of experimental variables. Of these the most interesting is the effect of pH. Over the pH range 6.3–8.2, PAP formation increases with pH. However, over this range, PAP formed by kidney increases about sixfold, while PAP formed by liver increases only about 1.5-fold.

5.6. Implications of the Reactive Metabolite Hypothesis

Since the dose–response relationship is an essential factor in the study of drug toxicity and since dose usually refers to that of the parent drug, in the case of toxicity mediated by the formation of reactive metabolites it is essential to recognize the limitations of the coventional yardstick for the measurements of dose–response. As general rules it may be proposed that there is no necessary relationship between the dose of parent drug and the amount of reactive metabolite that is produced, excreted, or bound to macromolecules. Hence there is no necessary relationship between the dose of parent drug and ultimate toxicity. For most tissues, when other factors remain constant, the local concentrations of parent drug and/or its metabolites are approximately proportional to the dose of the parent drug. However, this does not hold for the kidney, particularly the distal nephron, for which a physiologic function of the tubule such as the reabsorption of

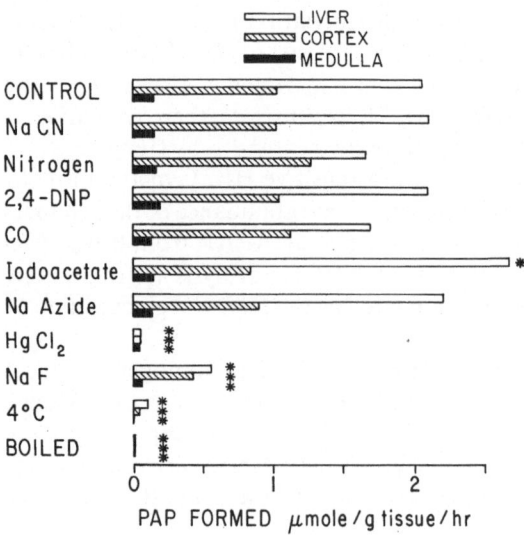

FIGURE 4. Deacetylation of APAP to PAP. 13,000 × gravity supernates of tissue homogenates incubated at 37°C for 1 hr in phosphate buffer (pH 7.2) and 11 mM APAP. Metabolic inhibitors added at conventional concentrations required for maximal positive effect. Nitrogen indicates substitution for oxygen in gas phase. Concentration of NaF was 10 mM. Tissue from Fischer male rats. $n = 3$ or greater. *, $p < 0.05$ compared to control.

water may be a more important determinant of drug concentration than initial dosage.

The excretion of potential toxic metabolites varies unpredictably with dosage. Dubach and Raaflaub (1969) observed about a 30-fold increase in the fractional abundance of 2-OH phenetidine that was excreted in the urine as the dose of phenacetin was increased 30-fold. (In the measurement of urinary excretory products, fractional abundance is the amount of a given compound, or metabolite, in relation to the total of parent drug plus all metabolites.) Since the total amount of a compound that is excreted is the product of fractional abundance times total drug and metabolites, the absolute amount of 2-OH phenetidine increased in linear fashion about 1000-fold. A different situation has been observed by Gemborys and Mudge (unpublished) in the hamster. At low doses of APAP the fractional abundance of the 3-S-adducts of APAP is about 34%. This falls to 10% as dose of APAP is increased. As a consequence the absolute amount of 3-S-adducts that are excreted remains relatively constant over a major portion of the dose-response curve.

The enzymatic production of reactive metabolites must be subject to a number of metabolic influences. Some of these have been studied in acute experiments, particularly the effects of compounds that either induce or inhibit the formation of drug-metabolizing enzymes. In each instance a positive experiment would distort the relationship between the dose of the parent drug and the amount of the ultimate reactive metabolite which is produced. As another example, consider the production of PAP from APAP by kidney. This reaction is so strongly pH dependent *in vitro* that, if the same findings held for the *in vivo* situation, changes in acid–base balance might well be more important than changes in initial dosage.

The reactive metabolite mechanism contains an additional complication when the binding of drug metabolites to tissues is taken into account. Consider the formation of PAP from APAP. This has by now been demonstrated for the mouse, hamster, and rat. There is good evidence that PAP may have two separate ultimate fates—either to be bound to tissue protein, or to be excreted in the urine, principally as conjugates. There is also good evidence that in the experimental animal the conversion of APAP to PAP is most readily demonstrable at doses of APAP too toxic to be administered to humans. In the event that in humans, with reasonable doses of APAP, it is not possible to detect the urinary excretion of PAP, does this mean that in humans PAP is not generated from APAP? Not at all. According to the tissue binding mechanisms, it is quite possible that PAP might be bound to tissue protein instead of being excreted. In a sense, this would be a "pharmacologic Catch-22."*

5.7. The Role of PAP in Analgesic Nephropathy

PAP has long been recognized as a cytotoxic agent (Bernheim *et al.,* 1937) and more recently as a specific nephrotoxin with relatively little toxicity for other

---

*"Catch-22" defies precise definition, but is essentially an impossible situation created by conflicting bureaucratic regulations. Taking the absence of a compound in one place as evidence for its presence elsewhere is a not unreasonable example.

organs (Calder *et al.,* 1979). The possible involvement of PAP in the pathogenesis of analgesic nephropathy depends primarily on the demonstration that APAP can be deacetylated to PAP. This reaction has been denied in the literature to date but has been demonstrated in the studies reported here. As a hypothesis, it might be proposed that APAP is deacetylated within the kidney to PAP and that the PAP thus formed has at least three possible fates: (1) to be reacetylated back to APAP (Smith and Williams, 1949); (2) to be covalently bound to kidney protein (Calder *et al.,* 1979); and (3) to undergo phenolic conjugation and thereby to be excreted at a high rate (Smith and Williams, 1949).

An additional fourth possibility must be recognized, namely that the phenolic conjugates might be deconjugated back to PAP. In Fig. 1 the formation of the phenolic conjugates is depicted as undirectional. This is an erroneous oversimplification in view of the fact that many mammalian tissues possess glucuronidase activity (Levvy and Conchie, 1966). On the basis of our preliminary experiments, glucuronidase activity is the same for the kidney medulla as for the cortex. This might have significant implications for the pathogenesis of analgesic nephropathy. The PAP that is generated by the kidney from APAP is overwhelmingly converted *in vivo* to inactive phenolic conjugates. For reasons previously discussed, these conjugates become highly concentrated in the tubular fluid of the distal nephron. If one assumed that local glucuronidase had access to compounds within the tubular fluid and if one also assumed the usual relationship between substrate concentration and the generation of endproducts, then it would follow that, since at the same substrate concentration glucuronidase activity is the same for the cortex as the medulla, when the higher concentration of PAP-glucuronide in the distal tubule is taken into account, the potential of the distal nephron to generate nephrotoxic PAP would be far higher than that of the proximal nephron. By reference to Fig. 1 it may be noted that four types of phenolic conjugates are excreted in the urine. Of these the conjugates of *N*-OH-APAP appear only after the administration of *N*-OH-APAP. Of the remaining three categories, it is only the conjugates of PAP that would give rise to a highly toxic compound upon deconjugation. Since the degree of access of glucuronides to tissue enzymes of the papilla is not known, the above mechanism is speculative.*

One of the arguments against the involvement of PAP in the pathogenesis of analgesic nephropathy is that in acute experiments the renal lesions are confined to the cortex. However, this argument is not convincing. For many toxic agents the localization and distribution of lesions vary depending on dosage and chronicity of exposure. As examples, consider the toxicity of a number of heavy metals.

5.8. Metabolic Pathways from APAP to Toxic Metabolites

Before considering these pathways in detail the importance of species and strains should be emphasized. For many species, as well as for many metabolites,

---

*The renal tubule is not completely impermeable to all glucuronides. Partial reabsorption of the glucuronide of APAP has been demonstrated by clearance experiments in the dog (Duggin and Mudge, 1975). However, the extent of reabsorption in the papilla is not known.

systematic studies are not available. Nevertheless, as an example, consider the excretion of the 3-*S*-adducts of APAP. These are derived from the metabolism of 3-*S*-glutathione APAP and are a measure of the amount of the electrophilic metabolite that is formed via the oxidative pathway from APAP to APBQI (Fig. 1). In humans the fractional abundance of these metabolites in the urine is 3% (Mrochek *et al.,* 1974). In the hamster, at a comparable dose of APAP, the fractional abundance is 34% (Gemborys and Mudge, unpublished). Clearly the results obtained in one species cannot be applied to another without reservations.

   As indicated in Fig. 1, if one starts from APAP it is possible to reach a reactive quinone by two separate routes. If APAP is initially oxidized via the cytochrome P-450 mixed function oxidase system it is converted to the quinone APBQI. However, if APAP is first deacetylated by a nonoxidative enzymatic step it is converted to PAP, which in turn can undergo nonenzymatic auto-oxidation to the quinone PBQI. The binding of radioactive APAP to kidney microsomes is enhanced by NADPH, while the binding of radioactive PAP is depressed by it (McMurtry *et al.,* 1978; Calder *et al.,* 1979). This apparent paradox may be explained by two separate roles of NADPH, which ultimately have opposite effects as measured by protein binding (Fig. 5). The conversion of APAP to APBQI has a requirement for NADPH; the conversion of APAP to PBQI does not. The second role of NADPH is its effect on the oxidative status of glutathione (GSH). Once either APBQI or PBQI has been formed it can react either with GSH or with protein. GSH is active in the reduced form but is inactive when oxidized to GSSG. Thus, as GSH levels increase, more reactive metabolite is bound to it and less to protein. When one starts with PAP this is the single effect of NADPH. When one starts with APAP this same effect undoubtedly persists but is overshadowed by the role that NADPH plays in the generation of the reactive metabolite itself.

   In a review article Duggin (1980) reported unpublished experiments in which APAP became bound to proteins in the medulla of rabbits by an NADPH-independent system which was not further characterized. The summarized findings are completely consistent with the deacetylation of APAP to PAP and the subsequent formation of the reactive quinone PBQI. Binding was inhibited by

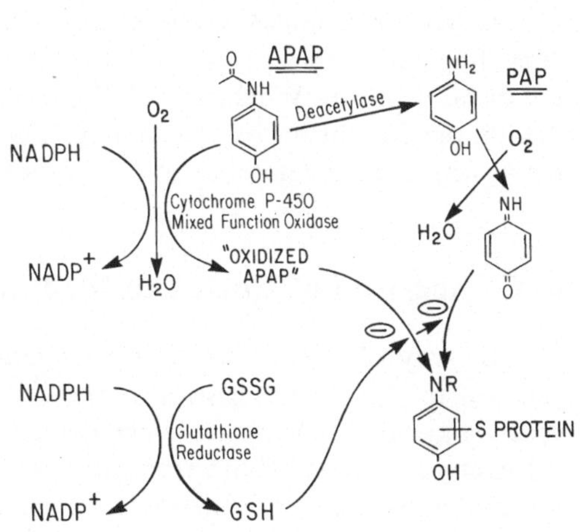

FIGURE 5. Schematic summary of the dual roles of NADPH in covalent binding of APAP to tissue protein. Encircled minus signs indicate competition from GSH resulting in decreased protein binding. For discussion, see text.

fluoride. As shown in Fig. 4, fluoride inhibits the generation of PAP from APAP. Binding was also inhibited by acetyl Co-A, which would reacetylate the PAP that had been formed. Finally, binding was inhibited by the addition of PAP, which would dilute the radioactive PAP generated from APAP and thereby reduce the radioactivity in the compound bound to protein. An additional way to establish this mechanism would be to compare the binding of acetyl and ring labeled radioactive APAP.

As indicated in Fig. 1, some reactions (heavy dashed lines) can occur nonezymatically in simple aqueous solutions. These include the dehydration of N-OH-APAP to APBQI and its subsequent reaction with sulfhydryls, and the auto-oxidation of APAP to PBQI and its subsequent reaction with sulfhydryls. The conjugation of the reactive metabolite of APAP (presumably APBQI) with GSH is also catalyzed by liver cytosol enzymes (Rollins and Buckpitt, 1979). It is not known whether the other nonenzymatic reactions are also subject to enzymatic catalysis, but this question warrants study.

5.9. Salicylates

The pharmacology of salicylates in relation to analgesic nephropathy has been recently reviewed by Duggin (1980) and will be discussed here only to make a limited number of comparisons to the studies on APAP. Both aspirin and salicylate are excreted by a pH-dependent mechanism, while phenacetin and APAP are not. This means that both urinary pH and the rate of urine flow may influence the concentration of salicylate in the urine. Aspirin rapidly acetylates proteins, particularly those involved in prostaglandin metabolism, and in a sense aspirin thus becomes covalently bound to protein. However, such binding appears to have consequences far different from those induced by APAP. It is probable that both aspirin and salicylate are concentrated within the cells of the distal tubule, but definitive data are not available. By analogy to the minor reactive metabolites of APAP, the role of gentisic acid, a metabolite of aspirin, warrants further study. This is capable of forming quinones and becomes strongly bound to proteins, including connective tissue (Roof and Turner, 1955).

Based on clinical observations probably the most important question regarding pathogenesis is the possible interaction between salicylates and APAP in the distal nephron. Wheldrake (1975) has shown that salicylate may interfere with membrane formation by the kidney. This type of observation warrants further study with special reference to drug combinations and chronic dosage.

6. Comparison of Cortex to Medulla

By way of summary, the trace metabolite hypothesis may be reviewed in relation to the characteristics of the cortex and medulla. It is proposed that a reactive metabolite becomes concentrated within the cells of the distal nephron and is thereby cytotoxic. With chronic drug abuse, individual episodes of tissue damage are probably minor, but their effect is cumulative and is aggravated by both high drug intake and/or dehydration. The recuperative capacity of the medulla compared to the cortex is not known.

There are at least three types of mechanism by which the concentration of a cytotoxic agent might become higher in the cells of the distal than proximal nephron: (1) the metabolite might be formed at an extrarenal site, filtered at the glomerulus, or secreted across the tubule, and become concentrated distally by the reabsorption of water; (2) the same mechanism might hold if the compound were formed within the kidney itself, provided it could gain access to the tubular fluid; and (3) both the cortex and medulla might have a similar intrinsic capacity to generate the metabolite, but since product formation is also determined by substrate concentration, the latter might become the overriding determinant and thus permit greater production in the distal than proximal tubule. It is theoretically possible that the distal tubule might have an enzymatic function not possessed by the proximal, but there are no examples which would be relevant to the present question.

All data suggest that the reactive metabolite is, or can be, generated within the kidney itself. As to intrinsic enzymatic capacity, for most reactions there is greater activity in the cortex than medulla. This certainly applies to the mixed function oxidase system (Zenser *et al.,* 1978; Duggin, 1980) as well as to deacetylase. From preliminary studies, glucuronidase activity in cortex and medulla are equal, but tissue localization is not known and its role is conjectural. Unfortunately, most studies on drug metabolism have failed to distinguish between cortex and medulla and, in addition, have been of an acute rather than chronic nature.

Considered in its entirety, the available evidence strongly suggests that PAP is the most likely candidate for the critical role of cytotoxic metabolite. This applies to the disorder resulting from chronic abuse. With acute overdosage, which may result in acute renal as well as hepatic failure, the oxidative metabolite (APBQI) might produce the renal lesion. The development of analgesic nephropathy involves complex interactions among a number of physiologic and pharmacologic factors. From the nature of the disorder it is not essential that more than trace amounts of the putative cytotoxic agent be involved. PAP, which has now been identified as a trace metabolite, has the appropriate characteristics for each step of the pathophysiologic mechanism which has been proposed. It remains to be seen to what extent further documentation can be obtained in the human disease itself.

Acknowledgments

The original investigations reported herein were supported by U.S. Public Health Service Grants AM 06818 and AM 16960.

References

Bernheim, F., Bernheim, M. L. C., and Michael, H. O., 1937, The action of *p*-aminophenol on certain tissue oxidations, *J. Pharmacol. Exp. Ther.* **61**:311.

Brodie, B. B., and Axelrod, J., 1949, The fate of acetophenetidin (phenacetin) in man and methods for the estimation of acetophenetidin and its metabolites in biological material, *J. Pharmacol. Exp. Ther.* **95**:58.

Burry, A., Cross, R., and Axelsen, R., 1977, Analgesic nephropathy and the renal concentrating mechanism, in: *Pathology Annual, Part 2* (S. S. Sommers and R. P. Rosen, eds.), Appleton-Century-Crofts, New York, p. 1.

Calder, I. C., Williams, P. J., Woods, R. A., Funder, C. C., Green, C. R., Ham, K. N., and Tange, J. D.,1975, Nephrotoxicity and molecular structure, *Xenobiotica* 5:303.

Calder, I. C., Yong, A. C., Woods, R. A., Crowe, C. A., Ham, K. N., and Tange, J. D., 1979, The nephrotoxicity of *p*-aminophenol. II. The effect of metabolic inhibitors and inducers, *Chem.-Biol. Interactions* 27:245.

Carpenter, H. M., III, 1978, Mechanisms of the nephrotoxicity of acetaminophen studied in the mouse kidney cortex slice system, PhD Thesis, Dartmouth College, Hanover, New Hampshire.

Dubach, U. C., 1978, Nephropathies due to analgesics, *Contrib. Nephrol.* 10:75.

Dubach, U. C., and Raaflaub, J., 1969, Neue aspekte zur frage der nephrotoxizitat von phenacetin, *Experentia* 25:956.

Duggin, G. G., 1980, Mechanisms in the development of analgesic nephropathy, *Kidney Int.* 18:553.

Duggin, G. G., and Mudge, G. H., 1975, Renal tubular transport of paracetamol and its conjugates in the dog, *Br. J. Pharmacol.* 54:359.

Duggin, G. G., and Mudge, G. H., 1976a, Analgesic nephropathy: Renal distribution of acetaminophen and its conjugates, *J. Pharmacol. Exp. Ther.* 199:1.

Duggin, G. G., and Mudge, G. H., 1976b, Phenacetin: renal tubular transport and intrarenal distribution in the dog, *J. Pharmacol. Exp. Ther.* 199:10.

Gemborys, M. W., Mudge, G. H. and Gribble, G. W., 1980, Mechanism of decomposition of *N*-hydroxyacetaminophen, a postulated toxic metabolite of acetaminophen, *J. Med. Chem.* 23:304.

Hinson, J. A., 1980, Biochemical toxicology of acetaminophen, in: *Reviews in Biochemical Toxicology, Volume 2* (E. Hodgson, J. R. Bend, and R. M. Philpot, eds.) Elsevier/North-Holland, New York, p. 103.

Hinson, J. A., Pohl, L. R., and Gillette, J. R., 1979, *N*-Hydroxyacetamimophen: A microsomal metabolite of *N*-hydroxyphenacetin but apparently not of acetaminophen, *Life Sci.* 24:2133.

Hinson, J. A., Pohl, L. R., Monks, T. J., Gillette, J. R., and Guengerich, F. P., 1980, 3-Hydroxyacetaminophen: A microsomal metabolite of acetaminophen. Evidence against an epoxide as the reactive metabolite of acetaminophen, *Drug Metab. Dispos.* 8:289.

Jones, D. P., Sundby, G., Ormstad, K., and Orrenius, S., 1979, Use of isolated kidney cells for study of drug metabolism, *Biochem. Pharmacol.* 28:929.

Levvy, G. A., and Conchie, J., 1966, β-Glucuronidase and the hydrolysis of glucuronides, in: *Glucuronic Acid, Free and Combined* (G. J. Dutton, ed.), Academic Press, New York, p. 301.

Margetts, G., 1976, Phenacetin and paracetamol, *J. Internat. Med. Res.* 4:(Suppl. 4):55.

McMurtry, R. J., Snodgrass, W. R., and Mitchell, J. R., 1978, Renal necrosis, glutathione depletion, and covalent binding after acetaminophen, *Toxicol. Appl. Pharmacol.* 46:87.

Miner, D. J., and Kissinger, P. T., 1979, Evidence for the involvement of *N*-acetyl-*p*-quinoneimine in acetaminophen metabolism, *Biochem. Pharmacol.* 28:3285.

Mrochek, J. E., Katz, S., Christie, W. H., and Dinsmore, S. R., 1974, Acetaminophen metabolism in man, as determined by high-resolution liquid chromatography, *Clin. Chem.* 20:1086.

Mudge, G. H., Gemborys, M. W., and Duggin, G. G., 1978, Covalent binding of metabolites of acetaminophen to kidney protein and depletion of renal glutathione, *J. Pharmacol. Exp. Ther.* 206:218.

Nelson, S. D., Forte, A. J., and Dahlin, D. C., 1980, Lack of evidence for *N*-hydroxyacetaminophen as a reactive metabolite of acetaminophen *in vitro, Biochem. Pharmacol.* 29:1617.

Nery, R., 1971, The possible role of *N*-hydroxylation in the biological effects of phenacetin, *Xenobiotica* 1:339.

Rollins, D. E., and Buckpitt, A. R., 1979, Liver cytosol catalyzed conjugation of reduced glutathione with a reactive metabolite of acetaminophen, *Toxicol. Appl. Pharmacol.* 47:331.

Roof, B. S., and Turner, J. C., 1955, Protein interactions of gentisic acid and certain of its oxidation products, *J. Clin. Invest.* 34:1647.

Rosner, I., 1976, Experimental analgesic nephropathy, *CRC Crit. Rev. Toxicol.* 4:331.

Ross, B., Tange, J., Enslie, K., Hart, S., Smail, M., and Calder, I., 1980, Paracetamol metabolism by the isolated pufused rat kidney, *Kidney Int.* **18**:562.

Smith, J. N., and Williams, R. T., 1949, Studies in detoxication 23. The fate of aniline in the rabbit, *Biochem. J.* **44**:242.

Smith, P. K., 1958, *Acetophenetidin, A Critical Bibliographic Review,* Interscience, New York.

Welch, R. M., Conney, A. H., and Burns, J. J., 1966, The metabolism of acetophenetidin and *N*-acetyl-*p*-aminophenol in the cat, *Biochem. Pharmacol.* **15**:521.

Wheldrake, J. F., 1975, The effect of aspirin (acetyl salicylate) on macromolecule turnover in rat kidney and liver, *Experentia* **31**:559.

Zenser, T. V., Mattammal, M. B., and Davis, B. B., 1978, Differential distribution of the mixed-function oxidase activities in rabbit kidney, *J. Pharmacol. Exp. Ther.* **207**:719.

20

Endemic Balkan Nephropathy

PHILIP W. HALL, III

1. Introduction

Endemic Balkan nephropathy (EBN) is a unique form of chronic interstitial renal disease which is endemic to isolated rural populations of Bulgaria, Romania, and Yugoslavia. The area is confined within a 200-mile radius of Belgrade. The involved villages lie along river valleys in multiple areas of Romania and Yugoslavia where flooding periodically occurs (Wolstenholme and Knight, 1967). The disease is clinically characterized by the insidious onset of a normocytic, normochromic anemia, azotemia, and persistent proteinuria usually of <1 g/24 hr. There is no increase in the incidence of hypertension in the involved populations and the disease slowly progresses to end-stage renal failure over a period of 5–10 years following the onset of azotemia. While there is no uniform agreement concerning the early pathologic changes, it is generally accepted to be a chronic interstitial disease as it appears in the uremic patient. There is marked tubular atrophy with focal areas of tubular regeneration, interstitial fibrosis, and minimal round cell infiltration. The tissue loss is predominently cortical and most marked in the cortex opposite the hilum (Radonic *et al.,* 1966).

The diagnosis of EBN will be made by the clinician if the patient presents with the majority of the findings summarized in Tables I and II. The differential diagnosis of a patient presenting with these findings, without the geographic exposure, might be nonobstructive pyelonephritis, reflux nephropathy, or

PHILIP W. HALL, III • Department of Medicine, Case Western Reserve University and Cleveland Metropolitan General Hospital, Cleveland, Ohio 44109.

TABLE I. Typical Findings in Endemic Balkan Nephropathy

History
  Lived in an endemic village >20 years, native or immigrant
  As likely male as female
  Farm workers almost exclusively
  Negative history for hypertension, edema, or any form of urinary tract disease
  Multiple family members may be involved

Complaints
  Vague, symptomatic only, consisting of fatigue, backache, and headache

Physical findings
  Yellow skin pigmentation in advanced disease
  Negative for hypertension, edema, or organomegaly

analgesic abuse nephropathy. The laboratory and physical findings would be most unusual for chronic glomerulonephritis or vascular disease. The nature of the proteinuria and the absence of hypertensive changes would mediate against these diagnoses. If no reflux or anatomical abnormalities can be demonstrated, the diagnosis of pyelonephritis would be most unlikely. If analgesic abuse can be ruled out, one is left with the diagnosis of chronic interstitial disease, etiology unknown. This is the situation reported to exist in greater than 10% of the cases of chronic interstitial nephritis studied at the University of Pennsylvania Hospital (Murray and Goldberg, 1975).

2. Epidemiologic Investigations

Epidemiologic methods have been employed to define the geographic distribution, the natural history of the disease, the relationship to climate, altitude, season

TABLE II. Typical Findings in Endemic Balkan Nephropathy—Diagnostic Studies

Urinalysis: trace 1+ proteinuria, <1 g/24 hr; occasional granular casts; no RBC or WBC casts

Blood studies: mild normocytic normochromic anemia of chronic renal failure

Chemistries: typical for any form of chronic azotemic renal failure; increased anion gap acidosis; normal serum proteins, Na, K, Cl

Renal function studies
  a. Increased α-amino nitrogen and amino acid excretion; minimal glycosuria
  b. Tubular proteinuria with increased excretion of B_2M, retinol binding protein, lysozyme, ribonuclease, 3S gamma globulin, and IgG kappa light chains
  c. Diminished ability to dilute and concentrate; pitressin resistant

Radiologic studies: symmetric, bilaterally small kidneys without evidence of calyceal distortion, scarring, or obstruction of the year, occupation, etc. The disease is found in the highest prevalence in rural areas where the population density is the lowest. Such evidence points against an infectious agent. A geographic area is defined to be "endemic" if population surveys for proteinuria, using the sulfosalicyclic acid (SSA) test, show an incidence >7% and the public health records indicate a high death rate from kidney disease. In many of these villages, the incidence of proteinuria has been reported as high as 34% (Stojimirovic, 1974).

The SSA screening test is sensitive but not specific for EBN. The diffuse tubuloinsterstitial pathology described in EBN suggested to us that the nature of the protein spilled in the urine might be of differential diagnostic aid. The initial studies, utilizing electrophoretic analysis of urine protein concentrates, demonstrated the common occurrence of tubular proteinuria in village populations in Bosnia, Yugoslavia where EBN is highly endemic (Hall *et al.,* 1967). We repeatedly screened a small village population (403 individuals) 3–6 times/year for 10 years. Children of <10 years of age showed intermittent proteinuria. The percent of the population involved increased with age (Fig. 1). Twelve of the 403 people died of kidney disease over this period of observation (mortality rate 298/100,000/year). The deaths occurred in people 36–79 years of age (M. Vasiljevic, N. Popovic, J. Gaon, and P. W. Hall, unpublished observations). To further establish the extent of the involvement of the disease, we investigated a village population in the Vratza district of Bulgaria (Ts. Dimitrov, I. Dinev, T. Sattler, and P. W. Hall, unpublished observations). The incidence of tubular proteinuria, as detected by radioimmunoassay, increased from 5% of those individuals between 30–40 years of age to 60% of those >69 years of age (Fig. 2). An elevated serum creatinine concentration (SCr)

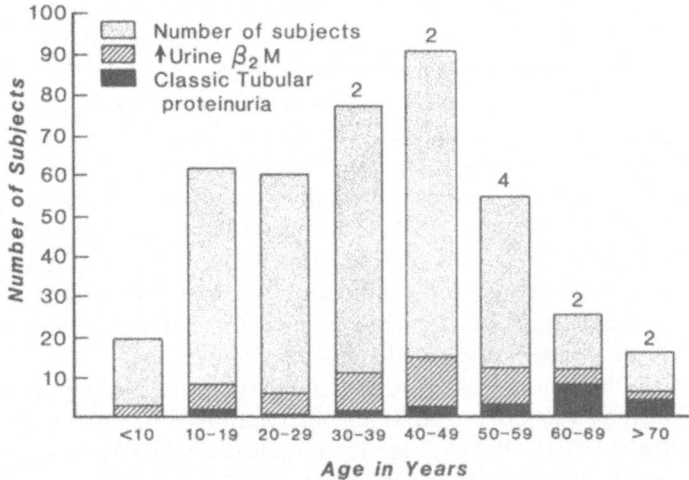

FIGURE 1. The relationship of age (abscissa) to the prevalence of B_2 microglobulinuria and classic tubular proteinuria (ordinate) in a population of 403 individuals living in an "endemic" village in Bosnia, Yugoslavia. The numbers at the top of the bars from the fourth decade on are deaths that occurred in these groups during the 7-year followup.

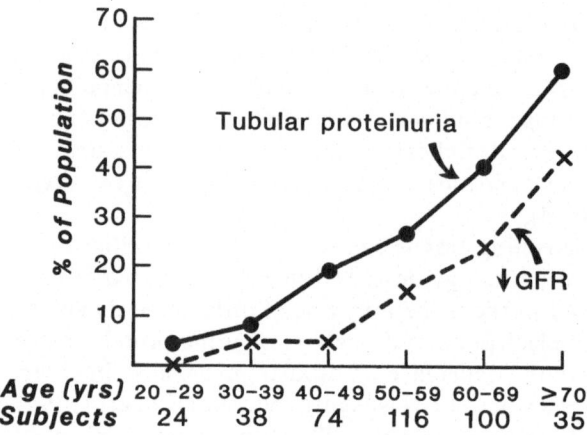

FIGURE 2. The relationship of age (abscissa) to the incidence of tubular proteinuria and decreased glomerular filtration rate (GFR) (ordinate) in a population of 387 adults living in the Vratza district of Bulgaria where EBN is prevalent.

occurred in 5% of the population between 40 and 50 years of age and increased to 40% of those over 69 years of age.

Epidemiologic studies, reviewed elsewhere (Hall and Dammin, 1978), have demonstrated that immigrants acquire this disease after a 20-year exposure to an endemic focus. Followup studies of identical twins, where one twin left the endemic area by age 18, demonstrated that continued exposure in the remaining twin resulted in the development of the disease, while the twin who emigrated to a "nonendemic" area remained free of the disease (Danilovic and Stojimirovic, 1967). Essentially all involved individuals, male and female, farmed the land. In one survey of an entire village population, we were able to find only 7 of 35 individuals over the age of 65 who did not have tubular proteinuria; all were men and all were nonfarmers; wives of these men actually farmed (A. Gaon, S. Schwab, J. Gaon, and P. W. Hall, unpublished observations). The study also revealed that the most involved families were the socioeconomically poorest families in the village. This most likely was due to the fact that their farm lands, in contrast to the farm lands of the less involved families, were periodically flooded, making soil fertilization useless and resulting in poor crop production. Attempts to correlate variations in the incidence of proteinuria, clinical appearance of the disease, or death rates with seasons of the year have not yielded significant patterns.

In order to determine what part genetics plays in the development of EBN, a collaborative study was done examining the relationship between HLA antigens and EBN (Minev et al., 1978). These investigators studied 180 patients with the clinically established diagnosis of EBN whose residence was the Vratza district of Bulgaria. This group was compared to a healthy group of 1264 individuals from the same village. A significant increase in the frequency of the occurrence of HLA B_{18} was found in the patients as compared to healthy individuals. These data can be interpreted to suggest that a predisposition to develop EBN may exist in certain families. Although the disease seems to be a progressive one once azotemia is established, we observed many individuals with tubular proteinuria who did not show progression of renal disease over a period of more than 10 years. It would be of interest to determine whether or not such individuals had HLA B_{18}.

3. Possible Etiologic Factors

The epidemiologic studies point to an environmental rather than an inherited etiology to EBN. The natural history of the disease, the pathology, and the epidemiology suggest a toxic agent. Using such reasoning, several substances have been considered. The list includes silica, cadmium, lead, uranium, aristolochic acid, ochratoxin, citrinin, and coronavirus. On the basis of the evidence collected, a case can be made for each of the above agents. When the available data are reviewed, however, none pass the critical test and many can be readily discarded.

Lead was the first element to be incriminated as an etiologic agent in EBN (Danilovic, 1958). Bread consumed by inhabitants of some villages was found to contain 5–10 times normal lead concentrations. The bread was made of wheat ground in mills where cracks in the millwheel were filled with a lead compound. Such milling is no longer the custom in most areas where the disease is still prevalent. Two different clinical presentations have been described in patients with lead poisoning when considered from the standpoint of kidney disease. That seen in children consists of a Fanconi-like syndrome with amino aciduria, glycosuria, and renal tubular acidosis. It presents as an acute illness in association with other systemic manifestations. The second type of renal disease is more chronic in nature and is characterized by a course similar to that of hypertensive nephrosclerosis and EBN. There is a gradual onset of uremia which is asymptomatic. The pathology described in such cases is primarily vascular and glomerular in location. The pattern of proteinuria seen in this disease is that of a glomerular one (Vacca *et al.*, 1980). Studies using chelating agents have failed to show any difference in lead excretion when comparing a normal to an "endemic" population (Gaon *et al.*, 1962). These data make it most unlikely that lead is the causative agent in EBN.

The epidemiology, clinical pattern, and laboratory findings in EBN suggest to some investigators that cadmium may be the causative agent. Cadmium nephropathy is characterized by subtle alterations in proximal tubular functions which can only be demonstrated by research techniques. These include altered handling of glucose, amino acids, and low-molecular-weight proteins by the proximal tubule. Long-term studies on populations exposed to chronic cadmium demonstrate that such a nephrotoxicity does not lead to the development of chronic end-stage renal disease in a significant percent of an exposed population (Kazantzis, 1979). Cadmium concentrations in hair samples, limited autopsy and biopsy material, and water supplies obtained from Yugoslavian populations at risk to develop EBN have been found to be within normal limits (Fajgelj *et al.*, 1975; M. Piscator, M. Vasiljevic, and P. W. Hall, unpublished data).

The chronic interstitial fibrosis and tubular atrophy characteristic of EBN can be induced by radioactivity. An evaluation of the concentrations of radioactive elements in the water and soil samples from villages in Bulgaria where the disease is prevalent showed that the concentrations of radon, radium, thorium, and uranium were within the accepted range of normal by international standards (Karamik-hailova *et al.,* 1960). These findings were confirmed in Yugoslavia (J. Peric, D. Stefanovic, Z. Radovanovic, and P. W. Hall, unpublished observations). There are too few data available to determine whether or not radioactivity plays a significant role in the development of EBN.

Silicic acid, released from silicate minerals, induces a proliferative inflammatory lesion in the interstitial tissue of the kidney with subsequent necrosis (Newberne and Wilson, 1970). The pathology seen in the experimental animal studies (Markovic and Lebedev, 1979) is that of lymphocytic and plasma cell infiltration around blood vessels, glomeruli, and tubules, with sclerosis of the interstitial tissue. This lesion was produced by feeding mice drinking water containing 50 mg of quartz plus 100 mg of quartzite/liter. Silicon has been described as inducing renal disease in humans. Neither the clinical course nor the histologic lesions described by light and electron microscopy are typical for those reported in EBN (Hauglustaine *et al.,* 1980). Silicate concentrations in renal tissue have been reported from only three cases reviewed by these authors. In each case it was found to be approximately ten times the normal level. No data are available on the silicate concentrations in the kidneys of patients with EBN. Analysis of water samples obtained from Bulgarian and Yugoslavian endemic foci, as reviewed elsewhere (Hall and Dammin, 1978), has not revealed concentrations of silicate approaching those used to induce the disease in experimental animals. This element has widespread distribution throughout the world. Not all forms of the element ingested or inhaled are toxic (Newberne and Wilson, 1970). Biopsy analysis of silicate concentrations, using emission spectroscopy, should yield valuable data to establish whether or not silica plays a role in endemic nephropathy. Although the silicon hypothesis is an intriguing one, there seems to be little evidence to support it as the causative agent.

The major pathology in EBN appears to be confined to the kidney. Preliminary data concerning involvement of the kidney in other species with lesions similar to human disease are very limited. A chronic interstitial nephropathy has not been observed in several species of domestic animals, including dog, sheep, cow, horse, pig, and chicken, nor in frogs (personal observations). That the disease is species- and organ-specific and that it requires several years to develop meet several of the criteria describing a slow virus diseae. This was first suspected to be the case in EBN when viral-like particles were seen by electron microscopy in the renal tubules of patients dying from EBN (Georgescu *et al.,* 1970). Subsequent attempts to isolate and cultivate a virus in culture were unsuccessful (Georgescu *et al.,* 1977). Numerous cytoplasmic vesicles in the proximal tubule containing particles which were characteristic of coronavirus were seen in seven cases of EBN using electron microscopic techniques (Apostolov *et al.,* 1975). The glomerular mesangial reaction and segmental thickening of the glomerular basement membrane with subendothelial membranous deposits observed by these investigators are quite different than the electron microscopic findings reported by others. Examination of renal biopsy material from 33 cases failed to reveal any recognizable lesion in the glomerulus by light or electron microscopy (Hall and Dammin, 1978). The data supporting a viral etiology remain controversial and relate only to the electron microscopic findings. A significant part of the problem leading to the confusion rests in variations in the diagnosis and selection of cases for microscopic study.

Ochratoxin A is a nephrotoxic mycotoxin metabolite occurring in foodstuffs. This toxin has been isolated from certain strains of *Aspergillus ochraceus,* which are commonly found contaminating wheat and grain stores. It has been established to be the cause of porcine nephropathy, which is comparable, in many ways, to EBN (Krogh, 1974). The toxin itself has been isolated in increased amounts in foodstuffs collected from several villages in Yugoslavia and Bulgaria where the disease is endemic (Barnes *et al.,* 1977; Krogh *et al.,* 1977). In addition, ochratoxin has been isolated from the serum of patients with EBN (Hult *et al.,* 1979). The renal pathology is characterized by extensive interstitial fibrosis and proximal tubular degeneration (Krogh *et al.,* 1977). Animal experiments have demonstrated that liver glycogen stores are depleted and serum glucose concentrations increased within hours after the administration of ochratoxin A (Suzuki *et al.,* 1975). Renal glycogen is also diminished (Meisner and Selanik, 1979). Meisner's study demonstrated that ochratoxin A diminished renal phosphoenolypruvate carboxy-kinase activity (PEPCK). The addition of ochratoxin A to fresh renal cortical slices depressed the transport of both organic anions and cations (Berndt and Hayes, 1979). Several studies, reviewed elsewhere (Hall and Dammin, 1978), have demonstrated an increase in amino acid excretion in patients with endemic nephropathy. Polyuria, glycosuria, proteinuria, and decreased urine osmolality are common findings in the animal with ochratoxin A-induced nephropathy and also in EBN (Berndt *et al.,* 1980). These investigators point out that repeated small doses of ochratoxin A have an cumulative effect. As PEPCK and ochratoxin are both found in the proximal tubule of the experimental animal, these tubular functional abnormalities may be a representation of the toxic effect of ochratoxin A. Measurement of the PEPCK in renal biopsy material obtained from appropriately selected cases of EBN should yield valuable information concerning the relationship between ochratoxin A and EBN.

The progression of EBN that occurs in emigrants who have left the endemic foxus is suggestive of a self-perpetuating disease having an autoimmune component. There are multiple brief conflicting reports concerning the detection of IgA, IgG, and IgM in kidney tissue (Hall and Dammin, 1978). No immune deposits could be detected, using a fluorescent tagged antibody, in the 33 renal biopsy cases from an endemic focus in Yugoslavia (Hall and Dammin, 1978). Doichinov was unable to detect C_3 or C_4 component of complement, IgA, IgG, or IgM in the glomeruli or tubules of ten renal biopsies taken from EBN patients from Bulgaria (personal communication). In a preliminary report, attempts to identify the presence of tubular antigen in biopsy material from patients with EBN were unsuccessful (Radonic *et al.,* 1977). Low titer antibodies to smooth muscle antigen and cytoplasmic antibodies have been reported in the sera of patients from Romania with tubular proteinuria (Moraru *et al.,* 1977). These investigators postulated that these antigen–antibody complexes subsequently become trapped in the glomerulus and lead to the development of the late glomerular alterations seen in EBN. Although there are some isolated reports to the contrary, available data do not support the concept of a major immunologic component in this disease.

There are similarities between EBN and analgesic abuse nephropathy. Both produce a chronic interstitial disease which takes a comparatively long time to develop. One very unusual and unique feature of both diseases is the associated increase in the prevalence of papillary transitional cell tumors of the renal pelvis and ureters but not the bladder. Histologic evaluation was made on tissue obtained from patients undergoing surgery for papillary transitional cell tumors. Carcinoma-*in-situ* was found in multiple sites in the transitional epithelium of the upper urinary tract. These patients were long-term residents of epidemic foci (Petkovic *et al.,* 1976). As early as 1960 it was suggested that the coincidental occurrence of the endemic nephropathy and the tumors was the result of a toxic and "blastogenic action at the same time, the latter being manifested after a prolonged period of time, more than 30 years" (Petrinska-Venkova, 1960). Data on the use of analgesics by the populations involved are not available. A combination analgesic containing phenacetin, aspirin, and caffeine has been found in many households of several village populations in endemic foci in Yugoslavia (personal observation). The availability (in 1976) of such analgesic compounds to the rural farm populations makes analgesic abuse a possibility. However, papillary necrosis, common in analgesic-induced disease, has not been seen in EBN patients. Further studies are needed to determine what role analgesics play in the development of EBN.

4. The Relationship of EBN to Papillary Transitional Cell Tumors

The coincident increased incidence of tumors in patients suffering from both analgesic abuse nephropathy and EBN suggests that some similar mechanism for induction of both diseases may exist. Phenacetin derivatives are most likely the carcinogenic agent in analgesic-associated cases (Bengtsson *et al.,* 1978). As these investigators pointed out, there is a close chemical relationship between *n*-hydroxylated metabolites of phenacetin and known carcinogenic amines which have been established to cause bladder cancer. Multiple metabolites of phenacetin are known to be carcinogenic. The increased prevalence of the carcinogenic changes in the renal pelvis and ureter in contrast to the bladder is similar in both analgesic abuse and EBN (Gonwa *et al.,* 1980; Petkovic, 1978). This finding suggests that the carcinogenic agent may only be active in the urine for a brief period as it flows through the pelvis and ureter and is inactivated by the time it reaches the bladder. An alternative explanation is that the metabolic activities of the renal pelvis and ureteral epithelial cells are different from those of the bladder cells, making them more susceptible to the mutagenic or carcinogenic agent. Embryologically the tissues are derived from different origins. The ureter and renal pelvis develop from the metanephric kidney tubules, while the bladder develops from an outpouching of the cloaca. No extensive search for mutagenic or carcinogenic compounds in the "endemic" environment has been conducted.

The relationship between cancer and renal tubular functional abnormalities is well recognized. Renal tubular acidosis has been described as a complication in chronic myelogenous leukemia and multiple myeloma. There is experimental evidence to indicate that an overload of low-molecular-weight proteins, particularly Bence-Jones proteins, results in proximal tubular functional abnormalities (Clyne *et al.,* 1979). We demonstrated, in prospective studies, the sequential changes in protein excretion that occur in a population destined to develop EBN (Hall and Vasiljevic, 1973). The urine concentrations of several low-molecular-weight proteins, including retinol binding protein, lysozyme, ribonuclease, 3S gamma globulin, IgG kappa light chains, and beta$_2$ microglobulin (B_2M), were determined 3–6 times/year for more than 5 years. The first of those proteins to be increased in the urine was B_2M. This protein continued to be spilled in increased quantities for a long period of time in some instances without the additional changes in proteinuria or development of kidney disease. Prior to a change in the SCr, the next protein to be detected elevated was the IgG kappa light chain. This was followed by a generalized increase in all the other low-molecular-weight proteins measured.

Beta$_2$ microglobulin is a light-chain-like protein having a molecular weight of 11,800 daltons. It occupies a position in the HLA molecule similar to that of the light chains on the immunoglobulin molecules. Amino acid sequencing studies have demonstrated that there are homologous regions on the constant fractions of IgG light and heavy chains to those of the B_2M molecule (Peterson *et al.,* 1972). Loading experimental animals with B_2M is associated with a marked tubular proteinuria consisting of several low-molecular-weight proteins, suggesting that this protein may play a role in inducing tubular injury (Hall *et al.,* 1980).

The similarities between light-chain nephropathy and EBN led us to consider the possibility that B_2M itself, if presented in increased loads to the kidney, may cause proximal tubular damage. In order to determine whether the B_2 microglobulinuria of EBN was representative of injury to the proximal tubule or an increased load of B_2M presented to the tubule, it was necessary to measure serum B_2M concentrations (SB_2M) and compare those with SCr in an "endemic" population. Such a study (Sattler *et al.,* 1977) showed that 13.4% of a "healthy" population living in a village where EBN is endemic had elevated SB_2M and normal SCr. This finding suggested that B_2M production may be increased in such populations. Elevated SB_2M has been reported in the sera of cancer patients, as reviewed elsewhere (Hall and Dammin, 1978). The source of the increased B_2M production could be the papillary transitional cell tumors. We have examined renal and tumor tissue obtained from patients operated on for papillary transitional cell tumors for the presence of B_2M in the tissue. All patients came from an endemic area of Bulgaria. Beta$_2$ microglobulin was found in the proximal tubular cytoplasm as well as in the stroma of the tissue, using the fluorescent conjugated antibody technique (Doichinov and Hall, 1978). Renal function studies were carried out on a group of such patients. We compared SB_2M, SCr, and fractional excretion of B_2M (FeB_2M) in three groups: (1) a suitable control population; (2) a "healthy" group living in a highly "endemic" area; and (3) a group of tumor patients operated on for papillary transitional cell tumors, all of whom came from the same area as those of group 2. The SB_2M/SCr ratios of those healthy individuals living in an endemic area were significantly higher than the control population ($p < 0.01$) and were as equally elevated in the tumor group. The FeB_2M was higher in the endemic healthy population than in the control and was highest in the tumor-operated patients (Hall *et al.*, 1981). These data can be interpreted to indicate that B_2M production was increased in both "healthy" individuals exposed to an endemic focus and in tumor patients. The findings of an increased fractional excretion in these groups suggested that tubular damage had occurred in addition to overloading.

It is interesting to speculate about the role of B_2M in the development of EBN. The protein itself can induce renal tubular functional abnormalities in experimental animals. The kidneys of a significant percent of the population at risk to develop EBN are presumably subjected to an increased load of this protein. Similar proximal tubular functional abnormalities have been attributed to an increased production of light chains in such diseases as multiple myeloma and myelomonocytic leukemia (Muggia *et al.*, 1969, Maldonado *et al.*, 1975). The fact that chronic interstitial nephropathy is not seen in patients with these diseases could be explained by the rapidity with which the patients succumb to their underlying malignancy. In contrast, papillary transitional cell tumors, which appear to contain large amounts of B_2M, are slow growing and much less malignant. As a consequence, people live for many years with carcinoma-*in-situ* lesions and papillomas without evidence of metastases. Several cases of bilateral tumors occurring sequentially have been reported (Petrinska-Venkova, 1960; Petkovic, 1978). The occurrence of B_2 microglobulinuria, elevated SB_2M concentrations, EBN, and papillary transitional cell papillomas in the same geographically isolated populations and often in the same patient seems more than a chance happening.

5. Prevention and Therapy

Without the etiology of EBN defined, preventative and therapeutic measures are limited. The available evidence indicates that the disease can be prevented if the individual has less than 15–20 years of exposure. A group of 110 individuals were relocated from a highly "endemic" area to a nonendemic area in Bulgaria and followed for 15 years (Dinev, 1974). Among the emigrants, 40% developed evidence of EBN over the followup period. There were 26 deaths during that time, 18 of which had EBN established at autopsy. Individuals relocated who had spent less than 10 years in the endemic focus prior to relocation showed no evidence of the disease during the followup period. Although this appears to be a satisfactory preventative measure, it obviously is not a practical one. Improvement in water supply, sanitary conditions, and other indicators of the standard of living has occurred over the past 15–20 years. Preliminary data (Fajgelj *et al.*, 1981) indicate that the prevalence of EBN is decreasing in certain endemic villages in Bosnia, Yugoslavia. These investigators imply that this is due to an improvement in the overall standard of living.

The therapy for EBN is supportive only. Multiple hemodialysis centers are now operating in the involved countries. Transplantation has proven to be effective in early reports. Several centers are now performing transplants in Yugoslavia. The first transplant of a patient with endemic nephropathy was performed in Romania recently (Pasare *et al.,* in press). Suitable candidates for transplant in Bulgaria are being sent to Moscow. Followup studies are available on seven transplants performed on patients with EBN from Nis, Yugoslavia who received their cadaver transplants in Lyon, France (Touraine *et al.,* in press). Six patients underwent transplants and one had a second transplant. The followup period was 4–46 months. All results, including biopsies, indicate that the course followed by these patients is typical of that of the transplant population in general. The individuals are currently living in endemic areas of Yugoslavia.

6. Summary and Conclusions

In summary, EBN is a chronic interstitial renal disease, the pathology of which is principally tubular atrophy and interstitial fibrosis of a uniform nature with minimal cellular infiltrate. It does not appear to have an immune component. It does appear to be an acquired disease and not a genetically inherited one. The ochratoxin A-induced animal disease is the model most closely approximating the human disease. The epidemiology fits very well with a slow virus etiology, but supporting data are morphologic only and controversial. Radiation, analgesic abuse, and silica have been implicated, but definitive studies have not been conducted to determine that any of these are causative agents. The coincidental occurrence of B_2 microglobulinuria, an elevated SB_2M, EBN, and papillary transitional cell tumors in the same population leads to the interesting speculation that the renal disease may be a "light-chain nephropathy." The solution of the problem of what causes EBN may have a direct bearing on the etiology of 5–15% of the patients currently being treated for end-stage renal disease in the United States who carry the diagnosis of chronic interstitial disease.

Acknowledgments

This work was supported in part by HEW Contract HSM 110-72-124, HEW Contract 86-68-107, and HEW Grant AM 19556.

This work obviously represents the collaborative research efforts of many principal investigators. I would like to express my special gratitude to Prof. J. Gaon, Dr. M. Vasiljevic, Dr. M. Popovic, the late Prof. V. Danilovic, the late Prof. A. Puchlev, Dr. Z. Radovanovic, Dr. D. Doichinov, Dr. Ts. Dimitrov, Dr. I. Dinev, Dr. A. Hrabar, Prof. S. Petkovic, Dr. G. Dammin, Dr. M. Chung-Park, Prof. M. Piscator, Engineer I. Peric, and Prof. D. Stefanovic, as well as to the late Dr. C. H. Rammelkamp for his continuous guidance and constructive criticism. I wish to acknowledge also the expert secretarial assistance provided by Heather Bailey.

References

Apostolov, K., Spasic, P., and Bojanic, N., 1975, Evidence of a viral etiology in endemic (Balkan) nephropathy, *Lancet* **1975** (December 27):1271.

Barnes, J. M., Austwick, P. K. C., Carter, R. L., Flynn, F. V., Peristianis, G. C., and Aldridge, W. N., 1977, Balkan (endemic) nephropathy and a toxin-producing strain of *penicillium verrucosum* var *cyclopium:* An experimental model in rats, *Lancet* **1977** (March 26):671.

Bengtsson, U., Johansson, S., and Angervall, L., 1978, Malignancies of the urinary tract and their relation to analgesic abuse, *Kidney Int.* **13**:107.

Berndt, W. O., and Hayes, A. W., 1979, *In vivo* and *in vitro* changes in renal function caused by ochratoxin A in the rat, *Toxicology* **12**:5.

Berndt, W. O., Hayes, A. W., and Phillips, R. D., 1980, Effects of mycotoxins on renal function: Mycotoxic nephropathy, *Kidney Int.* **18**:656.

Clyne, D. H., Pesce, A. J., and Thompson, R. E., 1979, Nephrotoxicity of Bence Jones proteins in the rat: Importance of protein isoelectric point, *Kidney Int.* **16**:345.

Danilovic, V., 1958, Chronic nephritis due to ingestion of lead-contaminated flour, *Br. Med. J.* i:27.

Danilovic, V., and Stojimirovic, B., 1967, Endemic nephropathy in Kolubara Serbia, in: *The Balkan Nephropathy,* Ciba Foundation Study Group No. 30 (G. E. W. Wolstenholme and J. Knight, eds.), Churchill, London, p. 44.

Dinev, I., 1974, Results of long-term observations of patients and normal individuals who have emigrated from the village of Karash and settled in Sofia suburb villages, in: *Endemic Nephropathy, Proceedings of the 2nd International Symposium on Endemic Nephropathy* (A. Puchlev, I. V. Dinev, B. Milev, and D. Doichinov, eds.), Publishing House of the Bulgarian Academy of Sciences, Sofia, Bulgaria, p. 138.

Doitchinov, D., and Hall, P. W., 1978, Beta$_2$ microglobulin (B$_2$M) distribution in tumors and renal tissue of patients with Balkan nephropathy (BEN) and carcinoma of the kidney, in: *7th International Congress of Nephrology.*

Fajgelj, A., Popovic, N., and Durovic, N., 1975, Cadmium as a possible etiological factor of endemic nephropathy, *Medski Arh.* **29**:241.

Fajgelj, A., Dedic, I., Pasic, I., Mulaomerovic, N., Gaon, J., and Filipovic, A., 1981, Trogodisnji (1975–1977) republicki program za otkrivanje I suzbijanje endemske nefropatije u sr Bosni I Hercegovini, in: *1st Yugoslav Congress of Nephrology,* Galenika, Belgrade, Yugoslavia, p. 42.

Gaon, J., Griggs, R. C., Vasiljevic, M., and Albegovic, S., 1962, Investigation of chronic endemic nephropathy in Yugoslavia. I. Lead as a possible etiologic agent, *Acta Med. Iugoslav.* **16**:347.

Georgescu, L., Litvac, B., Manescu, N., Petrovici, A., Schwarzkopf, A., and Zosin, C., 1970, Particules virales dans le rein de la nephropathie endemique Balkanique, *Sem. Hop.* **46**:3526.

Georgescu, L., Litvac, B., Diosi, P., Plavosin, L., and Herzog, G., 1977, Viruses in Balkan nephritis, *Am. Heart J.* **94**:805.

Gonwa, T. A., Corbett, W. T., Schey, H. M., and Buckalew, V. M., 1980, Analgesic-associated nephropathy and transitional cell carcinoma of the urinary tract, *Ann. Intern. Med.* **93**:249.

Hall, P. W., and Dammin, G. J., 1978, Balkan nephropathy, *Nephron* **22**:281.

Hall, P. W., and Vasiljevic, M., 1973, Sequential development of tubular proteinuria in Balkan nephropathy, in: *Protides of the Biological Fluids—21st Colloquium* (H. Peeters, ed.), Pergamon Press, Oxford, p. 501.

Hall, P. W., Gaon, J., Griggs, R. C., Piscator, M., Popovic, N., Vasilovic, M., and Zimonjic, B., 1967, The use of electrophoretic analysis of urinary protein excretion to identify early involvement in endemic (Balkan) nephropathy, in: *The Balkan Nephropathy,* Ciba Foundation Study Group No. 30 (G. E. W. Wolstenholme and J. Knight, eds.), Churchill, London, p. 72.

Hall, P. W., Ricanati, E. S., Vacca, C. V., and Chung-Park, M., 1980, Renal metabolism of B$_2$-microglobulin. Experimental animal studies, *Vox Sang.* **38**:343.

Hall, P. W., Doichinov, D., Dinev, I., Dimitrov, Ts., Radovanovic, Z., and Dammin, G., 1981, Balkan nephropathy: Relationship with transitional cell cancer, presented at the VIIIth International Congress of Nephrology, Greece.

Hauglustaine, D., Damme, B. V., Daenens, P., and Michielsen, P., 1980, Silicon nephropathy: A possible occupational hazard, *Nephron* **26**:219.

Hult, K., Hokby, E., Gatenbeck, S., Plestina, R., and Ceovic, S., 1979, Ochratoxin A and Balkan endemic nephropathy IV. Occurrence of ochratoxin A in humans, *Separatdruck aus Chem. Rundschou* **35**:32.

Karamikhailova, E., Nikolov, K., Doichinova, K., and Mikhailova, V., 1960, On the radioactivity of the water sources in the villages affected by endemic nephritis, in: *Endemic Nephritis in Bulgaria* (A. Puchlev, ed.), Medizina i Fizkultura, Sofia, Bulgaria.

Kazantzis, G., 1979, Renal tubular dysfunction and abnormalities of calcium metabolism in cadmium workers, *Environ. Health Persp.* **28:**155.

Krogh, P., 1974, Mycotoxic porcine nephropathy: A possible model for Balkan endemic nephropathy, in: *Endemic Nephropathy, Proceedings of the 2nd International Symposium on Endemic Nephropathy* (A. Puchlev, I. V. Dinev, B. Milev, and D. Doichinov, eds.), Publishing House of the Bulgarian Academy of Sciences, Sofia, Bulgaria, p. 266.

Krogh, P., Hald, B., Plestina, R., and Ceovic, S., 1977, Balkan (endemic) nephropathy and foodborn ochratoxin A: Preliminary results of a survey of foodstuffs, *Acta Pathol. Microbiol. Scand. B* **85:**238.

Maldonado, J., Velosa, J., Kyle, R., Wagoner, R., Holley, K., and Salasa, R., 1975, Fanconi syndrome in adults. A manifestation of a latent form of myeloma, *Am. J. Med.* **58:**354.

Markovic, B., and Lebedev, S., 1979, The Balkan endemic nephropathy and cancer of the urinary tract, *Etiol. Lab. Work Prevent.* **1**(1):1179.

Meisner, H., and Selanik, P., 1979, Inhibition of renal gluconeogenesis in rats by ochratoxin, *Biochem. J.* **180:**681.

Minev, M., Mikhaylov, T., Kastelan, A., Nyulassy, S., and Menzel, G., 1978, HLA system and Balkan endemic nephropathy, *Tissue Antigens* **11:**50.

Moraru, I., Karlsson, A., Lenkei, R., Dan, M.-E., Melencu, M., Cristian, R., and Dobre, I., 1977, Beta$_2$-microglobulin excretion and autoimmune phenomena in endemic Balkan nephropathy, *Rev. Roum. Med. Interne.* **15**(4):311.

Muggia, F., Heinemann, H., Farhangi, M., and Osserman, E., 1969, Lysozymuria and renal tubular dysfunction in monocytic and myelomonocytic leukemia, *Am. J. Med.* **47:**351.

Murray, T., and Goldberg, M., 1975, Chronic interstitial nephritis: Etiologic factors, *Ann. Intern. Med.* **82:**453.

Newberne, P. M., and Wilson, R. B., 1970, Renal damage associated with silicon compounds in dogs, *Proc. Natl. Acad. Sci.* **65:**872.

Pasare, G., Melencu, M., and Pirvulescu, C., in press, Le premier cas de transplant de rein chez un malade de nephropathie endemique, in: *IVth Symposium on Balkan Nephropathy*, Institute of Nephrology and Hemodialysis, Faculty of Medicine, Nis.

Peterson, P. A., Cunningham, B. A., Berggard, I., and Edelman, G. M., 1972, B$_2$-Microglobulin. A free immunoglobulin domain, *Proc. Natl. Acad. Sci.* **69:**1697.

Petkovic, S. D., 1978, Treatment of bilateral renal pelvic and ureteral tumors. A review of 45 cases, *Eur. Urol.* **4:**397.

Petkovic, S., Mutavdzic, M., Petronic, V., and Markovic, V., 1968, Geographical distribution of urothelium in Yugoslavia, *Urologia Suppl. (Treviso)* **35:**425.

Petkovic, S., Dammin, G., and Isvaneski, M., 1976, Metaplastic and neoplastic alterations by renal pelvic and ureteral tumours on normal looking mucosa, *Proc. Serb. Acad. Sci. Arts, Div. Med. Sci.* **27:**141.

Petrinska-Venkova, S., 1960, Morphological studies on endemic nephritis, in: *Medicine & Physcultura* (A. Puchlev, ed.), State Publisher, Sophia, Bulgaria, p. 72.

Radonic, M., Radosevic, Z., and Zupanic, V., 1966, Endemic nephropathy in Yugoslavia, in: *Symposium on Geographic Pathology of Renal Diseases,* International Academy of Pathology Monograph No. 6, Williams & Wilkins, Maryland, p. 503.

Radonic, M., Richet, G., Morel-Maroger, L., and Ceovic, S., 1977, Odredivanje antitijela protiv bazalne membrane tubular u bolesnika s endemskom nefropatijom, in: *Proceedings of the 3rd Symposium on Endemic Nephropathy,* Galenika, Belgrade, Yugoslavia, p. 196.

Sattler, T. A., Dimitrv, Ts., and Hall, P. W., 1977, Relationship between endemic (Balkan) nephropathy and urinary-tract tumours, *Lancet* **iii:**278.

Stojimirovic, B., 1974, Recent data on the incidence and distribution of endemic nephropathy along the Kolubara River, in: *Endemic Nephropathy, Proceedings of the 2nd International Symposium on Endemic Nephropathy* (A. Puchlev, I. V. Dinev, B. Milev, and D. Doichinov, eds.), Publishing House of the Bulgarian Academy of Sciences, Sofia, Bulgaria, p. 297.

Suzuki, S., Satoh, T., and Yamakazi, M., 1975, Effects of ochratoxin A on carbohydrate metabolism in rat liver, *Toxicol. Appl. Pharmacol.* **32:**116.

Touraine, J. L., Malik, M. C., Stefanovic, V., and Traeger, J., in press, Kidney transplantation in endemic nephropathy, in: *IVth Symposium on Balkan Nephropathy,* Institute of Nephrology and Hemodialysis, Faculty of Medicicine, Nis.

Vacca, C. V., Hall, P. W., and Hines, J. D., 1980, Heavy metal nephrotoxicity—Lead differentiated from cadmium and mercury, *Am. J. Clin. Pathol.* **73:**308.

Wolstenholme, G. E. W., and Knight, J. (eds.), 1967, *The Balkan Nephropathy,* Ciba Foundation Study Group No. 30, Churchill, London.

21

Nephrotoxicity of Natural Products
Mycotoxin-Induced Nephropathy

WILLIAM O. BERNDT

1. Introduction

No doubt a number of naturally produced chemical compounds have dramatic effects on renal function. Although it is conceivable that such effects may be beneficial as well as detrimental, in general, most of the focus of the biomedical and toxicologic community has been on the potential for the production of nephrotoxicity.

The mycotoxins, secondary fungal metabolites, are a diverse group of chemical substances produced by a diverse group of fungi. The diversity of the chemical nature of these compounds affords an excellent opportunity to study a potentially broad range of effects pertaining to renal function. As toxic mold metabolites, the mycotoxins have the potential to cause various diseases either after their ingestion or after contact to the skin. The diseases that may result from exposure to these substances are called mycotoxicoses. This description does not distinguish among effects on the nervous system, the liver, or the kidney, but simply describes the occurrence of a disease state resulting from the ingestion of or exposure to a

WILLIAM O. BERNDT • Department of Pharmacology and Toxicology, University of Mississippi Medical Center, Jackson, Mississippi 39216. Present address: Department of Pharmacology, College of Medicine, University of Nebraska Medical Center, Omaha, Nebraska 68105.

mycotoxin. Overt fungal infestation is not essential for the production of mycotoxicoses.

Humanity's contact with fungal products is not a new experience. Although the best documentation of an early mycotoxicosis comes from the ergotism experience of the Middle Ages, there are suggestions that even earlier problems occurred (Bagger, 1931). Indeed, the likelihood is rather great that mycotoxin contamination of food and animal feed has been with us for many, many years. In societies where modern food storage techniques have been employed, fungal contamination of food has not been considered a serious problem. Indeed, much of the major mycotoxicologic work in the past several decades has been directed at animal feed problems. Nonetheless, a generalization to humans from studies in domestic animals is important, because there have been discrete occurrences of human involvement. For example, the products of *Aspergillus flavus*, the aflatoxins, can be transferred directly to humans or transferred to humans through the mediation of farm animals. Aflatoxin B_1, for example, can contaminate human food supplies directly, or humans might receive aflatoxin M_1 (a carcinogenic metabolite of B_1) from milk. The cow converts the B_1 compound to the M_1 metabolite and passes the metabolite into the milk. Similarly, human consumption of other fungal products may come from the ingestion of contaminated meat and meat products, as has been suggested by Krogh (1976) to be the situation for ochratoxin A in pork.

Although clearly documented examples of human fungal intoxication (aside from ergotism) are relatively hard to find in the literature, there are some interesting and outstanding examples. Epidemiologic data suggest that aflatoxin ingestion by human populations in several countries have led to severe hepatic diseases, including hepatocarcinoma. These studies from Uganda, Taiwan, Kenya, and other countries have involved the occurrence of unusual outbreaks of diseases among various populations of residents. In general, it was possible to demonstrate a cause-and-effect relationship between the disease and the consumption of a common type of food that was mold-contaminated and contained measurable quantities of the aflatoxins. In addition, surveys of various food products for the fungi or the aflatoxins themselves have yielded positive results in some studies. Details of these epidemiologic investigations have been summarized by Hayes (1979).

Studies in laboratory animals clearly document a variety of effects of mycotoxins other than the hepatotoxic or carcinogenic effects of the aflatoxins. Several chemically unrelated fungal metabolites have been demonstrated to produce tumors in experimental animals: patulin, penicillic acid, luteoskyrin. Rubratoxin B has been shown to be mutagenic. Various prenatal effects of mycotoxins have been reported. For example, fetocidal effects and/or teratogenesis have been observed with a number of toxins: aflatoxin B_1, ochratoxin A, rubratoxin B, T-2 toxin, etc. (LeBreton *et al.,* 1964; Hayes, 1978). Various organ-specific actions of mycotoxic substances also have been reported: for example, effects of rubratoxins on hepatic function, penicillic acid effects on the cardiovascular system, citreovirdin effects on the nervous system, and so forth. Hence, not only do the mycotoxins represent a diverse chemical group of substances, they also present a great diversity of toxicologic actions. In addition to all of the above, effects on renal function have been well documented.

Several mycotoxins, including aflatoxin B_1 and rubratoxin B, but most significantly citrinin and ochratoxin A, have effects on renal function. Although renal function studies have not been done with aflatoxin B_1 or rubratoxin B, Hayes and Williams (1977) demonstrated morphologic changes in dog kidney that were similar to those reported in the natural outbreaks of contamination by these substances. Renal damage caused by rubratoxin B was primarily tubular in nature. Details of aflatoxin effects on renal structure or function have not been defined.

Citrinin, produced by *Penicillium citrinum, P. verdicatum,* and other molds, has a well-documented history of nephrotoxic effects (Berndt and Hayes, 1977). Ochratoxin A, produced by *Aspergillus ochraceus,* also produces alterations in renal function (Berndt and Hayes, 1979). Both substances have been implicated in the production of porcine nephropathy (Krogh *et al.*, 1973; Krogh, 1976). This disease is characterized by polyuria, glucosuria, proteinuria, and a reduced urine osmolality. These effects can be produced in pigs either by feeding grain contaminated with the appropriate mold, feeding the mold directly, or by the administration of citrinin and/or ochratoxin A. There is little doubt concerning the cause-and effect-relationship of citrinin and/or ochratoxin A in the production of this disease, and there is a reasonable probability that ochratoxin A may produce a similar syndrome in certain avian species (Elling *et al.*, 1975).

Whether or not a human renal dysfunction occurs associated with mycotoxin ingestion remains to be seen. The endemic Balkan nephropathy may be such a disease (Barnes, 1967), but absolute data confirming a cause-and-effect relationship between toxin ingestion and the disease are lacking. The similarity between Balkan nephropathy and porcine nephropathy, however, is unmistakable (Berndt *et al.,* 1980). Krogh *et al.* (1977) have found a significant contamination by ochratoxin A of various foodstuffs in one endemic village as well as a significant occurrence of ochratoxin A in the blood of affected individuals. On the other hand, Barnes *et al.,* (1977) reported that the most frequent mold contaminant from those same food samples was a non-ochratoxin A or citrinin producer. Further, young rats fed culture extracts of that mold, *P. verrucosum,* developed only modest histologic changes in their nephrons, a situation quite different from animals fed ochratoxin A or citrinin. On the other hand, *P. viridicatum* also is reported to be an ochratoxin A and citrinin producer and is as common a contaminant of cereals and vegetables as *P. citrinum* and *A. ochraceus.* The exact role of fungal toxins, however, in the production of the endemic Balkan nephropathy is quite unsettled.

A first step in a better understanding of mycotoxin-induced nephropathy is the development of an animal model suitable for laboratory studies. Barnes *et al.,* (1977) and Berndt and Hayes (1977) have used the laboratory rat as such a model. Berndt, Hayes, and colleagues employed the Sprague-Dawley rat in a series of studies on citrinin-induced alterations in renal function (Berndt and Hayes, 1977; Berndt *et al.*, 1980; Phillips *et al.*, 1980b), as well as to study effects produced by ochratoxin A (Berndt and Hayes, 1979). All of these investigations involved the acute administration of citrinin and both whole animal (anesthetized and unanesthetized) and *in vitro* experiments were undertaken.

2. Effects of Citrinin on Renal Function and Structure

A sustained disruption of renal function occurs in the unanesthetized rat after the administration of a single dose of citrinin intraperitoneally. Depending upon the dose, renal dysfunction persists for 3 or 4 days, with the occurrence of full recovery by 7 or 8 days. For animals that survive for 36 hr after administration of the toxin, the renal dysfunction is characterized by a "high-output" renal failure. In this syndrome the daily urine volume is three or four times the normal and the urine osmolality sustained at less than one-third of normal (Fig. 1). Accompanying these observations is an increase in urinary glucose excretion as well as a marked proteinuria. The BUN is elevated in a dose- and time-dependent manner. Animals such as these maintained in metabolism cages will lose weight initially, but subsequently will eat and gain weight at a more or less normal rate even while renal dysfunction persists. Many animals that die after the administration of citrinin do so within 36–48 hr of the injection with anuric acute renal failure. This overall pattern of response agrees well with that reported in the literature for porcine nephropathy (Krogh, 1976; Krogh *et al.*, 1973) and with reports of a similar disease in poultry (Elling *et al.*, 1975).

A more detailed examination of renal function in the anesthetized rat also has been undertaken (Phillips *et al.*, 1980). Despite the elevated 24-hr urine output, the anesthetized rat that had received citrinin 24–48 hr before the clearance experiment showed a greatly reduced urine flow assessed on a ml/min basis. In addition, both the para-aminohippurate (PAH) clearance and the glomerular filtration rate

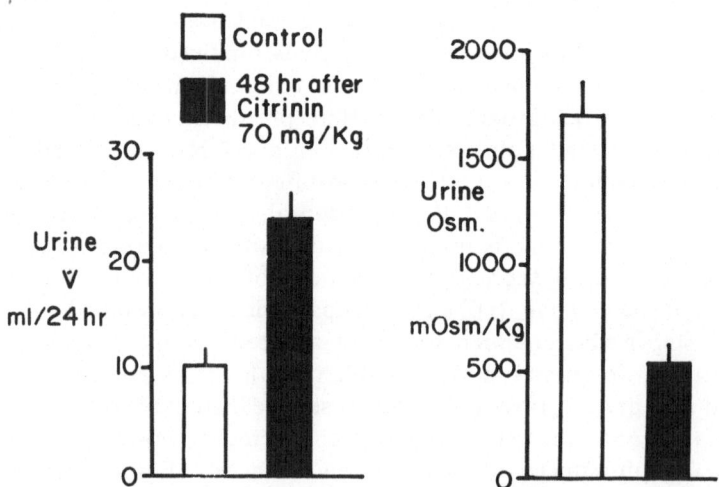

FIGURE 1. Effect of a nephrotoxic dose of citrinin on urine volume and urine osmolality in the male rat. Height of the bar is the mean and the vertical line the standard error for *n* = 4 experiments.

TABLE I. Effect of Citrinin on Renal Function Parameters in the Anesthetized Rat.[a]

	Values as percent of control	
	Citrinin, acute 50 mg/kg	Citrinin preRx 24 hr before exp., 50 mg/kg
Urine flow	89	22[b]
C_{In}	65[b]	8[b]
C_{PAH}	61[b]	1[b]
Na^+ excretion	81	2[b]
K^+ excretion	24[b]	15[b]
Percent change in filtered K^+ reabsorbed	35[b]	−80[b]

[a] Measurements made 120 min after start of infusion containing PAH and inulin.
[b] Significantly different from control, $p < 0.05$.

(inulin clearance) were reduced (Table I) whether the rats were pretreated with citrinin or it was administered acutely. An explanation for the discrepancy in urine flow between the animals housed in metabolism cages (Fig. 1) and those animals used for clearance experiments (Table I) is uncertain. Possibly the oliguria observed in the clearance experiments results from an interaction of anesthesia with the already compromised renal failure.

In addition to effects on filtration and blood flow, citrinin pretreatment causes a very dramatic reduction in sodium excretion and potassium excretion. The normally observed potassium reabsorption seen in the anesthetized rat was converted to potassium secretion after a relatively large dose of citrinin administered 24–48 hr before the clearance experiment. Whether or not this reflected a direct loss of potassium from the diseased renal tissue into the tubular fluid or a loss of potassium from the extracellular stores has not been determined. In any event, the absolute magnitude of the potassium loss was relatively modest after citrinin administration. That is to say, total potassium excretion was greatly reduced, although the amount of potassium appearing in the urine was too large to be accounted for by glomerular filtration alone.

Acute administration of citrinin to the anesthetized rat will produce effects within 2 hr that are similar to those described above. The magnitude of the effect (Table I) is considerably less than that observed in animals pretreated with the nephrotoxin, but significant reductions in glomerular filtration, renal blood flow, and potassium excretion are observed. By 2 hr, urine flow and sodium excretion remain unaffected.

The decreased potassium excretion noted with citrinin given either acutely or in a pretreatment regimen is somewhat reminiscent of that produced by the organomercurial diuretics or amiloride. These compounds appear to act in the distal nephron of the kidney to reduce potassium secretion, particularly if the secretory mechanism is very active. A similar mechanism could be involved in the reduction of potassium excretion following the acute administration of citrinin, although at this time there is no direct evidence for that mechanism. The net tubular secretion of potassium observed in the pretreated rats probably involves a complicated mechanism. There is no doubt that during citrinin-induced nephrotoxicity the integrity of the proximal tubular membrane is lost (see below). Because a large quantity of filtered potassium is reabsorbed along the proximal segment of the nephron, one might anticipate an increased delivery of potassium to distal segments. This, coupled with normal or near normal potassium secretion along the distal segment, might account for the net secretion of potassium. This could occur, however, only if the acute inhibitory effect of citrinin on potassium excretion would have disappeared by the 24-hr time period when the clearance experiment was performed on the pretreated animals.

Electron microscopic studies of renal tissue removed from rats pretreated with citrinin support entirely the time course of the events observed in the clearance experiments. Lockard *et al.* (1980) examined tissues at 24-hr intervals for 4 days after the administration of a nephrotoxic dose of citrinin. At each time period studied, the most prominent changes observed were in the cells of the proximal convoluted tubule. Distal tubular damage was not observed and other components of the nephron appeared relatively normal morphologically. By 48 hr, multifocal areas of necrosis were observed in the proximal convoluted segments. In addition, extensive mineral deposits were present in the necrotic areas. By 72 hr, initial regeneration of the proximal tubular cells was apparent in some of the necrotic areas of the cortex. By 96 hr, tubular regeneration was prominent. The time course of morphologic changes agrees well with the renal function studies. For example, the morphologic disruption was observed by 24 hr but peaked only at 48 and 72 hr. Maximal functional disturbances were observed during the 48- to 72-hr time period. The regeneration of nephron cells by 72–96 hr is compatible with the beginning of recovery of normal function observed over those times.

Two observations are worth noting with respect to the morphologic studies. First, the overall character of the disruption was very similar to that observed with other nephrotoxic substances. That is to say, vacuolization, swollen mitochondria, necrosis, etc. is a commonly reported sequence of events observed with a variety of nephrotoxins. Second, the damage observed in the rat kidney was confined to the proximal convoluted tubule. There was no evidence of damage to the straight part of the proximal tubule, let alone distal tubular areas. Studies in the mouse and dog by Carlton and colleagues (Carlton *et al.*, 1973; Jordan *et al.*, 1977) suggested the occurrence of distal tubular lesions. Although these investigators did find effects on the proximal tubule, these effects were observed only after very large doses of the nephrotoxin. Studies in the rat (Lockard *et al.*, 1980) do not substantiate these studies. This is an interesting species difference and one which might be pursued profitably in an attempt to better understand the nephrotoxic phenomena.

3. Effects of Citrinin on Renal Transport Mechanisms

In addition to the obvious effects on renal function overall, citrinin also has effects on specific renal transport processes. Whether administered as a pretreatment or added directly to fresh renal cortex slices, citrinin can reduce the transport (renal slice accumulation) of PAH (Fig. 2). Under acute conditions, it is possible that the effect on PAH transport is a competitive one, since citrinin is an organic anion and shows clearance values consistent with renal tubular secretion (Phillips *et al.*, 1980b). However, a competitive effect of citrinin on TEA (tetraethylammonium) transport is considerably less likely, although an interaction of organic anion and organic cation transport has been reported for some substances (Koschier and Berndt, 1977). After pretreatment, virtually all of the administered citrinin had disappeared by 48 hr (see below; Phillips *et al.*, 1979) and yet the effects of citrinin on PAH and TEA transport still persisted. These observations suggest that the action of citrinin in blocking organic anion and cation transport was related to some sustained, metabolic, or structural alteration related to the nephrotoxicity. This does not exclude the possibility that some component of the transport inhibition may relate to a competitive effect, such as has been reported with dinitrophenol (Berndt and Grote, 1968), but it is likely that the major, sustained effect was related to whatever was the cause of the nephrotoxicity. Not all substances are affected as dramatically as PAH and TEA. The nonmetabolizable, neutral amino acid α-aminoisobutyrate (AIB) was transported normally in tissues removed from animals pretreated with citrinin. When citrinin was added directly to

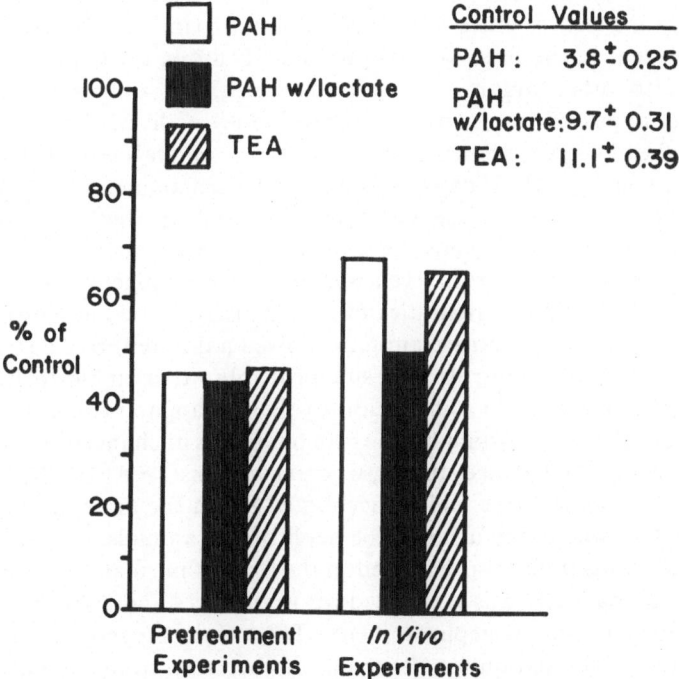

FIGURE 2. Effect of citrinin on renal transport of organic ions by rat renal cortex slices. In the pretreatment experiments, citrinin, 70 mg/kg, was administered IP 3 days before the slice experiment. For the *in vitro* experiments, citrinin, 5×10^{-5} M, was added to fresh renal cortex slices; $n = 3$.

fresh renal slices, AIB transport was modestly enhanced. No depression of AIB transport was observed in any experimental protocol (Berndt and Hayes, 1977).

The dramatic effects of citrinin on PAH and TEA transport corroborate entirely the morphologic studies suggesting a site of action in the proximal convoluted tubule. Although considerable PAH transport occurs in the pars recta or straight part of the proximal tubule in the rat, there is also no doubt that significant organic anion transport occurs in the convoluted proximal segment of the nephron as well (Roch-Ramel and Weiner, 1980). The morphologic observations suggest damage to the convoluted segment and the PAH and TEA transport studies fully substantiate this observation.

4. Pharmacokinetics and Metabolism of Citrinin

Carbon 14-labeled citrinin has been used to study the pharmacokinetics and tissue distribution of this fungal metabolite. After the intravenous administration to the rat of the [^{14}C]citrinin in a non-nephrotoxic dose, the radioactivity disappeared from the blood compartment by a double exponential process. Approximately 10% of the injected radioactivity was in the plasma 15–30 min after injection and this decreased to less than 1% by 12 hr; only trace quantities remained by 48–72 hr. Only the liver and the kidney contained large amounts of radioactivity. Fifteen minutes after injection approximately 30% of the injected dose was present in the liver, with approximately 8% in the kidney. The possible distribution to other organs also was examined, but only small amounts of radioactivity appeared in organs other than kidney or liver.

When the above experiments were repeated in animals rendered nephrotoxic (citrinin, 55 mg/kg, IP) 3 days before the pharmacokinetic study, a biphasic disappearance of ^{14}C from plasma also was observed. However, the pharmacokinetic parameters (Table II) were changed from those calculated for animals with intact renal function. The serum levels were maintained longer when renal function was compromised and the quantities of radioactivity in the kidney and liver also were sustained. The nephrotoxic animals showed a different rate constant for the alpha phase than did controls, a result probably attributable to loss of renal function. Other methods of elimination of the ^{14}C apparently were maintained, however, since the rate constant for the beta phase was unchanged. This latter point was supported by the balance studies, where it was observed that although 80% of administered radioactivity could be accounted for in the urine of normal rats by 48 hr, only 32% was in the urine of the nephrotoxic animals. On the other hand, 15% of total radioactivity was excreted in the feces of normal rats, whereas nearly 50% of the radioactivity was accounted for in the feces and colon contents of the rats with citrinin-induced nephrotoxicity. These data suggest that although the normal route of elimination for citrinin is via the urine, in the absence of normal renal function other excretory mechanisms are available. Studies with the isolated perfused liver (H. M. Mehendale and W. O. Berndt, unpublished) indicate that relatively large amounts of radioactivity appear in the bile after addition of

TABLE II. Pharmacokinetic Parameters for
the Disappearance of ^{14}C from the Serum of
Rats after Administration of $[^{14}C]$citrinin:
$$(Cp)_t = Ae^{-\alpha t} + Be^{-\beta t}$$

	Control[a]	Nephrotoxic[b]
A	10.5	29.4
B	0.54	11.4
α, hr^{-1}	0.27	1.18
β, hr^{-1}	0.046	0.049

[a] $[^{14}C]$ Citrinin, 5 mg/kg, given IV on day of experiment.
[b] Citrinin, 55 mg/kg, given IP, 3 days before the pharmacokinetic experiment. On the day of the experiment, 5 mg/kg
$[^{14}C]$citrinin was administered IV.

$[^{14}C]$citrinin to the perfusate. Possibly under normal conditions citrinin and/or its metabolites undergo enterohepatic circulation with ultimate disposition through the urine.

Through the use of high-performance liquid chromatography (Phillips *et al.*, 1980a), it has been possible to demonstrate that small amounts of citrinin are metabolized by the rat. At least two polar metabolites of citrinin were observed in the serum, bile, and urine in animals that received the mycotoxin. Quantitatively, in acute experiments these compounds account for a very small percentage (less than 1%) of the total citrinin administered if the dose is non-nephrotoxic. With nephrotoxic doses of the order of 50–70 mg/kg, as much as 50% of the radioactivity excreted in urine appeared as metabolites. The chemical nature of the metabolites is unknown. Incubation with glucuronidase or sulfatase suggested that the metabolites were neither glucuronides nor sulfates. Because the metabolites are more polar than citrinin, they could be mercapturic acid derivatives resulting from glutathione conjugation, or oxidation products, e.g., alcohols or acids. Studies with the isolated perfused kidney and HPLC analysis (Berndt, unpublished) failed to demonstrate renal metabolism of citrinin. Probably, therefore, the liver is the primary organ of metabolism.

5. Effects of Citrinin on Biochemical Processes in the Kidney

The mechanisms underlying the production of nephrotoxicity by citrinin are obscure. Because citrinin exerts some effects under purely *in vitro* conditions, it is possible, if not likely, that the nephrotoxic response observed in the intact animal is attributable to disruption of biochemical events in the intrarenal distribution of blood. Further, these data suggest that citrinin itself has the potential for altering renal function. Of course, this does not preclude the possibility that a metabolite is active. Even though the kidney itself does not appear to metabolize citrinin, a reactive metabolite might be made in a distant organ, for example, the liver. Whether the parent compound or metabolite is active, various biochemical functions of the kidney might be the target of the toxic species. For example, the effects of citrinin on nonprotein sulfhydryl groups (glutathione predominantly, GSH) has been examined. In addition, the possibility of "covalent binding" of citrinin or a citrinin metabolite by renal tissue also has been investigated.

The administration of a nephrotoxic dose of citrinin to the rat has significant effects on both renal (Fig. 3) and hepatic glutathione content. Within 1 hr after administration of the nephrotoxin, renal glutathione falls to approximately 50% of its normal value. Recovery proceeds promptly, with control values being achieved in 12 hr and values greater than control seen by 3 days. As prompt a reduction in liver glutathione occurs after citrinin as in the kidney, but recovery is somewhat slower and no overshoot has been observed. In any event, liver glutathione content reaches control values by 24–48 hr after citrinin. Another nephrotoxin, ochratoxin A, has relatively minor effects on renal or hepatic glutathione. The difference between ochratoxin A and citrinin is most striking at the early time periods. Although the magnitude of the reduction in kidney glutathione is not as great as that produced by other nephrotoxins (Mudge *et al.*, 1978), the pattern of response and the temporal relationships are not unlike those seen with other nephrotoxic compounds, such as acetaminophen. That is, the very early depletion of glutathione followed by relatively prompt recovery, with the toxicity expressed only after a considerable time lag, i.e., hours to days, is the standard pattern observed for agents acting through some form of reactive intermediate.

To examine this possibility further, attempts were made to determine the extent, if any, of covalent binding by renal tissue of [^{14}C]citrinin administered to

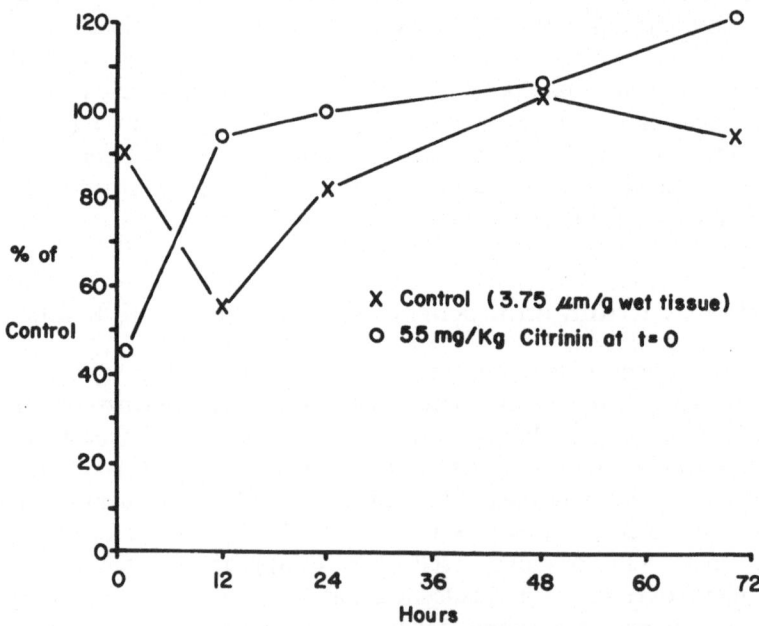

FIGURE 3. Effect of citrinin on renal cortical nonprotein sulfhydryl groups (NPSG, GSH); *n* = 3.

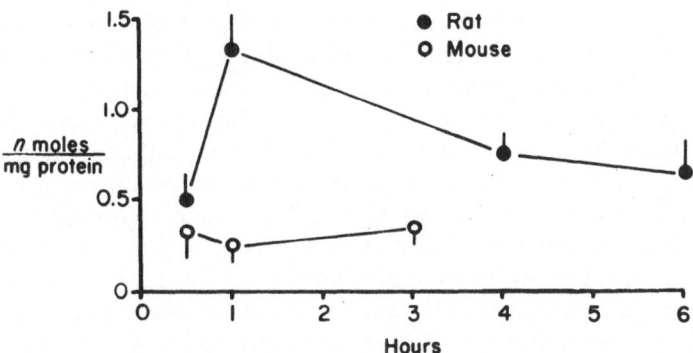

FIGURE 4. Covalent binding of ^{14}C after administration of [^{14}C]citrinin by rat or mouse kidney cortex; $n = 4$.

rats in a nephrotoxic dose. The results of some of those studies are shown in Fig. 4. The mouse also was examined for covalent binding, since citrinin has a distinctly different effect on renal transport in the mouse (Berndt and Hayes, 1980). Rather than a depression of PAH transport such as was noted with the rat, in the mouse citrinin produced a stimulation of PAH transport and, except in extremely high doses, did not affect the PAH acetylation reaction. Other controls that have been done (data not presented) were those in which citrinin binding to a non-target tissue (liver) was undertaken. In all of these studies, covalent binding has been described as that radioactivity that persists on a TCA-insoluble material after numerous extractions with organic solvents selected because they dissolve authentic citrinin readily.

Although some binding of ^{14}C was observed with rat liver tissue after the administration of [^{14}C]citrinin, the extent of this binding could only be called "background" when compared to the large values seen in kidney. Binding at the level of 0.1–0.4 nmole/mg protein was observed in liver and no peak was observed over the same time course studied for kidney. The renal cortical binding of label persists for 48–72 hr, although after 24 hr only trace amounts of irreversibly bound label was seen. Mouse renal tissue binds ^{14}C poorly by comparison to rat renal tissue, a finding which might be viewed as completely consistent with the different effect of the nephrotoxin in the two species.

Although the covalent binding experiments do not prove the presence of a reactive intermediate related to citrinin nephrotoxicity, the patterns of response are at least consistent with those observed with other nephrotoxins thought to produce their effects by such an action (Mudge *et al.*, 1978). The fact that no metabolism of citrinin has been observed with the isolated perfused kidney suggests that if a reactive compound is involved in the nephrotoxicity, then this must be produced in the liver or some other organ and transported to the kidney. Although the chemical nature of such an intermediate has not been established, it is possible that the substance might be a breakdown product of glutathione conjugation or a product of citrinin oxidation which escaped conjugation. In either case the metabolite produced in the liver might be expected to have the stability to allow synthesis in a remote organ and transport to the kidney intact. If a glutathione conjugate is involved, and the breakdown of this compound leads to the formation of a toxic metabolite, the kidney might well be the organ expected to show toxicity, since it is this organ that plays a very large role in the formation of mercapturic acid derivatives from glutathione conjugates. The enzyme γ-glutamyltranspetidase is present in high concentrations in rat kidney and is the first enzyme involved in the mercapturic acid synthetic pathway after glutathione conjugation has occurred. Direct evidence, however, for either the presence of a glutathione conjugate of citrinin or for the role of such a compound in the nephrotoxicity is not yet available.

The morphologic studies of Lockard *et al.* (1980) also revealed the presence of mineral deposits in cells and tubular lumina of rats treated with nephrotoxic doses of citrinin. Such deposits might result from the precipitation of calcium, which in turn, if of sufficient magnitude at an intracellular site, might result in mitochondrial dysfunction. Intracellular accumulation might have resulted from an effect of citrinin on calcium-ATPase in kidney tissue, an action which might underlie the subsequent nephrotoxicity. *In vitro* studies to examine this possibility are presently underway (Berndt and Hayes, 1980). The uptake of ^{45}Ca was measured in fresh renal cortex slices to which was added either citrinin or ochratoxin A in a concentration of 10^{-5} M. Data from these experiments indicate that either fungal toxin could increase calcium flux into the renal cortex slices. Total tissue calcium measurements demonstrated a significant increase in tissue calcium of a magnitude comparable to the increase in ^{45}Ca uptake. Although these data are preliminary, they suggest that these fungal toxins may exert at least a part of their effect on renal tissue through an action on calcium transport. These observations also are consistent with the fact that citrinin inhibits calcium-ATPase in renal cortical tissue (unpublished observations).

Since both citrinin and ochratoxin A can alter calcium transport, this effect has a certain appeal as an explanation for the effects produced on renal function by these compounds. Although covalent binding studies have not been undertaken with ochratoxin A, the absence of an effect of this compound on tissue glutathione content suggests that the reactive intermediate hypothesis may not be tenable as a general mechanism of action for these nephrotoxic fungal products. Possibly, the effect of these compounds on calcium transport represents an action which initiates a toxic response expressed subsequently by a variety of biochemical and physiologic changes in renal tissue. Certainly the renal function response of the rat is different after ochratoxin A than after citrinin, although both substances produce marked alterations in the normal physiologic behavior. The exact nature of those responses may be dictated by alterations in certain biochemical processes (e.g., glutathione depletion) which resulted from disruption of calcium transport. Exactly how the calcium transport process is linked to these other biochemical changes is uncertain at this time and is under investigation.

6. Importance of Laboratory Studies

All of the information discussed above resulted from laboratory experiments. Exactly how these studies can be compared to human or animal experience in the field remains to be seen. For a number of reasons, direct comparisons may be inappropriate. For example, in the field setting the "test" animal may have consumed fungal toxins in relatively low concentrations over relatively long periods of time. The above studies were conducted under conditions where relatively large, acute doses were administered. The role of repeated administration of ochratoxin A is particularly important since even in the laboratory setting it was not possible to produce a nephrotoxic response by the administration of a single dose of this fungal product. Single doses large enough to produce any observable effect resulted in severe diarrhea and usually death. Single doses small enough to avoid diarrhea and death had no effect on renal function. On the other hand, daily, small doses over a few days produced renal dysfunction without the demise of the animal. Chronic studies with very small doses have not been undertaken but may be very important in establishing a more "field-like" nephrotoxic response in the laboratory rat. Because of the pharmacokinetic pattern exhibited by citrinin, it is not entirely clear what the role of repeated daily administration would be for that compound. Attempts to elicit renal dysfunction in the rat with the administration of small doses on a daily basis over 1–2 weeks failed. This is not surprising considering that citrinin is eliminated virtually completely within 24 hr.

In the field, it would no doubt be very unusual for the "test" animals to be exposed to a single fungal product. Fungal contamination of feed usually involves more than one mold and even if only one mold is involved that mold will usually produce more than one product. Hence, the "pure toxicity" described in the laboratory setting, although necessary to understand specific events, may be considerably less realistic as a model for the field situation than might have been predicted at the outset. The effects of chronic administration of very small doses of mixtures of toxins may be necessary in the laboratory setting if a genuine model for the field situation is to be developed. Obviously, the interaction of one fungal product with another (i.e., possible synergistic effects) may be very important. Not only are multiple biochemical actions possible, but one toxin may affect the metabolism or excretion of another.

This chapter has attempted to describe one situation where a natural product may have a distinct undesirable effect on a normal physiologic function. Because exposure to products of this sort may be beyond our control, at least under some circumstances, an effort should be made to minimize those exposures that are known to occur and to better understand the consequences of those exposures. Whether or not fungal products as contaminants in the human food supply have a major role in renal disease in humans is unknown. However, the potential for such an effect certainly exists.

Acknowledgments

Work from the author's laboratory was supported by USPHS research grants ES 01643 and ES 02191. One predoctoral trainee was supported on USPHS training grant ES 00475.

References

Bagger, G., 1931, *Ergot and Ergotism,* Guerney and Jackson, London.

Barnes, J. M., 1967, Possible nephrotoxic agents, in: *Ciba Foundation Study Group* (G. E. W. Wolstenholme and J. Knight, eds.), Little, Brown, and Co., Boston, Massachusetts, p. 110.

Barnes, J. M., Austwide, P. K. C., Carter, R. L., Flynn, F. U., Peristianis, G. C., and Aldridge, W. N., 1977, Balkan (endemic) nephropathy and a toxin producing strain of *Penicillium verrucosum,* var. *cyclopium:* An experimental model in rats. *Lancet* **1977**(March 26):671.

Berndt, W. O., and Grote, D., 1968, The accumulation of C^{14}-dinitrophenol by slices of rabbit kidney cortex, *J. Pharmacol. Exp. Ther.* **164**:223.

Berndt, W. O., and Hayes, A. W., 1977, Effects of citrinin on renal tubular transport functions in the rat, *J. Environ. Pathol. Toxicol.* **1**:93.

Berndt, W. O., and Hayes, A. W., 1979, *In vivo* and *in vitro* changes in renal function caused by ochratoxin A in the rat, *Toxicology* **12**:5.

Berndt, W. O., and Hayes, A. W., 1980, Effects of citrinin on renal transport by mouse kidney, *Pharmacologist* **22**:157.

Berndt, W. O., and Hayes, A. W., 1981, Effects of nephrotoxic fungal toxins on renal calcium transport, *Toxicologist* **1**:9.

Berndt, W. O., Hayes, A. W., and Phillips, R. D., 1980, Effects of mycotoxins on renal function: Mycotoxic nephropathy, *Kidney Int.* **18**:656.

Carleton, W. W., Szczech, G. M., and Tuite, J., 1973, Toxicosis produced in dogs by cultures of *Penicillium citrinum, Toxicol. Appl. Pharmacol.* **25**:457.

Elling, F., Hald, B., Jacobsen, C., and Krogh, P., 1975, Spontaneous cases of nephropathy in poultry associated with ochratoxin A, *Acta Pathol. Microbiol. Scand. A* **83**:739.

Hayes, A. W., 1978, Mycotoxin teratogenicity, in: *Toxins: Animal, Plant, Microbial* (P. Rosenberg, ed.), Pergamon Press, New York, p. 739.

Hayes, A. W., 1979, Biological activities of mycotoxins, *Mycopathologia* **65**:29.

Hayes, A. W. and Williams, W. L., 1977, Acute toxicity of aflatoxin B_1 and rubratoxin B in dog, *J. Environ. Pathol. Toxicol.* **1**:59–70.

Jordan, W. H., Carleton, W. W., and Sansing, G. A., 1977, Citrinin mycotoxicosis in the mouse, *Food Cosmet. Toxicol.* **15**:29.

Koschier, F. J., and Berndt, W. O., 1977, The relationship between 2,4,5-T and the renal base transport system, *Biochem. Pharmacol.* **26**:1709.

Krogh, P., 1976, Mycotoxic nephropathy, *Adv. Vet. Sci. Comp. Med.* **20**:147.

Krogh, P., Hald, B., and Pedersen, E. J., 1973, Occurrence of ochratoxin A and citrinin in cereals associated with mycotoxic porcine nephropathy, *Acta Pathol. Microbiol. Scand. B* **81**:689.

Krogh, P., Hald, B., Pestina, R., and Ceovic, S., 1977, Balkan (endemic) nephropathy and foodborne ochratoxin A: Preliminary results of a survey of foodstuffs, *Acta Pathol. Microbiol. Scand. B* **85**:238.

LeBreton, E., Frayssinet, C., Lafarge, C., and DeRecondo, A. M., 1964, Aflatoxine—Mecanisme de l'action, *Food Cosmet. Toxicol.* **2**:675.

Lockard, V. G., Phillips, R. D., Hayes, A. W., Berndt, W. O., and O'Neal, R. M., 1980, Citrinin nephrotoxicity in rats: A light and electron microscopic study, *Exp. Mol. Path.* **32**:226.

Mudge, G. H., Gemborys, M. W., and Duggin, G. G., 1978, Covalent binding of metabolites of acetominophen to kidney protein and depletion of renal glutathione, *J. Pharmacol. Exp. Ther.* **206**:218.

Phillips, R. D., Berndt, W. O., and Hayes, A. W., 1979, Distribution and excretion of ^{14}C-citrinin in rats, *Toxicology* **12**:285.

Phillips, R. D., Hayes, A. W., and Berndt, W. O., 1980a, High performance liquid chromatographic analysis of the mycotoxin citrinin and its application to biologic fluids, *J. Chromatog.* **190**:419.

Phillips, R. D., Hayes, A. W., Berndt, W. O., and Williams, W. L., 1980b, Effects of citrinin on renal function and structure. *Toxicology* **16**:123.

Roch-Ramel, F., and Weiner, I. M., 1980, Renal excretion of urate: Factors determining the action of drugs, *Kidney Int.* **18**:665.

22

Lead Nephrotoxicity

RICHARD P. WEDEEN

1. Historical Perspective

Lead poisoning and gout are two of the most ancient diseases known to medicine. The distinctive colic followed by palsy was characterized by Nikander in the second century B.C., while Hippocrates had described podagra two centuries earlier. The dramatic symptoms of acute lead intoxication resulted in early recognition, but the delayed effect of prolonged low-dose lead absorption presented considerable diagnostic difficulty. The late sequelae of chronic lead exposure remain the subject of continuing controversy. Neither Sir George Baker nor Tanguerel des Planches, the major contributors to knowledge of lead toxicity in the eighteenth and nineteenth centuries, noted that renal disease or gout were complications of chronic lead intoxication. The legacy of controversy surrounding these issues is the subject of this chapter.

Confusion concerning lead nephropathy is rooted in the growth of understanding of renal diseases in the nineteenth century. Kidney disease in lead poisoning was first recorded by Lancereaux (1862). Lead nephropathy was an incidental finding in an artist who habitually held his paint brushes in his mouth. Lancereaux noted a remarkable atrophy of the renal cortex and tubular fibrosis in the post-mortem examination of the kidneys of this patient. A year later he described four cases of interstitial nephritis in plumbism, but this report was

RICHARD P. WEDEEN • Veterans Administration Medical Center, East Orange, New Jersey 07019; and College of Medicine and Dentistry of New Jersey, Newark, New Jersey 07103.

overshadowed by a similar publication by Ollivier entitled "De l'albuminurie saturnine" (Ollivier, 1863). The crucial difference between the two descriptions of lead nephropathy was that Lancereaux emphasized the absence of proteinuria, while Ollivier considered the renal disease of lead simply another form of Bright's disease. The great French physician, Jean Charcot, credited Ollivier with the discovery of lead nephropathy, with the result that the interstitial nephritis of lead became inextricably confused with proteinuric diseases of the renal glomerulus. The absence of sustained proteinuria in lead nephropathy, except during the uremic phase, caused many modern physicians to doubt the existence of lead-induced renal disease in adults. By the beginning of the twentieth century, recognition of occupational lead nephropathy virtually disappeared from the American medical literature, although its existence in Europe was repeatedly confirmed (Emmerson, 1973).

Lead nephropathy seemed only of historical interest in the English-speaking nations until a curious epidemic of renal failure was recognized in young adults in Queensland, Australia at the turn of the century. After decades of arduous epidemiologic studies it was established that the epidemic of "nephritis" among Queensland youth was the consequence of excessive lead absorption in childhood (Emmerson, 1973; Henderson and Inglis, 1957). Subsequently, a transient Fanconi syndrome was identified among American children during the acute lead intoxication associated with pica. Chronic renal disease was not, however, identified as a sequel to childhood lead poisoning in the United States (Burde and Choate, 1972). The contrast between the American and Australian followup studies remains to be satisfactorily explained. Transition from acute proximal tubule dysfunction to chronic interstitial nephritis has been observed in experimental animals but not in humans (Aviv *et al.*, 1980). While there is some evidence of a proximal tubule transport defect in adults with lead nephropathy, it is not clear whether this finding distinguishes lead nephropathy from other forms of interstitial nephritis (Hong *et al.*, 1980).

In the United States, lead nephropathy in adults was recognized only among illicit whiskey consumers in the southern states (Morgan and Hartley, 1976). Outside of the "moonshine" belt, even overt lead intoxication was often mistaken for cholecystitis, pancreatitis, alcoholism, schizophrenia, or porphyria. In the Criteria for a Recommended Standard—Occupational Exposure to Inorganic Lead issued by the National Institute of Occupational Safety and Health (1972), lead nephropathy was relegated to the distant past and the underdeveloped countries of the world. The marked contrast between the American and European experience with occupational lead nephropathy was attributed to the advanced safety standards in American industry. Another possible explanation was that in the absence of symptomatic plumbism, the Americans had been using inappropriate diagnostic criteria for the identification of lead nephropathy.

Failure to identify lead nephropathy among American lead workers in the past can be attributed to three diagnostic misconceptions: (1) renal disease was identified primarily by the presence of proteinuria, which is characteristically absent in the early stages of interstitial nephritis; (2) the laboratory diagnosis of lead poisoning was based on the presence of blood lead concentrations greater than 80 μg/dl; and (3) histopathologic features of interstitial nephritis were called "pyelonephritis," fostering the impression that bacterial invasion of the renal parenchyma was the cause of renal damage in lead workers (Heptinstall, 1974). The resulting confusion was compounded by the inability to separate a variety of intercurrent renal diseases, including nephrosclerosis, gout nephropathy, and glomerulonephritis.

2. Occupational Lead Nephropathy

It was within this setting that we began to examine lead workers for excessive lead stores in 1973. It soon became apparent that the blood lead concentration was unreliable as an index of past lead absorption (Vitale *et al.,* 1975). We relied instead on the EDTA lead-mobilization test, which had been validated by Emmerson in Australia as an indicator of patients with lead nephropathy (1973). EDTA (CaNa$_2$ edetate) chelates lead stored primarily in bone and soft tissues and causes it to be excreted in the urine. Despite minor variations in the method of testing, it has been found that adults without unusual lead exposure excrete less than 600 μg Pb/day after parenteral administration of 1 or 2 g EDTA (Wedeen *et al.,* 1975).

Traditional tests of blood lead, hemoglobin concentrations, or basophilic stippling are not sufficiently sensitive to detect cumulative toxicity from chronic low-dose lead exposure. Moreover, sensitive assays for defects in heme synthesis have not been adequately correlated with specific lead-induced organ damage. Heme synthesis defects have been correlated primarily with blood and urine lead concentrations, which, unfortunately, do not accurately reflect hazardous cumulative lead stores. Urine and blood leads indicate current lead absorption rather than the risk of lead nephropathy.

Use of the EDTA lead-mobilization test has made it possible to identify preclinical occupational lead nephropathy among industrially exposed workers. One hundred and forty lead workers received 1 g of EDTA IM twice, 8–12 hr apart, during collections of a 24-hr urine. Five milliliters of EDTA (1 g) was mixed with 1 ml of 1% procaine to reduce pain at the injection site (Wedeen *et al.,* 1979). One hundred thirteen of the 140 men excreted more than 1000 μg Pb during the next 24 hr. Fifty-six of these men were not studied further, either because they refused additional tests or they had evidence of other disease which might compromise renal function, such as hypertension, gout, or renal stones. It is important to note that in this phase of our studies, patients with hypertension or gout were specifically excluded in order to isolate lead nephropathy. In the remaining 57 EDTA-positive men, glomerular filtration rate (GFR) was measured by iothala-mate-I[125] clearance. Twenty-one of these 57 men were found to have a GFR of less than 90 ml/min/1.73 m^2 body surface area, the generally accepted lower limit of normal for adult men under 55 years of age. At the time of evaluation none had proteinuria or elevated blood lead concentrations according to then current criteria. After excluding five men over 54 years of age and one with hypertension, 15 remained with compelling evidence of occupational lead nephropathy. Twelve of this group had never had any symptoms or laboratory findings suggestive of lead poisoning. In six of 12 renal biopsies no abnormalities were found, while in the remaining six, focal interstitial nephritis was evident. Acid-fast intranuclear inclusions characteristic of acute lead intoxication were absent. Goyer's work suggests that these lead-containing inclusions are removed from the kidney during EDTA testing (Goyer and Wilson, 1975). Immunofluorescent microscopy revealed tubular or glomerular immune deposits in seven of eight biopsies examined. The histologic studies thus confirmed the diagnosis of interstitial nephritis in half of the men and raised the possibility that immunologic mechanisms mediate lead nephropathy.

Following long-term chelation therapy over 6–50 months, four of eight patients showed an increase in GFR of 20% or more (Fig. 1). The increased GFR was accompanied by comparable improvement in effective renal plasma flow (PAH clearance). Treatment consisted of 1 g EDTA (with procaine) IM three times a week until the EDTA lead-mobilization test had returned to less than 1000 μg Pb/day. The improvement in renal function following long-term chelation therapy confirmed the etiologic diagnosis.

Since these observations were reported, subclinical occupational lead nephropathy has been found in both the United States (Baker *et al.,* 1979; Lilis *et al.,* 1979) and Europe (Vangelista *et al.,* 1978). The United States Supreme Court has upheld the notion, promulgated in the Final Standard—Occupational

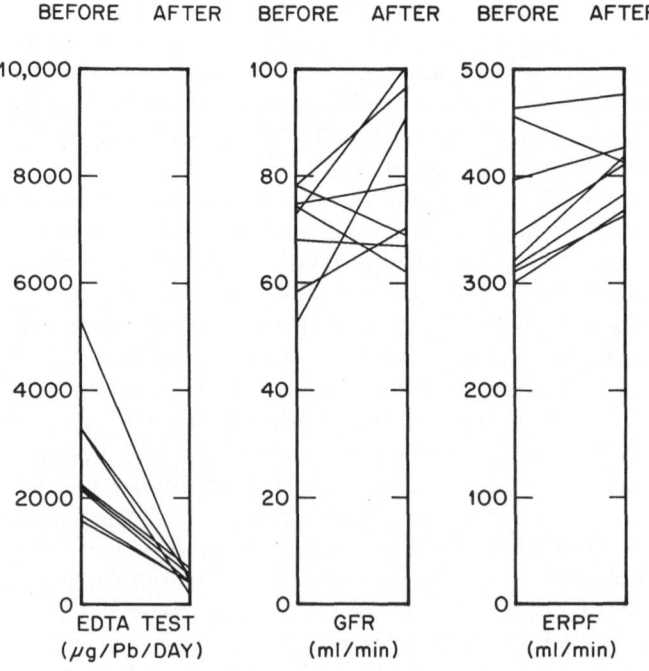

FIGURE 1. The effect of long-term chelation therapy on subclinical occupational lead nephropathy. Glomerular filtration rates (GFR) and effective renal plasma flow (ERPF) increased by 20% or more in four of eight patients given long-term chelation therapy.

Exposure to Lead (United States, 1978), that regulatory responsibility includes prevention of asymptomatic lead nephropathy.

3. Gouty Nephropathy

As long as the existence of occupational lead nephropathy was doubted in this country, it was necessary to exclude lead workers with gout or hypertension from the study. Long suspected as complications of lead poisoning, gout and hypertension would have confused the diagnosis, since, according to the generally accepted view, both are common causes of renal disease (Editor, 1981). The relationship of lead to gout has, however, been the subject of considerable controversy for over 200 years. In one of the first applications of modern chemistry to medicine, in 1767, George Baker proved that the cause of the Devonshire colic was lead admixed with cider. Baker's announcement enraged local farmers, whose livelihoods were seriously threatened by the disclosure. In the ensuing polemics, James Hardy, a staunch defender of Baker's thesis, noted that gout as well as colic and palsy were consequences of Devonshire cider (Hardy, 1780). Hardy's interesting observation seems to have been obscured by his rhetoric. Physicians paid little attention to the role of lead in gout until Alfred Baring Garrod devised a method for detecting hyperuricemia (Garrod, 1859).

Garrod's measurement of serum uric acid drastically altered medical understanding of the gouty diathesis. Gout had long been considered a protean disease which included a vast array of nonarticular symptoms, including colic, palsy, and encephalopathy, referred to as "irregular" gout. With the exception of kidney disease, the nonarticular manifestations of gout would not today be considered part of the arthritic syndrome. Similarly, after centuries of preoccupation with alcoholic beverages, the notion that drinking causes gout has gradually faded from the medical literature. The long association of gout with wine and "irregular" symptoms could be explained if lead were a sporadic contaminant of alcoholic drinks.

Kidney disease associated with gout also remains a subject of considerable controversy. While the history of ureteral stones in gout can be traced to Galen, modern recognition of intrarenal urate deposits probably should be credited to Castelnau (1843). He noted the characteristic white striations in the outer renal medulla due to intraluminal precipitation of uric acid. Microtophi in the renal cortex are also considered characteristic of the gouty kidney (Charcot, 1881). It has generally been assumed that interstitial nephritis in gout is due to one of these mechanisms, even when no urate or uric acid deposits can be identified in the tissue.

Despite the growing consensus on this view of gout nephropathy, it seemed possible that lead, rather than uric acid, might sometimes be responsible for the gouty kidney even in contemporary patients. Up until the last decade, kidney disease had been observed to be the major cause of death in patients with gout. Yet, long-term followup studies in both California and New York failed to provide evidence that either hyperuricemia or gout cause progressive renal failure (Fessel,

1979; Yu *et al.*, 1979). Moreover, Garrod was aware that at least one-fourth of his patients were lead workers who at one time or another had suffered symptomatic lead poisoning. What Garrod did not know was that the remainder of his patients may also have had excessive body lead burdens. England's most popular beverages in the eighteenth and nineteenth centuries were imported port and madeira. These wines had long been suspected of being the major cause of gout and colic, and were accused of adulteration with lead at the turn of the nineteenth century (Ellwanger, 1897). More recently, Ball (1971) tested some of the old Portuguese imports and found them heavily laced with lead. Ball concluded that the epidemic of gout in eighteenth and nineteenth century London may have had more of an assist from lead then even Garrod suspected. In the lead nephropathy of the "moonshine" belt and in Australia, gout is extremely common, whereas in other forms of renal failure, gout is distinctly rare (Emmerson, 1973).

In order to test the hypothesis that unsuspected lead poisoning may sometimes contribute to the renal disease of gout, we recently examined 44 gout patients at the Veterans Administration Medical Center in East Orange, New Jersey (Batuman *et al.*, 1981a). Excessive body lead stores were detected with the three-day EDTA lead-mobilization test, a modification of the one-day test for patients with compromised renal function (Emmerson, 1973). None of these men had ever experienced symptoms of lead poisoning, but each gave a typical history of acute gouty arthritis of the great toe. The mean serum uric acid was 9.5 ± 0.3 (SEM) mg/dl. Half of these men had renal failure manifested by a serum creatinine of 1.5 mg/dl or greater. In 12 patients in whom joint fluid was obtained, the presence of uric acid crystals confirmed the diagnosis.

There was a remarkable correlation between the degree of renal failure and the amount of mobilizable lead in these gout patients ($r = 0.38$, $p < 0.02$). Those with renal disease (mean serum creatinine 3.0 ± 0.4 mg/dl, creatinine clearance 54 ± 9 ml/min) excreted a mean of 806 ± 90 μg Pb during the three-day EDTA test (Table I). Gout patients without renal disease, on the other hand (mean serum creatinine 1.3 ± 0.1 mg/dl, creatinine clearance 84 ± 6 ml/min), excreted only 470 ± 52 μg Pb during the EDTA test. The gout patients with renal failure did not differ from those without renal failure with respect to age, blood pressure, history of lead exposure, serum uric acid, blood lead, or zinc protoporphyrin concentration. That the excessive excretion of lead in gout patients with renal failure was not due to renal disease per se was demonstrated in ten renal failure patients without gout who had no known exposure to lead. These non-lead, non-gout renal failure control patients had a mean serum creatinine of 2.9 ± 0.3 μg/dl, a mean creatinine clearance of 45 ± 9 ml/min, and excreted 424 ± 72 μg Pb over 3 days. Thus renal failure by itself does not lead to an increase in mobilizable lead.

In addition to accounting for the appearance of interstitial nephritis in gout, these studies help explain the blurring of distinctions between glomerular and interstitial disease; proteinuria increased with the progression of renal failure (Fig. 2). When the serum creatinine concentration was plotted against the 24-hr urine protein excretion a linear regression was obtained with an r value of 0.72 and a p of less than 0.001. Half of the gout nephropathy patients with creatinines over 2 mg/dl excreted more than 500 mg and five excreted more than 1 g of protein per day. It is

TABLE I. Clinical Characteristics of Gout Patients with and without Renal Failure[a]

	Number of patients	Age, yr	Gout, yr	High BP	Lead exposure	Serum uric acid, mg/dl	Serum creatinine mg/dl	Creatinine clearance, ml/min	24-hr urine protein mg/day	Blood lead, μg/dl	ZPP,[b] μg Pb/dl	EDTA test, μg Pb/3 days
No renal disease	22	53 ± 2	8 ± 1	12	16	9 ± 0.4	1.3 ± 0.1	84 ± 6	235 ± 50	24 ± 3	27 ± 3	470 ± 52
Renal disease	22	57 ± 2	8 ± 2	17	10	10 ± 0.5	3.0 ± 0.4	54 ± 9	760 ± 227	26 ± 3	43 ± 9	806 ± 90
p value		NS	NS	NS	NS	NS	<0.001	<0.01	<0.05	NS	NS	<0.005

[a] The only significant difference apart from renal disease is the excessive mobilizable lead in the gout patients with renal disease. [From Batuman et al. (1981a); reprinted by permission of The New England Journal of Medicine.]
[b] ZPP, Blood zinc protoporphyrin concentration.

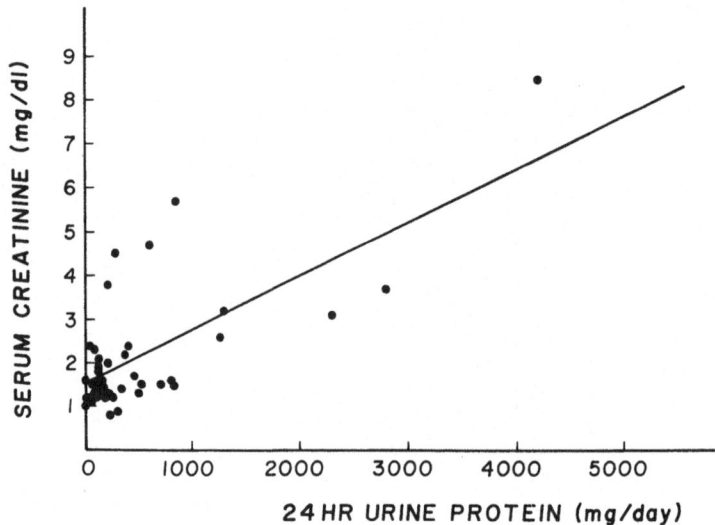

FIGURE 2. The correlation of proteinuria with serum creatinine in 44 gout patients. Proteinuria increases with renal failure. [From Batuman *et al.* (1981a); reprinted by permission of *The New England Journal of Medicine.*]

thus understandable that the interstitial nephritis of gout and lead could be confused with glomerular disease when only patients with symptomatic renal failure are examined.

These data would seem to confirm the view expressed by Lancereaux (1881) that "the nephritis and arthritis of saturnism are precisely the identical lesion of nephritis and arthritis of gout." Our findings support the conclusion that lead may sometimes contribute to gouty nephropathy even in contemporary patients. While the source of lead in the New Jersey veterans we studied is not precisely known, both moonshine consumption and transient employment in the lead industry were commonly reported. The role of environmental sources of lead remains undefined in these men.

4. Nephrosclerosis

Since the EDTA lead-mobilization test permitted identification of lead workers and gout patients with previously unrecognized lead nephropathy, it seemed of interest to determine if lead could also sometimes be responsible for renal disease in hypertension. A preliminary survey of hypertensives without renal failure which we undertook some years ago had indicated that mobilizable lead in hypertensives was not different from that in normotensive controls. When, however, we examined EDTA test results in hypertensive veterans with and without renal disease a very different picture emerged (Batuman *et al.,* 1981b). Among 21 men with sustained diastolic blood pressures greater than 90 mm Hg, 11 with renal failure (mean serum creatinine 3.8 ± 0.4 mg/dl) excreted a mean of 763 ± 95 μg Pb/3 days, while ten with serum creatinines less than 1.6 mg/dl excreted only 336 ± 68 μg Pb/3 days ($p < 0.005$, Fig. 3).

Renal failure was not responsible for the excessive mobilizable lead in the hypertensives with kidney disease, since eight normotensive patients with comparable renal failure excreted only 387 ± 79 μg Pb/3 days. The EDTA test correlated directly with the serum creatinine levels ($r = 0.69$, $p < 0.001$) in the hypertensive patients, which was even more significant than the correlation of mobilizable lead with renal failure in gout patients. Furthermore, the mean serum uric acid concentration in hypertensives with renal failure (mean 9.9 ± 0.5 mg/dl) was significantly greater than that of patients with comparable renal failure but without hypertension (7.1 ± 0.9 mg/dl, $p < 0.02$). Thus hyperuricemia is more severe in hypertensives with lead nephropathy than in comparable normotensive renal failure patients.

These data suggest that unsuspected lead nephropathy may sometimes be responsible for hypertension and nephrosclerosis, just as it is sometimes reponsible for gout and gout nephropathy. The histologic appearance of the hypertensive kidney is often indistinguishable from that of lead nephropathy. Indeed, anteriolar changes typical of hypertension have been observed in kidneys with lead nephropathy in the absence of hypertension (Wedeen et al., 1975). If the EDTA lead-mobilization test is performed in hypertensive patients with renal failure, much of what has heretofore been considered "essential hypertension with

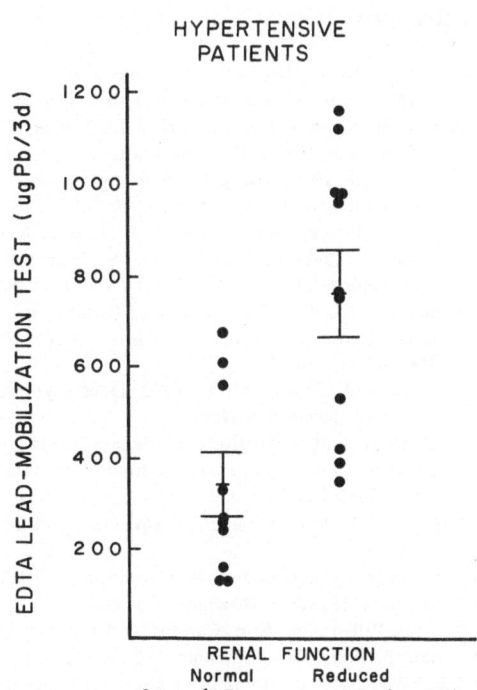

FIGURE 3. EDTA lead-mobilization tests in 21 veterans with hypertension. Mobilizable lead was significantly greater in hypertensives with renal failure than in those with normal renal function.

nephrosclerosis" may in the future be termed "lead nephropathy." The possibility that a lead-induced immune response mediates both hypertensive nephrosclerosis and lead nephropathy warrants careful evaluation.

In conclusion, our data suggest that occupational lead nephropathy and the contribution of lead to gout nephropathy and nephrosclerosis may have been underestimated in modern times. The diagnostic oversight appears to stem from the earliest descriptions of Bright's disease and may, in pa.t, be attributable to the ease with which proteinuria is detected in the clinical laboratory. The confusion is readily compounded by inappropriate diagnostic criteria for lead poisoning and by etiologic judgments based on nonspecific histologic findings in the kidney. Lead may not only be far more important as a toxicant than has generally been supposed, but the kidney may be the organ most sensitive to its noxious effects using currently acceptable diagnostic criteria.

Acknowledgments

This work was supported by funding from the Veterans Administration Medical Research Service. Acknowledgement is made to Vecihi Batuman, M.D., and Elaine Landy, R. N., without whom these studies could not have been performed.

References

Aviv, A., John, E., Bernstein, J., Goldsmith, D. I., and Spitzer, A., 1980, Lead intoxication during development: Its late effects on kidney function and blood pressure, *Kidney Int.* **17**:430.

Anon, 1981, Hypertension and uric acid, *Lancet* **1**:365.

Baker, E. L., Landrigan, P. J., Barbour, A. G., Cox, D. H., Folland, D. S., Ligo, R. N., and Throckmorton, J., 1979, Occupational lead poisoning in the United States: Clinical and biochemical findings related to blood lead levels, *Br. J. Ind. Med.* **36**:314.

Ball, G. V., Two epidemics of gout, 1971, *Bull. Hist. Med.* **45**:401.

Batuman, V., Maesaka, J. K., Haddad, B., Tepper, E., Landy, E., and Wedeen, R. P., The role of lead in gout nephropathy, 1981a, *N. Engl. J. Med.* **304**:520.

Batuman, V., Shobha, S., Landy, E., Maesaka, J. K., and Wedeen, R. P., Role of lead in hypertensive renal disease, 1981b, in: *VIIIth International Congress of Nephrology,* Athens, Greece, Abstract HK-096, p. 334.

Burde, B., and Choate, M. S., 1972, Does asymptomatic lead exposure in children have any latent sequelae?, *Pediatrics* **81**:1088.

Castelnau, N. F., Observations et reflexions sur la goutte et le rhumatisme, et specialement sur quelques accidents graves qui peuvent se manifester dans le cours de ces deux affections, 1843, *Arch. Gen. Med. Paris* **3**:285.

Charcot, J. M., 1881, *Clinical Lectures on Senile and Chronic Diseases,* The New Sydenham Society, London.

Editor, 1981, Hypertension and uric acid, *Lancet* **1**:365.

Ellwanger, G. H., 1897, *Meditations on Gout with a Consideration of its Cure Through the Use of Wine,* John Wilson and Son, Cambridge University Press.

Emmerson, B. T., 1973, Chronic lead nephropathy, *Kidney Int.* **4**:1.

Fessel, W. J., Renal outcomes of gout and hyperuricemia, 1979, *Am. J. Med.* **67**:74.

Garrod, A. B., 1859, *The Nature and Treatment of Gout and Rheumatic Gout,* Walton and Maberly, London.

Goyer, R. A., and Wilson, M. H., 1975, Lead-induced inclusion bodies. Results of ethylenediaminetetraacetic acid treatment, *Lab. Invest.* **32:**149.

Hardy, J., 1780, *An Answer to the Letter Addressed by Francis Riolloy, Physician of Newbury to Dr. Hardy, on the Hints Given Concerning the Origin of the Gout, in his Publication on the Colic of Devon,* T. Cadell, London.

Henderson, D. A., and Inglis, J. A., 1957, The lead content of bone in chronic Brights disease, *Aust. Ann. Med.* **6:**145.

Heptinstall, R. H., 1974, *Pathology of the Kidney,* 2nd ed. Little, Brown & Co., Boston, Massachusetts.

Hong, C. D., Hanenson, I. B., Lerner, S., Hammond, P. B., Pesce, A. J., and Pollak, V. E., 1980, Occupational exposure to lead: Effects on renal function, *Kidney Int.* **18:**489.

Lancereaux, E., 1862, Note relative a un cas de paralysie saturnine avec alteration des cordons nerveux et des muscles paralyses, *Gaz. Med. Paris* **17:**709.

Lancereaux, E., 1881, Nephrite et arthrite saturnines; coincidence de ces affections; parallele avec la nephrite et l'arthrite goutteuses, *Arch. Gen. Med. Paris* **6:**641.

Lilis, R., Valciukas, J., Fischbein, A., Andrews, G., and Selikoff, I. J., 1979, Renal function impairment in secondary lead smelter workers: Correlations with zinc protoporphyrin and blood lead levels, *J. Environ. Pathol. Toxicol.* **2:**1447.

Morgan, J. M., and Hartley, M. W., 1976, Etiologic factors in lead nephropathy, *South Med. J.* **69:**1445.

National Institute of Occupational Safety and Health, 1972, Criteria for a recommended standard— Occupational exposure to inorganic lead, U.S. Department of Health, Education and Welfare, Washington, D.C.

Ollivier, A., 1863, De l'albuinurie saturnine, *Arch. Gen. Med. Paris* **2:**530.

United States Department of Labor, Occupational Health and Safety Administration, Occupational Exposure to Lead: Final Standard, 1978, Federal Register, p. 52952.

Vangelista, A., Caudarella, R., and Bonomini, V., Lead poisoning nephropathy, 1978, in: *VIIth International Congress of Nephrology,* Montreal, Canada, Abstract, p. D32.

Vitale, L. F., Joselow, M. M., and Wedeen, R. P., 1975, Blood lead—an inadequate measure of occupational exposure, *J. Occup. Med.* **17:**155.

Wedeen, R. P., Maesaka, J. K., Weiner, B., Lipat, G. A., Lyons, M. M., Vitale, L. F., and Joselow, M. M., Occupational lead nephropathy, 1975, *Am. J. Med.* **59:**630.

Wedeen, R. P., Mallik, D. K., and Batuman, V., 1979, Detection and treatment of occupational lead nephropathy, *Arch. Intern. Med.* **139:**53.

Yu, T. F., Berger, T. L., Dorph, D. J., and Smith, H., 1979, Renal function in gout, V. Factors influencing the renal hemodynamics, *Am. J. Med.* **67:**766.

23

The Rat as an Animal Model of Lead Nephropathy

P. B. HAMMOND, C. D. HONG, E. J. O'FLAHERTY, S. I. LERNER, and I. B. HANENSON

1. Current Significance of Lead Nephropathy

Nephropathy is only one of a number of toxic effects of lead observed in humans. Others include anemia, peripheral neuropathy, and an array of signs and symptoms attributed to deranged function of the central nervous system. These effects have been reported principally in children of pre-school age and in adults occupationally exposed to lead. The frequency of occurrence and severity of adverse effects in these two populations has diminished over the years. There is still cause for concern, however, particularly in regard to occupational lead nephropathy. The death rate due to chronic nephritis among male lead workers in the first quarter of this century was twice as high as among non-lead-exposed males of the same social class (Lane, 1964). Even among men employed from 1947 to 1970 excess mortality due to nephritis appears to have occurred (Cooper and Gaffey, 1975). Recent studies of renal function in workers indicate that occupational lead nephropathy is still a major problem (Wedeen *et al.*, 1979; Hong *et al.*, 1980).

Lead nephropathy is not solely an occupational disease. Numerous cases have also been identified among moonshine whisky drinkers (Morgan *et al.*, 1966;

P. B. HAMMOND, E. J. O'FLAHERTY, and S. I. LERNER • Department of Environmental Health   C. D. HONG and I. B. HANENSON • Department of Medicine, University of Cincinnati Medical Center, Cincinnati, Ohio 45267.

Sandstead *et al.,* 1970) and among both older children and adults as a sequel to childhood plumbism (Henderson, 1958). The continuing significance of these latter two types of adult lead nephropathy is questionable. Lead contamination of moonshine whisky seems to be on the decline due to changes in moonshine still technology (Gerhardt *et al.,* 1980), and adult nephropathy due to childhood plumbism was an episodic occurrence unique to Queensland, Australia. Tepper (1963) reported a followup study of 165 Americans who had childhood plumbism more than 20 years earlier. He concluded that chronic renal disease in his series was not a sequel to childhood plumbism. Chisolm [as reported by Goyer (1971)] also found no evidence of renal disease as a sequel to childhood plumbism. The likelihood that adult nephropathy occurs as a sequel to childhood plumbism is further diminished by the fact that there has been a substantial reduction in the incidence of clinical lead poisoning among children in recent years.

The occurrence of renal involvement as a feature of childhood plumbism (rather than as a sequel) has been reported. Chisolm (1968) reported the occurrence of aminoaciduria, glycosuria, and hypophosphatemia in four of 15 lead-intoxicated young children. All these children had very high levels of exposure as judged by the fact that they all had blood lead concentrations (PbB) of 100 μg/dl or higher at the time of initial evaluation. Renal tubular disease was evident only among the four children with PbB over 245 μg/dl. Unfortunately, no evaluation of glomerular function was reported. Such high levels of lead exposure as were reported are rarely encountered today, at least in the experience of the pediatric services of the Cincinnati Children's Hospital.

In summary, lead nephropathy is currently an occupational problem, first and foremost. Thus, this is the problem that is in most urgent need of resolution. The ideal animal model of lead nephropathy should most closely mimic the characteristics of this form of the disease.

2. The Characteristics of Occupational Lead Nephropathy: Review of Past Studies

The salient features of occupational lead nephropathy have been described by numerous investigators. No two studies have been precisely the same with regard to the renal parameters investigated. Some have focused primarily on glomerular (filtration) function, whereas others have been largely limited to examination of possible tubular dysfunction.

2.1. Glomerular Filtration and Renal Plasma Flow

The most thorough studies of glomerular filtration in terms of numbers of subjects and precision of methodology are those reported by Wedeen's group (Wedeen *et al.,* 1975, 1979). Glomerular filtration rate (GFR) was measured by the clearance of radiolabeled iothalamate. Effective renal plasma flow (ERPF) was measured by the clearance of *p*-aminohippuric acid (PAH). Both GFR and ERPF were substantially depressed in the absence of elevated serum urea nitrogen (SUN) and serum creatinine (SCr). These findings have been confirmed by Hong *et al.* (1980) in a more limited series of subjects using inulin instead of iothalamate to measure GFR, but using the same technique to measure ERPF. Cramer *et al.* (1974) also studied GFR and ERPF in seven workers. GFR was depressed, but not ERPF. In other studies the clearances of urea and/or creatinine have been shown to be depressed in a substantial fraction of the cases studied (Richet *et al.*, 1966; Lilis *et al.*, 1968; Radocevic *et al.*, 1961). Thus, depression of GFR is a consistent manifestation of occupational lead nephropathy, probably accompanied by reduced ERPF.

2.2. Tubular Function

Studies of renal tubular function following lead exposure are not as extensive or as easily interpreted as studies of glomerular filtration function. The conventional parameters of tubular function which have been studied are maximal p-aminohippuric acid secretion (Tm_{PAH}), maximal glucose reabsorption (Tm_G), and the excretion of α-amino acids. The Tm_{PAH} has been shown to be reduced (Wedeen *et al.*, 1975, 1979; Hong *et al.*, 1980). This finding can not be taken to signify a direct effect of lead on tubular secretory processes. While Tm_{PAH} has been shown to be reduced in men with occupational lead exposure, this effect could be secondary to reduced GFR (Hong *et al.*, 1980). These same investigators have shown, however, that TM_G sometimes is reduced to a degree disproportionately greater than the reduction in GFR.

Aminoaciduria has been observed, albeit not consistently, in some studies (Goyer *et al.*, 1972; Clarkson and Kench, 1956). In other studies, aminoaciduria was not found (Cramer *et al.*, 1974; Hammond *et al.*, 1980). It is possible that reduction in GFR masked a reduction in reabsorptive capacity which might have been manifested as aminoaciduria had the filtered load of amino acids been normal.

Some attention has been given to the effect of occupational lead nephropathy on the renal handling of uric acid. Unlike the case with lead nephropathy due to consumption of moonshine whisky, gout is not a generally recognized feature of occupational lead nephropathy. Nonetheless, elevated serum uric acid concentration is an inconstant feature of occupational lead nephropathy (Wedeen *et al.*, 1975; Richet *et al.*, 1966).

2.3. Hypertension

Studies early in this century suggested that an increased incidence of hypertension occurred with excessive lead exposure. More recent studies have indicated, however, that hypertension is not an essential feature of lead nephropathy (Wedeen *et al.*, 1975; Radocevic *et al.*, 1961; Hong *et al.*, 1980).

2.4. Renal Pathology

To varying degrees, all major elements of the kidney are involved in occupational lead nephropathy, namely tubular epithelium, glomeruli, renal vessels, and interstitium. With particular reference to the pathogenesis of reductions in glomerular filtration, observations to date suggest that morphologic alteration of glomeruli probably is not the critical determinant of functional deficits. Thus, Wedeen *et al.* (1979) observed normal glomeruli in ten of 12 cases and focal tubular atrophy and interstitial disease in six of 12. In a more extensive study, Richet *et al.* (1966) found that interstitial and vascular lesions occurred earlier than glomerular lesions. This progression is depicted graphically in Fig. 1. However, tubular atrophy and other abnormalities of the proximal tubule are more prevalent than glomerulovascular or interstitial lesions (Wedeen *et al.*, 1975; Cramer *et al.*, 1974; Galle and Morel-Maroger, 1965).

3. Occupational Lead Nephropathy: Current Study

Studies of renal function in occupationally exposed workers are in progress at the University of Cincinnati under the direction of Dr. C. D. Hong. Certain recent findings are reported here which serve to confirm earlier work (Hong *et al.*, 1980; Wedeen *et al.*, 1979) and which expand upon earlier findings as to the degree of renal dysfunction among severely intoxicated workers.

3.1. Characteristics of the Subjects

The subjects worked in a secondary lead smelter. All had been exposed for many years and had abnormally high SUN and SCr (Table I). Although PbB values determined over the previous 6 years were not remarkably high, they may have been considerably higher in earlier years. The protocol for the study of renal function has been described previously (Hong *et al.*, 1980).

3.2. Results of the Study

As in the previous study (Hong *et al.*, 1980), the major findings were (1) reduced inulin clearance, (2) reduction in Tm_G disproportionately greater than the

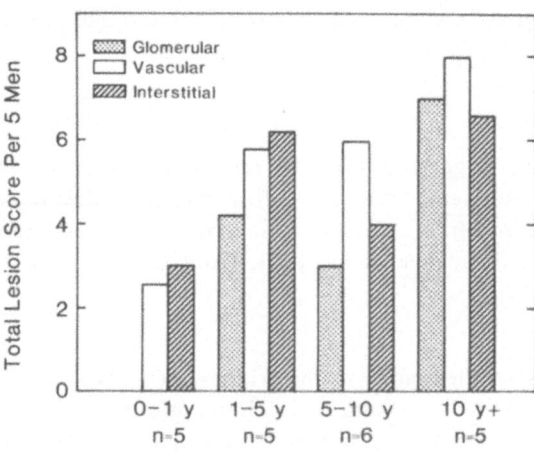

FIGURE 1. Frequency of lesions observed in a series of cases of occupational lead nephropathy as related to duration of employment. [From Richet *et al.* (1966).] Total lesion scores per five subjects are based on a scale of $\pm = 0.5$, $+ = 1$, $++ = 2$, $+++ = 3$ from the author's report; n indicates number of subjects in each category of duration of employment.

TABLE I. Characteristics of Subjects[a]

Subject	Age	Years of exposure	Average μg Pb/dl blood	SUN, mg/dl	SCr, mg/dl	SUA mg/dl
1	50	31	69	59	3.7	7.3
2	57	22	75	32	2.4	5.6
3	63	29	56	32	2.0	9.4
4	57	18	66	39	3.9	8.9
5	54	20	76	55	3.1	8.0
6	63	39	83	36	2.6	6.9
7	50	28	72	55	3.1	8.0
8	64	23	51	34	2.8	8.6
9	52	35	66	44	3.7	8.0

[a] SUN, Serum urea nitrogen. SCr, Serum creatinine. SUA, Serum uric acid.

fall in GFR (five of nine subjects), and (3) reduction in Tm_{PAH} not disproportionately greater than the fall in GFR. These findings are presented graphically along with the earlier findings in less severely affected individuals (Fig. 2). In some individuals the Tm_{PAH} appears disproportionately high in relation to GFR. The combined data from this study and others show a striking degree of consistency in C_{PAH} vs. C_{IN} over a wide range of C_{IN} (Fig. 3). Clearly, the filtration fraction C_{IN}/C_{PAH} does not decrease as C_{IN} falls. If anything, there is a tendency for C_{IN}/C_{PAH} to increase as C_{IN} falls. These observations are consistent with the postulate that reduced filtration is probably secondary to reduced blood flow rather than to an increase in the filtration barrier in glomeruli.

It is of interest that only three of the nine subjects had even slightly elevated serum uric acid (>8 mg/dl) and that none had signs or symptoms of gout. Yet their renal disease was probably, on the average, more severe than in the series of moonshine whisky drinkers studied by Morgan *et al.* (1966), in which six of 13 had gout. Relative severity of renal disease in the two series can only be compared on the basis of elevation in SCr, the only parameter of renal function which was reported in both studies. The SCr values were substantially more elevated in our subjects [Table I vs. Morgan *et al.* (1966)].

4. The Rat as an Animal Model of Lead Nephropathy

4.1. Need for a Model

Many questions remain concerning lead nephropathy. Thus, for example, there is considerable uncertainty as to the relative importance of duration as compared to intensity of lead exposure in regard to both onset and reversibility of the disease. Similarly, the importance of the age at which exposure occurs is not known. There also is essentially no information available concerning the interaction of lead exposure with other environmental insults, such as alcohol consumption and exposure to other environmental nephrotoxic substances, e.g., cadmium. Finally, the pathogenesis of the disease is still poorly understood. It is still not known whether the primary site of toxic insult is the renal vasculature, the glomerulus, or

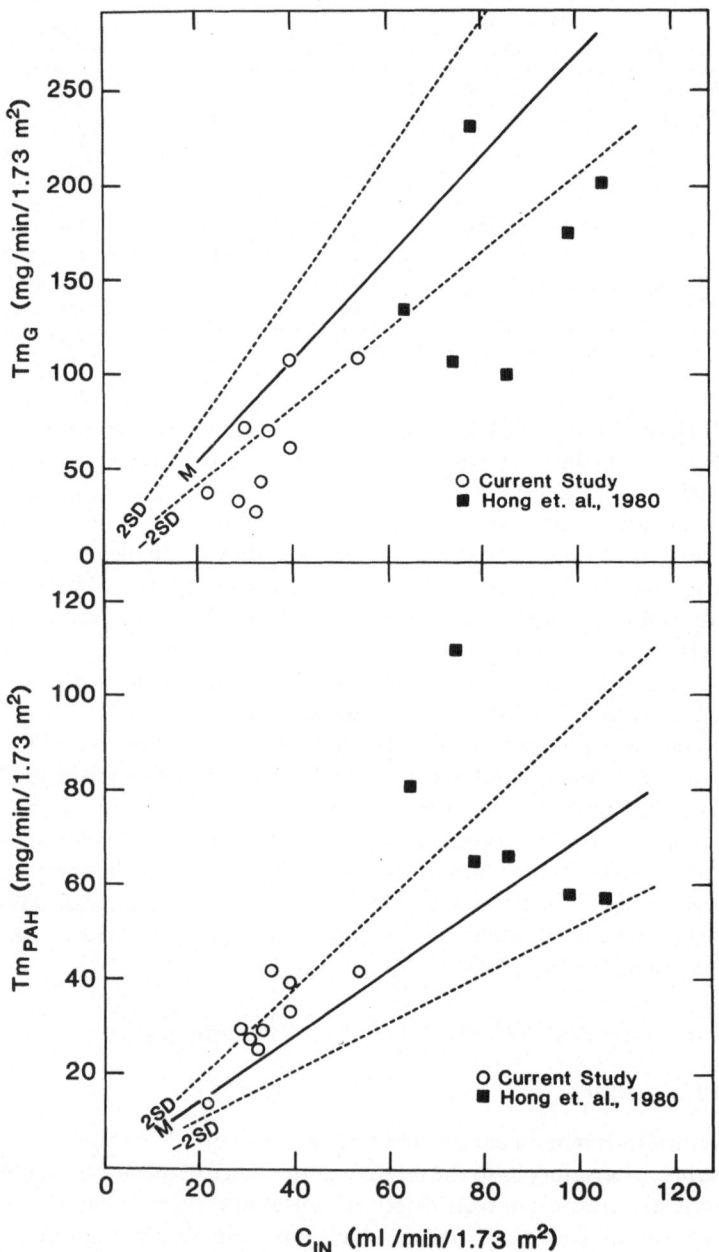

FIGURE 2. The relationship of tubular maximum for *p*-aminohippurate (Tm$_{PAH}$) and of tubular maximum for glucose (Tm$_G$) to inulin clearance (C$_{IN}$) in the same series of cases of occupational lead nephropathy. Statistical data (mean ± 2 SD) adopted from Smith *et al.* (1943).

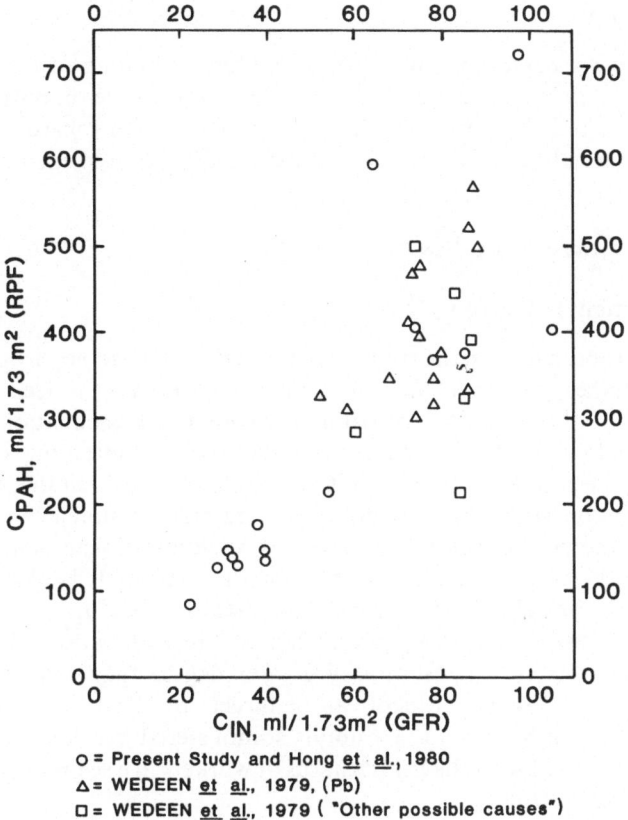

FIGURE 3. The relationship of *p*-aminohippurate clearance (C_{PAH}) to inulin clearance (C_{IN}).

the renal tubule. A suitable animal model would be immensely helpful in the resolution of these and other questions concerning lead nephropathy.

4.2. Criteria for Judging Suitability of the Model

The ideal animal model should closely mimic the features of the disease as they occur in humans. There actually may be several forms of lead nephropathy. The disease in children, heavy consumers of alcohol, and occupationally exposed workers may be distinctly different. Primary consideration should be directed, however, toward the disease as it occurs in lead-exposed workers. It is probably the most prevalent form and, of equal or greater importance, it is the only form that has been reasonably well characterized.

From a functional standpoint the ideal model would be a mature animal exhibiting a dose-related reduction in GFR and ERPF. These features are seen in humans with evidence of some concurrent decrement in tubular function. It would also be desirable that the general health status of the model not be seriously compromised.

4.3. Why the Rat?

There is no *a priori* reason for believing that the rat, among all available animal species, is most suitable. Attention is directed here toward the rat only because it is the only animal that has been studied in any detail with regard to the salient functional abnormalities described in occupationally exposed men.

4.4. Past and Current Studies

4.4.1. Experimental Design

The rat has been used in the study of many aspects of lead nephropathy. Only a few of the studies, however, have examined the effects of lead on the two parameters of renal function most strikingly depressed in occupationally exposed men, GFR and/or ERPF. Even among these few studies the experimental protocols have been quite variable. No two studies have utilized the same route of lead administration at comparable dosages in animals of similar age.

It is well known that the absorption and retention of lead is quite variable, depending on the route of administration or the vehicle. It is also known that dietary factors, principally the levels of calcium and iron, influence the gastrointestinal absorption of lead, and that suckling rats absorb lead from the gastrointestinal tract to a substantially greater degree than weanling rats. The problem of varying dosage protocol can be largely resolved by using PbB as the index of exposure rather than the amount administered per day. This index was found to correlate well with the health status of workers in a secondary lead smelter (Hammond *et al.,* 1980).

The immaturity of the rat kidney during the first three weeks of life, as to both functional and anatomical development (Falk, 1955; Horster and Lewy, 1970), also is a consideration in experimental design. It therefore is probably not appropriate to evaluate the effects of lead on renal function resulting from exposure during the suckling period.

TABLE II. Studies of Renal Function in Rats[a]

Study	Age of exposure, days	Study days	μg Pb/dl blood $\pm$ SE	GFR	RBF	Tm$_G$
1. Johnson and Kleinman (1979)	0–50	35–50	120 $\pm$ 21	nc	nc	—
2. Aviv *et al.* (1980)	30–63	42–63	155 $\pm$ 38	—	—	—
		84	56 $\pm$ 8	↓	↓	—
		175	24 $\pm$ 4	↓	↓	—
3a. Present work	0–540	480–540	156 $\pm$ 24	nc	—	nc
3b. Present work	42–210	180–210	89 $\pm$ 13	nc	—	nc
3c. Present work	42–540	480–540	87 $\pm$ 11	nc	—	—

[a] ↓, Decrease; nc, no change.

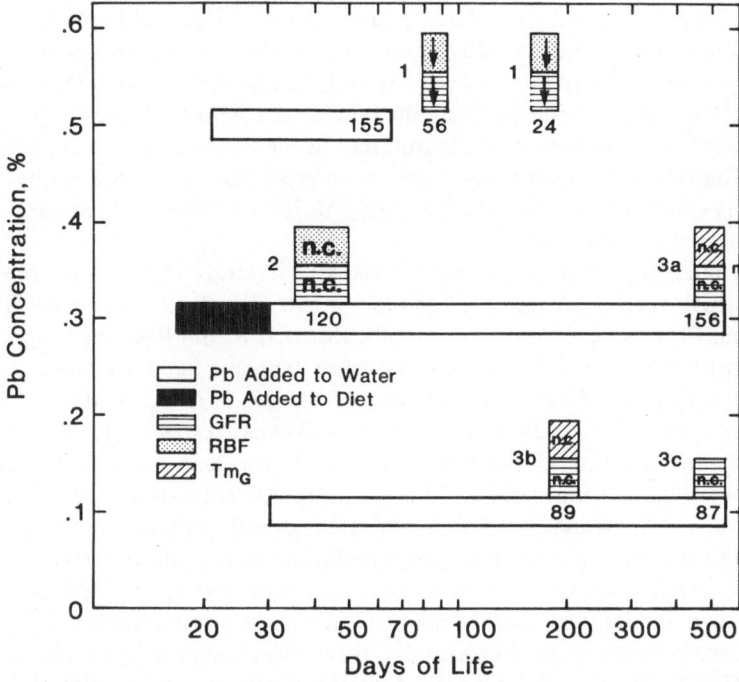

FIGURE 4. Graphic summary representation of studies concerning renal function as related to lead exposure in rats. Horizontal bars represent periods of lead exposure and route of administration. Vertical bars represent time at which studies of renal function were conducted and the results, in which nc indicates no change in function and ↓ indicates decreased function. Numbers within the horizontal bars and below the vertical bars indicate blood lead concentrations (μg Pb/dl) at the times specified on the horizontal axis. Numbers to the left of the vertical bars correspond to reference citations as in Table II.

4.4.2. Summary of Results

There are two published reports of effects of lead on GFR and renal blood flow (Aviv *et al.,* 1980; Johnson and Kleinman, 1979) (studies 1 and 2, Table II and Fig. 4). In the first study (Johnson and Kleinman, 1979), the animals were exposed to lead from birth, initially via the dams' milk, then via the diet, and, finally, via the drinking water. The only determination made of PbB was at the time that renal function was studied (35–50 days of age). There was no effect of lead on either GFR or ERPF. Fractional excretion of amino acids and phosphate was not altered. The only renal effect observed was an elevation in 24-hr urinary volume excretion and in fractional excretion of sodium following extracellular volume expansion. It should be noted in passing that the regimen of lead used did not result in reduction of body weight. Some of the rats left over from this study have since been maintained on the lead regimen to days 480–540, at which time GFR and Tm$_G$ have been determined and are reported here (study 3a, Table II and Fig. 4). There were no functional alterations related to lead exposure, even after this more extended period of time.

The PbB, already 120 ± 21 μg/dl in the earlier study, increased further to 156 ± 24 μg/dl. In a similar study initiated somewhat earlier in our laboratory and not previously reported, lead effects on GFR and Tm_G were examined (studies 3b and 3c, Table II and Fig. 4). In this study the effects of a semisynthetic diet vs. conventional laboratory chow were being compared, hence the designation of two separate groups. There were no lead-related effects on either GFR or Tm_G. In both of the above long-term studies the terminal body weights of the lead-exposed animals were 9% less than controls.

The final study reported in Table II and Fig. 4 (study 2) is the only one known to us in which there occurred a clearly demonstrable effect of lead on GFR and renal blood flow (RBF) (Aviv *et al.*, 1980). Both GFR and RBF were significantly and proportionally reduced 3 and 16 weeks after exposure had terminated, in spite of the fact that PbB had declined substantially from 155 ± 38 μg/dl at termination of exposure to 56 ± 8 and 24 ± 4 at 3 and 16 weeks, respectively, postexposure. Interstitial and vascular lesions were found in the lead-exposed group in the absence of any glomerular lesions. These findings are consistent with those usually reported in occupational lead nephropathy. Blood pressure was significantly elevated in the lead-exposed group, an uncommon finding among recently studied cases of lead nephropathy, but one commonly reported in the older literature.

The Aviv model falls short of the ideal in some respects. Body weight gains of lead-exposed animals were substantially lower than in controls, by 20% at 84 days and by 26% at 175 days. It is also somewhat doubtful that these animals had truly adult kidneys. Thus, it has been shown that during much of the exposure period used by the Aviv group (21–63 days of age) single-nephron glomerular perfusion rate and both whole-kidney glomerular perfusion rate and RBF are increasing per unit kidney weight (Aperia and Herin, 1975). Moreover, no effects of lead were noted in our laboratory on either GFR on Tm_G in mature rats attaining comparable PbB values for a prolonged period of exposure. This again suggests that the Aviv model involved maturational factors. A more careful analysis of dose-age-effect interrelationships is needed. The recent work reviewed here has at least defined more clearly the boundaries of the exposure conditions within which a good model might be successfully developed.

Acknowledgments

Supported by Grant #LH-229, International Lead Zinc Research Organization, and Grant #ES-00159, National Institute of Environmental Health Sciences. The human renal function studies were made possible by use of the General Clinical Research Center facility, supported by NIH grant RR-00068.

References

Aperia, A., and Herin, P., 1975, Development of Glomerular perfusion rate and nephron filtration in rats 17–60 days old, *Am. J. Physiol.* **228**:1319.

Aviv, A., John, E., Bernstein, J., Goldsmith, D. I., and Spitzer, A., 1980, Lead intoxication during development: Its late effects on kidney function and blood pressure, *Kidney Int.* **17**:430.

Chisolm, J. J., 1968, The use of chelating agents in the treatment of acute and chronic lead intoxication in childhood, *J. Pediatr.* **73**:1.

Clarkson, T. W., and Kench, J. E., 1956, Urinary excretion of amino acids by men absorbing heavy metals, *Biochem. J.* **62**:361.

Cooper, W. C., and Gaffey, W. R., 1975, Mortality of lead workers, *J. Occup. Med.* **17**:100.

Cramer, K., Goyer, R. A., Jagenburg, R., and Wilson, M. H., 1974, Renal ultrastructure, renal function, and parameters of lead toxicity in workers with different periods of lead exposure *Br. J. Ind. Med.* **31**:113.

Falk, G., 1955, Maturation of renal function in infant rats, *Am. J. Physiol.* **181**:157.

Galle, P., and Morel-Maroger, L., 1965, Les lesions renales du saturnisme humain et experimental, *Nephron* **2**:273.

Gerhardt, R. E., Crecelius, E. A., and Hudson, J. B., 1980, Trace element content of moonshine, *Arch Environ. Health* **35**:332.

Goyer, R. A., 1971, Lead and the kidney, *Curr. Top. Pathol.* **55**:147.

Goyer, R. A., Tsuchiya, K., Leonard, D. L., and Kahyo, H., 1972. Aminoaciduria in Japanese workers in the lead and cadmium industries, *Am. J. Clin. Pathol.* **57**:635.

Hammond, P. B., Lerner, S. I., Gartside, P. S., Hanenson, J. B., Roda, S. B., Foulkes, E. C., Johnson, D. R., and Pesce, A. J., 1980, The relationship of biological indices of lead exposure to the health status of workers on a secondary lead smelter, *J. Occup. Med.* **22**:475.

Henderson, D. A., 1958, The aetiology of chronic nephritis in Queensland, *Med. J. Aust.* **1**:377.

Hong, C. D., Hanenson, I. B., Lerner, S., Hammond, P. B., Pesce, A. J., and Pollak, V. E., 1980, Occupational exposure to lead: Effects on renal function, *Kidney Int.* **18**:489.

Horster, M., and Lewy, J. E., 1970, Filtration fraction and extraction of PAH during neonatal period in the rat, *Am. J. Physiol.* **219**:1061.

Johnson, D. R., and Kleinman, L. I., 1979, Effects of lead exposure on renal function in young rats, *Toxicol. Appl. Pharmacol.* **48**:361.

Lane, R. E., 1964, Health control in inorganic lead industries, *Arch. Environ. Health* **8**:243.

Lilis, R., Gavrilescu, N., Nestorescu, B., Dumitriu, C., and Roventa, A., 1968, Nephropathy in chronic lead poisoning, *Br. J. Ind. Med.* **25**:196.

Morgan, J. M., Hartley, M. W., and Miller, R. E., 1966, Nephropathy in chronic lead poisoning, *Arch. Intern. Med.* **118**:17.

Radocevic, Z., Saric, M., Beritic, T., and Knezevic, J., 1961, The kidney in lead poisoning, *Br. J. Ind. Med.* **18**:222.

Richet, G., Albahary, C., Morel-Maroger, L., Guillaume, P., and Galle, P., 1966, Les alterations renales dans 23 cas de saturnisme professionnel, *Bull. Soc. Med. Hop. Paris* **117**:441.

Sandstead, H. H., Michelakis, A. M., and Temple, T. E., 1970, Lead intoxication. Its effect on the renin–aldosterone response to sodium deprivation, *Arch. Environ. Health* **20**:356.

Smith, H. W., Goldring, W., Chasis, H., Ranges, H. A., and Bradley, S. E., 1943, Application of saturation methods to the study of glomerular and tubular function in the human kidney, *J. Mt. Sinai Hosp. (NY)* **10**:59.

Tepper, L. B., 1963, Renal function subsequent to childhood plumbism, *Arch. Environ. Health* **7**:82.

Wedeen, R. P., Maesaka, J. K., Weiner, B., Lipat, G. A., Lyons, M. M., Vitale, L. F., and Joselow, M. M., 1975, Occupational lead nephropathy, *Am. J. Med.* **59**:630.

Wedeen, R. P., Mallik, D. K., and Batuman, V., 1979. Detection and treatment of occupational lead nephropathy, *Arch. Intern. Med.* **139**:53.

IV

PATHOPHYSIOLOGIC MECHANISMS OF TOXICITY INDUCED BY ENVIRONMENTAL TOXICANTS

JERRY B. HOOK, Section Editor

24

Environmental Toxicities and Hazards
Introduction

JERRY B. HOOK

Many environmental and industrial chemicals are known to be nephrotoxic. This section will focus on certain nephrotoxic metals and halogenated aliphatic hydrocarbons. Also included are chapters on pollutants such as polychlorinated and polybrominated biphenyls and pesticides like Kepone. These agents, unlike nephrotoxic hydrocarbons or metals, produce few acute effects on the kidney, but may potentiate the nephrotoxic effects of other chemicals.

Occupational exposure to cadmium has only recently been correlated with nephropathy. In Chapter 25 Nordberg presents a review of the metabolism of Cd and stresses the importance of binding of Cd to metallothionein, a low-molecular-weight protein, in the transport of Cd to the kidney. Nordberg treats the relationship between whole-body accumulation of Cd and its excretion in urine and feces mathematically and has constructed a model which can be used to determine the maximally tolerated concentration of cadmium in the renal cortex.

The next two chapters, by Goyer and Fowler, deal with more extensive studies on Cd and other metals. In Chapter 26 Goyer describes the pathogenesis of Cd nephropathy in experimental animals and outlines similarities between these and human nephrotoxicity. The critical concentration of 200 μg/g described by Nordberg is supported by these animal studies. Goyer divides Cd nephropathy into

JERRY B. HOOK • Center for Environmental Toxicology, Michigan State University, East Lansing, Michigan 48824.

two phases. The first phase includes Cd accumulation in the renal cortex accompanied by an increased synthesis of metallothionein without any detectable Cd in the plasma or any discernible changes in the kidney. Phase II is the toxic phase of Cd nephropathy characterized by increased urinary excretion of Cd, renal tubular dysfunction, and detectable plasma concentrations of the metal.

In Chapter 27 Fowler examines the effects of some metal and organic nephrotoxicants on specific cell types and organelle systems. He presents data which indicate that Hg^{++} and Cr^{+6} exert cellular toxicity by direct interaction with the brush border membranes of the kidney proximal tubular cells. The nuclear function of the proximal tubule cells are altered by metals such as Pb, Cd, and Hg. He suggests that these nephrotoxicants are capable of influencing proximal cell nuclear function *in vivo* either directly or secondarily by inducing cell death and increasing cell turnover, thereby increasing normal changes in nuclear DNA. Lead and other agents produce marked changes in cell mitochondria ultrastructure and function. The effect of nephrotoxicants on renal lysosomal and microsomal systems are also reported by Fowler. Thus it is clear that chemicals are capable of inducing toxicity by affecting one or more intracellular organelle systems in the proximal tubule. Subsequent work will determine which effects are primary etiologic events and which are secondary to toxicity.

The next set of chapters deals with nephrotoxic effects seen with halogenated hydrocarbons and the potentiation of these effects by environmental pollutants. In Chapter 28 Kluwe discusses the mechanism of acute nephrotoxicity of the aliphatic halogenated hydrocarbons, which appears to be dependent on toxicant distribution and the innate susceptibility of the kidney. The injury caused by these chemicals is localized to the proximal tubules in the pars recta of rats. This localization is correlated with the S_3 cells. Kluwe proposes that the action of these nephrotoxicants is via the formation of reactive intermediates and is correlated with their localization in the S_3 cells because these cells contain the highest concentration of renal microsomal enzymes. His review stresses the importance of the kidney as a metabolic organ and intrarenal metabolism to a reactive intermediate as a necessary component of the mechanism of nephrotoxicity.

The concept of the kidney as a metabolic organ is further supported by the data presented in Chapter 29 by Hook and Serbiá. The environmental pollutants PBBs and PCBs do not produce any direct effect on renal tissue or function, but have been found to be potent inducers of renal and hepatic microsomal enzymes. The enhanced toxicity seen with chloroform after PBB treatment further substantiates the concept of the intrarenal production of a reactive intermediate or metabolite.

In Chapter 30 Hewitt *et al.* review the potentiation of chloroform-induced nephrotoxicity by another class of compounds—the ketones. Mirex and Kepone, two structurally related insecticides, were found to have different effects. Whereas Kepone enhanced chloroform nephrotoxicity in rats and mice, Mirex did not. Multiple doses of Mirex did, however, lead to enhanced chloroform-induced nephrotoxicity, suggesting that the carbonyl moiety, the difference in structure between Mirex and Kepone, may be a prime determinant of the potentiating capability of Kepone. Further studies with other ketones provided additional evidence that ketones or chemicals capable of being converted to ketones increase the susceptibility of the kidney to the toxic actions of haloalkanes. However, the mechanism of toxicity remains unresolved.

25

Metabolism of Cadmium

GUNNAR F. NORDBERG

1. Introduction

During the last few years cadmium has been increasingly recognized as a dangerous environmental pollutant. Specifically, the subject of discussion has been the effect of cadmium on the kidney tubule, as it is the earliest health effect ("critical effect") after long-term exposure (G. F. Nordberg, 1974; Friberg *et al.*, 1974; Task Group on Metal Toxicity, 1976). Such renal effects occur with long-term accumulation of cadmium in the renal cortex when a certain tissue level (150–300 μg/g) is reached. An approximate estimation of the potential risk of renal effects can thus be made on the basis of calculations of the accumulation of cadmium in the kidney. In order to make adequate estimations of renal accumulation from data on intake, a detailed understanding is needed of the absorption, tissue distribution, retention, and excretion, i.e., the metabolism of cadmium.

There has been a considerable improvement in our understanding of cadmium metabolism during the the last 10 years. Perhaps the most important factor is the involvement of metallothionein, a low-molecular-weight protein, in the transport and intracellular binding of cadmium. A considerable portion of the present chapter will therefore be devoted to these aspects.

GUNNAR F. NORDBERG • Department of Environmental Hygiene, University of Umea, Umea, Sweden.

2. Isolation and Characterization of Metallothionein

The study of the role of metallothionein in Cd metabolism requires that a pure and well-characterized protein be used. A few aspects of the isolation and characterization of metallothionein will therefore be given.

2.1. Rabbit Metallothionein

Liver tissue from cadmium-exposed rabbits was homogenized; the homogenate was treated with rivanol and potassium bromide and subsequently with ethanol and chloroform in order to precipitate high-molecular-weight proteins and hemoglobin (G. F. Nordberg et al., 1972). After precipitation and centrifugation procedures, the supernatant was removed for gel chromatography on a Sephadex G-75 column. Further purification was obtained through isoelectric focusing (Fig. 1). Two forms of metallothionein (isoelectric points 3.9 and 4.5) were thus obtained. Both had a very high cysteine content (27 and 29%, respectively) and were without aromatic amino acids. This demonstrated that a high purity of metallothionein had been obtained. The molecular weight of the two forms, as calculated from their amino acid composition (after adding the metal content), amounted to 6000–7200.

2.2 Mouse Metallothionein

Similar purification methods, but using ultracentrifugation instead of the precipitation methods with rivanol, were used for the isoloation of mouse liver metallothionein (M. Nordberg et al., 1975). A high purity was also obtained in this case. In the mouse liver, one main metallothionein form, with an isoelectric point of 4.2, which fulfilled the criteria for high purity of metallothionein (high cysteine content 34.6% and absence of aromatic amino acids), was obtained. This protein was used in some of the studies of transport and effects of metallothionein.

A considerable amount of information about metallothionein has recently been published. The amino acid sequence in the protein has been identified for some forms (M. Nordberg and Kojima, 1979). The breakdown of metallothionein into different forms is still not fully clarified. Isoelectric focusing has been a valuable and successful method used for this purpose by several investigators (M. Nordberg and Kojima, 1979; Cherian, 1974). Other authors have used ion-exchange chromatography and have also been able to separate two principal forms of metallothionein (Kojima and Kägi, 1978).

The molecular weight of metallothionein has been the subject of much debate. Kägi (1970) reported a molecular weight of 6600 from amino acid analysis. This molecular weight was also obtained by gel filtration of oxidized apometallothionein in the presence of guanine HCl or urea (random coil formation) (Kägi et al., 1974). Native protein separation on a calibrated Sephadex G-75 column gave a molecular weight of approximately 12,000 for metallothionein from chicken and rats (Weser et al., 1973). Based on estimations from amino acids analyses, G. F. Nordberg et al. (1972) and M. Nordberg et al. (1975) obtained a molecular weight of 6000–7000 for metallothionein with varying metal content. This molecular weight has also

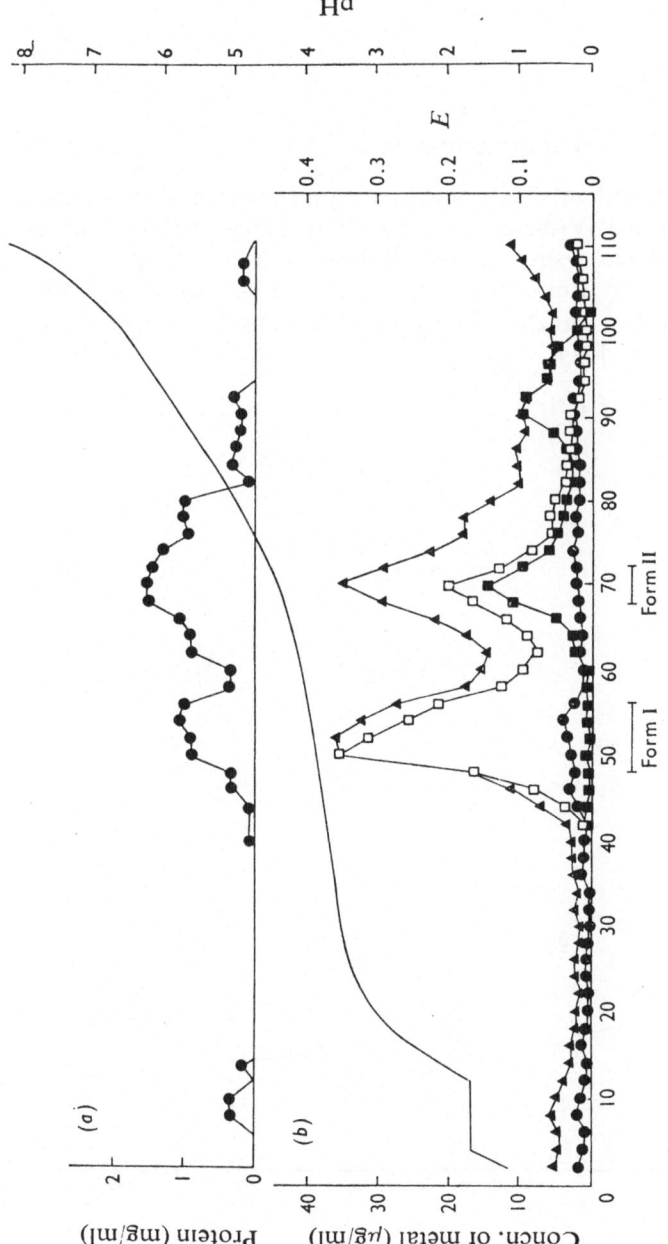

FIGURE 1. Results of isoelectric focusing of rabbit liver metallothionein obtained through precipitation steps and gel-chromatography. Fractions (2 ml) were collected and measurements performed concerning (a) concentration of protein and (b). $\square$—$\square$, Concentration of Cd; ■—■, concentration of Zn; ●—●, E_{280}; ○—○, E_{250}; —, pH at + 4° C. [From G. F. Nordberg *et al.* (1972).]

been obtained in sequence studies. The difference in the molecular weight found with gel chromatography and that found with amino acid analyses may be explained by the nonglobular shape of metallothionein (M. Nordberg and Kojima, 1979).

3. Cadmium Uptake

3.1. Uptake via the Gastrointestinal Tract

In experiments with rats, mice, and monkeys, an absorption percentage of 0.5–3% has been reported (Friberg *et al.,* 1974). In human volunteers, absorption between 5 and 7% has been reported (Rahola *et al.,* 1972). The absorption in women with low body iron stores is up to 20% (Flanagan *et al.,* 1978). Other factors of importance for gastrointestinal absorption are calcium and protein intake (Friberg *et al.,* 1974).

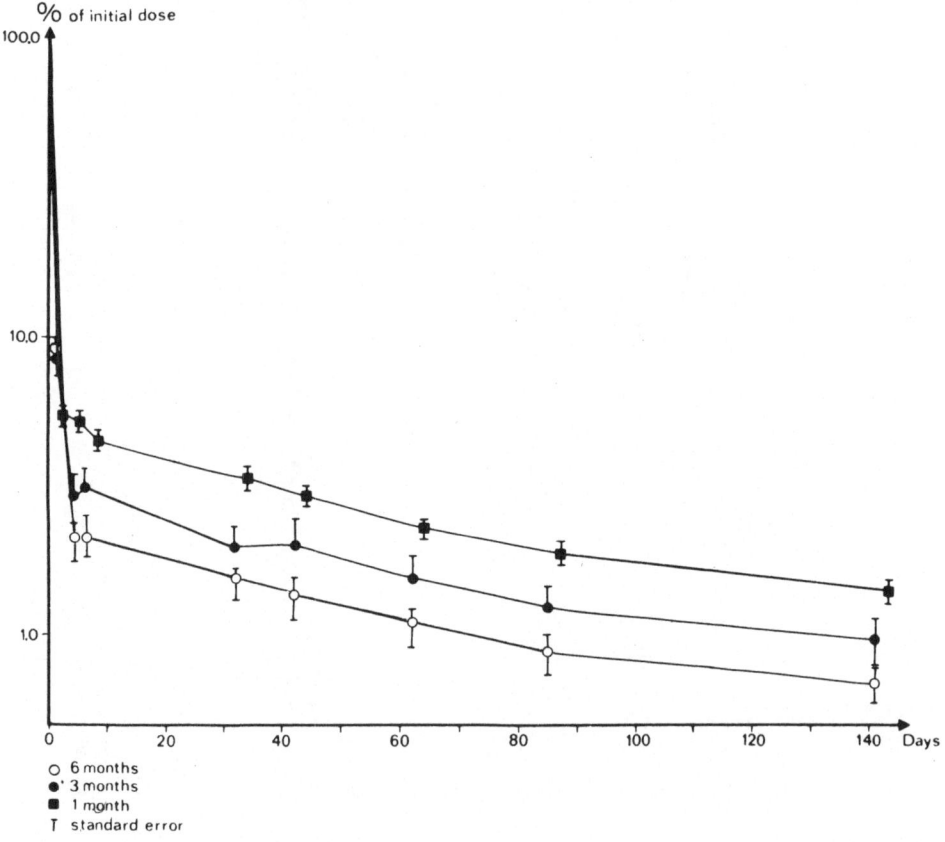

FIGURE 2. Whole-body retention of cadmium in mice of various ages during 5 months after exposure to a single oral dose of $^{109}CdCl_2$ (corrected for radioactive decay). [From Engström and Nordberg (1979).]

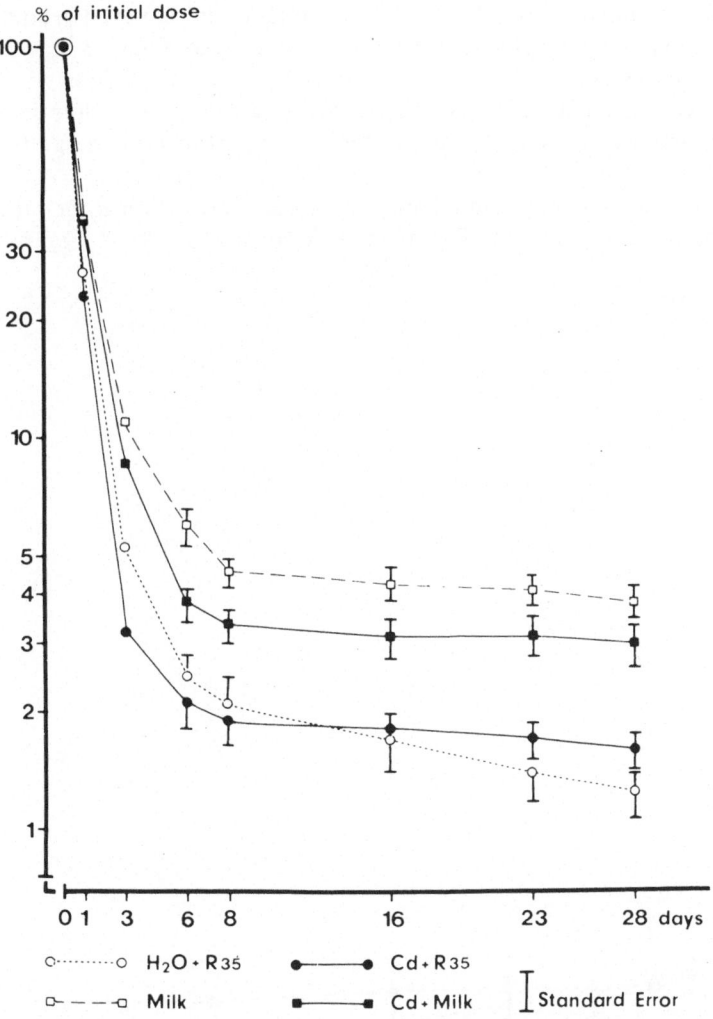

FIGURE 3. Whole-body retention of cadmium in mice consuming various diets after a single oral dose of ^{109}Cd. H_2O + R35: pelleted diet and deionized drinking water. Milk: milk diet, no pellets. Cd + R35: pelleted diet + drinking water with 50 ppm Cd. Cd + Milk: milk diet with 50 ppm Cd, no pellets. [From Engström and Nordberg (1978).]

Engstrom and Nordberg (1978, 1979) showed that age, dose, pretreatment with cadmium, as well as diet are of importance for cadmium retention after several days. Figure 2 shows the results of retention studies in mice given radioactive $^{109}CdCl_2$ by gavage and studied by repeated whole-body countings for several months. Young male mice (1 month old) absorbed 5%, 3-month-old mice 2.9%, and 6-month-old mice only 2.1% of the dose. The biologic half-time was the same in the three groups. Several authors have reported a high absorption of several metals, including Cd, in neonatal animals (Matsusaka *et al.*, 1969; Kello and Kostial, 1977). This may be explained by the milk diet consumed by neonates relative to older animals. Engstrom and Nordberg (1978) demonstrated that even in animals of the same age, milk diet did influence gastrointestinal uptake. Figure 3 shows the results of these experiments.

The experiments by Engstrom and Nordberg (1978, 1979) thus demonstrated that both age and milk diet were factors of importance in the gastrointestinal uptake of cadmium.

Absorption and retention of oral ^{109}Cd was also shown to be influenced by pretreatment with cadmium (Engstrom and Nordberg, 1979). Figure 4 demon-

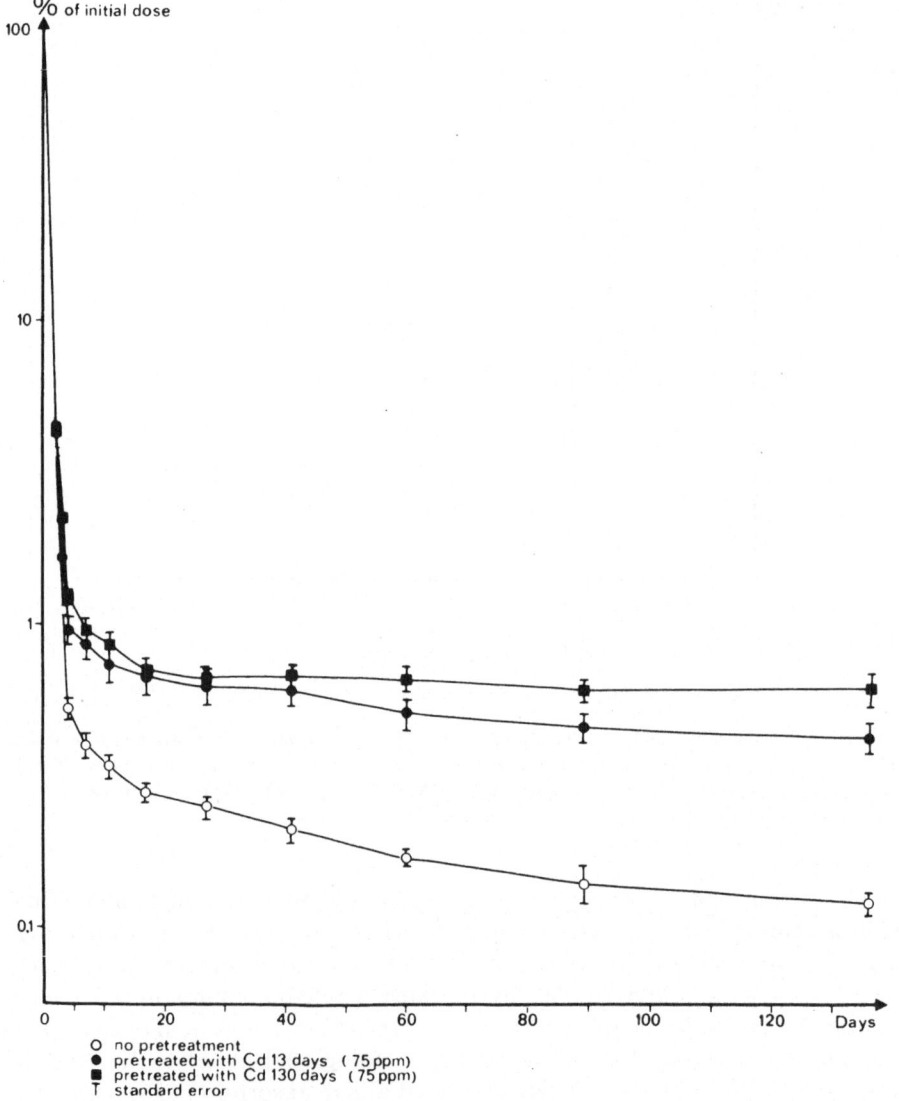

FIGURE 4. Whole-body retention of cadmium in mice given various schedules of pretreatment before a single oral dose of ^{109}Cd. [From Engström and Nordberg (1979).]

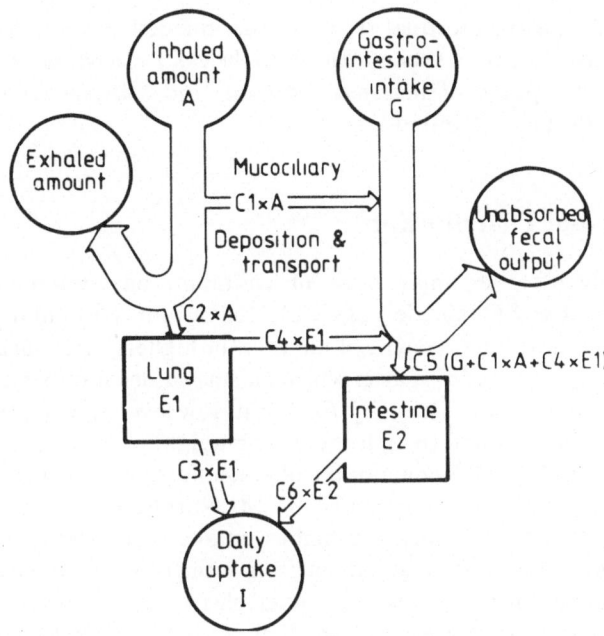

FIGURE 5. Model describing the uptake of cadmium after ingestion or inhalation. [From G. F. Nordberg and Kjellström (1979).]

strates the difference in retention between mice without pretreatment and those given pretreatment for 13 or 130 days with nonradioactive cadmium. In addition to a greater absorption, pretreatment also gave rise to a longer biologic half-time of cadmium, 78 versus 292 days. The resulting difference in body reten-tion after 140 days was 4–6 times greater. The liver retention was particularly affected by the pretreatment. Similar effects on the biologic half-time were also found with Cd exposure subsequent to [109]Cd pretreatment. Engstrom and Nordberg (1980) also demonstrated that dosage was another factor of importance in gastrointestinal absorption. In mice at low doses (1 μg/kg), 0.5% was absorbed, at intermediate doses (15–750 μg/kg), 1.4% was absorbed, and at high doses (37 mg/kg), 3.2% was absorbed.

3.2. Uptake by the Respiratory Tract and Total Absorption

Uptake and clearance of various cadmium compounds from the respiratory tract have been studied only to a limited extent. Whereas soluble Cd compounds would be expected to be cleared from the lung quickly, some compounds that are found in industrial air may be retained for years; high concentrations were found in autopsy specimens from workers who died up to 19 years after retirement (G. F. Nordberg *et al.*, 1978; Brune *et al.*, 1980). General knowledge about the deposition of inhaled particles would also be applicable to cadmium particles and it can be assumed that a model according to Fig. 5 can be used (G. F. Nordberg and Kjellström, 1979). In this model a certain proportion of an inhaled amount is transferred to the gastrointestinal tract via the mucociliary escalator. The total absorption is thus the sum of such a transfer to the gastrointestinal tract in addition to gastrointestinal uptake of ingested cadmium and a direct transfer from the pulmonary tissues into the blood.

4. Transport and Distribution

Immediately after a single dose of cadmium has entered the systemic circulation (e.g., after SC injection of $CdCl_2$), cadmium is bound in blood plasma to proteins with a molecular weight of albumin or higher. Clearance from plasma takes place during the first 48 hr, after which plasma values stabilize at a level that is about $1/1000$ of the initial value (Fig. 6). At this low level approximately one-half of plasma cadmium is bound to high-molecular-weight proteins and the other half to metallothionein (Fig. 7). Renal uptake of systemically injected metallothionein-bound cadmium has been documented by M. Nordberg and Nordberg (1975). Figure 8 demonstrates the markedly higher uptake of injected metallothionein-bound cadmium in the kidney in comparison with the corresponding amount of injected cadmium chloride. The efficient renal clearance and tubular reabsorption of Cd-MT gives importance even to small concentrations of Cd-MT in plasma. Thus, even a small amount of cadmium present in the plasma in the metallothionein-bound form, as small as 0.1 ng Cd/ml, can eventually serve in the important function of transporting cadmium to the renal cortex.

During the initial phase of distribution of cadmium (i.e., the first 24 hr), cadmium is mainly bound to albumin in plasma. During this phase Cd is distributed to a large extent to the liver, where it will be bound to metallothionein. Cadmium will also enter the blood cells. Uptake in blood cells may occur either directly from plasma or via the bone marrow. A direct transfer into blood cells seems to be the most likely explanation for the relatively fast uptake (Fig. 6). In blood cells cadmium will also be bound to metallothionein (M. Nordberg, 1978). When the blood cells are turned over, this metallothionein will become available for transport.

After the initial distribution phase to the liver, there is a slow redistribution of cadmium from the liver and other tissues to the kidney. In view of the very efficient transport of cadmium to the kidney when bound to metallothionein, it is likely that this mechanism is of importance for the redistribution process. Whereas the liver has the highest concentration of cadmium immediately after a single injection of cadmium chloride, the kidney will have the highest concentration for a longer postinjection time period, as shown in Fig. 9. In this figure it may also be noted that cadmium is not taken up in the central nervous system, but a considerable accumulation may occur in the testicles. Uptake of cadmium in the testicles has been studied in more detail, as demonstrated by the autoradiogram in Fig. 10. It is seen that cadmium is localized in the interstitial tissue in the testicles, i.e., in the testosterone-producing Leydig cells. This is of interest and may be related to the possible influence of cadmium on testosterone balance, which has been

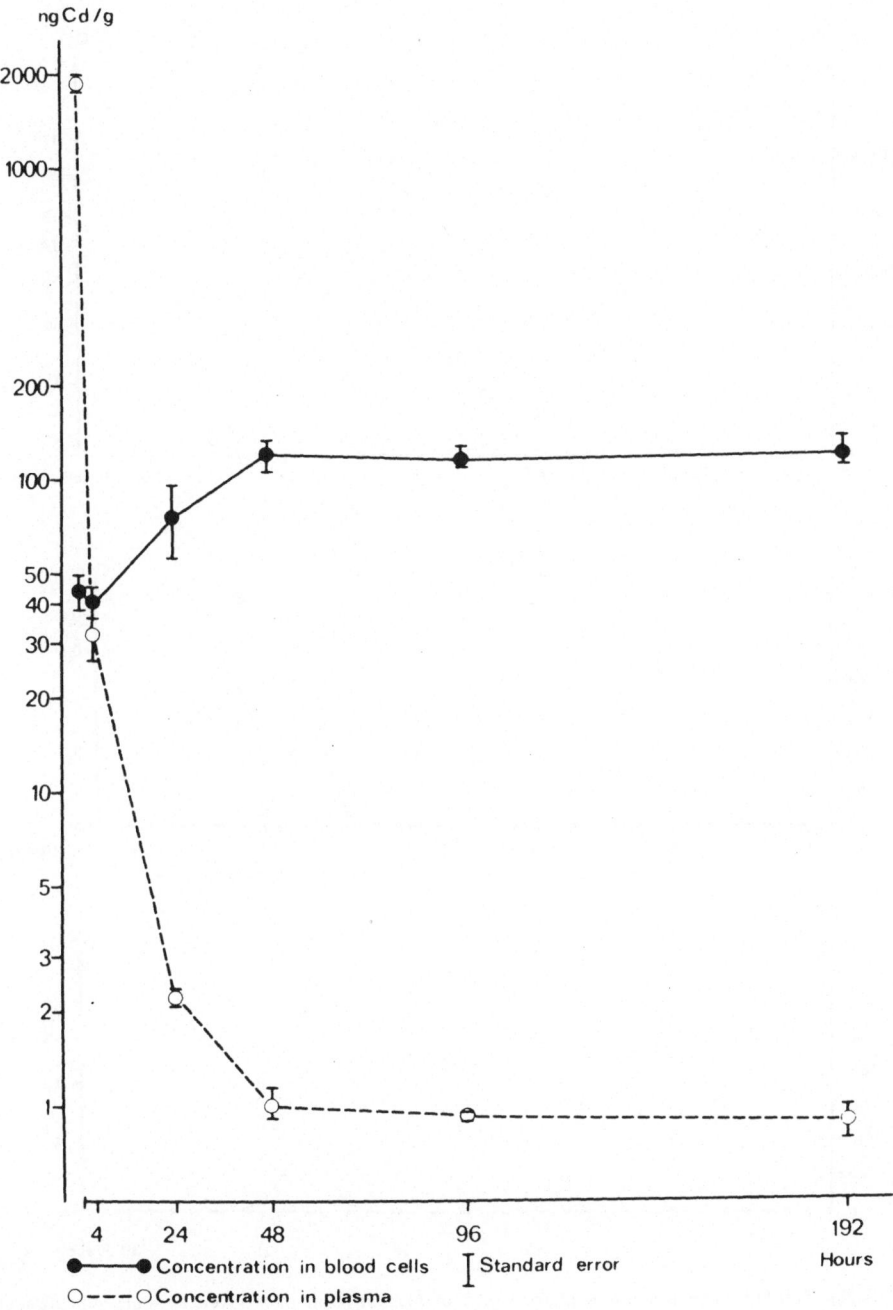

FIGURE 6. Concentrations of cadmium in plasma and blood cells in mice given a single subcutaneous injection of [109]CdCl₂ (1 mg/kg). [From M. Nordberg (1978).]

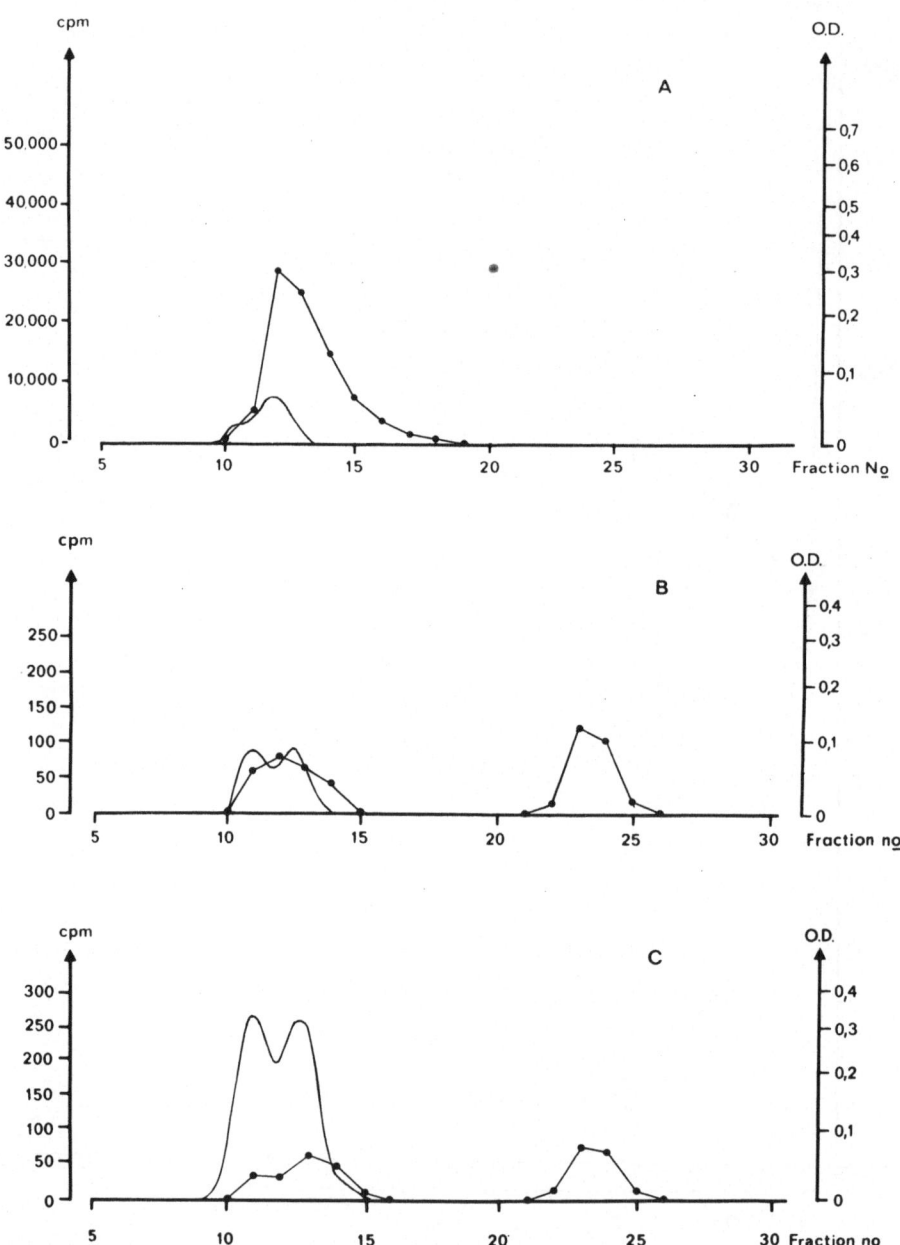

FIGURE 7. Elution profiles on Sephadex G75 of plasma from mice given a single subcutaneous injection of $^{109}CdCl_2$ (1 mg Cd/kg). (—) OD_{254nm}, (●—●) ^{109}Cd, cpm. A: 20 min after injection. B: 96 hr after injection. C: 192 hr after injection. [From M. Nordberg (1978).]

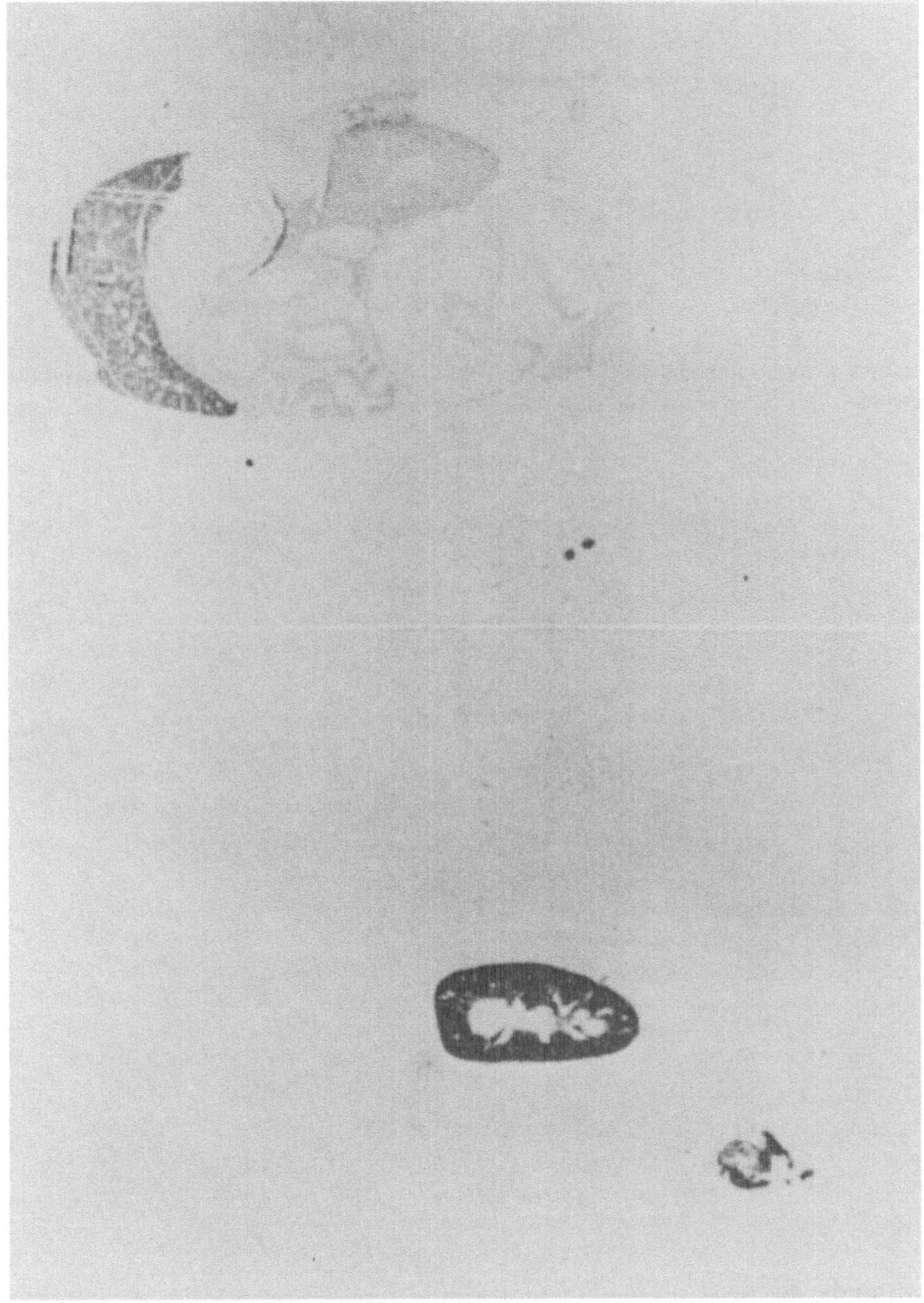

FIGURE 8. Whole-body autoradiogram of mice. (top) Twenty minutes after IV injection of ^{109}CdCl$_2$. Dark areas correspond to high concentration of cadmium. The highest Cd concentration is seen in the liver. (bottom) Twenty minutes after IV injection of ^{109}Cd-MT. Accumulation is seen almost exclusively in the kidney. [From M. Nordberg and Nordberg (1975).]

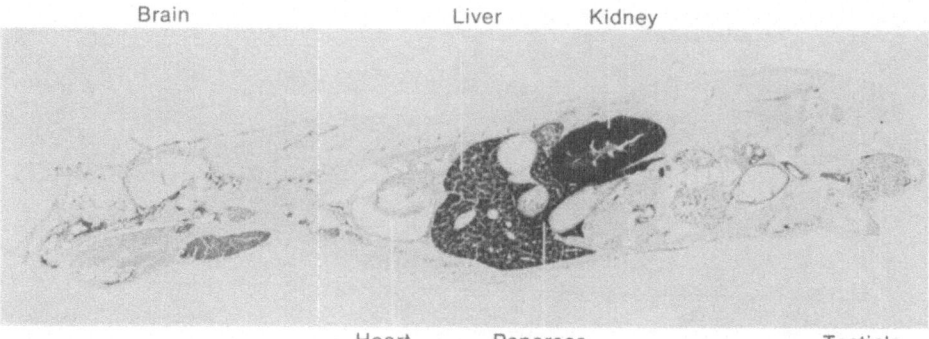

FIGURE 9. Autoradiographic distribution of [109]Cd in a mouse 112 days after a single intravenous injection.

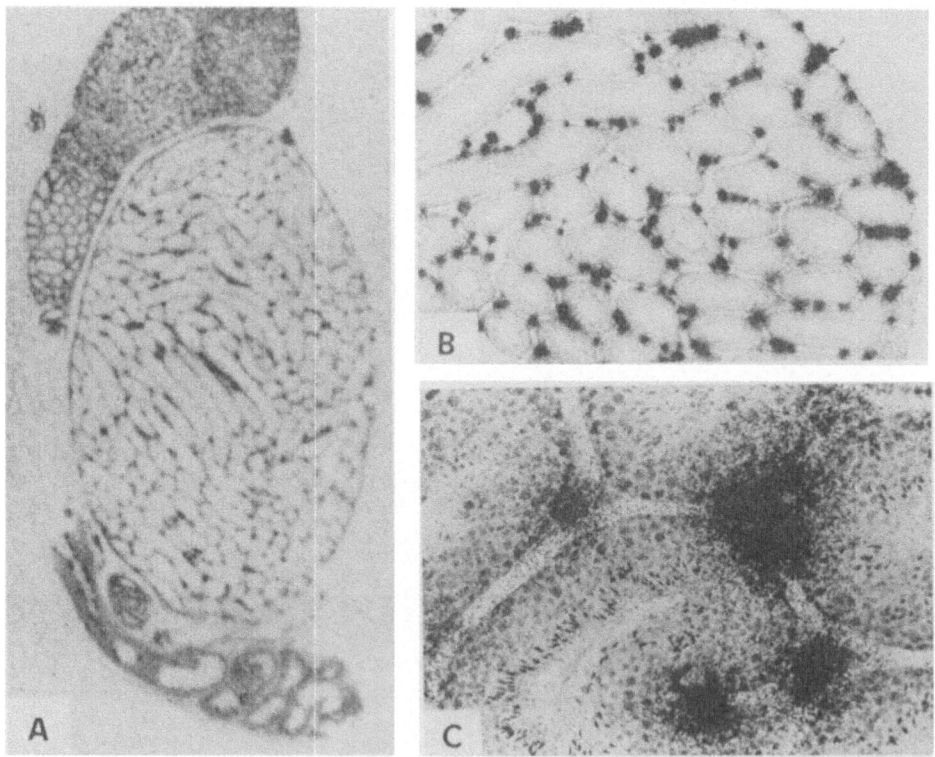

FIGURE 10. Autoradiograms from testicles of mice. A: Apposition autoradiogram from testis and epididymis (freeze-sectioned 5 μm) from a mouse given repeated injections of [109]CdCl$_2$ during 5 months (×9). B: Microautoradiogram (4 μm section + autoradiogram) of a testicle from a mouse SC-injected with [109]CdCl$_2$ and killed 4 hr later. Preparation by freeze-drying and embedding in paraffin wax *in vacuo*. Htx-eosin (×22). C: Micoroautoradiogram (4 μm section + autoradiogram) of a testicle from a mouse IV-injected with [109]CdCl$_2$ and killed 1 day after injection. Preparation as in B (×140).

demonstrated in experimental animals (G. F. Nordberg, 1975). This may also be related to the high incidence of prostate cancer that has been recorded in workers with high occupational exposure to cadmium (Piscator, 1981).

5. Excretion

5.1. Relationship between Fecal Excretion and Body Burden of Cadmium

It is difficult to measure the excretion of cadmium by the fecal route in humans, because of the interference of unabsorbed Cd in the feces. An experiment with mice employing injected radioactive cadmium was reported by G. F. Nordberg (1972). Mice were repeatedly injected subcutaneously with various doses of $^{109}CdCl_2$ and the feces and urine collected in metabolism cages. The body retention of ^{109}Cd was monitored by whole-body counting of the animals. The results from these measurements are shown in Fig. 11. Although there is a large interindividual scattering of the data, a statistically significant correlation is found between body burden and fecal excretion of cadmium. It may be noted that the line intersects the ordinate at the fecal excretion rate of about 120 ng/day, indicating that this part of fecal excretion is dependent on daily dose and not on body burden of cadmium. In another experiment, a ten times lower dose of cadmium was given in the same way and similar measurements were made. The results are shown in Fig. 12. It can be

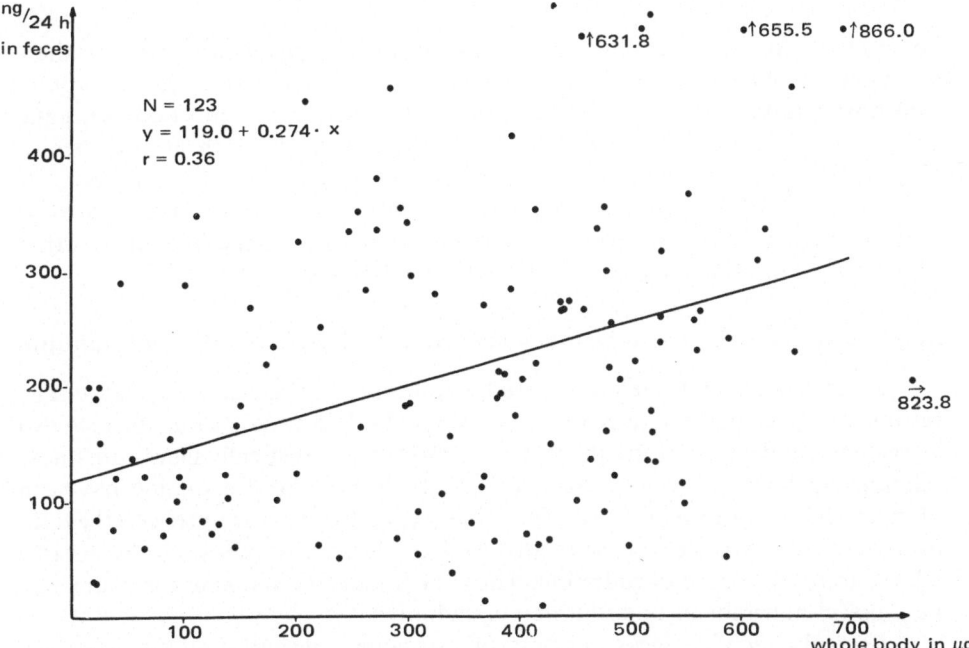

FIGURE 11. Individual values for fecal excretion of cadmium in relation to body burden in mice given SC injection of Cd 0.25 mg/kg for 5 days per week. [From G. F. Nordberg (1974b).]

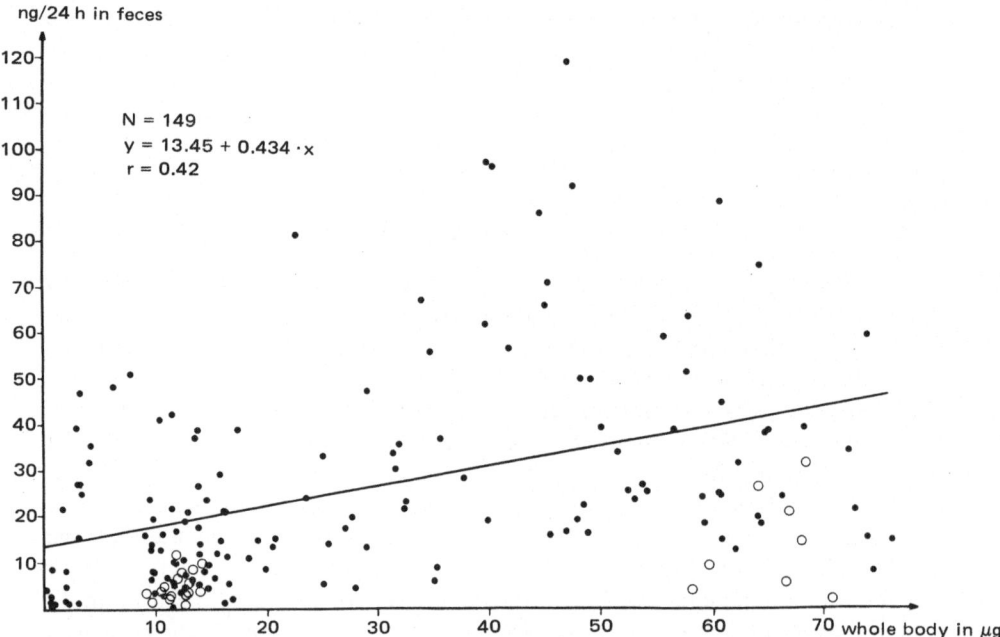

FIGURE 12. Individual values for fecal excretion of cadmium in relation to body burden in mice given SC injections of Cd 0.025 mg/kg for 5 days per week. [From G. F. Nordberg (1974b).]

noted in this figure as well that the line calculated from the values intersects the y axis at approximately 13 ng/day, i.e., one-tenth of the value in the previous figure. This observation is thus in accordance with a dependence on daily dose, since the daily dose in this case was ten times lower than in the experiment described in Fig. 11. Some of the animals in the experiment corresponding to Fig. 12 were left without exposure for 3 weeks. Observations after this time interval are indicated by unfilled circles. These observations consistently fall below the line, and further support a relation of the fecal excretion to daily exposure.

5.2. Relationship between Urinary Excretion and Body Burden of Cadmium

Urine was also collected in the study on fecal excretion described above. The results from such observations are shown in Fig. 13. For urinary values there is also a great interindividual scatter in the data. However, a statistically significant linear relationship between urinary excretion and body burden of cadmium has been observed (G. F. Nordberg, 1972). The urinary excretion (at variance with the fecal excretion) does not seem to be related to daily dosage. It appears to be directly related to body burden of cadmium. Thus, the urinary excretion of cadmium may be a useful indicator of body burden of cadmium.

Recently, *in vivo* measurements of cadmium concentrations in liver and kidney of occupationally exposed human beings have been performed (Roels *et al.*, 1979) and correlated to simultaneously measured cadmium concentrations in urine. These studies in humans have confirmed the conclusions from the animal studies. Concentrations of cadmium in urine are thus used both in environmental and industrial health studies as an indicator of body accumulation (mainly renal accumulation) of cadmium.

6. Mathematical Model for Cadmium Metabolism

Based on the data on cadmium metabolism presented in the previous sections of this chapter, the flow of cadmium among various body compartments can be described. In order to use this information for calculations of cadmium accumulation in the kidney cortex and in other tissues, the flow of cadmium was further described by a number of differential equations.

A detailed description of the model has been presented by G. F. Nordberg and Kjellström (1979) and Kjellström and Nordberg (1978). The first part of this model describes pulmonary and gastrointestinal uptake of cadmium and has been presented in Fig. 7. Total daily uptake I was derived from this part of the model. The pathways for cadmium after uptake are shown in Fig. 14. As previously discussed, cadmium is mainly bound to albumin in plasma (blood compartment B1) immediately after uptake. Cadmium bound in this way will be transported mainly to the liver (C12 = 0.25) and other tissues (C9 = 0.44). Part (C7 = 0.25) of the daily uptake is bound in plasma directly to metallothionein. A part (C11 = 0.27) of

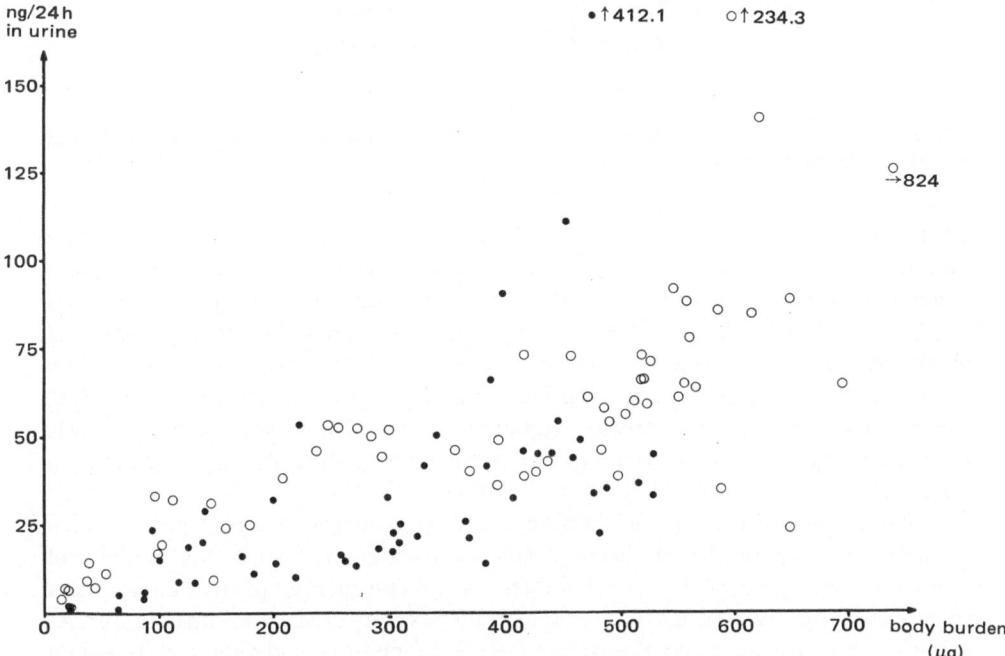

FIGURE 13. Individual values for urinary excretion of cadmium related to body burden in mice given 0.25 mg Cd/kg for 5 days per week. [From G. F. Nordberg (1974b).]

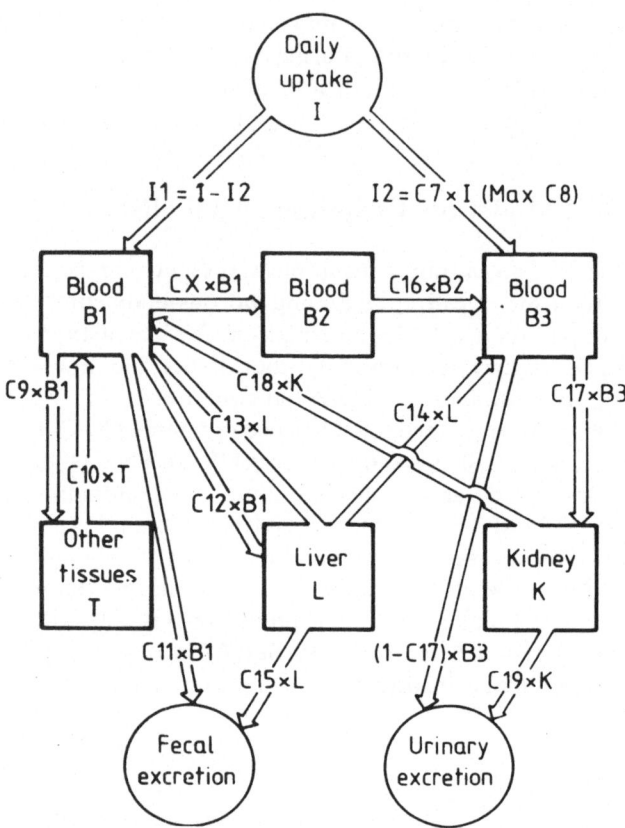

FIGURE 14. Flow diagram of cadmium among various body compartments after uptake from lung or gastrointestinal tract according to Fig. 5.

B1 is excreted in feces and represents that part of fecal excretion that is dependent on daily dose. A small part (CX = 0.04) of B1 is transferred to blood cells (B2). From the blood cells cadmium-metallothionein is released in relation to blood cell turnover (C16 = 0.012). B3 is free metallothionein-bound cadmium in plasma that originates also from the liver (C14 = 0.00016). Such cadmium will mainly (C17 = 0.95) be transported to the kidney and partly be excreted in the urine. Since renal tubular reabsorptive capacity is diminished with age, C17 was made partly age-dependent. Urinary excretion in this model is mainly dependent on renal accumulation C19 × K (C19 = 0.00014).

Values for cadmium accumulation in the renal cortex by age in persons with normal cadmium intake in Sweden have been calculated with this model and compared with empirically found values. As shown in Fig. 15, this comparison turns out favorably, as does that for the urinary excretion of cadmium (Fig. 16). As discussed by Nordberg and Kjellström (1979), other empirical data, e.g., liver and blood values in persons undergoing surgery for gallbladder operations, also fit with the values calculated with this model. Whereas the model seems to give reasonably

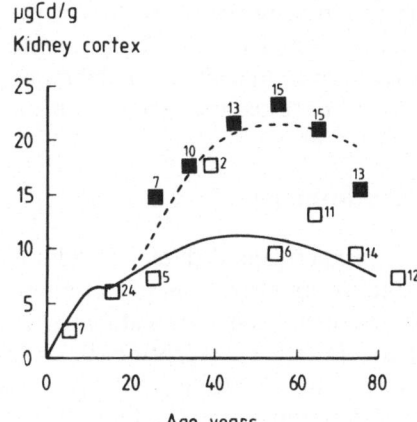

FIGURE 15. Calculated and empirically found concentrations in kidney cortex by age (Sweden); (—) calculated, nonsmokers; (--) calculated, smokers; (□) observed, nonsmoker; (■) observed, smokers. Values show number of observations.

good results for liver and kidney values under long-term exposure conditions, it is still questionable whether cadmium accumulation in other compartments can be described as well.

Since the critical effect of cadmium is renal tubular dysfunction, calculations of intakes giving rise to certain concentrations in the kidney cortex are of particular interest. The daily intake by the oral route required to reach 200 μg/g in kidney cortex was calculated to be 440 μg. For industrial exposure during 25 years, the present model arrives at a concentration in industrial air of approximately 40 μg/m^3 to reach 200 μg/g in kidney cortex.

Although these estimates are probably quite reasonable, because they also fit with epidemiologic evidence concerning dose–response relationships, it should be remembered that the present model is based on a number of assumptions concerning pathways for the cadmium flow in the human organism which have only been partly confirmed by experimental data. For example, the model is based on linear kinetics with regard to uptake. Experimental data on cadmium

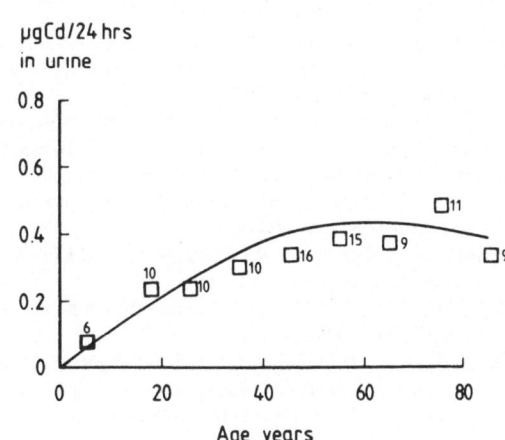

FIGURE 16. Calculated and empirical concentrations of cadmium in urine by age (Sweden): (—) calculated excretion; (□) observed.

metabolism reviewed in this chapter indicate that the gastrointestinal uptake of cadmium and biologic half-time do not display fully linear kinetics. In order to make the mathematical model more generally valid, such deviations from linearity should be taken into account in the future.

7. Summary

Experimental data on cadmium metabolism, i.e., uptake, transport, tissue binding, and excretion of cadmium, have been presented. Several factors, such as dose, diet, and pretreatment, were identified as being of importance for gastrointestinal absorption. Binding of cadmium in tissues and transport has been shown to be related to the presence of the low-molecular-weight protein metallothionein in blood and tissues.

A selective uptake in the kidney of metallothionein-bound cadmium present in plasma has been demonstrated. Relationships between whole-body accumulation of cadmium and excretion in urine and feces were further established.

The available evidence concerning metabolism was used to formulate a multicompartment model for cadmium metabolism in human beings. The concentrations of cadmium in major tissue compartments as calculated by the model agree reasonably well with the empirically found values and the model may thus be used to calculate approximate intake levels that would be required to reach the critical concentration of cadmium in the renal cortex.

References

Brune, D., Nordberg, G., and Wester, P. O., 1980, Distribution of 23 elements in the kidney, liver and lungs of workers from a smeltery and refinery in north Sweden exposed to a number of elements and of a control group, *Sci. Total Environ.* **16**:13.

Cherian, M. G., 1974, Isolation and purification of cadmium binding proteins from rat liver, *Biochem. Biophys. Res. Commun.* **61**:920.

Engstrom, B., and Nordberg, G. F., 1978, Effects of milk diet on gastrointestinal absorption of cadmium in adult mice, *Toxicology* **9**:195.

Engstrom, B., and Nordberg, G. F., 1979, Factors influencing absorption and retention of oral [109]Cd in mice; Age, pretreatment and subsequent treatment with non-radioactive cadmium, *Acta Pharmacol. Toxicol.* **45**:315.

Engstrom, B., and Nordberg, G. F., 1980, Dose dependence of gastrointestinal absorption and biological half-time of cadmium in mice, *Toxicology* **13**:215.

Flanagan, P. R., McLellan, J. S., Haist, J., Cherian, M. G., Chamberlain, M. J., and Valberg, L. S., 1978, Increased dietary cadmium absorption in mice and human subjects with iron deficiency, *Gastroenterology* **74**:841.

Friberg, L., Piscator, M., Nordberg, G. F., and Kjellström, T., 1974, in: *Cadmium in the Environment,* 2nd ed., CRC Press, Cleveland, Ohio.

Kägi, J. H. R., 1970, Hepatic metallothionein, in: *Int. Congr. Biochem. Abstr. 8th* (Switzerland, September 1970), p. 130.

Kägi, J. H. R., Himmelhoch, S. R., Whanger, P. D., Bethune, J. L., and Vallee, B. L., 1974, Equine hepatic and renal metallothioneins. Purification, molecular weight, amino acid composition and metal content, *J. Biol. Chem.* **249**:3537.

Kello, D., and Kostial, K., 1977, Influence of age on whole-body retention and distribution of ^{115m}Cd in the rat, *Environ. Res.* **14**:92.

Kjellström, T., and Nordberg, G. F., 1978, A kinetic model of cadmium metabolism in the human being, *Environ. Res.* **16**:248.

Kojima, Y., and Kägi, J. H. R., 1978, Metallothionein, *Trends Biochem. Sci.* **3**:90.

Matsusaka, N., Inaba, J., Ichikawa, R., Ikeda, M., and Ohkubo, Y., 1969, Some special features of nuclide metabolism in juvenile mammals, in: *Proceedings of the Ninth Annual Hanford Biology Symposium,* Richland, Washington, p. 217.

Nordberg, G. F., 1972, Cadmium metabolism and toxicity: Experimental studies on mice with special reference to the use of biological materials as indices of retention and the possible role of metallothionein in transport and detoxification of cadmium, *Environ. Physiol. Biochem.* **2**:7.

Nordberg, G. F., 1974a, Health hazards of environmental cadmium pollution, *Ambio* **3**(2):55.

Nordberg, G. F., 1974b, in: *Cadmium in the Environment,* 2nd ed. (L. Friberg, M. Piscator, G. F. Nordberg, and T. Kjellström, eds.), CRC Press, Cleveland, Ohio.

Nordberg, G. F., 1975, Effects of long-term cadmium exposure on the seminal vesicles of mice, *J. Reprod. Fert.* **45**:165.

Nordberg, G. F., and Kjellström, T., 1979, Metabolic model for cadmium in man, *Environ. Health Persp.* **28**:211.

Nordberg, G. F., Nordberg, M., Piscator, M., and Vesterberg, O., 1972, Separation of two forms of rabbit metallothionein by isoelectric focusing, *Biochem. J.* **126**:491.

Nordberg, G. F., Wester, P. O., and Brune, D., 1978, Tissue levels of 25 elements in smelter workers—A preliminary communication, in: *Proceedings of International Symposium on the Control of Air Pollution in the Working Environment,* Liber Tryck, Stockholm, p. 261.

Nordberg, M., 1978, Studies on metallothionein and cadmium, *Environ. Res.* **15**:381.

Nordberg, M., and Kojima, Y., 1979, Metallothionein and other low molecular weight metal-binding proteins, in: *Metallothionein Experientia,* Supplement 34 (J. H. R. Kägi and M. Nordberg, eds.), Birkhäuser, Basel, p. 41.

Nordberg, M., and Nordberg, G. F., 1975, Distribution of metallothionein-bound cadmium and cadmium chloride in mice: Preliminary studies, *Environ. Health Persp.* **12**:103.

Nordberg, M., Nordberg, G. F., and Piscator, M., 1975, Isolation and characterization of a hepatic metallothionein from mice, *Environ. Physiol. Biochem.* **5**:396.

Piscator, M., 1981, Role of cadmium in carcinogenesis with special reference to cancer of the prostate, *Environ. Health Persp.* **40**:107.

Rahola, T., Aaran, R. K., and Miettinen, J. K., 1972, Half-time studies of mercury and cadmium by whole body counting, in: IAEA Symposium on the Assessment of Radioactive Organ and Body Burden, in: *Assessment of Radioactive Contamination in Man,* IAEA, Vienna, p. 553.

Roels, H., Bernard, A., Buchet, J. P., Goret, A., Lauwerys, R., Chettle, D. R., Harvey, T. C., and Al-Haddad, I., 1979, Critical concentration of cadmium in renal cortex and urine, *Lancet* **1**:221.

Task Group on Metal Toxicity, 1976, Effects and dose–response relationships of toxic metals, A consensus report from an international meeting organized by the Subcommittee on Toxicology of Metals, in: *Effects and Dose–Response Relationships of Toxic Metals* (G. F. Nordberg, ed.), Elsevier, Amsterdam, p. 3.

Weser, U., Rupp, H., Doney, F., Linnemann, F., Vaelter, W., Voetsch, W., and Jung, G., 1973, Characterization of Cd, Zn-thionein(metallothionein) isolated from rat and chicken liver, *Eur. J. Biochem.* **39**:127.

26

Cadmium Nephropathy

ROBERT A. GOYER

1. Introduction

It is well established now that cadmium accumulates in the kidney with age and if exposure is excessive cadmium will induce a progressive form of chronic renal disease. Unlike many other nephrotoxins, including other heavy metals, such as lead and mercury, there are virtually no acute effects of inorganic cadmium salts on the kidney, except perhaps for some nonspecific effects that have been seen in animals given near lethal doses.

Renal effects are dependent on increasing concentration of cadmium bound to metallothionein. The nephropathy occurs when cadmium concentration reaches a level that has been widely referred to as the critical concentration of cadmium. Cadmium nephropathy has only been recognized in the past 30–40 years, but it has been intensely studied, particularly in persons with heavy occupational exposures (Friberg, 1948; Adams *et al.*, 1969; Lauwerys *et al.*, 1979).

2. Characteristics of Cadmium Nephropathy

Some of the characteristics of cadmium nephropathy are listed in Table I. The effects of cadmium on proximal renal tubular function are manifested by increased cadmium in the urine, proteinuria, aminoaciduria, glucosuria, and decreased renal

ROBERT A. GOYER • National Institute of Environmental Health Sciences, Research Triangle Park, North Carolina 27709.

TABLE I. Cadmium Nephropathy Characteristics

Increase in urinary cadmium excretion
Renal tubular dysfunction
    Low-molecular-weight proteinuria
    Aminoaciduria
    Glucosuria
    Tubular reabsorption of phosphate
Interstitial inflammation and fibrosis
Critical concentration of cadmium in renal cortex

tubular reabsorption of phosphate. Morphologic changes are nonspecific and consist of tubular cell degeneration in the initial stages, progressing to an interstitial inflammatory reaction and fibrosis. The nephropathy occurs when cadmium concentration reaches a level that has been widely referred to as the critical concentration of cadmium.

2.1. Proteinuria

The proteinuria is principally a tubular proteinuria consisting of low-molecular-weight proteins whose tubular absorption has been impaired by cadmium injury to proximal tubular lining cells. The predominant protein is a β_2-microglobulin. This protein was first isolated and characterized from human urine by Berggard and Bearn (1968) and is normally absorbed by proximal renal tubular lining cells. Although increased urinary excretion might be expected to reflect a nonspecific index of proximal renal tubular cell injury, its presence in urine has been most closely related in a quantitative way with cadmium nephropathy (Kjellstrom *et al.*, 1977), and with current availability of radioimmune assay procedures, it is probably, next to cadmium determination per se, the single most useful index of cadmium-induced renal disease. It is widely used to monitor excessive cadmium exposure in populations at risk. β_2-Microglobulin has a molecular weight of 11,800 daltons and is thought to be a component of immunoglobulins. It is produced on the surface of lymphocytes, but is recoverable from the surface of nearly all cells. It may also be a component of a number of cell surface antigens.

Serum levels increase in a variety of pathologic conditions, particularly disease processes where there is increase in cell turnover rate, as in neoplastic disorders, particularly of the reticuloendothelial system, such as lymphosarcoma, and in various inflammatory and immunologic processes (Franklin, 1975). Low serum levels are present when there is increased renal loss as might occur in cadmium toxicity. However, Tsuchiya *et al.* (1979) in Japan find increased serum levels in the early stages of cadmium nephropathy, suggesting that cadmium may in fact stimulate β_2-microglobulin synthesis. Therefore, changes in serum or urine β_2-microglobulin are not specific for cadmium toxicity and must be correlated with excessive cadmium excretion for purposes of diagnosis. Nevertheless, it does provide a good index to monitor the status of cadmium toxicity. A number of other low-molecular-weight proteins have been identified in the urine of workers with excessive cadmium exposure, such as retinol-binding protein, lysozyme, ribonuclease, and immunoglobulin light chains (Lauwreys *et al.*, 1979).

Lauwerys *et al.* (1979) have also identified high-molecular-weight proteins in the urine, such as albumin and transferin, indicating that some workers may actually have a mixed proteinuria and suggesting a glomerular effect as well. The nature of the glomerular lesion in cadmium nephropathy has not been studied extensively, but Japanese workers (Kawamura *et al.*, 1978) have demonstrated decreased inulin clearance in rats given excessive cadmium.

2.2. Aminoaciduria

Aminoaciduria in cadmium toxicity is generalized, reflecting increased excretion of amino acids normally reabsorbed by proximal tubular lining cells. The severity of the aminoaciduria increases in cadmium workers with increasing levels of cadmium exposure (Tsuchiya, 1978). In addition, particularly large increases in proline and hydroxyproline excretion have been noted in patients with chronic cadmium toxicity with bone disease or Itai-Itai disease, but this probably reflects the changes in bone metabolism found in these people. Glucosuria and decreased tubular reabsorption of phosphate parallel the occurrence of low-molecular-weight proteinuria and aminoaciduria, reflecting the proximal tubular cell effect.

2.3. Clinical Occurrence of Tubular Dysfunction

Proximal tubular dysfunction may be symptom-free for a number of years, but Kazantzis (1979) reports that followup of nine of 12 workers, initially investigated in 1962, has shown that tubular dysfunction may progress. Five of the nine showed progressive reduction in glomerular function, six had hypercalcuria, two had renal calculi, and one developed osteomalacia and evidence of distal tubular dysfunction.

Although most of the data available to date related to cadmium exposure and cadmium nephropathy have been obtained from workers with occupational exposure, there is some evidence now that persons in the general population with nonoccupational exposure to cadmium may also have cadmium-related renal tubular dysfunction. Among inhabitants of cadmium-polluted areas of Japan where dietary content of cadmium is increased, the prevalence of proteinuria and glucosuria is higher than in control areas and there is some association between increased excretion of low-molecular-weight proteins in urine and level of cadmium pollution (Shigematsu *et al.*, 1978) Also, Lauwerys *et al.* (1980) in Belgium found that a group of women living near a nonferrous metal smelter had a higher body burden as reflected by an increased excretion of cadmium in urine and a higher prevalence of signs of renal dysfunction than women from a control area.

2.4. Critical Concentration of Cadmium

With this awareness that cadmium-induced nephropathy may even occur in persons in the general population, it becomes of major public health importance to know what is the maximum level of cadmium exposure that a person can be exposed to without risk of renal tubular dysfunction and cadmium nephropathy. Also, the concept of a critical concentration of cadmium has very important implications with regard to establishing maximum levels of cadmium that human populations may be exposed to with some margin of safety. A World Heath Organization (WHO) Task Force established the critical concentration of cadmium in the renal cortex of humans to be about 200 μg/g of renal cortex as early as 1972 on the basis of appearance of renal tubular dysfunction and a limited number of cadmium measurements from autopsy and biopsy material (World Health Organization, 1979).

Kjellstrom *et al.* (1977) have established a metabolic model relating daily intake of cadmium and concentration of cadmium in renal cortex. The geometric average intake of cadmium was 14 μg cadmium per day, corresponding to a concentration of cadmium in the renal cortex at about age 50 of around 10 μg/g. The WHO Task Force estimated that daily ingestion of 200–300 μg cadmium/day would be required to reach a kidney cortex concentration of 200 μg cadmium/g at age 50 for a 70-kg man. The model proposed by Kjellstrom agrees quite well with this recommendation. More recently, Roels *et al.* (1979) from Brussels in collaboration with a group from Birmingham in England measured renal cortical cadmium concentrations by a neutron activation technique in workers with various degrees of renal tubular dysfunction and concluded that the "critical" or "effect" level of cadmium in the kidney cortex was probably between 200 and 250 μg/g. This figure is also in agreement with the WHO Task Force Report (World Health Organization, 1979) and the Kjellstrom *et al.* model (1977).

3. Clinical Recognition

The most important measure of excessive cadmium exposure is increased cadmium excretion in the urine. In persons in the general population, without excessive cadmium exposure, urine cadmium excretion is both small and constant, that is, it is usually on the order of only 1 or 2 μg/day, or less than 1 μg/g creatinine. With excessive exposure to cadmium as might occur in workers, there may be a delay before urine cadmium excretion is increased. Nevertheless, increased urine cadmium does parallel increased renal cadmium content, so that urine cadmium measurement does provide a good index of cadmium exposure and body burden of cadmium. Nogawa (1979) determined the urinary concentration of cadmium corresponding to a 1% prevalence rate of a number of abnormal urinary findings. Tubular proteinuria, as indicated by measurable excretion of β_2-microglobulin, occurred at the 1% prevalence rate with urinary cadmium concentration of 3.2 μg/g of creatinine. This was at a slightly lower urine cadmium level than other signs of renal tubular dysfunction. Followup studies of workers with cadmium exposure and correlations between cadmium in kidney and cadmium in urine suggest that the "critical level" of urine concentration of cadmium is 10–15 μg cadmium/g urinary creatinine; that is, renal tubular dysfunction is likely to occur (World Health Organization, 1980).

4. Pathogenesis

4.1. Role of Metallothionein

The pathogenesis of cadmium nephropathy has been studied in a number of animal models, including the mouse, rat, rabbit, and monkey. These studies are important not only to support human dose–response data, but to provide some understanding of the cellular mechanisms contributing to cadmium nephropathy and, it is hoped, in the long term to provide a scientific basis for prevention or treatment of this disorder.

One of the most challenging questions regarding the metabolism of cadmium has been the role of metallothionein in cellular metabolism and potential toxicity. Nordberg *et al.* (1975) made the very interesting observation that the intravenous injection of cadmium bound to metallothionein protects mice from the testicular necrosis that occurs following injection of a similar dose of inorganic cadmium. This supports the notion that the binding of cadmium to metallothionein has a protective effect on cadmium toxicity. However, in these same studies they also observed what may be regarded as a paradoxical effect, the cadmium-metallothionein-produced necrosis of renal tubular lining cells. However, injection of a similar dose of cadmium as cadmium chloride had no effect on the kidney, nor did the injection of a similar dose of zinc-metallothionein. From these observations, it was hypothesized that the cadmium-metallothionein complex per se is toxic to cells. This question has been pursued further by Cherian *et al.* (1976) and Goyer *et al.* (1978), as well as in studies done or in progress in other laboratories (Shaikh and Hirayama, 1979; Suzuki *et al.*, 1979).

4.2. Comparative Toxicology of Cadmium Chloride and Cadmium-Metallothionein

The distribution of cadmium in various organs has been compared after the injection of cadmium chloride and cadmium-metallothionein (Cherian *et al.*, 1976). When cadmium chloride is injected nearly all (60–70%) of the cadmium is recovered in the liver. This is similar to what occurs when cadmium chloride is ingested orally. With time cadmium moves from the liver to the kidney. Only a small fraction of the injected cadmium appears in the urine. When cadmium-metallothionein is injected, a major fraction becomes deposited in the kidney. This is accompanied by renal tubular cell injury, loss of cadmium from the kidney, and the appearance of cadmium in the urine. The effects of an injection of only 200 μg of cadmium complexed to metallothionein on rat renal tubular cells are very dramatic. Four hours after the injection of the cadmium-metallothionein complex the normal luminal surface of microvilli is lost because of the distention and blebbing of the cell. A profuse increase in microvacuoles occurs. This phenomena is not unique to cadmium-metallothionein; it has been observed that the renal tubule absorbs a number of other macromolecules, polysaccharides, highly anionic polycarboxylic, polyamino acids such as the chelating agent EDTA, and other small proteins (Jacques, 1975).

In addition, mitochondria become swollen, may contain calcium deposits, and become disrupted. By 24 hr, there is massive cell necrosis, and the cytoplasm shows extensive mitochondrial calcification in the necrotic cells.

Cadmium contents of the kidney and liver were compared through the experiment and there was little increase in liver cadmium. Marked increase in kidney cadmium occurred initially and then decreased, presumably due to loss from necrotic tubular cells. This accounts for the increase in urine cadmium content 24 hr after the injection of cadmium-metallothionein. Fowler and Nordberg (1978) showed that much of the cadmium injected with metallothionein accumulates in the lysosomal system of the cells. This suggests that cadmium-metallothionein is dissociated, since cadmium-metallothionein induced by inorganic cadmium is located in the cytoplasm and not lysosomes. Suzuki *et al.* (1979) prepared metallothionein with different ratios of cadmium and zinc and found that renal tubular necrosis is related to the cadmium content, not the amount of metallothionein. Although alternative explanations of the cell injury induced by cadmium-metallothionein administration have been presented (Gonick, 1975; Fowler and Nordberg, 1978), the morphologic evidence of cell injury and ruptured membranes, accompanied by an apparent separation of cadmium from the cadmium-metallothionein complex, suggests that the problem may occur at the level of the cadmium-metallothionein interaction at the cell surface. *In vitro* studies of cadmium-metallothionein in renal cell cultures should provide more specific insight (Cherian, 1980).

4.3. *In Vivo* Model of Cadmium Nephrotoxicity

Another question is how do these observations relate to the pathogenesis of cadmium nephrotoxicity following exposure to inorganic salts of cadmium. In a later study the events occurring during the parenteral administration of cadmium chloride to rats were followed (Goyer *et al.,* 1978).

The weekly urinary excretion of cadmium during the injection of cadmium chloride (0.6 mg/kg per day) 5 days per week for 8 weeks is shown in Figure 1. There is virtually no increase in urinary cadmium or discernible morphologic or functional change in the kidney until after the fifth or sixth week. At this time there is simultaneous onset of morphologic evidence of tubular cell damage, increased urinary excretion of cadmium, and occurrence of signs of renal tubular dysfunction. The cell damage has the same morphologic appearances as seen by injecting cadmium-metallothionein complex intravenously, namely increased vacuolation and cell volume, disruption of mitochondria, and cell necrosis.

Plasma of these rats were examined by Sephadex gel chromatography for cadmium-metallothionein complex each week from the onset of the experiment. After 3 weeks of cadmium chloride injection, no cadmium-metallothionein was present, but a clearly discernible peak consistent with cadmium-metallothionein was found in the 6-week sample. This occurred just before the large increase in urine cadmium during the seventh week, as shown in Fig. 1. The peaks of cadmium-metallothionein present in plasma after the seventh and eighth weeks were even larger than found in the sixth-week analysis.

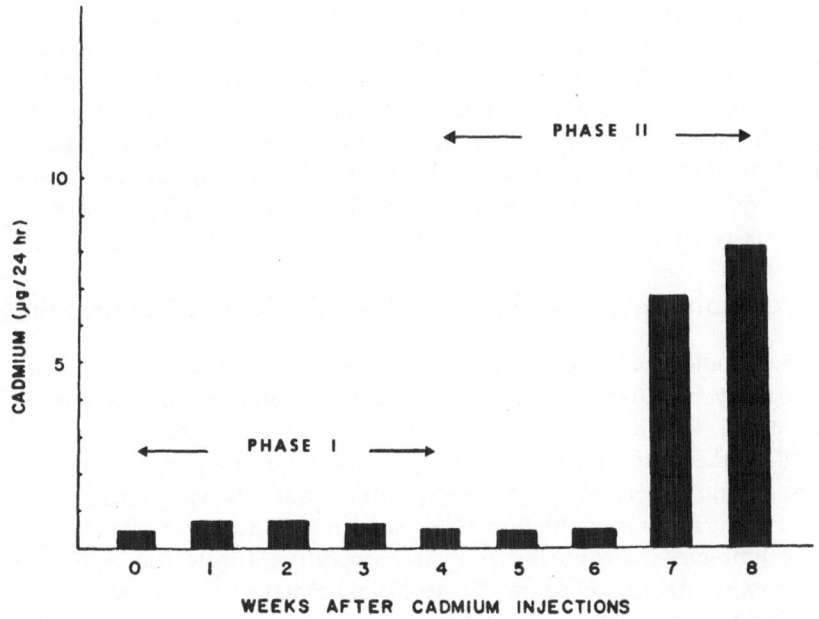

FIGURE 1. Weekly urinary excretion of cadmium during the injection of cadmium chloride (0.6 mg/kg per day) 5 days per week for 8 weeks. [From Goyer *et al.* (1978).]

Cadmium and zinc concentrations in liver and kidney were measured weekly throughout the experiment (Table II). Levels in control animals were similar to the levels found in the experimental group before injection of cadmium and remained so throughout the experiment. The animals receiving injections of cadmium chloride showed continuous increases of cadmium and zinc in liver and kidney. Presumably the increase in zinc reflects some zinc binding to increased intracellular

TABLE II. Concentrations of Cadmium and Zinc in Rat Liver and Kidney Following Intraperitoneal Injection of Cadmium Chloride[a]

Weeks	No. of animals	Liver		Kidney	
		μg Cd/g	μg Zn/g	μg Cd/g	μg Zn/g
1	9	48.3 ± 6.4	59.0 ± 8.5	16.1 ± 2.5	29.9 ± 2.5
2	9	50.8 ± 9.9	82.3 ± 7.9	21.3 ± 3.4	36.1 ± 1.4
3	12	71.1 ± 7.1	74.4 ± 10.9	46.5 ± 8.8	34.8 ± 7.7
4	15	114.3 ± 13.0	96.5 ± 12.7	74.8 ± 10.5	44.7 ± 3.6
5	19	103.7 ± 9.6	50.2 ± 5.8	95.3 ± 10.7	25.6 ± 1.3
6	9	202.5 ± 31.5	81.9 ± 14.5	113.5 ± 16.9	34.6 ± 7.0
7	16	250.0 ± 13.0	95.4 ± 4.4	203.6 ± 6.5	44.1 ± 7.6
8	12	365.5 ± 43.7	113.5 ± 14.2	203.5 ± 11.2	59.4 ± 9.8

[a]The results are expressed as mean ± SEM. Rats were injected with 0.6 mg cadmium/kg body weight daily for 5 days per week and killed at various time intervals during the 8 weeks of the experiment.

metallothionein induced by the cadmium injections. The renal concentration of cadmium was measured for cadmium content of the whole kidney. If these concentrations are increased by 50% to reflect cadmium in the renal cortex, the concentration of cadmium in the cortex at the sixth week was 150–200 $\mu g/g$, the time in the experiment when cadmium-metallothionein was found in plasma, and evidence of tubular cell injury was detectable. This corresponds quite well with studies by others and the proposed hypothesis that about 200 μg cadmium/g of renal cortex is the critical concentration of renal cadmium.

5. Correlations between Experimental and Human Nephrotoxicity

These studies and similarities of this model to experiences with cadmium nephrotoxicity in humans suggest that cadmium nephropathy has two phases. During phase I, cadmium accumulates in the renal cortex accompanied by increased synthesis of metallothionein. There is little or no increase in cadmium in the urine and no discernible morphologic or functional changes. This phase occurs in people in the general population during a normal lifetime exposure to cadmium from the ambient environment. It is also operative during the early stages of excessive exposure to cadmium. No cadmium-metallothionein is detectable in plasma during this period.

The transition to phase II, or the toxic phase of cadmium nephropathy, is characterized by increased cadmium in the urine, evidence of renal tubular dysfunction, and detectable metallothionein in plasma, at least in the rat model. It is likely that when persons with this stage of nephropathy are removed from excessive cadmium exposure the changes are reversible. However, if allowed to progress, loss of tubular cells may occur with decrease in renal concentration of cadmium and eventual interstitial fibrosis and permanent damage.

References

Adams, R. G., Harrison, J. F., and Scott, P., 1969, The development of cadmium-induced proteinuria, impaired renal function and osteomalacia in alkaline battery workers, *Q. J. Med.* **38**:425.

Berggard, I., and Bearn, A. G., 1968, Isolation and properties of a low molecular weight β_2-globulin occurring in human biological fluids, *J. Biol. Chem.* **243**:4095.

Cherian, M. G., 1980, The synthesis of metallothionein and cellular adaptation to metal toxicity in primary kidney epithelial cell cultures, *Toxicology* **17**:225.

Cherian, M. G., Goyer, R. A., and Richardson, L. D., 1976, Cadmium-metallothionein-induced nephropathy, *Toxicol. Appl. Pharmacol.* **7**:700.

Fowler, B. A., and Nordberg, G. F., 1978, The renal toxicity of cadmium metallothionein: Morphometric and x-ray microanalytical studies, *Toxicol. Appl. Pharmacol.* **46**:609.

Franklin, E. C., 1975, β_2-Microglobulin—small molecule—big role? *N. Engl. J. Med.* **293**:1254.

Friberg, L., 1948, Proteinuria and kidney injury among workmen exposed to cadmium and nickel dust, *J. Ind. Hyg. Toxicol.* **30**:32.

Gonick, H., Indraprast, S., and Neustein, H., 1975, Cadmium induced experimental Fanconi syndrome, *Curr. Probl. Clin. Biochem.* **4**:111.

Goyer, R. A., Cherian, M. G., and Richardson, L. D., 1978, Renal tubular effects of cadmium, in: *Cadmium, Proceedings, First International Cadmium Conference,* International Lead Zinc Research Organization, New York, p. 183.

Jacques, P. J., 1975, The endocytic uptake of macromolecules, in: *Pathobiology of Cell Membranes* (B. F. Trump and A. U. Arstilla, eds.) Volume 1, Academic Press, New York, p. 255.

Kawamura, J., Yoshida, O., Nishino, K., and Itokawa, Y., 1978, Disturbances in kidney functions and cadmium and phosphate metabolism in cadmium-poisoned rats, *Nephron* **20**(2):101.

Kazantzis, G., 1979, Renal tubular dysfunction and abnormalities of calcium metabolism in cadmium workers, *Environ. Health. Persp.* **28**:155.

Kjellstrom, T., Ervin, P. E., and Rahnster, B., 1977, Dose–response analysis of cadmium-induced tubular proteinuria, A study of urinary β_2-microglobulin excretion among workers in a battery factor, *Environ. Res.* **13**:303.

Lauwerys, R. R., Roels, H. A., Buchet, J. P., Bernard, A., and Stanescw, D., 1979, Investigations on the lung and kidney function in workers exposed to cadmium, *Environ. Health Persp.* **28**:137.

Lauwerys, R. R., Roels, H., Bernard, A., and Buchet, J. P., 1980, Renal response to cadmium in a population living in a nonferrous smelter area in Belgium, *Int. Arch. Occup. Environ. Health* **45**:271.

Nogawa, K., Kobagashi, E., and Honda, R., 1979, A study of the relationship between concentrations in urine and renal effects of cadmium, *Environ. Health Persp.* **28**:161.

Nordberg, G. F., Goyer, R. A., and Nordberg, M., 1975, Comparative toxicity of cadmium-metallothionein and cadmium chloride on mouse kidney, *Arch. Pathol.* **99**:192.

Roels, H., Bernard, A., Buchet, J. P., and Goret, A., Lauwreys, R., Chettle, D. R., Harvey, T. C., and Haddad, I. A., 1979, Critical concentration of cadmium in renal cortex and urine, *Lancet* **1**:221.

Shaikh, A. A., and Hirayama, K., 1979, Metallothionein in the extracellular fluids as an index of cadmium toxicity, *Environ. Health Persp.* **28**:267.

Shigematsu, I., Kawaguchi, T., and Yamagawa, H., 1978, Reanalysis of the results of health examinations conducted on the general population in cadmium-polluted areas of Japan, in: *Cadmium Studies in Japan, A Review* (K. Tsuchiya, ed.), Elsevier/North-Holland Biomedical Press, New York, p. 253.

Suzuki, K. T., Takenaka, S., and Kubota, K., 1979, Fate and comparative toxicity of metallothionein with differing cadmium–zinc ratios in rat kidney, *Arch. Contam. Toxicol.* **8**:85.

Suzuki, Y., 1977, Significance of cadmium excretion in the urine by long term cadmium administration to rats, II. Changes of cadmium distribution in the urine, *Japan. J. Ind. Health* **19**:200.

Tsuchiya, K., 1978, Epidemiological studies on low-molecular-weight proteins and amino acids in urine of cadmium workers and itai-itai disease patients, in: *Cadmium Studies in Japan, A Review* (K. Tsuchiya, ed.), Elsevier/North-Holland Biomedical Press, New York, p. 276.

Tsuchiya, K., Iwao, S., Sugita, M., and Sakurai, H., 1979, Increased urinary β_2-microglobulin in cadmium exposure: Dose–effect relationship and biological significance of β_2-microglobulin, *Environ. Health Persp.* **28**:147.

World Health Organization, 1979, Environmental Health Criteria, Interim report, EHE/EHC, 72, 20, World Health Organization, Geneva.

World Health Organization, 1980, Technical Report No. 677, Recommended health-based limits in occupational exposure to heavy metals, World Health Organization, Geneva.

27

Ultrastructural and Biochemical Localization of Organelle Damage from Nephrotoxic Agents

BRUCE A. FOWLER

1. Introduction

The kidney is a frequent target organ for a variety of toxic agents due to its metabolic capacity for concentrating toxicants and/or metabolites during the excretory process and to the exquisite sensitivity of the numerous metabolic processes performed by this organ under normal circumstances and after injury. Toxic metals such as mercury, cadmium, chromium, and lead all show extensive kidney accumulation and toxicity to a host of renal metabolic processes. Other classes of toxicants, such as halogenated organics (carbon tetrachloride, cyclodiene pesticides, TCDD, and brominated alkyl compounds), are also capable of extreme toxicity following accumulation and interaction with renal microsomal enzyme systems.

Mechanisms by which toxicants such as those listed above produce damage are mediated by a number of factors, including specificity for action on particular

BRUCE A. FOWLER • Laboratory of Pharmacology, National Institute of Environmental Health Sciences, Research Triangle Park, North Carolina 27709.

renal cell types and interaction with specific essential organelle systems. Table I shows the major cell types of the kidney and attempts to indicate those cells that are most frequently involved in toxic processes. It is clear that only some cell types are readily involved in toxic processes, while the majority are usually spared. The metabolic basis for this predilection of effects appears to be in part determined by blood transport of toxicants, since cells of the renal blood vasculature are frequently involved. Kidney cells with primary reabsorptive functions, such as proximal tubule cells, are the other major group usually involved in toxic processes, ostensibly as a result of uptake, concentration, and/or metabolism of toxicants.

In addition, susceptibility to toxicity also appears to be determined by the sensitivity of particular intracellular organelle systems to a given toxic agent. Some compounds appear to preferentially exert toxicity via action on cell membranes and attendant metabolic processes, whereas others selectively affect nuclear, lysosomal, mitochondrial, or endoplasmic reticulum functions. Still others appear to simultaneously affect a number of systems. The following review will examine the effects of some metal and organic nephrotoxins on specific cell types and organelle systems. It is hoped that this approach will illustrate some of the primary mechanisms of known nephrotoxin action and lead to a better understanding of our current knowledge of toxic pathogenesis in the kidney.

TABLE I. Cell Types of the Kidney

Vasculature:	Arteries/arterioles/ veins
	1. Smooth muscle cells[a]
	2. Arterial endothelial cells
	3. Cells of the juxtaglomerular apparatus
	4. Endothelial cells of the peritubular capillaries[a]
Glomeruli:	Bowman's capsule
	5. Parietal epithelial cells
	Capillaries
	6. Fenestrated endothelial cells
	7. Podocytes[a]
	8. Mesangial cells[a]
Proximal tubule:	
	9. Proximal convoluted (S_1) segment[a]
	10. Proximal convoluted descending (S_2) segment[a]
	11. Proximal straight (S_3) segment[a]
Loop of Henle:	
	12. Cells of the thin limb
Distal tubule:	
	13. Distal tubule cells
Collecting tubules:	
	14. Collecting tubule cells
Collecting ducts:	
	15. Collecting duct cells

[a] Cell types frequently involved in toxic or pathologic processes.

2. Organelle Systems

2.1. Cell Membrane

Ultrastructural and biochemical studies have demonstrated that agents such as inorganic mercury (Hg^{++}) and chromate (Cr^{6+}) exert toxic effects on the brush border membrane of kidney proximal tubule cells. Gritzka and Trump (1968) and Ganote et al. (1975) performed time-course studies which identified loss of the proximal tubule cell brush border as among the earliest structural lesions to occur following acute administration of Hg^{++}. Ganote et al. (1975) also demonstrated concomitant increases in tissue content of water and Ca^{++} in the inner cortex but an absence of effects on para-aminohippurate accumulation and renal slice oxygen consumption until later in the toxic process. On the basis of these data, the authors suggested that cellular membrane damage was a primary event in Hg^{++}-induced proximal tubule cell necrosis. Other studies by Evan and Dail (1974), using Cr^{6+} to induce proximal tubule cell necrosis, also demonstrated swelling and loss of microvilli as the earliest ultrastructural lesions and associated this effect with lysozymuria. These authors felt that the observed lysozymuria was due to decreased reabsorption of this low-molecular-weight protein and as such provided a correlative functional parameter for the observed brush border damage.

More recent comprehensive studies by Kempson et al. (1977), using Hg^{++}, have examined the relationship between ultrastructural damage to the brush border of proximal tubule cells and urinary excretion of marker enzymes. These authors observed decreases in the brush border marker enzymes alkaline phosphatase, leucine aminopeptidase, 5-nucleotidase, maltase, and Mg^{++}-ATPase but no change in the basolateral membrane marker Na^+–K^+-ATPase. These findings were correlated in time with ultrastructural damage to the brush border and a 42-fold maximal increase in the urinary excretion of alkaline phosphatase, suggesting loss of the brush border into the urinary filtrate.

Data from the above clearly indicated that Hg^{++} and Cr^{6+} may exert cellular toxicity by direct interaction with the brush border membranes, but the molecular mechanisms underlying this lesion are unknown and await further study.

2.2. Nuclear Effects

Metals such as lead (Pb), cadmium (Cd), and mercury (Hg) have also been shown to alter nuclear function in kidney proximal tubule cells. Lead accumulation in tubule cell nuclei under acute or chronic circumstances has been shown to cause formation of inclusion bodies (Fig. 1) which contain the highest intracellular concentrations of Pb in these cells (Goyer et al., 1970; Moore and Goyer, 1974; Fowler et al., 1980). Karyomegaly is also frequently observed in tubule cells of lead-treated animals, often at relatively low Pb dose levels (Fowler et al., 1980). Alterations in renal tubule cell nucleic acid synthesis have been reported in a series of studies by Choie and Richter (1974a,b). These authors observed a 45-fold increase in DNA synthesis in proximal tubule cell nuclei of mice given a single

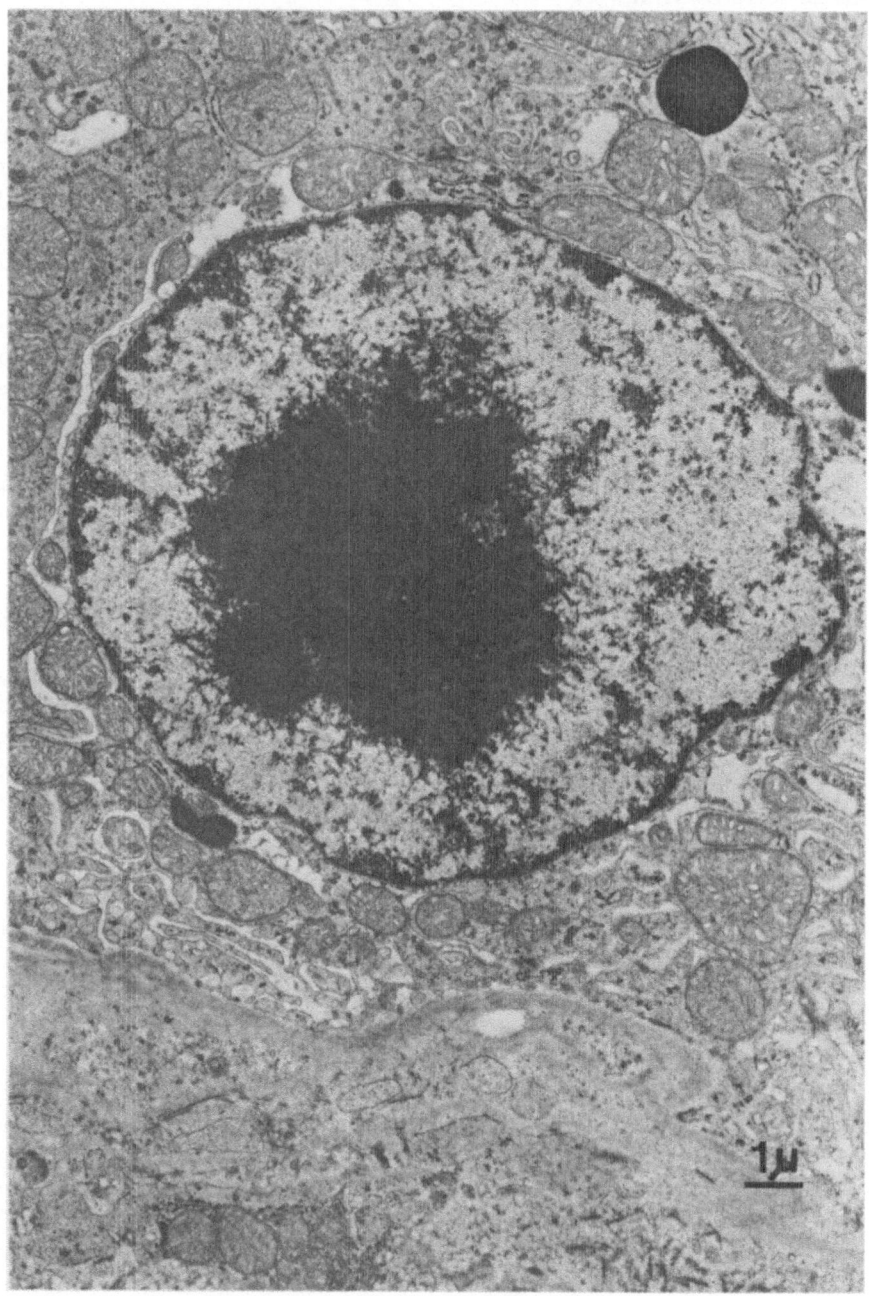

FIGURE 1. Proximal tubule cell from a rat exposed to lead in drinking water, showing large intra-nuclear inclusion body.

injection of Pb at a dose of 5 μg/g, indicating that marked *in vivo* changes in tubule cell nucleic acid metabolism may be produced by relatively low-level Pb exposure.

In contrast, Brubaker *et al.* (1973) observed a 36% decrease in renal DNA synthesis under conditions of subacute exposure to methyl mercury, demonstrating that inhibition of tubule cell nuclear function may also occur at well-tolerated dose levels of this environmental agent. In addition, injection of cadmium-thionein, a low-molecular-weight cadmium-binding protein, has been shown to cause a significant inhibition of renal RNA synthesis, presumably due to the toxicity of the cadmium ion to the proximal tubule cells (Squibb *et al.*, 1982a).

Another type of effect which should be considered in relation to the action of agents such as metals on kidney tubule cell nuclear function is the production of cell death and replacement during acute toxicity. Cuppage and co-workers (Cuppage and Tate, 1967; Cuppage *et al.*, 1967, 1969, 1972) have reported biphasic increases in the incorporation of [^{3}H]thymidine during the tubule cell replacement period following Hg^{++}-induced necrosis, indicating that changes in renal DNA synthesis may also occur as a renal nuclear response to acute injury.

Major points to be derived from these studies are that common environmental nephrotoxins such as lead, cadmium, and mercury are capable of greatly influencing renal proximal tubule cell nuclear function *in vivo* either directly at low-level exposures or secondarily by inducing cell death and replacement with concomitant changes in normal nuclear DNA synthesis. The value of this information rests with an enhanced understanding and perspective of how metals or other agents may influence normal tubule cell replication.

2.3. Mitochrondria

Studies by a number of authors have demonstrated that renal proximal tubule cell mitochrondria are highly susceptible to toxic injury both *in vivo* and *in vitro*. Exposure to agents such as arsenate (Brown *et al.*, 1976), methyl mercury (Fowler and Woods, 1977; Woods and Fowler, 1977), and lead (Goyer, 1968; Goyer and Krall, 1969; Fowler *et al.*, 1980) has been shown to produce marked changes in proximal tubule cell mitochondria ultrastructure (Fig. 2) and function under conditions of prolonged oral exposure. The biochemical mechanisms of mitochondrial toxicity from these agents appear to result from differential effects on mitochondrial membranes, sensitivity of specific respiratory substrates, and marker enzymes in key pathways. The following discussion will attempt to summarize current knowledge of nephrotoxin action at these intramitochondrial loci.

Mitochondrial membranes play a key role in the functional integrity of this organelle and the primary effects of toxic metals such as lead on membranes of renal mitochondria have received attention from several investigators (Goyer and Rhyne, 1973; Fowler *et al.*, 1981a,b). Lead has been shown to have a great affinity for renal mitochondria *in vivo* (Barltrop *et al.*, 1971, Oskarsson *et al.*, 1981) and for the mitochondrial membranes in particular (Fowler *et al.*, 1981b), with loss of respiratory control, impairment of conformational behavior, and altered membrane marker enzyme activities as among the most striking effects (Goyer and

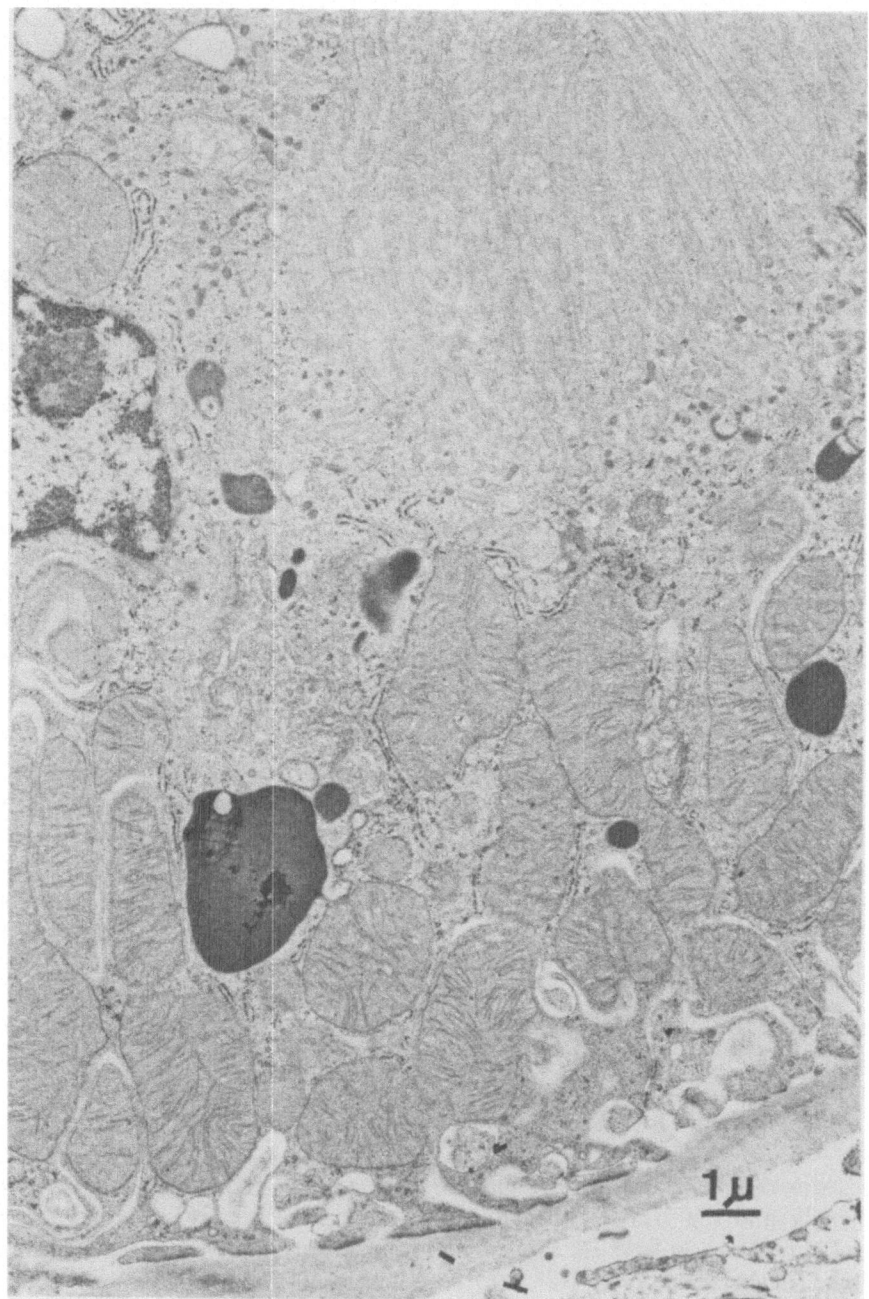

FIGURE 2. Electron micrograph of renal proximal tubule cell from a rat exposed to lead in drinking water, showing *in situ* swollen condition of the mitochondria.

Krall, 1969; Fowler *et al.,* 1981a). Subsequent studies (Fowler *et al.,* 1981b) have shown that the marked physical alteration of the renal mitochondrial membranes by lead (Fig. 3) is associated with a loss of membrane charging and ethidium bromide binding.

The NAD-linked substrate respiration of renal mitochondria has been found to be generally more highly sensitive to toxicant action than that for succinate. Arsenate (Brown *et al.,* 1976), methyl mercury (Fowler and Woods, 1977), and lead (Goyer, 1968; Goyer and Krall, 1969) have all been shown to preferentially inhibit respiration of NAD-linked substrates such as pyruvate/malate and α-ketoglutarate in relation to *in situ* swelling. The mechanism of this preferential substrate effect is usually attributed to inhibition of the lipoic acid moiety (Ulmer and Vallee, 1969). More recent studies (Matlib and Srere, 1976) have shown, however, that simple physical swelling of mitochondria by phosphate will produce a similar preferential inhibition of NAD-linked substrate respiration which is apparently due to disruption of the relationship between the matrix pyruvate dehydrogenase complex and the inner membrane electron transport chain. Such observations again point to the importance of understanding the relationship between mitochondrial physical integrity and biochemical dysfunction.

The other area of renal mitochondrial toxicity which is of particular concern following chronic exposure to metals like lead concerns effects on key cellular pathways leading to potential decreases in essential metabolic products such as heme. Chronic lead feeding studies (Rhyne and Goyer, 1971) have shown a decrease in renal mitochondrial content of cytochrome aa_3. More recent studies (Fowler *et al.,* 1980) have demonstrated that chronic lead exposure leads to inhibition of the renal mitochondrial heme biosynthetic pathway enzymes δ-ALA synthetase (ALAS) and ferrochelatase, thus providing a mechanistic explanation for the observed decrease in mitochondrial cytochrome content. Other studies (Woods and Fowler, 1977) have demonstrated an inhibition of renal mitochondrial ferrochelatase with concomitant coproporphyrinuria and secondary δ-ALAS induction following exposure to methyl mercury at a shorter time period.

The overall metabolic significance of the nephrotoxic action of metals on the mitochondrial heme pathway rests not only with understanding overt cellular toxicity due to depletion of cytochrome-mediated respiration, but also more insidious effects on microsomal enzyme systems responsible for the metabolism of carcinogenic organic compounds. In addition, the inhibitory effects of metals on renal mitochondrial heme metabolism have been found (Woods and Fowler, 1977; Fowler and Woods, 1977) to provide a useful index of ongoing nephrotoxicity due to an enhanced urinary excretion of porphyrins which occurs prior to changes in standard tests of renal function or lysozymuria.

2.4. Lysosomes

The effects of nephrotoxins on the renal lysosome system have recently received extensive attention from a number of authors. Mercury as either methyl mercury (Fowler *et al.,* 1974, 1975b) or Hg^{++} (Madsen and Christensen, 1978) has been found to accumulate in renal lysosomes and to alter the specific activities of a

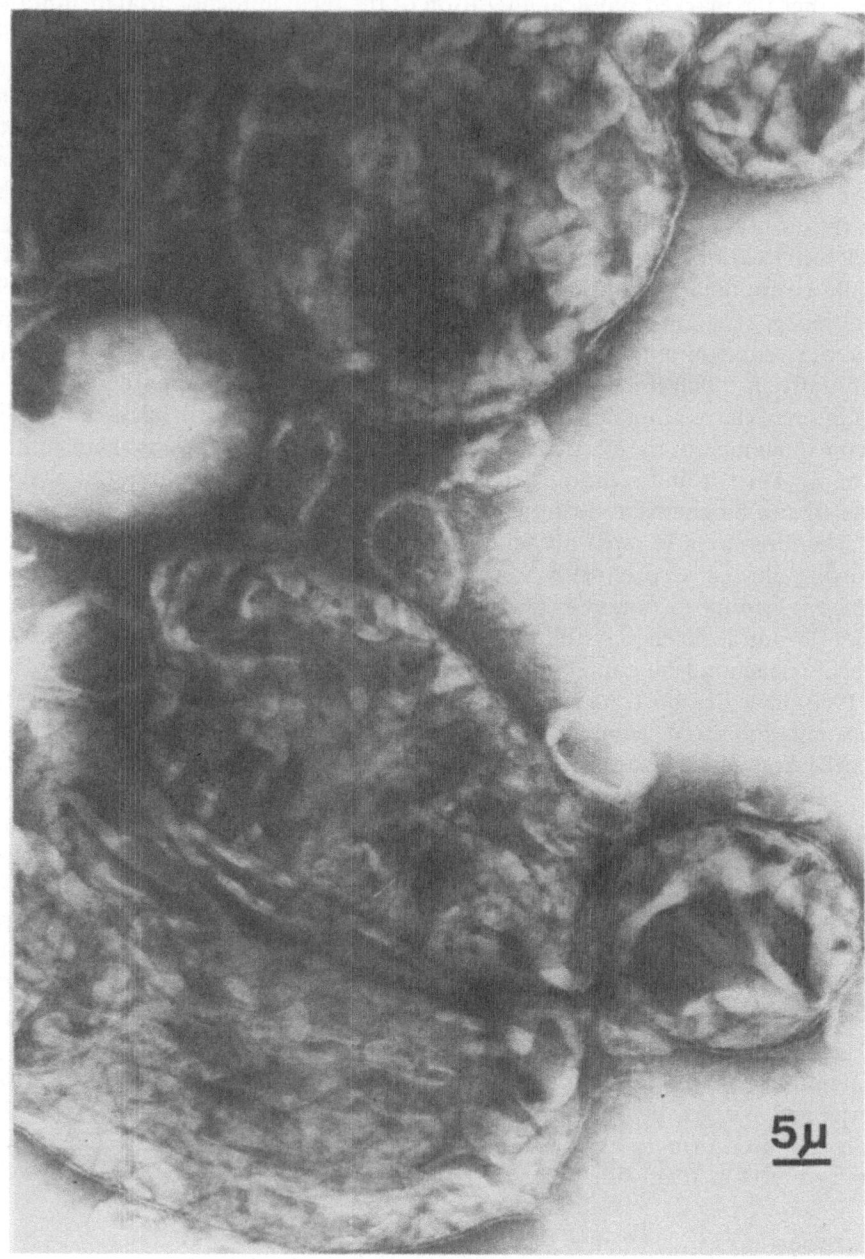

FIGURE 3. Ammonium molybdate-stained mitochondria incubated with Pb (10^{-4} M), showing cristae and membrane distortion.

number of marker enzyme activities and proteolytic functions. Studies by Mego and Barnes (1973) and Mego and Cain (1975) have demonstrated that Hg^{++} and Cd^{++} inhibit the proteolytic capability of renal lysosomes to metabolize ^{125}I-albumin. *In vivo* studies by Squibb *et al.* (1979, 1981, 1982) have shown that renal accumulation of Cd-thionein following injection results in a temporary uptake and degradation of a portion of the injected Cd dose by the renal proximal tubule cell lysosome system (Squibb *et al.*, 1979) with an apparent decreased fusion of primary lysosomes with pinocytotic vesicles (Fig. 4) and a 50% reduction of cathepsin D activity. Injection of maleate (Christensen and Maunsbach, 1980) has also been found to inhibit lysosomal proteolysis in kidney tubule cells and it has also been suggested that the underlying mechanism involves failure of pinocytotic vesicle fusion with primary lysosomes. In both of the above studies, the functional effect of these findings has been the development of tubular proteinuria (Fig. 5).

2.5. Microsomes

Studies by a number of workers have focused attention on the effects of nephrotoxic agents on renal microsomal enzyme systems and the structure of endoplasmic reticulum in proximal tubule cells in relation to the metabolism of these compounds. A number of organic compounds, such as carbon tetrachloride (Striker *et al.*, 1968), paraquat (Fowler and Brooks, 1971), dieldrin (Fowler, 1972), and TCDD (Fowler *et al.*, 1977), have been shown to cause a proliferation of smooth endoplasmic reticulum which is primarily localized in cells of the terminal (S_3) segments of the proximal tubule (Fig. 6). In one of these studies (Fowler *et al.*, 1977) this intracellular effect was associated with induction of microsomal monooxygenase enzymes, while in another (Striker *et al.*, 1968) an inhibitory effect was observed. These results and those of others (Wattenberg and Leong, 1962) have suggested that renal microsomal enzyme activity is primarily localized in one segment of the proximal tubule. Studies by others (Kluwe and Hook, 1980) have demonstrated that a variety of organic nephrotoxins that must be metabolically activated preferentially exert toxicity on cells of the S_3 segment, suggesting that the locus of metabolic activation may be the same as that of nephrotoxicity.

Other nephrotoxic agents, such as methyl mercury (Fowler *et al.*, 1975b), have been shown to inhibit renal microsomal acid hydrolase activities, which may be in part due to the known inhibitory effects of this agent on renal protein synthesis (Brubaker *et al.*, 1973). Inducers of renal heme oxygenase, such as tin (Kappas and Maines, 1976), nickel, and platinum (Maines and Kappas, 1977), are also capable of inhibiting heme-dependent renal microsomal enzymes, thus demonstrating that nephrotoxicity from agents such as metals may occur secondarily to alteration of intracellular heme metabolism.

From the above studies, it is clear that *in situ* ultrastructural changes in the endoplasmic reticulum are associated with changes in renal microsomal enzyme function and hence these morphologic changes may be used to identify loci of toxic metabolism or injury for agents whose toxic mechanisms involve interaction with this system.

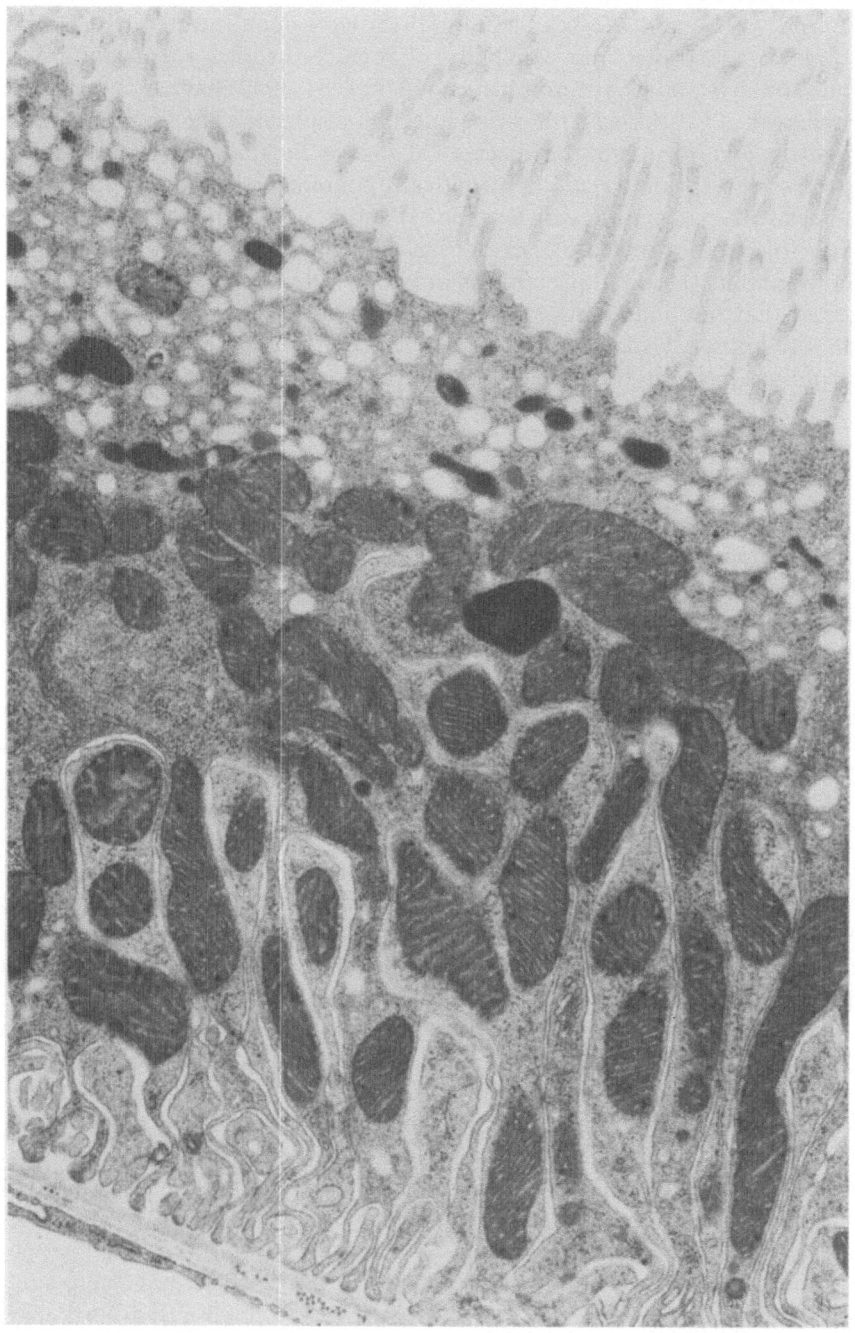

FIGURE 4. Renal proximal tubule cell from a rat injected with Cd-thionein (0.6 mg Cd/kg), showing small, dense apical lysosomes and proliferated pinocytotic vesicles.

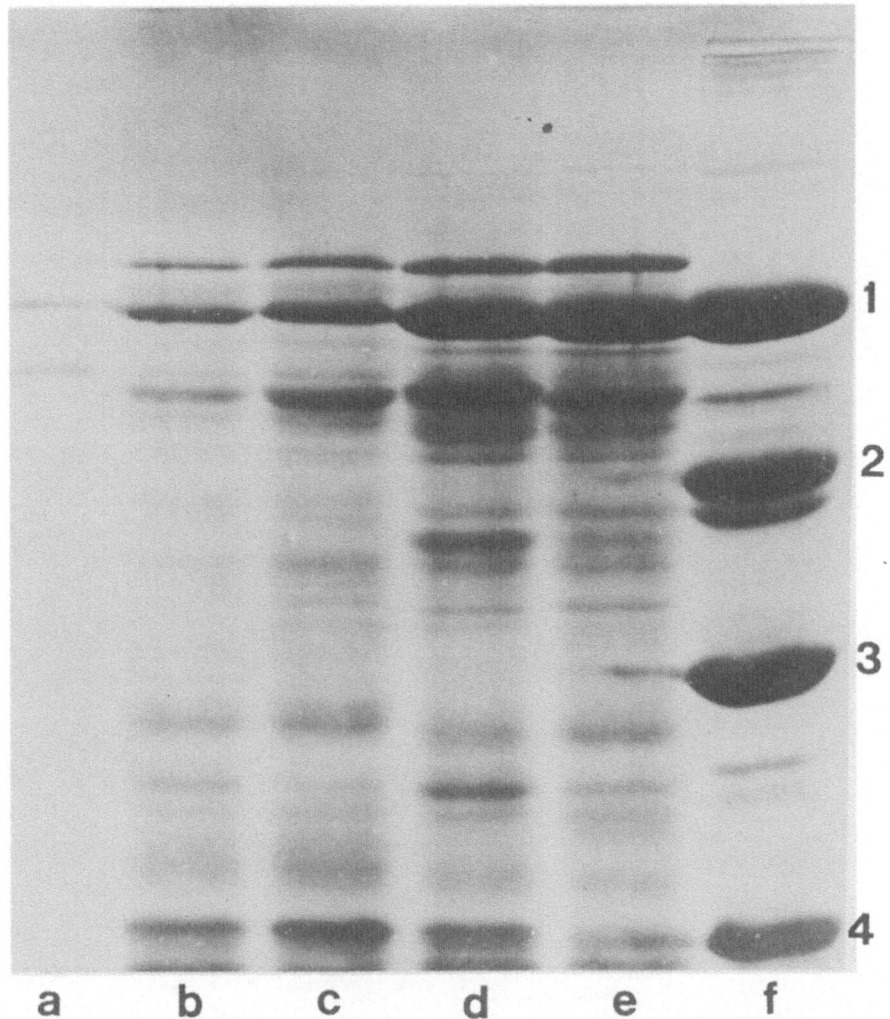

FIGURE 5. SDS gel electrophoresis of urinary proteins from rats injected with Cd-thionein at dose of 0.6 mg Cd/kg, showing a striking increase in low-molecular-weight protein excretion. (a) Control; (b,c) urine collected over an 8-hr period after Cd-thionein treatment; (d,e) urine collected over a 24-hr period after Cd-thionein treatment; (f) molecular-weight standards: (1) albumin, 67,000; (2) ovalbumin, 43,000; (3) chymotrypsinogen A, 25,000; (4) ribonuclease, 13,700.

3. Nephrotoxin Interactions

A relatively new area of investigation concerns interactions between nephrotoxic agents in the kidney. Studies by Mahaffey and Fowler (1977) and Mahaffey *et al.* (1981) on Pb × Cd × As interactions focused considerable attention on antagonism between Pb and Cd with respect to decreased numbers of lead inclusion bodies in animals given both Pb and Cd. This effect was associated with a

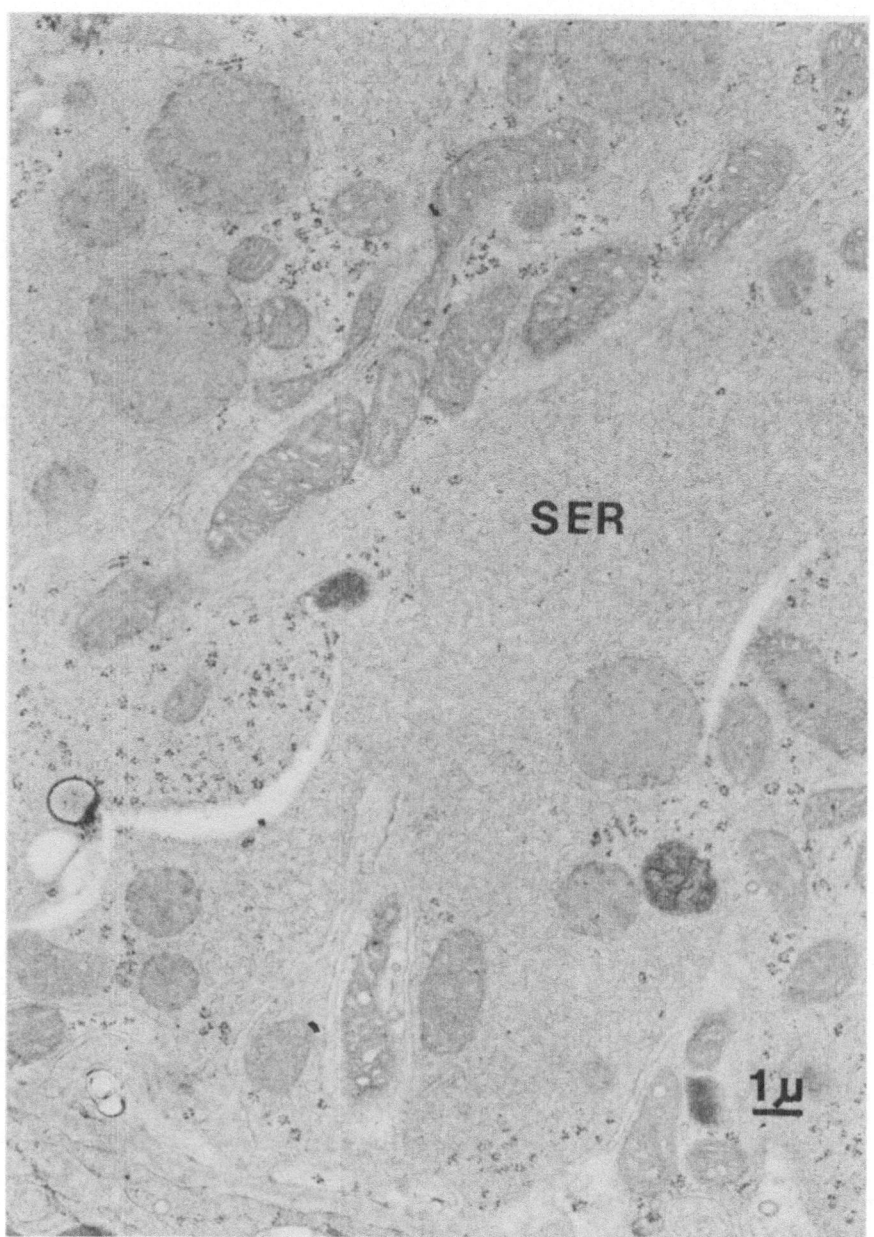

FIGURE 6. Renal proximal tubule cell of the S_3 segment from a rat injected with TCDD (5 μg/kg) and killed 7 days later, showing SER proliferation.

60% reduction in total renal lead concentrations but no change in lead-induced porphyrinuria (Mahaffey *et al.*, 1981). Other studies concerning Hg–Se interactions in the kidney (Carmichael and Fowler, 1979) have demonstrated the presence of Hg–Se intranuclear inclusions which appear to play a role in decreasing the nephrotoxic effects of Hg^{++} on renal proximal tubule cells. Studies by Drummond and Kappas (1980) have demonstrated the protective effects of concomitant exposure to Mn^{++} and Zn^{++} in preventing Sn^{++} and Ni^{++} induction of renal heme oxygenase and associated decreases in cytochrome P450 levels. In other studies, agents that induce microsomal enzymes, such as dieldrin (Fowler, 1972) or phenobarbital (Fowler *et al.*, 1975a), have been found to reduce renal tubule cell damage following prolonged exposure to methyl mercury by enhancing renal excretion of the Hg^{++} ion or stimulating *in vivo* dealkylation of the methyl mercury molecule.

A major point to be emphasized from the above-cited studies is that significant interactions between agents have been documented in the kidney which markedly alter the known intracellular responses produced by administration of a single element. Most of these interactions are associated with pronounced morphologic changes in important subcellular organelle systems, which suggest a mechanism for the interaction. The practical importance of such information rests with indicating the need for evaluation of nephrotoxic agents in multi-element or -factor exposure situations, since usual cellular responses may be greatly altered.

4. Summary—Integrated Assessment of Cellular Toxicity

4.1. Cell Types

One of the major considerations in assessing mechanisms of nephrotoxin action is the fact that only certain renal cell types are usually involved. Cells of the renal blood vasculature and in particular those of the peritubular capillaries frequently manifest damage in the form of interstitial fibrosis. Agents such as cadmium (Stowe *et al.*, 1972; Fowler *et al.*, 1975c) and lead (Goyer, 1971) are examples of nephrotoxins that exert toxicity to the renal blood vasculature. At present, a mechanistic explanation for interstitial fibrosis is lacking, but this response presumably occurs due to cell death and replacement with attendant scarring. Involvement of capillary endothelial cells in this process occurs due to the role played by the peritubular vasculature in transport of toxins and/or their metabolites to and from the tubule cells. Podocytes and mesangial cells of the glomerulus are the other cells of the blood vasculature most frequently involved in toxic processes due to deposition of immune or protein complexes following glomerular deposition.

Cells of the proximal tubule are most frequently involved in toxic processes due to their reabsorptive functions and the presence of metabolic activation systems. In addition, nephrotoxic agents have also been shown to preferentially act on different segments of the proximal tubule. Toxins such as cadmium-thionein preferentially damage cells of the pars convoluta (S_1) segment following glomerular filtration and reabsorption by these cells. Mercuric chloride has been classically used to damage terminal pars recta (S_3) segments of the proximal tubule (Gritzka and Trump, 1968), while methyl mercury exposure has been shown to damage both the second and third segments of the proximal tubule (Fowler *et al.*, 1974).

Metabolic activation of organic nephrotoxins (Kluwe and Hook, 1980) has been shown to cause preferential damage to S_3 proximal tubule cells and it has been suggested (Kluwe and Hook, 1980) that this is due to the presence of metabolic activation systems in these cells (Fowler *et al.*, 1977; Wattenberg and Leong, 1962).

4.2. Organelle Systems

It should be clear from the discussion above that nephrotoxic agents may exert toxicity via action on one or a number of intracellular organelle systems. The value of this knowledge rests with expediting eluciation of toxic mechanisms by focusing attention on the primary intracellular sites of action. In addition, this knowledge should also provide insight into the mechanisms behind clinical manifestations of renal injury, such as decreased tubular reabsorption of proteins, amino acids, glucose, and electroytes, as well as aiding development of new biologic indicators of toxicity (i.e., porphyrinurias). It is also worth noting that primary damage to one organelle system will disrupt cellular homeostasis by altering interrelationships with other organelles, leading to secondary loss of function due to a loss in normal interorganelle relationships. This means that investigations of the primary locus of intracellular toxicity must also consider whether the observed effects are primary or secondary.

References

Barltrop, D., Barrett, A. J., and Dingle, J. T., 1971, Subcellular distribution of lead in the rat, *J. Lab. Clin. Med.* **77**:705.

Brown, M. M., Rhyne, B. C., Goyer, R. A., and Fowler, B. A., 1976, The intracellular effects of chronic arsenic administration on renal proximal tubule cells, *J. Toxicol. Environ. Health* **1**:507.

Brubaker, P. E., Klein, R., Herman, S. P., Lucier, G. W., Alexander, L. T., and Long, M. D., 1973, DNA, RNA, and protein synthesis in the brain, liver and kidneys of asymptomatic methyl mercury treated rats, *Exp. Mol. Pathol.* **18**:263.

Carmichael, N. G., and Fowler, B. A., 1979, Effects of separate and combined chronic mercuric chloride and sodium selenate administration in rats: Histological, ultrastructural and X-ray microanalytical studies of liver and kidney, *J. Environ. Pathol. Toxicol.* **3**:399.

Choie, D. D., and Richter, G. W., 1974a, Cell proliferation in the mouse kidney induced by lead. I. Synthesis of deoxyribonucleic acid, *Lab. Invest.* **30**:647.

Choie, D. D., and Richter, G. W., 1974b, Cell proliferation in the mouse kidney induced by lead. II. Synthesis of ribonucleic acid and protein, *Lab. Invest.* **30**:652.

Christensen, E. I., and Maunsbach, A. B., 1980, Proteinuria induced by sodium maleate in rats: Effects on ultrastructure and protein handling in renal proximal tubule, *Kidney Int.* **17**:771.

Cuppage, F. E., and Tate, A., 1967, Repair of the nephron following injury with mercuric chloride, *Am. J. Pathol.* **51**:405.

Cuppage, R. E., Cunningham, N., and Tate, A., 1969, Nucleic acid synthesis in the regenerating nephron following injury with mercuric chloride, *Lab. Invest.* **26**:122.

Cuppage, F. E., Chiga, M., and Tate, A., 1972, Cell cycle studies in the regenerating rat nephron following injury with mercuric chloride, *Lab. Invest.* **26**:122.

Drummond, G. S., and Kappas, A., 1980, Metal ion interactions in the control of haem oxygenase induction in liver and kidney, *Biochem. J.* **192**:637.

Evan, A. P., and Dail, W. G., Jr., 1974, The effects of sodium chromate on the proximal tubules of the rat kidney: Fine structural damage and lysozymuria, *Lab. Invest.* **30**:704.

Fowler, B. A., 1972, The morphologic effects of dieldrin and methyl mercuric chloride on pars recta segments of rat kidney proximal tubules, *Am. J. Pathol.* **69**:163.

Fowler, B. A., and Brooks, R. E., 1971, Effects of the herbicide paraquat on the ultrastructure of mouse kidney, *Am. J. Pathol.* **63**:505.

Fowler, B. A., and Woods, J. S., 1977, Ultrastructural and biochemical changes in renal mitochondria during chronic oral methyl mercury exposure: The relationship to renal function, *Exp. Mol. Pathol.* **27**:402.

Fowler, B. A., Brown, H. W., Lucier, G. W., and Beard, M. E., 1974, Mercury uptake by renal lysosomes of rats ingesting methyl mercury hydroxide: Ultrastructural observations and energy dispersive X-ray analysis, *Arch. Pathol.* **98**:297.

Fowler, B. A., Lucier, G. W., and Mushak, P., 1975a, Phenobarbital protection against methyl mercury nephrotoxicity, *Proc. Soc. Exp. Biol. Med.* **149**:75.

Fowler, B. A., Brown, H. W., Lucier, G. W., and Krigman, M. R., 1975b, The effects of chronic oral methyl mercury exposure on the lysosome system of rat kidney. Morphometric and biochemical studies, *Lab. Invest.* **32**:313.

Fowler, B. A., Jones, H. S., Brown, H. W., and Haseman, J. K., 1975c, The morphologic effects of chronic cadmium administration on the renal blood vasculature of rats given a low and normal calcium diet, *Toxicol. Appl. Pharmacol.* **348**:233.

Fowler, B. A., Hook, G. E. R., and Lucier, G. W., 1977, Tetrachlorodibenzo-*p*-dioxin induction of renal microsomal enzyme systems: Ultrastructural effects on pars recta (S_3) proximal tubule cells, *J. Pharmacol. Exp. Ther.* **20**:712.

Fowler, B. A., Kimmel, C. A., Woods, J. S., McConnell, E. E., and Grant, L. D., 1980, Chronic low level lead toxicity in the rat. III. An integrated toxicological assessment with special reference to the kidney, *Toxicol. Appl. Pharmacol.* **56**:59.

Fowler, B. A., Squibb, K. S., Oskarsson, A., Taylor, J. A., and Carver, G. T., 1981a, Lead-induced alteration of renal mitochondrial membrane structure and function, *Toxicologist* **1**:19 (abstract).

Fowler, B. A., Taylor, J. A., and Okarsson, A., 1981b, Compartmental binding of Pb in rat kidney mitochondria: The relationship to respiratory function and ethidium bromide binding. *Fed. Proc.* **40**:828 (abstract).

Ganote, C. E., Reimer, K. A., and Jennings, R. B., 1975, Acute mercuric chloride nephrotoxicity: An electron microscopic and metabolic study, *Lab. Invest.* **31**:633.

Goyer, R. A., 1968, The renal tubule in lead poisoning. I. Mitochondrial swelling and aminoaciduria, *Lab. Invest.* **19**:71.

Goyer, R. A., 1971, Lead and the kidney: Current topics in pathology, *Ergebnisse Pathol.* **55**:147.

Goyer, R. A., and Krall, R., 1969, Ultrastructural transformation in mitochondria isolated from kidneys of normal and lead intoxicated rats, *J. Cell. Biol.* **41**:393.

Goyer, R. A., and Rhyne, B. C., 1973, Toxic changes in mitochondrial membranes and mitochondrial function, in: *Pathobiology of Cell Membranes I* (B. F. Trump and A. U. Arstilla, eds.), Academic Press, New York, p. 383.

Goyer, R. A., May, P., Cates, M. M., and Krigman, M. R., 1970, Lead and protein content of isolated intranuclear inclusion bodies from kidneys of lead-poisoned rats, *Lab. Invest.* **22**:245.

Gritzka, T. L., and Trump, B. F., 1968, Renal tubular lesions caused by mercuric chloride. Electron microscopic observations: Degeneration of the pars recta, *Am. J. Pathol.* **52**:1225.

Kappas, A., and Maines, M. D., 1976, Tin: A potent inducer of heme oxygenase in kidney, *Science* **192**:60.

Kempson, S. A., Ellis, B. G., and Price, R. G., 1977, Changes in rat renal cortex, isolated plasma membranes and urinary enzymes following the injection of mercury chloride, *Chem.-Biol. Interact.* **18**:217.

Klein, R., Herman, S. P., Bullock, B. C., and Talley, F. A., 1973, Early functional and pathological changes in rat kidney during methyl mercury intoxication, *Arch. Pathol.* **96**:83.

Kluwe, W. M., and Hook, J. B., 1980, Effects of environmental chemicals on kidney metabolism and function, *Kidney Int.* **18**:648.

Madsen, K. M., and Christensen, E. F., 1978, Effects of mercury on lysosomal protein digestion in the kidney proximal tubule, *Lab. Invest.* **38**:165.

Mahaffey, K. R., and Fowler, B. A., 1977, Effects of concurrent administration of dietary lead, cadmium, and arsenic in the rat, *Environ. Health Persp.* **19**:165.

Mahaffey, K. R., Capar, S. G., Gladen, B. C., and Fowler, B. A., 1981, Concurrent exposure to lead, cadmium and arsenic: Effects on toxicity and tissue metal concentrations in the rat, *J. Lab. Clin. Med.* **98**:463.

Maines, M. D., and Kappas, A., 1977, Enzymes of heme metabolism in the kidney: Regulation by trace metals which do not form heme complexes, *J. Exp. Med.* **146**:1286.

Matlib, M. A., and Srere, P. A., 1976, Oxidative properties of swollen rat liver mitochondria, *Arch. Biochem. Biophys.* **174**:705.

Mego, J. L., and Barnes, J., 1973, Inhibition of heterolysosome formation and function in mouse kidneys by injection of mercuric chloride, *Biochem. Pharmacol.* **22**:373.

Mego, J. L., and Cain, J. A., 1975, An effect of cadmium on heterolysosome formation and function in mice, *Biochem. Pharmacol.* **24**:1227.

Moore, J. F., and Goyer, R. A., 1974, Lead-induced inclusion bodies: Composition and probable role in lead metabolism, *Environ. Health Persp.* **7**:121.

Oskarsson, A., Squibb, K. S., and Fowler, B. A., 1981, Subcellular binding of ^{203}Pb in rat kidneys—Detection of a low molecular-weight cytosolic lead-binding component, *Toxicologist* **1**:81.

Rhyne, B. C., and Goyer, R. A., 1971, Cytochrome content of kidney mitochondria in experimental lead poisoning, *Exp. Mol. Pathol.* **14**:386.

Squibb, K. S., Ridlington, J. W., Carmichael, N. G., and Fowler, B. A., 1979, Early cellular effects of circulating cadmium-thionein on kidney proximal tubules, *Environ. Health Persp.* **28**:287.

Squibb, K. S., Pritchard, J. B. and Fowler, B. A., 1982, The renal metabolism and toxicity of metallothionein, in: Foulkes, E. C. (ed.), *Biological Roles of Metallothionein: Proc. U.S.A.-Japan Workshop on Metallothionein,* Elsevier/North Holland 1982, p. 181.

Squibb, K. S., Taylor, J. A., and Fowler, B. A., 1981, Early biochemical effects of cadmium-thionein in the rat kidney, *Fed. Proc.* **40**:836 (abstract).

Stowe, H. W., Wilson, M., and Goyer, R. A., 1972, Clinical and morphologic effects of oral cadmium toxicity in rabbits, *Arch. Pathol.* **94**:389.

Striker, G. E., Smuckler, E. A., Kohnan, P. W., and Nagle, R. B., 1968, Structural and functional changes in rat kidney during CCl₄ intoxication, *Am. J. Pathol.* **53**:769.

Ulmer, D. D., and Vallee, B. L., 1969, Effects of lead on biochemical systems, in: *Trace Substances in Environmental Health—II. Proceedings University of Missouri Annual Conference on Trace Substances in Environmental Health, 2nd* (D. D. Hemphill, ed.), University of Missouri, Columbia, Missouri, p. 7.

Wattenberg, L. W., and Leong, J. L., 1962, Histochemical demonstration of reduced pyridine nucleotide dependent polycyclic hydrocarbon metabolizing systems, *J. Histochem. Cytochem.* **10**:412.

Woods, J. S., and Fowler, B. A., 1977, Renal porphyrinuria during chronic methyl mercury exposure, *J. Lab. Clin. Med.* **90**:266.

28

Mechanisms of Acute Nephrotoxicity

Halogenated Aliphatic Hydrocarbons

WILLIAM M. KLUWE

1. Scope

1.1. Incidence of Chemical Injury to the Kidney

1.1.1. Humans

Kidney disease is unquestionably a major contributing factor to health problems in the United States, with renal failure alone estimated to cause up to 100,000 deaths yearly (DHEW-NIH, 1978). Although the proportion of kidney disease directly caused, or contributed to, by chemical exposures is unknown, the common association of some analgesics, antibiotics, and heavy metals with renal dysfunction indicates the innate susceptibility of the human kidney to chemical insult. Occupational and environmental exposures to chemicals other than heavy metals are less clearly associated with nephrotoxicity, possibly because of inadequate patient histories and worker exposure records, insensitive diagnostic tests, and the lack of followup studies. The large population exposed to halogenated aliphatic hydrocarbon compounds at work or in the general

WILLIAM M. KLUWE • National Toxicology Program, Research Triangle Park, North Carolina 27709.

environment, in particular, necessitates an assessment of the effects of these potentially nephrotoxic compounds on human renal function.

1.1.2. Experimental Animals

The toxicologic characterization of chemicals occurs principally in experimental animals, where target organs are identified, dose–response relationships are documented, and reversibility of the effects is determined. Broad toxicology screening studies in rodents have demonstrated that the kidney is commonly the primary target organ for a diverse array of industrial and agricultural chemicals. Evidence of nephrotoxicity is generally based upon histologic observation of morphologic aberrations in the renal tissue, often in the absence of functional abnormalities. (The reserve capacity of the kidney and rapid cellular repair may preclude the diminution of function despite the loss of functioning renal cells.) Screening studies in animals, unfortunately, neither indicate mechanisms of chemical toxicity nor frequently suggest the risk of human exposure to such. A descriptive screening study, therefore, is only the initial step in toxicologic evaluation and more sophisticated methodologies must be employed to extrapolate accurately from animal studies to human risk evaluation.

1.2. Spectrum of Nephrotoxic Halogenated Aliphatic Chemicals: Uses and Likelihood of Human Exposure

Halogenated aliphatic chemicals, principally those with short alkyl chains and one or more chlorine (Cl) or bromine (Br) substituents, constitute a class of chemicals that are especially toxic to the kidneys of experimental animals (Kluwe, 1981). A partial list of such compounds is contained in Table I, the length of which is more likely limited by the number of toxicology studies published in the biomedical literature than by the propensity of this class of chemicals to cause

TABLE I. Nephrotoxic Halogenated Aliphatic Hydrocarbons

Chemical	Use
Allyl chloride (3-chloro-1-propene)	Intermediate
Carbon tetrachloride	Intermediate, solvent
Chloroform	Intermediate, solvent
α-Chlorohydrin (3-chloropropan-1,2-diol)	Rodenticide
Chloroprene (2-chlorobutadiene)	Intermediate
Dibromochloropropane	Pesticide
Ethylene dibromide	Intermediate, pesticide
Ethylene dichloride	Intermediate
Epichlorohydrin (3-chloro-1,2-epoxypropane)	Intermediate
Hexachloro-1,3-butadiene	By-product
1,1,2-Trichloroethane	Degreaser
Trichloroethylene	Degreaser
Tris(2,3-dibromopropyl) phosphate	Flame retardant
Vinyl chloride	Intermediate
Vinylidene chloride	Intermediate

kidney injury. Compounds such as those in Table I are used as pesticides, solvents, and flame retardants, and as intermediates in plastic formulations and other synthetic processes. In addition to worker exposure, however, the ultimate release of these agents into the general environment often results in their ubiquitous distribution. Organochlorine compounds also arise from the chlorination of sewage during chemical sterilization (Bunn et al., 1975). Extensive human exposures to a variety of halogenated aliphatic hydrocarbon compounds, therefore, are likely to occur both at work and in the home.

2. Susceptibility of the Kidney to Chemical Insult

The precise reason for the susceptibility of the rodent kidney to organohalide compounds is unknown, but an attempt will be made here to describe reasonable hypotheses and their supporting evidence.

2.1. Toxicant Distribution

It is generally assumed that the magnitude of injury produced by a chemical is directly proportional to the amount of toxicant present at the target site and the length of time the toxicant is in contact with the target tissue. The high rate of renal perfusion—20–25% of cardiac output—assures adequate delivery of blood-borne halogenated aliphatic chemicals to the kidney. While this phenomenon alone may increase the likelihood of a nephrotoxic response, the limited pharmacokinetic studies to date do not indicate that organohalide compounds are selectively accumulated by the kidney or that their half-lives in the kidney are longer than in other soft tissues. Transient periods of high chemical concentrations in the kidney often occur in conjunction with the renal excretion of water-soluble metabolites of halogenated aliphatic compounds. The location of the chemical in this situation, however, is intraluminal rather than intracellular and probably constitutes little or no risk for cellular injury, since the hydrophilic metabolites would not be readily reabsorbed by the tubular cells. The nephrotoxic actions of halogenated chemicals, therefore, do not appear to be dependent solely on toxicant distribution.

2.2. Innate Susceptibility of the Kidney

If not due to toxicant distribution, then the susceptibility of the kidney can be presumed to be an innate property of the affected cells.

2.2.1. Intrarenal Localization of Injury—Proximal Tubules

Glomerular and distal tubular lesions are relatively infrequent following acute or subacute halogenated aliphatic chemical exposures; injury is restricted rather exclusively to the proximal tubule. Cytotoxicity is manifest morphologically as a vacuolar degeneration or necrosis (depending on the severity of intoxication) of the proximal tubular epithelium. Functionally, energy-dependent secretive and reabsorptive activities of the proximal tubule are depressed. The tubular functions generally return to normal within 4–10 days of chemical exposure, commensurate with the histologic appearance of cellular regeneration. The time sequence of injury and repair, therefore, is relatively brief and a good correlation exists between the morphologic lesions and functional effects.

Selective susceptibility of the proximal tubular cells to chemical injury is not a phenomenon exclusive to halogenated compounds—many other types of chemicals also damage the proximal tubules rather selectively (Schreiner and Maher, 1965). The cells in this portion of the nephron, therefore, appear to be particularly susceptible to chemical insult. Possible reasons include the high-energy requirements of proximal tubular cells and their extensive exposure to toxicants freely filtered at the glomerulus and reabsorbed along with fluids and electrolytes as the urine is formed (Hook, 1980). The latter property may be of lesser importance than the former with regard to organohalide nephropathies, since the serum half-lives of lipophilic halogenated aliphatic chemicals are generally quite short and their hydrophilic metabolites are poorly reabsorbed in the kidney.

2.2.2. Intratubular Localization of Injury—Pars Recta

A specific pattern of renal tubular injury is produced in rats treated with certain halogenated aliphatic hydrocarbons, particularly the olefinic and poly-halide compounds. Anatomically, the injured proximal tubular cells are restricted to those in the outer medullary stripe (Fig. 1), an area of the kidney populated largely by the pars recta or straight, descending segments of the proximal tubules. Proximal tubular cells in the cortex frequently undergo mild degenerative changes, but generally do not become necrotic unless very high doses are used. Cell injury is initially observed at the inner medullary border; as the dose of toxicant is increased, the area of severe injury expands toward the cortical aspects of the kidney, encompassing cells of the pars convoluta as well as the pars recta.

A specific type of proximal tubular cell, the S_3 cell is abundant in the pars recta. For at least one halogenated aliphatic nephrotoxicant, 1,2-dibromo-3-chloropropane (DBCP), it can be shown that S_3 cells undergo a variety of degenerative ultrastructural changes, while S_2 cells adjacent to the injured S_3 cells are relatively unaffected (W. M. Kluwe and B. A. Fowler, unpublished). This finding suggests that S_3 cells are specific cellular targets for DBCP in the kidney. Subtle cytologic differences between pars recta (S_3) and pars convoluta (S_1, S_2) cells exist; microvilli are slightly longer and the mitochondria are smaller, less numerous, and more randomly arrayed in S_3 cells. No attempt has been made to correlate these cytologic differences between proximal tubular cell types with susceptibility to organohalide nephrotoxicity. Functionally, pars convoluta cells are more active in sodium reabsorption than are pars recta cells, while Tune *et al.* (1969) have demonstrated that pars recta cells are more active in organic anion secretion. The toxicologic significance of the high capacity of S_3 cells for anion secretion is evidenced by the finding that organic anion transport by kidney cells (*in vitro*) is one of the most sensitive indicators of halogenated aliphatic hydrocarbon cytotoxicity (Table II). Therefore, S_3 proximal tubular cells appear to be specific

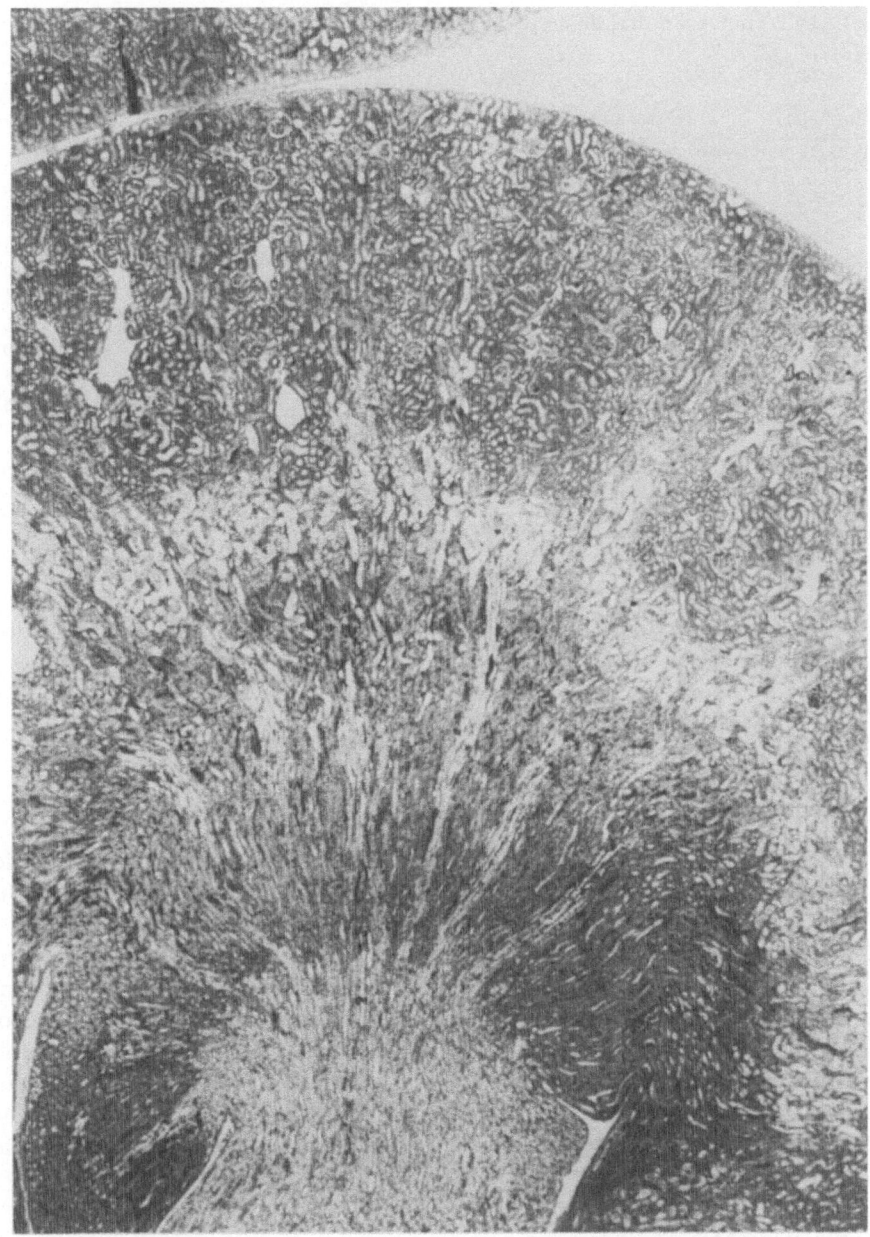

FIGURE 1. Intrarenal localization of proximal tubular injury. Male Fischer 344 rats were treated subcutaneously with 100 mg/kg 1,2-dibromo-3-chloropropane and killed 72 hr later. Proximal tubules in the outer medullary stripe have become necrotic and the epithelial cells have sloughed off, leaving dilated, desquamated tubules. Granular epithelial cell casts are evident in distal tubules in the inner medulla. × 16.

TABLE II. Organohalide-Induced Loss of Tubular Cell Functions *in Vitro*[a]

Organohalide	Concentration, M	Percent of control		
		PAH S/M	$[K]_{ICW}$	QO_2
DBCP	1×10^{-5}	96 ± 8	101 ± 4	102 ± 6
	3.2×10^{-5}	74 ± 11^b	96 ± 11	106 ± 5
	1×10^{-4}	55 ± 4^b	72 ± 14^b	98 ± 14
	3.2×10^{-4}	22 ± 6^b	66 ± 11^b	73 ± 5^b
EDB	1×10^{-4}	98 ± 11	100 ± 14	98 ± 6
	3.2×10^{-4}	68 ± 14^b	92 ± 7	108 ± 11
	1×10^{-3}	43 ± 5^b	65 ± 7^b	64 ± 9^b
HCBD	3.2×10^{-4}	96 ± 10	108 ± 6	104 ± 9
	1×10^{-3}	73 ± 8^b	94 ± 12	94 ± 12
	3.2×10^{-3}	38 ± 9^b	64 ± 6^b	60 ± 8^b

[a] 1,2-Dibromo-3-chloropropane (DBCP), 1,2-dibromoethane (EDB), hexachloro-1,3-butadiene (HCBD) or vehicle (control) were added to incubates of renal cortical slices and the following parameters measured after 60 min: active accumulation of *p*-aminohippuric acid (PAH S/M), intracellular concentration of potassium ($[K]_{ICW}$), and rate of oxygen consumption (QO_2; $\mu l\ O_2/min/mg$).
[b] Significantly lower than control, $p < 0.05$.

targets for certain organohalide nephrotoxicants both functionally and morphologically.

3. Metabolic Activation of Halogenated Aliphatic Chemicals

3.1. Toxic Metabolites

The toxic effects of organohalide compounds on the liver have been evaluated more completely than those on the kidney. It is apparent from these studies that most halogenated aliphatic hydrocarbon compounds are relatively innocuous to cells in their parent forms. When acted upon by mammalian enzymes, however, cytotoxic metabolites can be produced. Such toxification reactions are commonly referred to as "metabolic activation." Although specific pathways of biotransformation may be species-dependent, the conversion of lipophilic organohalide compounds to hydrophilic products is universally necessary for their elimination from the mammalian body. It is likely, therefore, that in most cases humans do not differ qualitatively from experimental animals in response to organohalide compounds, though quantitative differences in response may be marked.

3.2. Reactive Metabolites

3.2.1. Chemical Structures

Enhanced hydrophilicity of lipophilic chemicals is achieved by ionization or increased polarization of the molecule. Many halogenated aliphatic chemicals are oxidized to polar products, often resulting in the loss of one or more halide ions and the formation of an electron-deficient center (Fig. 2). Molecules with adjacent halide substituents and halogenated olefinic compounds are especially susceptible to epoxidation, resulting in the formation of chemically unstable intermediates (Fig. 2; Jones *et al.*, 1979; Pohl *et al.*, 1979; Reichert *et al.*, 1979). The end products or the intermediates of these oxidative reactions, then, are reactive, electrophilic moieties capable of bonding covalently to adjacent molecules with nucleophilic centers. Potential targets for alkylation are membrane proteins and lipids, nucleic acids, and polypeptides.

3.2.2. Potential Effects

Chemical addition (alkylation, acylation) to macromolecules could be detrimental to cell function and conceivably produce cellular death. Membrane perturbation, enzyme inactivation, and misinterpretation of informational macromolecules are theoretical consequences of electrophilic addition. Precise mechanisms of reactive metabolite nephrotoxicity have not yet been elucidated, but good correlations between covalent bonding of metabolites to renal macromolecules and tissue necrosis have been demonstrated (Ilett *et al.*, 1973; Mitchell *et al.*, 1977).

3.3. Renal Xenobiotic Metabolism

Halogenated aliphatic hydrocarbons and other xenobiotic (foreign) compounds are metabolized by cytochrome P-450-dependent microsomal monooxygenases. These enzymes are components of the smooth endoplasmic reticulum membrane and metabolize endogenous as well as exogenous substrates. The importance of microsomal (and nonmicrosomal) metabolism in the expression of organohalide toxicities, including dose–response relationships, species differences,

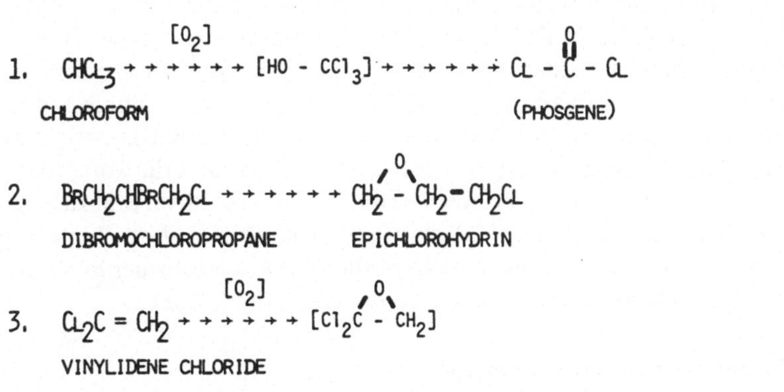

FIGURE 2. Conversion of halogenated aliphatic hydrocarbons to reactive metabolites. Structures in parentheses are hypothetical.

and thresholds for tissue injury, are becoming increasingly apparent as our understanding of chemical toxicity progresses.

3.3.1. Capacities

Hepatic tissue exhibits the highest specific activities of cytochrome P-450-dependent monooxygenases in the body. This property, along with the reception of portal blood flow, makes the liver the major site of xenobiotic metabolism. The capacity of the kidney for metabolism of xenobiotic chemicals is much less than that of the liver and probably contributes little to the overall rate of elimination of most foreign compounds from the body (through urine is a major route of excretion). The reactive nature of the suspected toxic metabolites and their specificities for producing injury at discrete anatomical sites, however, suggest that local (i.e., intrarenal) metabolism may be responsible for the nephrotoxic actions of halogenated aliphatic chemicals.

3.3.2. Localizations of Enzymatic Activities

Xenobiotic metabolizing enzyme activities are generally measured in whole-tissue homogenates or other preparations that do not differentiate among various cell types. Selective localization of an enzyme within a specific cell type, therefore, could result in an underestimation of its activities within an organ. The kidney, for example, contains several different cell types and, unlike the liver, their distributions within the organ are not nearly homogeneous. Recent studies indicate that renal monooxygenase activities and cytochrome P-450 concentrations are higher in the medulla than in the papilla, and higher in the cortex than in the medulla (Zenser *et al.,* 1978), roughly commensurate with the frequency distribution of proximal tubular cells in the kidney. Moreover, Fowler *et al.* (1977) have demonstrated that certain monooxygenase activities are greatest in the outer medullary stripe (pars recta) and that 2,3,7,8-tetrachlorodibenzo-*p*-dioxin, an inducer of renal xenobiotic metabolism, selectively increases the amount of smooth endoplasmic reticulum in the S_3 cells, but not in adjacent tubular cells. Proximal tubular cells, therefore, may have much higher specific activities of xenobiotic metabolizing enzymes than suggested by studies utilizing whole-kidney microsomes or other preparations. Moreover, S_3 cells, in particular, may have higher specific activities of some xenobiotic metabolizing enzymes than do other proximal tubular cells. The correlation between xenobiotic metabolism and susceptibility to the cytotoxicity of halogenated aliphatic hydrocarbons—both high in proximal tubular cells—supports the general hypothesis that certain nephrotoxic organohalide compounds are metabolically activated in the kidney.

3.3.3. Noncytochromal Oxidation

Cytochrome P-450-independent oxidation of xenobiotic compounds in the kidney has recently been described by Zenser *et al.* (1979). This reaction exhibits highest activities in the inner medullary portion of the kidney and is arachidonic acid-dependent. The proposed mechanism is the cooxidation of arachidonic acid and xenobiotic compounds by prostaglandin cyclooxygenase. Though present in the cortex and outer medulla, as well as in the inner medulla, arachidonate-dependent oxidation does not appear to be causally associated with the nephrotoxicity of halogenated aliphatic hydrocarbons, since the distribution of its activity in the kidney does not correlate with the sites of organohalide injury.

4. Metabolism of Halogenated Aliphatic Compounds to Nephrotoxicants: Examples

4.1. Chloroform (CHCl₃)

Of all the halogenated compounds, $CHCl_3$ has received the most intensive scrutiny with regard to the mechanism of its nephrotoxic effects. $CHCl_3$ is a potent acute nephro- and hepatotoxicant in mammals and a carcinogen in rodents. It produces renal epithelial cell tumors upon chronic exposure in rats, but its acute necrogenic effects on the kidney are more prominent in mice. $CHCl_3$ is enzymatically oxidized by microsomal monooxygenases to phosgene (Fig. 2), a poisonous, chemically reactive molecule believed to cause tissue injury via the acylation of essential cellular structures (Pohl *et al.,* 1979).

4.1.1. Xenobiotic Metabolism and CHCl₃ Nephrotoxicity

The correlation between metabolism and toxicity can be evaluated by comparing susceptibilities to nephrotoxicity before and after modification of the biotransformation pathways. Treatment of mice with polybrominated biphenyls (PBB) increases the activities of renal and hepatic microsomal enzymes, including the monooxygenases, in a dose-dependent manner (W. M. Kluwe and J. B. Hook, unpublished). Accordingly, susceptibility of the kidney (and liver) to $CHCl_3$ toxicity is enhanced by PBB pretreatment (Table III). Conversely, susceptibility to $CHCl_3$ nephrotoxicity is reduced upon pretreatment with piperonyl butoxide (Table III), an inhibitor of many renal and hepatic monooxygenase activities. The experimental evidence suggests, therefore, that $CHCl_3$ is metabolically activated in mice to nephrotoxic products.

TABLE III. Effects of Various Pretreatments on the Activities of Microsomal Enzymes and Organ Susceptibilities to Chloroform Toxicity in Male Mice

Pretreatment	Kidney		Liver	
	Microsomal enzyme activities	CHCl₃ toxicity	Microsomal enzyme activities	CHCl₃ toxicity
Polybrominated biphenyls[a]	Increased	Increased	Increased	Increased
Piperonyl butoxide[b]	Decreased	Decreased	Decreased	Decreased
Phenobarbital[c]	No change	No change	Increased	Increased

[a] Kluwe and Hook (1978).
[b] Kluwe and Hook (1981).
[c] Kluwe *et al.* (1978).

4.1.2. Site of Activation

The kidney could logically be assumed to be the site of metabolism of $CHCl_3$ to a nephrotoxic product, as the enzyme systems necessary for oxidation are present and the likelihood of transport of chemically unstable molecules such as phosgene within the body is slight. The effects of PBB and piperonyl butoxide on $CHCl_3$ hepatotoxicity were similar to those on $CHCl_3$ nephrotoxicity, however, necessitating further experimentation to eliminate the liver as a possible site of activation. Mice pretreated with phenobarbital, a selective inducer of hepatic, but not renal, xenobiotic metabolism, were found to be more susceptible than controls to $CHCl_3$ hepatotoxicity, indicating greater conversion of the parent compound to a toxic metabolite, presumably phosgene, in the liver (Table III). No change in susceptibility to $CHCl_3$ nephrotoxicity occurred, however, indicating that nephrotoxic metabolites were not formed in the liver. Strain and sex differences in susceptibility to $CHCl_3$ nephrotoxicity but not hepatotoxicity have also been documented (Kluwe, 1981), providing further experimental support for the independence of the renal and hepatic $CHCl_3$-induced lesions and for the hypothesis that the nephrotoxic metabolite is generated intrarenally.

4.2. Other Organohalide Compounds

The susceptibilities of rodents to the nephrotoxic effects of many other halogenated aliphatic chemicals, including carbon tetrachloride, trichloroethane, trichloroethylene, and dibromochloropropane, are also altered by stimulation or inhibition of renal and hepatic xenobiotic metabolism (Kluwe *et al.,* 1979; W. M. Kluwe, unpublished). While these reports suggest a cause and effect relationship between organohalide metabolism and renal proximal tubular injury, the data are insufficient in most cases to indicate sites of toxic metabolite generation or the structures of the toxic molecules.

5. Interactions between Toxification and Detoxification Reactions

5.1. Complexities of Halogenated Aliphatic Chemical Metabolism

The metabolism of organohalide compounds has been described up to this point in a rather cursory manner, due in large part to a lack of information pertaining to specific mechanisms of biotransformation and inadequate metabolite identifications. The large number of structurally different metabolites that have been isolated from the urine of treated animals, however, indicates the existence of multiple pathways of organohalide catabolism. As the intermediates or end products of the various pathways are likely to differ significantly in their nephrotoxic potencies, the species, strain, and sex differences in susceptibilities to halogenated aliphatic hydrocarbon nephrotoxicities are probably due to differences—generally quantitative—in toxicant pharmacokinetics rather than to fundamental differences in the mechanism of toxicant action. Animal models, therefore, may indicate which chemicals are potentially nephrotoxic in humans, but further evaluation will be needed to determine the probable exposure levels at which injury will ensue. An understanding of the comparative pharmacokinetics of the specific toxicant in humans and experimental animals may greatly increase the reliability of animal studies as predictions of human response. Two factors complicating the relationship of metabolism to chemical toxicity—competitive and sequential reactions—are discussed in the subsequent paragraphs.

5.2. Competing Pathways of Metabolism

Halogenated aliphatic chemicals can be substrates for two or more competing pathways of metabolism. If only one produces a toxic product, then quantitative tissue injury will be dependent on the relative activities of, or balance between, the competing pathways (as well as on the amount of chemical encountered). Vinylidene chloride (VDC), CCl_4, $CHCl_3$, and dibromochloropropane (DBCP) have all been shown, or have been hypothesized, to be substrates for competitive pathways of metabolism (Kluwe *et al.*, 1978; Reichert *et al.*, 1979; W. M. Kluwe, unpublished). Susceptibilities to their nephrotoxic effects, therefore, may differ substantially among species, depending on the relative activities of the toxification and detoxification reactions. Furthermore, the balance between two or more metabolic pathways may be dose dependent—predominance can be altered by saturation of one of the pathways, resulting in threshold doses for tissue injury.

5.3. Sequential Metabolism

The metabolism of halogenated aliphatic compounds may also occur sequentially, the end product being stable, water-soluble, and generally nontoxic. If an intermediate in a sequential metabolism sequence is a toxic moiety, then the degree of injury produced will be dependent only on the amount of the intermediate present, not on the amount of end-product formed. As with competing pathways of metabolism, thresholds for tissue injury may exist such as when the capacity of the detoxification reaction is exceeded.

5.3.1. Glutathione

An example of sequential metabolism is the activation of halogenated aliphatic chemicals to electrophilic intermediates and subsequent detoxification by conjugation with glutathione. The abundant cytosolic tripeptide glutathione exists intracellularly in both an oxidized (GSSG) and a reduced (GSH) form. Reduced glutathione is predominant due to the actions of the enzyme glutathione reductase. GSH has a nucleophilic center (the sulfhydryl group on the cysteine residue) and is thought to protect against cellular injury by bonding covalently to reactive, electrophilic molecules, thereby rendering them nontoxic.

5.3.1a. Conjugation of Glutathione to Halogenated Aliphatic Hydrocarbon Metabolites. The organohalide–glutathione conjugate is excreted (generally in urine) as a water-soluble mercapturic acid metabolite, necessitating replacement of the eliminated glutathione by *de novo* synthesis. Transient depletion of renal GSH occurs when large amounts of halogenated aliphatic hydrocarbon chemicals are converted to electrophilic products *in vivo* (Table IV) or *in vitro*. For many organohalide compounds, the appearance of dose-dependent tissue injury seems to coincide with a measurable decline in tissue GSH concentrations, suggesting that conjugation with GSH is a detoxification reaction with limited capacities, and that saturation of the system can result in tissue injury. Investigations into the role of GSH depletion in the development of organohalide nephrotoxicities are generally insufficient, at the present time, to suggest cause and effect relationships.

5.3.1b. Renal Glutathione Concentrations and Susceptibility to Chemical Injury. The importance of GSH conjugation in the detoxification of halogenated aliphatic hydrocarbon metabolites, however, is suggested by reports that animals depleted of GSH by treatment with diethyl maleate exhibit increased susceptibilities to the nephrotoxic effects of certain organohalide compounds. Susceptibility to $CHCl_3$ nephrotoxicity in mice, for example, is greatly enhanced by GSH·depletion (Table V). Since the abilities of experimental animals to conjugate organohalide toxicants to GSH varies widely (W. M. Kluwe, unpublished), investigations into possible correlations between such GSH conjugating abilities and susceptibilities to kidney injury could be utilized to explore mechanisms of nephrotoxicity. Renal GSH concentrations and conjugating activities, nonetheless, appear to be important determinants of organohalide nephropathies.

6. Summary

Animal studies suggest that halogenated aliphatic hydrocarbon compounds as a class possess significant nephrotoxic potential. Their propensities for causing injury to the human kidney upon occupational or environmental exposure, however, cannot be properly assessed until mechanisms of action are defined and

TABLE IV. Depletions of Renal Glutathione by Organohalide Compounds in Mice[a]

Dose, mg/kg	Renal glutathione, % of control			
	DBCP[b]	EDB[b]	HCBD[b]	$CHCl_3$[c]
33	98 ± 6	106 ± 8	84 ± 12	—
74	—	—	—	92 ± 8
100	78 ± 5[d]	80 ± 9[d]	64 ± 6[d]	—
330	53 ± 6[d]	42 ± 7[d]	40 ± 5[d]	—
370	—	—	—	62 ± 10[d]
1000	—	—	32 ± 3	—

[a] The animals were killed 2 hr after IP injection of 1,2-dibromo-3-chloropropane (DBCP), 1,2-dibromoethane (EDB), hexachloro-1,3-butadiene (HCBD), chloroform ($CHCl_3$), or vehicle (control).
[b] Kluwe *et al.* (1981).
[c] Kluwe and Hook (1981).
[d] Significantly lower than control, $p < 0.05$.

TABLE V. Effect of Diethyl Maleate on Susceptibility to Chloroform Nephrotoxicity[a]

Parameter	Pretreatment	CHCl₃, mg/kg:		
		50	150	450
[BUN], mg/dl	Vehicle	20 ± 2	26 ± 2	83 ± 22
	Diethyl maleate	28 ± 4	76 ± 18^b	126 ± 13^b
PAH S/M	Vehicle	12.6 ± 2.1	11.2 ± 1.6	4.9 ± 0.4
	Diethyl maleate	9.8 ± 2.6	5.4 ± 0.8^b	2.0 ± 0.4^b
[renal glutathione],	Vehicle		3.06 ± 0.52	
μmoles/g	Diethyl maleate (600 mg/kg)		0.59 ± 0.07^b	

[a] Male mice were killed 2 hr after intraperitoneal injection of diethyl maleate for measurement of glutathione, or received an injection of chloroform (CHCl₃) 90 min after diethyl maleate and were killed 24 hr later for determination of blood urea nitrogen concentration (BUN) and ability of renal slices to accumulate p-aminohippuric acid *in vitro* (PAH S/M). Data from Kluwe and Hook (1981).
[b] Significantly different from vehicle, $p < 0.05$.

possible differences in organohalide pharmacokinetics in humans and the rodent models are considered.

The intrarenal target of most halogenated aliphatic chemicals in rats appears to be the proximal tubular cells, perhaps because of the ability of this cell type to oxidize xenobiotic compounds to toxic products. Some organohalides, furthermore, affect S_3 cells specifically. Glutathione may protect proximal tubular cells from injury by reactive, electrophilic organohalide metabolites by direct conjugation with such. This protective system, however, can be saturated by depletion of tissue GSH stores. The precise mechanisms (e.g., subcellular targets) by which toxic metabolites of halogenated aliphatic hydrocarbons produce cellular dysfunction and death are still unknown.

The existence of competing and sequential pathways of organohalide metabolism complicates the characterization of dose–response relationships and may result in threshold doses for tissue injury. Extrapolation of animal data to human risk evaluation, therefore, must recognize the limitations of toxicologic screening studies and the importance of more mechanistic information.

References

Bunn, W. W., Haas, B. B., Deane, E. R. and Kleopfer, D., 1975, Formation of trihalomenthanes by chlorination of surface water, *Environ. Lett.* **10**:205.

DHEW-NIH, 1978, *Research Needs in Nephrology and Urology. Volume I. Report of the Coordinating Committee,* U.S. Department of Health, Education and Welfare Publication No. (NIH) 78-1481.

Fowler, B. A., Hook, G. E. R., and Lucier, G. W., 1977, Tetrachlorodebenzo-p-dioxin induction of renal microsomal enzyme systems: Ultrastructural effects on *pars recta* (S_3) proximal tubule cells of the rat kidney, *J. Pharmacol. Exp. Ther.* **203**:712.

Hook, J. B., 1980, Toxic responses of the kidney, in: *Casarett and Doull's Toxicology* (J. Doull, C. D. Klaassen, and M. O. Amdur, eds.), Macmillan, New York, p. 232.

Ilett, K. F., Reid, W. D., Sipes, I. G., and Krishna, G., 1973, Chloroform toxicity in mice: Correlation of renal and hepatic necrosis with covalent binding of metabolites to tissue macromolecules, *Exp. Mol. Pathol.* **19:**215.

Jones, A. R., Fakhouri, G., and Gadiel, P., 1979, The metabolism of the soil fumigant 1,2-dibromo-3-chloropropane in the rat, *Experientia* **35:**1432.

Kluwe, W. M., 1981, The nephrotoxicity of low molecular weight halogenated alkane solvents, pesticides and chemical intermediates, in: *Toxicology of the Kidney* (J. B. Hook and R. L. Dixon, eds.), Academic Press, New York, p. 179.

Kluwe, W. M., and Hook, J. B., 1978, Polybrominated biphenyl-induced potentiation of chloroform toxicity, *Toxicol. Appl. Pharmacol.* **45:**861.

Kluwe, W. M., and Hook, J. B., 1981, Potentiation of acute chloroform nephrotoxicity by the glutathione depletor diethyl maleate and protection by the microsomal enzyme inhibitor piperonyl butoxide. *Toxicol. Appl. Pharmacol.* **59:**457.

Kluwe, W. M., McCormack, K. M., and Hook, J. B., 1978, Selective modification of the renal and hepatic toxicities of chloroform by induction of drug-metabolizing enzyme systems in kidney and liver, *J. Pharmacol. Exp. Ther.* **207:**566.

Kluwe, W. M., Herrmann, C. L., and Hook, J. B., 1979, Effects of polychlorinated biphenyls and polybrominated biphenyls on the renal and hepatic toxicities of several chlorinated hydrocarbon solvents in mice, *J. Toxicol. Environ. Health* **5:**605.

Kluwe, W. M., McNish, R., Smithson, K., and Hook, J. B., 1981, Depletion by 1,2-dibromoethane, 1,2-dibromo-3-chloropropane, tris(2,3-dibromopropyl) phosphate and hexachloro-1,3-butadiene of reduced non-protein sulfhydryl groups in target and non-target organs. *Biochem. Pharmacol.* **30:**2265.

Mitchell, J. R., McMurtry, R. J., Statham, C. N., and Nelson, S. D., 1977, Molecular basis for several drug-induced nephropathies, *Am. J. Med.* **62:**518.

Pohl, L. R., George, J. W., Martin, J. L., and Krishna, G., 1979, Deuterium isotope effect in *in vivo* bioactivation of chloroform to phosgene, *Biochem. Pharmacol.* **28:**561.

Reichert, D., Werner, H. W., Metzler, M., and Henschler, D., 1979, Molecular mechanisms of 1,1-dichloroethylene toxicity: Excreted metabolites reveal different pathways of reactive intermediates, *Arch. Toxicol.* **42:**159.

Schreiner, G. E., and Maher, J. F., 1965, Toxic nephropathy, *Am. J. Med.* **38:**409.

Tune, B. P., Burg, M. B., and Patlak, C. S., 1969, Characteristics of *p*-aminohippurate transport in proximal renal tubules, *Am. J. Physiol.* **217:**1057.

Zenser, T. V., Mattammal, M. B., and Davis, B. B., 1978, Differential distribution of the mixed-function oxidase activities in rat kidney, *J. Pharmacol. Exp. Ther.* **207:**719.

Zenser, T. V., Mattammal, M. B., and Davis, B. B., 1979, Demonstration of separate pathways for the metabolism of organic compounds in rabbit kidney, *J. Pharmacol. Exp. Ther.* **208:**418.

29

Potentiation of the Action of Nephrotoxic Agents by Environmental Contaminants

JERRY B. HOOK and VICTORIA C. SERBIÁ

1. Introduction

The mammalian kidney, because of its high rate of perfusion, active transport capabilities, and concentrating functions, often is exposed to much higher concentrations of chemicals than are other organs. This ability of the kidney to concentrate foreign chemicals or xenobiotics usually serves to expedite the renal excretion of these compounds; however, high concentrations of xenobiotics may also predispose renal tissue to damage (Hook, 1980). Some of these chemicals may thus subsequently produce direct renal cytotoxicity. The therapeutic agent acetaminophen and widely used chemicals such as chloroform ($CHCl_3$), carbon tetrachloride (CCl_4), trichloroethylene, and 1,1,2-trichloroethane are examples of xenobiotics that produce direct tissue damage. These chemicals produce identifiable lesions that may be life-threatening; however, their effects are predictable and dose-related. Other chemicals cause much more subtle perturbations. The polybrominated biphenyls (PBBs) and polychlorinated biphenyls

JERRY B. HOOK and VICTORIA C. SERBIÁ • Center for Environmental Toxicology, Michigan State University, East Lansing, Michigan 48824.

(PCBs) are such chemicals. These environmental pollutants produce biochemical changes in the kidney that alone appear to have no functional correlation (that is, no functional nephropathy can be measured), yet they alter the response of the kidney to other chemical agents. Thus, although they produce no functional lesion, the insidiousness of their effect may present more of a hazard to man than the more predictable direct-acting nephrotoxicants (Kluwe and Hook, 1980b).

PBBs became of toxicologic concern following an agricultural accident in the state of Michigan that resulted in human consumption of PBBs in meat and dairy products (Kay, 1977). The particular formulation involved, Firemasters BP-6 and FF-1 (Velsicol Chemical Company, St. Louis, Michigan), were complex mixtures of brominated biphenyls (mostly hexabromobiphenyl). PBB molecules are nonpolar, lipophilic, and resistant to enzymatic degradation. They are, therefore, well absorbed following oral ingestion, rapidly partitioned into fatty tissues, and are excreted very slowly. Estimated tissue half-lives are as long as 7–10 years.

The forced feeding of large amounts of PBBs (25 g/day) to cows produced dilatation of renal tubules and collecting ducts along with elevations of serum urea nitrogen and urine protein concentrations (Moorehead et al.,1977). However, the cows were also emaciated. Thus the relative contributions of PBB treatment to the reported symptoms are difficult to distinguish from those symptoms secondary to body wasting. Lower doses (0.25 g/day) produced no kidney abnormalities. Rats ingesting approximately 5 mg/kg/day of PBBs for up to 3 months exhibited normal renal function. No changes were detected in: the clearance of para-aminohippurate (PAH) and the clearance of inulin; fractional sodium excretion; renal ammoniagenesis and gluconeogenesis; and the in vitro accumulation of PAH and NMN (Table I). Renal morphology was also unaffected. Renal microsomal mixed-function oxidase (MFO) activities, however, were sharply elevated (McCormack et al., 1978). PBBs induced renal enzymes in a selective manner; arylhydrocarbon hydroxylase (AHH) activity was increased several-fold (Table II), whereas epoxide hydratase (EH) activity was reduced or unaffected (McCormack et al., 1978). This observation was interesting, for in the liver the activities of both AHH and EH were increased following PBBs. Many aromatic hydrocarbons are believed to be enzymatically toxified by AHH to reactive epoxides and detoxified by EH to dihydrodiols; thus, these observations were judged to be of toxicologic concern. As described in Chapter 28 by Kluwe, the

TABLE I. Effect of PBB Feeding (3 Months) on Ammoniagenesis, Gluconeogenesis, and the Accumulation of PAH and NMN in Rats[a]

Treatment	Net production, μmoles/mg of wet tissue wt/hr		Slice/medium ratio	
	Ammonia	Glucose	PAH	NMN
Control	3.4	0.03	10.5	5.79
100 ppm of PBB	3.5	0.03	10.2	5.20

[a] Modified from McCormack et al. (1978).

TABLE II. Induction of Aryl
Hydrocarbon Hydroxylase (AHH)
Activity by Polybrominated
Byphenyls (PBB)[a]

Dietary PBB, ppm	AHH Activity	
	Kidney	Liver
0	0.08	12.1
1	0.09	14.6
25	0.12	26.7
100	0.25	53.2

[a] Enzyme activities (relative fluorescence units per mg
protein per minute) in 14,000g supernatant fraction of
homogenates of kidney and liver from ICR male mice
maintained for 2 weeks on diets containing various
concentrations of PBB. Modified from Kluwe and
Hook (1980b).

existence of multiple pathways of metabolism for halogenated hydrocarbons has
been shown in animals, suggesting that a balance between toxification and
detoxification reactions may be determined by the activities of specific enzymes. It
follows, then, that secondary treatment with an agent metabolized by these
enzymes may result in enhanced nephrotoxicity (i.e., the reduced activity of renal
EH as well as the increased activity of AHH may produce an imbalance of
toxification reactions and thereby alter target organ toxicity).

PCBs are similar to PBBs in many respects. They are complex mixtures of
halogen-substituted biphenyls; they are lipid-soluble, resistant to metabolic
degradation, and they are accumulated in fatty tissues. PCBs have been used as
flame retardants, plasticisers, and as heat-dispersing agents in electrical trans-
formers. Because of their many uses, PCBs are global contaminants, although their
current manufacture and use are restricted.

The acute toxicity of PCB, like that of PBB, is relatively low. Ingestion of
subtoxic amounts, however, stimulates the activities of many xenobiotic
metabolizing enzymes. Occupational exposure to PCB has been reported to affect
drug disposition in humans (Alvares et al., 1977). PCB induces cytochrome P-450
and microsomal enzyme activities in rodent kidney but is less potent, in this respect,
than is PBB (Table III).

The findings elucidated by these PBB and PCB studies indicate that these
environmental pollutants may upset the balance between toxification and
detoxification reactions by selectively inducing specific microsomal enzyme
activities, and thus the response to metabolically activated toxicants may be altered
by exposure to even small amounts of PBBs and PCBs. Although traditionally the
kidney has been thought to play a minor role in the activation of toxicants, it is
highly possible that metabolism of chemicals within the kidney may be an
important factor in determining the response of the kidney to chemical exposure
(Kluwe and Hook, 1980b).

Chloroform ($CHCl_3$) and carbon tetrachloride (CCl_4) are thought to be converted to toxic products via microsomal metabolism (Rechnagle and Glende, 1973; Sipes *et al.*, 1977). These compounds, therefore, have been used as models with which to ascertain the effects of PBB-induced enzyme stimulation on kidney metabolism and the subsequent response to these toxicants. This review will discuss the potentiation of action of these nephrotoxicants, as well as the potentiation of acetaminophen, a therapeutic agent also thought to be metabolically activated, by the polybrominated and polychlorinated biphenyls.

2. Potentiation of Chloroform Toxicity

Chloroform ($CHCl_3$) is a common organic solvent, an intermediate in the production of many other chemicals, and a trace contaminant of municipal water supplies—possibly originating as such by the chlorination of methanes during chemical sterilization of sewage (Bellar *et al.*, 1974; Bunn *et al.*, 1975). Chloroform is widely recognized to be hepatotoxic in experimental animals and in humans (Von Oettingen, 1964; Winslow and Gerstner, 1978). A large body of evidence indicates that hepatic injury by $CHCl_3$ is probably not due to $CHCl_3$ per se, but is produced by a $CHCl_3$ metabolite. Several investigators have suggested that the initial step leading to $CHCl_3$-induced tissue injury is biotransformation of $CHCl_3$ to the reactive intermediate phosgene ($COCl_2$) by enzymes of the mixed-function oxidase system (Pohl *et al.*, 1977; Mansuy *et al.*, 1977; Sipes *et al.*, 1977). Formation of $COCl_2$ has been postulated to proceed through an oxidative dechlorination mechanism which involves oxidation of the C–H bond of $CHCl_3$ to produce the trichloromethanol (CCl_3–OH) derivative, a highly unstable derivative which may spontaneously dehydrochlorinate to $COCl_2$ (Pohl *et al.*, 1977).

Chloroform is also nephrotoxic (Von Oettingen, 1964; Winslow and Gerstner, 1978). The magnitude of nephrotoxicity varies with species, strain, and sex (Von Oettingen, 1964; Hill *et al.*, 1975), although anesthesia, hepatic necrosis, and death may occur after $CHCl_3$ administration before any marked nephropathy is observable. Male mice are particularly sensitive to the nephrotoxicity of $CHCl_3$

TABLE III. Induction of Renal
Arylhydrocarbon Hydroxylase (AHH)
Activity by Polybrominated Biphenyls
(PBB) and Polychlorinated Biphenyls (PCB)[a]

Dietary concentration, ppm	AHH activity, % of control	
	PBB	PCB
25	150	85
100	310	—
200	466	292
400	—	478

[a]PBB or PCB was fed to mice (in the diet) for 14 days. Modified from Kluwe and Hook (1980b).

TABLE IV. Effects of Polybrominated
Biphenyls (PBB) on Chloroform-Induced
Nephrotoxicity in Mice[a]

Function	CHCl$_3$, ml/kg		
	0.000	0.005	0.025
BUN, mg/dl			
0 ppm PBB	21	27	26
1 ppm PBB	22	23	35
25 ppm PBB	22	22	47[b]
100 ppm PBB	21	28	81[b]
PAH, S/M			
0 ppm PBB	7.8	7.6	6.2
1 ppm PBB	7.5	5.8	4.4[b]
25 ppm PBB	8.1	5.5	4.1[b]
100 ppm PBB	8.4	5.4[b]	3.7[b]

[a] Elevations of BUN and reductions of PAH S/M are signs of kidney injury 48 hr after IP chloroform (S/M is slice-to-medium concentration ratio). Each value is the mean from six mice. Data modified from Kluwe and Hook (1978).
[b] Significantly different from control (same dietary concentration of PBB, 0.000 ml/kg chloroform), $p < 0.05$.

and the degree of sensitivity to the poison varies among several strains (Hill et al., 1975). The functional lesion in the kidney appears to be primarily due to proximal tubular damage, although structural alterations are seen in other portions of the nephron as well. Functionally, the nephropathy appears similar to other types of acute renal insult with polyuria, glucosuria, and proteinuria at low doses, leading to anuria and complete renal failure with higher doses (Hook et al., 1979). Inhibition of PAH uptake by renal cortical slices appears to be one of the most sensitive indices of toxicity (Watrous and Plaa, 1972).

Modulation of hepatic microsomal activities markedly alters the quantitative hepatotoxicity of CHCl$_3$ in rats, suggesting that liver injury is related to enzymatic metabolism of CHCl$_3$ (Scholler, 1970). Recently Kluwe et al. (1978) reported that in mice, induction of drug metabolizing enzymes also alters CHCl$_3$ nephrotoxicity. Polybrominated biphenyls increased both enzyme activities and CHCl$_3$ nephrotoxicity in a dietary concentration-dependent manner (Table IV). Treatment of mice with piperonyl butoxide, an inhibitor of drug metabolism, reduced enzyme activities in kidney and liver and quantitatively decreased CHCl$_3$ nephrotoxicity (Kluwe and Hook, 1981). These results therefore suggest that metabolites of CHCl$_3$ are responsible for both the renal and the hepatic lesions.

The experiments summarized on Table V also suggest that the intrarenal metabolism of CHCl$_3$ is responsible for kidney injury. Treatment with phenobarbital increased the activities of drug metabolizing enzymes in the liver, but not in the kidney. Similarly, phenobarbital enhanced CHCl$_3$ toxicity to the liver, but not to the kidney (Kluwe and Hook, 1978). Treatment with 3-methylcholanthrene increased enzyme activities in both organs, but reduced CHCl$_3$ toxicity to the kidney despite little or no effect on toxicity to the liver (Kluwe and Hook, 1978). If a single $CHCl_3$ metabolite that produces both kidney injury and liver injury was formed only in the liver, then the effects of enzyme inducing agents on $CHCl_3$ injury would be expected to be the same for both organs. Intraorgan metabolism of $CHCl_3$ to a reactive intermediate, on the other hand, could be differentially affected by inducing agents if the toxic intermediates were different in the two organs. Alternately, even if the toxic products in both organs were the same, organ-specific differences in response to the inducing agents could lead to organ-specific differences in toxicity. Thus, these results are consistent with the premise that the kidney, rather than the liver, is the site of activation of $CHCl_3$ to a nephrotoxic agent. The effects of enzyme inducing agents on the liver, however, may alter the pharmacokinetics of $CHCl_3$. The serum half-life of $CHCl_3$ in mice was sharply reduced by treatment with polybrominated biphenyls and polychlorinated biphenyls (W. M. Kluwe and J. B. Hook, unpublished), presumably because of increased hepatic $CHCl_3$ metabolism. A similar phenomenon may explain the effect of 3-methylcholanthrene to reduce $CHCl_3$ nephrotoxicity. As a result of hepatic metabolism, the delivery of $CHCl_3$ to the kidney could have been reduced. Thus, the overall inhibitory effect of 3-methylcholanthrene on $CHCl_3$ nephrotoxicity may have been a consequence of altered hepatic, rather than renal, metabolism.

Recently, Pohl *et al.* (1977, 1979) reported that $CDCl_3$ was metabolized to $COCl_2$ at approximately half the rate of $CHCl_3$ and that $CDCl_3$ was also less hepatotoxic than $CHCl_3$ in rats. Since the C–D bond is stronger than the C–H

TABLE V. Effects of Phenobarbital (PB) and 3-Methylcholanthrene (3MC) on Enzyme Activities and Acute $CHCl_3$ Toxicity[a]

	Activity, % of control			
	PB		3MC	
Enzyme	Kidney	Liver	Kidney	Liver
AHH	93	125	286[b]	346[b]
PCNMA	106	201[b]	87	113
BP-2-OH	100	137	ND	ND
BP-4-OH	100	331[b]	ND	ND

Parameter	Vehicle pretreated	PB	3MC
SGPT, units/ml	64	98[b]	69
BUN, mg/100 ml	76	72	31[b]
PAH S/M	3.9	3.2	10.0[b]
KW/BW, ×100	1.79	1.79	1.47[b]

[a] PB was administered for 4 days and 3MC for 3 days. The mice were then either killed for determination of enzyme activities or challenged with 0.25 mg/kg $CHCl_3$ and killed 24 hr later. ND, Not determined. Data modified from Kluwe *et al.* (1978).

[b] Values significantly different from control (enzyme activities) or vehicle pretreated (toxicity tests).

bond, these data suggest that cleavage of the C–H bond is the rate-limiting step in the activation of $CHCl_3$, and the subsequent formation of $COCl_2$ may lead to tissue damage. $CDCl_3$ has also been reported to be less hepatotoxic than $CHCl_3$ in mice, suggesting that mouse liver metabolizes $CHCl_3$ via the same mechanism as rat liver (Ahmadizadeh *et al.*, 1981). $CDCl_3$ is also less toxic to mouse kidneys than $CHCl_3$ (Ahmadizadeh *et al.*, 1981). Thus this deuterium isotope effect found in $CHCl_3$-induced nephrotoxicity suggests that the kidney may metabolize $CHCl_3$ in the same manner as the liver (i.e., via the same metabolite).

3. Potentiation of Other Halogenated Hydrocarbons

Carbon tetrachloride, CCl_4, is both nephrotoxic and hepatotoxic. The administration of CCl_4 to rodents results in liver necrosis (Štriker *et al.*, 1968) and produces a recognizable renal lesion in male rats (Von Oettingen, 1964; Hill *et al.*, 1975).

CCl_4 toxicity requires CCl_4 metabolism (Rechnagel and Glende, 1973). The first step in liver toxicity is the very rapid formation of the trichloromethyl free radical $\cdot CCl_3$ (Slater, 1972) by enzymes present in the endoplasmic reticulum of the liver (Rechnagle and Glende, 1973). This highly reactive intermediate can then interact with liver macromolecules. Prior treatment with a nonlethal dose of CCl_4 will protect the liver against a subsequent challenge, probably because the first dose decreases the drug metabolizing enzymes of the liver.

Striker *et al.* (1968) have shown in the kidney of rats that early signs of damage occur long after CCl_4 has been eliminated, suggesting that in the kidney, as in the liver, a metabolite of CCl_4 is responsible for inducing tissue damage. Presumably the metabolic activation of CCl_4, as with $CHCl_3$, is a renal as well as hepatic phenomenon.

Little effort has been made to relate metabolism to the nephrotoxic effects of other halogenated aliphatic hydrocarbons. The nephrotoxicities of carbon tetrachloride, trichloroethylene, and 1,1,2-trichloroethane in mice, however, are enhanced by ingestion of polybrominated biphenyls and polychlorinated biphenyls (Kluwe *et al.*, 1979). Exposure to PCB, like exposure to PBB, thus appears to enhance the metabolism of some chlorinated aliphatic hydrocarbons to nephrotoxic products. In contrast, however, PCB reduced the nephrotoxicity of chloroform in mice (Kluwe *et al.*, 1978). PCB stimulation of a microsomal xenobiotic mechanism may therefore reduce, as well as enhance, chlorinated hydrocarbon nephrotoxicity. The differences between the effects of PBB and PCB in chloroform toxicity suggest that these agents may have different effects on the overall metabolism of chloroform. That is, PBB and PCB, though they share many physical and chemical characteristics, may have qualitatively different effects on xenobiotic metabolism. PCB stimulated hepatic xenobiotic metabolism as well as renal metabolism, however, and the serum half-life of chloroform was significantly less in PCB-treated mice than in control mice (W. M. Kluwe and J. B. Hook, unpublished observations). The protective effects of PCB on chloroform nephrotoxicity may have been a consequence of increased hepatic chloroform metabolism with a resulting decrease in delivery of chloroform to the kidney. Therefore, although no definitive data are available to rule out the formation of a toxic material in the liver and subsequent transport to the kidney, these results suggest that renal metabolism is an important determinant in the nephrotoxicities of chloroform and related compounds.

4. Potentiation of Acetaminophen (APAP) Toxicity

N-Acetyl-para-aminophenol, commonly known as acetaminophen (APAP) or paracetamol, is an antipyretic analgesic which in human overdose situations has produced a characteristic, and sometimes fatal, massive centrilobular hepatic necrosis. In addition, a significant number of individuals also developed acute renal insufficiency and, in terminal cases, acute renal failure (Boyer and Rouff, 1971; Clark et al., 1973). Renal ultrastructural damage similar to that found with other forms of toxic nephropathy includes loss of luminal brush borders, mitrochondrial disarray, sloughing of cells, and disruption of tubular basement membranes (Kleinman et al., 1980).

The mechanism of APAP-induced hepatic necrosis has been extensively studied (Mitchell et al., 1973a,b; Jollow et al., 1973; Potter et al., 1973), but only recently has an appropriate animal model been identified to evaluate the mechanism involved in renal necrosis. Using the Fisher 344 rat, McMurty et al. (1978) produced acute renal tubular necrosis which was restricted to the pars recta of the proximal tubule. These authors postulated that the biochemical mechanism of APAP toxicity in the kidney was similar to that in the liver; APAP was converted by a cytochrome P-450-NADPH dependent reaction to a chemically reactive arylating agent which could covalently bind to cellular macromolecules producing pathologic changes (McMurty et al., 1978). However, binding occurred only after substantial loss of glutathione. The loss of glutathione was presumably due to a direct conjugation of the nucleophilic glutathione with reactive metabolites of APAP. Several pathways have recently been postulated for acetaminophen metabolism (Fig. 1). The reactive metabolite is believed to be N-acetylimidoquinone, which is apparently formed by a previously uncharacterized cytochrome P-450 mechanism (Hinson et al., 1981).

McMurty et al. (1978) have demonstrated a direct correlation between loss of renal glutathione, covalent binding of APAP metabolites to renal tissue, and APAP-related renal cortical necrosis, using cytochrome P-450 inducers and inhibitors, suggesting that conversion of APAP to an electrophilic metabolite by cytochrome P-450 is a critical step in the pathogenesis of acute APAP nephrotoxicity.

Due to the extensive capacity of the liver to metabolize APAP, it is difficult to unequivocally demonstrate that the reactive metabolite of APAP is produced within the kidney in the intact animal. Many in vitro renal preparations may not adequately reflect the capacity of the kidney to metabolize APAP to a reactive intermediate. Concentration of APAP and metabolites by the kidney in urine and tissue may also contribute to acute APAP nephrotoxicity. However, the ability of

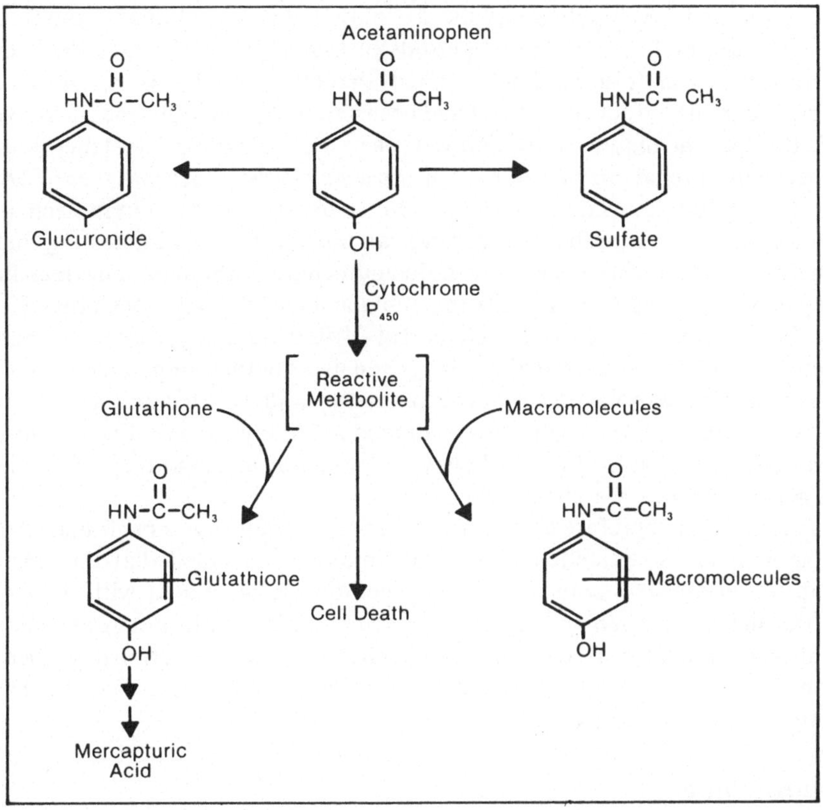

FIGURE 1. Pathways of acetaminophen metabolism. [From Gillette (1981). Reproduced with permission.]

APAP to deplete nonprotein sulfhydryls in the isolated perfused rat kidney was enhanced by prior treatment of the donor rats with PBBs and was reduced by piperonyl butoxide (Table VI). More recently, mercapturic acid conjugation of APAP (presumed to be indicative of the conjugation of a reactive intermediate with glutathione) by the isolated perfused rat kidney was enhanced threefold to fourfold by PBB, whereas formation of the nontoxic glucuronide and sulfate conjugates was decreased by 20% (J. F. Newton, W. E. Braselton, Jr., C.-H. Kuo, W. M., Kluwe, M. W. Gembroys, G. H. Mudge, and J. B. Hook, unpublished). It is clear, therefore, that metabolites of APAP can be produced directly within the kidney. Future experiments will determine the quantitative importance of renal metabolism in the activation of nephrotoxic agents in intact animals.

5. Duration of the Potentiation Effect

The question arises as to the duration of the interaction between the halogenated aromatic hydrocarbons and such nephrotoxic agents as chloroform.

These compounds are highly lipophilic, are only slightly metabolized, and thus are retained in the body for extended periods of time (Mathews *et al.*, 1977, 1978). McCormack *et al.* (1981) demonstrated the effect of PBBs on renal drug metabolism in the kidney of rats that had been exposed prenatally and up to 28 days of age through the maternal diet and were then weaned onto control diet. Marked enhancement of renal enzyme activity produced by PBBs which was seen at 29 days was still seen 300 days later. This marked increase in enzyme activity seen at 300 days after discontinuing the PBB diet was an unexpected and interesting finding. Furthermore, chemicals capable of inducing drug metabolizing enzymes in the kidney may be carried through milk to subsequent generations (McCormack *et al.*, 1981). Female rats received 10 or 100 ppm of PBBs in their diet; there was a marked induction of enzyme activity in the kidneys and livers of their pups at 28 days of age. Some of these animals were then allowed to mature and were fed only a control diet. They were then bred and their pups evaluated at 28 days of age. Even in this next generation, there was still marked enzyme induction in the kidneys. Only in the third generation was the effect eliminated.

These studies therefore demonstrate that PBBs may have a profound effect on enzyme activity for extended periods of time. It is assumed that the enhanced toxicity of chloroform seen in a single generation is correlated with the enzyme activities measured even though those specific activities are not responsible for activation of chloroform. Nevertheless, it is clear that the interaction described may extend far beyond the time of direct exposure and increase the risk of enhanced toxicity even to subsequent generations.

Acknowledgments

The authors' research described herein is supported by USPHS grant ES00560 and a grant from the Michigan Department of Agriculture. Michigan State

TABLE VI. Effects of Polybrominated Biphenyls (PBBs) and Piperonyl Butoxide Pretreatment on Depletion of Nonprotein Sulfhydryl Groups by Acetaminophen (APAP) in Isolated Perfused Rat Kidneys[a]

Pretreatment	APAP in perfusate	Percent depletion of nonprotein sulfhydryls		
		Cortex	Medulla	Papilla
None	3×10^{-8} M	11.6	10.5	38.8
PBB (150 mg/kg)	3×10^{-8} M	29.5^b	51.0^b	55.7
None	3×10^{-5} M	29.8	34.9	53.0
Piperonyl butoxide (600 mg/kg)	3×10^{-5} M	7.8^c	10.9^c	33.5

[a] Data are expressed as percent depletion of nonprotein sulfhydryl concentration from the nonperfused kidney of each animal. PBB and piperonyl butoxide alone had no effect. Values present means of three to six kidneys. Data modified from Kluwe and Hook (1980b).
[b] Significantly different from control (no pretreatment, 3×10^{-8} M APAP), $p < 0.05$.
[c] Significantly different from control (no pretreatment, 3×10^{-5} M APAP), $p < 0.05$.

University Agricultural Experimental Station Journal Article No. 10357. We express our appreciation to Elizabeth M. Madigan for typing the manuscript.

References

Ahmadizadeh, M., Kuo, C-H., and Hook, J. B., 1981, Nephrotoxicity and hepatotoxicity of chloroform in mice: Effect of deuterium substitution 8:105–111.

Alvares, A. P., Fishbein, A., Anderson, K. E., and Kappas, A., 1977, Alteration in drug metabolism in workers exposed to polychlorinated biphenyls, *Clin. Pharmacol. Ther.* 22:140.

Bellar, T. A., Lichtenberg, J. J., and Kroner, R. C., 1974, The occurrence of organohalides in chlorinated drinking waters, *J. Am. Water Works Assoc.* 66:703.

Boyer, T. D., and Rouff, S. L., 1971, Acetaminophen-induced necrosis and renal failure, *J. Am. Med. Assoc.* 218:440.

Bunn, W. W., Haas, B. B., Deane, E. R., and Klepfer, D., 1975, Formation of trihalomethanes by chlorination of surface water, *Environ. Lett.* 10:205.

Clark, R., Borirakchanyavat, V., Davidson, A. R., Thompson, R. P. H., Widdop, B., Goulding, R., and Williams, R., 1973, Hepatic damage and death from overdose of paracetamol, *Lancet* 1:66.

Gillette, J. R., 1981, An integrated approach to the study of chemically reactive metabolites of acetaminophen, *Arch. Intern. Med.* 141:375.

Hill, R. M., Clemens, T. L., Liu, D. K., Vesell, E. S., and Johnson, W. D., 1975, Genetic control of chloroform toxicity in mice, *Science* 190:159.

Hinson, J. A., Pohl, L. R., Monks, T. J., and Gilette, J. R., 1981, Acetaminophen-induced hepatotoxicity—Minireview, *Life Sci.* 29:107.

Hook, J. B., 1980, Toxicology of the kidney, in: *Casarett and Doull's Toxicology: The Basic Science of Poisons* (J. Doull, C. D. Klassen, and M. D. Amdur, eds.), Macmillan, New York, p. 232.

Hook, J. B., McCormack, K. M., and Kluwe, W. M., 1979, Biochemical mechanisms of nephrotoxicity, in: *Reviews in Biochemical Toxicology,* (E. Hodgson, J. R. Bend, and R. M. Philpot, eds.), Elsevier/North-Holland, New York, p. 53.

Jollow, D. J., Mitchell, J. R., Potter, W. Z., Davis, D. C., Gillette, J. R., and Brodie, B. B., 1973, Acetaminophen-induced hepatic necrosis. II. Role of covalent binding *in vivo, J. Pharmacol. Exp. Ther.* 187:195.

Kay, K., 1977, Polybrominated biphenyls (PBB) environmental contamination in Michigan, 1973–1976, *Environ. Res.* 13:74.

Kleinman, J. G., Breitenfield, R. V., and Roth, D. A., 1980, Acute renal failure associated with acetaminophen ingestion: Report of a case and review of the literature, *Clin. Nephrol.* 14:210.

Kluwe, W. M., and Hook, J. B., 1978, Polybrominated biphenyl-induced potentiation of chloroform toxicity, *Toxicol. Appl. Pharmacol.* 45:861.

Kluwe, W. M. and Hook, J. B., 1980a, Metabolic activation of nephrotoxic haloalkanes, *Fed. Proc.* 39:3129.

Kluwe, W. M., and Hook, J. B., 1980b, Effects of environmental chemicals on kidney metabolism and function, *Kidney Int.* 18:648.

Kluwe, W. M., and Hook, J. B., 1981, Potentiation of acute chloroform nephrotoxicity by the glutathione depletor diethyl maleate and protection by the microsomal enzyme inhibitor piperonyl butoxide, *Toxicol. Appl. Pharmacol.* 59:457.

Kluwe, W. M., McCormack, K. M., and Hook, J. B., 1978, Selective modification of renal and hepatic toxicities of chloroform by induction of drug metabolizing enzyme systems in the kidney and liver, *J. Pharmacol. Exp. Ther.* 207:566.

Kluwe, W. M., Herman, C. L., and Hook, J. B., 1979, Effects of dietary polychlorinated biphenyls and polybrominated biphenyls on the renal and hepatic toxicities of several chlorinated hydrocarbon solvents in mice, *J. Pharmacol. Exp. Ther.* 5:605.

Mansuy, D., Beune, P., Crestell, T., Lange, M., and Leroux, J. P., 1977, Evidence for phosgene formation during liver microsomal oxidation of chloroform, *Biochem. Biophys. Res. Commun.* 79:513.

Mathews, H. B., Kato, S., Morales, N. M., and Tuey, D. B., 1977, Distribution and excretion of 2,4,5,2',4',5'-hexabromobiphenyl, the major component of Firemaster BP-6, *J. Toxicol. Environ. Health* **3**:599.

Mathews, H., Fries, G., Gardner, A., Garthoff, L., Goldstein, J., Ku, Y., and Moore, J., 1978, Metabolism and biochemical toxicity of PCPs ana PBBs, *Environ. Health Persp.* **24**:147.

McCormack, K. M., Kluwe, W. M., Rickert, D. E., and Sanger, V. L., and Hook, J. B., 1978, Renal and hepatic microsomal enzyme stimulation and renal function following three month dietary exposure to polybrominated biphenyls, *Toxicol. Appl. Pharmacol.* **44**:539.

McCormack, K. M., Lepper, L. F., Wilson, D. M., and Hook, J. B., 1981, Biochemical and physiological sequelae to perinatal exposure to polybrominated biphenyls: A multigenaration study in rats. *Toxicol. Appl. Pharmacol.* **59**:300–313.

McMurty, R. J., Snodgrass, W. R., and Mitchell, J. R., 1978, Renal necrosis, glutathione depletion and covalent binding after acetaminophen, *J. Toxicol. Appl. Pharmacol.* **46**:87.

Mitchell, J. R., Jollow, D. J., Potter, W. Z., Davis, D. C., Gillette, J. R., and Brodie, B. B., 1973a, Acetaminophen-induced hepatic necrosis. I. Role of drug metabolism, *J. Pharmacol. Exp. Ther.* **187**:185.

Mitchell, J. R., Jollow, D. J., Potter, W. Z., Gillette, J. R., and Brodie, B. B., 1973b, Acetaminophen-induced hepatic necrosis. IV. Protective role of glutathione, *J. Pharmacol. Exp. Ther.* **187**:211.

Moorehead, P. D., Willett, L. B., Brumm, C. J., and Mercer, H. D., 1977, Pathology of experimentally induced polybrominated biphenyl toxicosis in pregnant heifers, *J. Am. Vet. Med. Assoc.* **170**:307.

Pohl, L. R., and Krishna, G., 1978, Deuterium isotope effect in bioactivation and hepatotoxicity of chloroform, *Life Sci.* **23**:1067.

Pohl, L. R., Bhooshan, B., Whittaker, N. F., and Kirshna, G., 1977, Phosgene: A metabolite of chloroform, *Biochem. Biophys. Res. Commun.* **79**:684.

Pohl, L. R., George, J. W., Martin, J. L., and Kirshna, G., 1979, Dueterium isotope effect in *in vivo* bioactivation of chloroform to phosgene, *Biochem. Pharmacol.* **28**:561.

Potter, W. Z., Davis, D. C., Mitchell, J. R., Jollow, D. J., Gillette, J. R., and Brodie, B. B., 1973, Acetaminophen-induced hepatic necrosis. III. Cytochrome P-450-mediated covalent binding *in vitro*, *J. Pharmacol. Exp. Ther.* **187**:203.

Rechnagel, O. R., and Glende, E. A., 1973, Carbon tetrachloride hepatotoxicity: An example of lethal cleavage, *CRC Crit. Rev. Toxicol.* **2**:263–297.

Scholler, K. L., 1970, Modification of the effects of chloroform on the rat liver, *Br. J. Anesthesiol.* **42**:602.

Sipes, I. G., Krishna, G., and Gillette, J. R., 1977, Bioactivation of carbon tetrachloride, chloroform and bromotrichloromethane: Role of cytochrome P-450, *Life Sci.* **20**:1541.

Slater, T. F., 1972, *Free Radical Mechaniams in Tissue Injury,* Pion, London.

Striker, G. E., Smuckler, E. A., Kohnon, P. W., and Nagle, R. B., 1968, Structural and functional changes in rat kidney during CCl_4 intoxication, *Am. J. Pathol.* **53**:769.

Von Oettingen, W. F., 1964, *The Halogenated Hydrocarbons of Industrial and Toxicological Importance,* Elsevier, Amsterdam, p. 77.

Watrous, W. M., and Plaa, G. L., 1972, Effect of halogenated hydrocarbons on organic ion accumulation by renal cortical slices of rats and mice, *Toxicol. Appl. Pharmacol.* **22**:528.

Winslow, S. G., and Gerstner, H. B., 1978, Health aspects of chloroform—A review, *Drug. Chem. Toxicol.* **1**:259.

30

Alteration of Chloroform-Induced Nephrotoxicity by Exogenous Ketones

WILLIAM R. HEWITT, ESTHER M. BROWN,
MICHEL G. CÔTÉ, GABRIEL L. PLAA, and
HIROAKI MIYAJIMA

1. Introduction

Chemical-induced potentiation of haloalkane toxicity is not a novel observation. While potentiation of liver injury is the most frequently examined aspect of this problem (Hewitt *et al.*, 1980a), a number of reports have indicated that the renal injury produced by various haloalkanes can be exacerbated by prior exposure to a number of different compounds. For example, Klaassen and Plaa (1966) demonstrated that a single 5 g/kg dose of ethanol significantly increased the chloroform ($CHCl_3$)- and 1,1,2-trichloroethane-induced depression of mouse

WILLIAM R. HEWITT • Department of Veterinary Anatomy–Physiology, College of Veterinary Medicine, and Department of Pharmacology, School of Medicine, University of Missouri–Columbia, Columbia, Missouri 65211. **ESTHER M. BROWN** • Department of Veterinary Anatomy–Physiology, College of Veterinary Medicine, University of Missouri—Columbia, Columbia, Missouri 65211. **MICHEL G. CÔTÉ and GABRIEL L. PLAA** • Département de Pharmacologie, Faculté de Médecine, Université de Montréal, Montréal, Québec, Canada H3C 3J7. **HIROAKI MIYAJIMA** • Département de Pharmacologie, Faculté de Médecine, Université de Montréal, Montréal, Québec, Canada H3C 3J7.

kidney phenolsulfonephthalein excretion. Subsequently, Watrous and Plaa (1971) found that two additional alcohols, isobutyl and isoamyl alcohol, potentiated $CHCl_3$-induced nephrotoxicity in mice. However, these investigators also found that seven other alcohols did not increase the renal damage produced by $CHCl_3$. Recently, Kluwe and Hook (1978) and Kluwe et al. (1979) found that mice fed polybrominated biphenyls were markedly more susceptible to the nephrotoxic effects of $CHCl_3$, carbon tetrachloride, trichloroethylene, and 1,1,2-trichloroethane. In contrast, pretreatment of mice with phenobarbital had no effect on $CHCl_3$-induced nephrotoxicity, whereas pretreatment with 3-methylcholanthrene, 2,3,7,8-tetrachlorodibenzo-p-dioxin, or polychlorinated biphenyls actually reduced the renal toxicity of $CHCl_3$ (Kluwe et al., 1978). Polychlorinated biphenyl pretreatment did potentiate trichloroethylene-induced renal dysfunction in mice, but did not produce a significant increase in carbon tetrachloride or 1,1,2-trichloroethane kidney injury (Kluwe et al., 1979).

Unfortunately, the diverse structures of these chemicals make it difficult to identify a structural feature(s) associated with the ability to potentiate haloalkane nephrotoxicity. In contrast, recent studies (Hewitt et al., 1980a) have indicated that one structural determinant, a carbonyl moiety, is common to several agents that potentiate haloalkane liver injury. Thus five ketonic chemicals (acetone, 2-butanone, 2-hexanone, 2,5-hexanedione, Kepone) and three chemicals that are metabolized to ketones (isopropanol, 2-butanol, n-hexane) have been reported by various investigators to potentiate the liver injury produced by one or more haloalkanes. Potentiation of liver injury has also been observed when haloalkanes are administered to animals in a state of metabolic ketosis produced by alloxan-induced diabetes or by 1,3-butanediol administration. These observations led to the formulation of a hypothesis suggesting that administration or generation of ketonic substances increases the susceptibility of the liver to the toxic action of haloalkanes. Interestingly, during the course of these studies, several of the compounds examined were found to potentiate $CHCl_3$-induced renal damage. This suggests that the hypothesis developed for liver may also be applicable to the kidney. That is, the susceptibility of the kidney to haloalkane toxicity may be enhanced by prior exposure to ketonic compounds or compounds that are metabolized to ketones in the body. The intent of this chapter is to review the studies that have led to this conclusion.

2. Alteration of $CHCl_3$-Induced Nephrotoxicity by Mirex and Kepone

Our interest in the ability of ketones to potentiate haloalkane-induced nephrotoxicity originated with a series of studies comparing the ability of two structurally related insecticides, Mirex and Kepone (chlordecone), to exacerbate the kidney damage produced by $CHCl_3$ in mice (Hewitt et al., 1979) and rats.

In the first series of experiments, male Swiss-Webster mice were challenged with $CHCl_3$ (0.1 ml/kg, PO) 18 hr after a single oral dose (50 mg/kg) of Mirex or Kepone. Kidney injury was estimated 24 hr after $CHCl_3$ administration using various functional parameters [e.g., depression of lactate-stimulated renal cortical slice accumulation of *p*-aminohippurate (PAH) and elevation of blood urea nitrogen (BUN) content] and histologic studies.

Treatment of mice with Mirex or Kepone alone did not produce an appreciable alteration in the parameters of renal injury examined (Table I). As expected, $CHCl_3$ produced a marked degree of renal damage in vehicle-pretreated mice, as evidenced by an approximate 59% decrease in slice PAH accumulation and a marked increase in the percentage of abnormal (degenerated + necrotic) tubules found upon histologic examination. $CHCl_3$ administration to mice pretreated with Mirex produced alterations in renal histology and function to the same extent observed in vehicle-pretreated mice. In contrast, Kepone pretreatment potentiated the nephrotoxic response to the challenge dose of $CHCl_3$. The combination of Kepone + $CHCl_3$ resulted in a significant elevation ($\sim 62\%$ increase) in BUN when compared to either treatment given alone. Kepone-induced potentiation of $CHCl_3$ kidney damage was also apparent histologically. A significant increase in the percentage of abnormal renal tubules was observed in mice treated with both Kepone and $CHCl_3$. Furthermore, necrotic tubules accounted for a greater proportion of the total (32% necrotic) in mice receiving Kepone + $CHCl_3$ than in vehicle-pretreated mice challenged with $CHCl_3$ (20% necrotic). However, administration of $CHCl_3$ to Kepone-pretreated mice did not further depress the ability of renal cortical slices to accumulate PAH.

Similar results were observed when $CHCl_3$ (0.5 ml/kg, PO) was administered to rats pretreated with a single 50 mg/kg dose of Mirex or Kepone (Table II). As observed with mice, Mirex and Kepone did not produce renal injury in rats when

TABLE I. Effects of Mirex and Kepone Pretreatment on CHCl₃-Induced Nephrotoxicity in Male Mice[a]

Pretreatment	CHCl₃ challenge (0.1 ml/kg, PO)	PAH S/M ratio	BUN, mg/100 ml	Abnormal tubules, %
Vehicle	No	16.78 ± 1.00 (15)	21 ± 1 (16)	0.8 ± 0.4 (5)
Vehicle	Yes	$6.85 \pm 0.75^{b,c}$ (14)	26 ± 2 (14)	$41.6 \pm 5.0^{b,c}$ (5)
Mirex (50 mg/ kg)	No	17.66 ± 0.80 (9)	18 ± 3 (10)	1.4 ± 0.2 (5)
Mirex (50 mg/ kg)	Yes	9.25 ± 1.22^{c} (8)	20 ± 3 (8)	37.0 ± 5.8^{c} (5)
Kepone (50 mg/kg)	No	18.19 ± 1.19 (15)	20 ± 1 (15)	1.8 ± 0.4 (5)
Kepone (50 mg/kg)	Yes	5.55 ± 0.52^{c} (14)	$42 \pm 7^{c,d}$ (14)	$57.8 \pm 4.8^{c,d}$ (5)

[a] Mice were challenged with CHCl₃ 18 hr following pretreatment (PO) with vehicle, Mirex, or Kepone. The animals were killed 24 hr later. Values are expressed as mean ± SE. Values in parentheses represent the number of mice studied. Data were compiled from Hewitt *et al.* (1979) and are used with the permission of Academic Press.
[b] Values significantly different from the control (vehicle alone) group, $p < 0.05$.
[c] Values significantly different from the corresponding group not given CHCl₃, $p < 0.05$.
[d] Values significantly different from the group receiving vehicle + CHCl₃, $p < 0.05$.

TABLE II. Effects of Mirex, Kepone, and Phenobarbital Pretreatment (Single Dose) on CHCl$_3$-Induced Nephrotoxicity in Male Sprague-Dawley Rats[a]

Pretreatment	CHCl$_3$ challenge (0.5 ml/kg PO)	PAH S/M ratio	BUN, mg/100 ml	Abnormal tubules, %
Vehicle	No	26.41 ± 0.86 (10)	13 ± 1 (10)	3.8 ± 0.4 (5)
Vehicle	Yes	26.20 ± 1.82 (10)	23 ± 2[b,c] (10)	42.4 ± 9.8[b,c] (5)
Mirex (50 mg/kg)	No	25.52 ± 1.42 (10)	12 ± 1 (10)	4.0 ± 0.7 (5)
Mirex (50 mg/kg)	Yes	25.78 ± 0.97 (10)	22 ± 2[c] (10)	49.8 ± 7.3[c] (5)
Kepone (50 mg/kg)	No	27.62 ± 1.16 (10)	14 ± 1 (10)	4.2 ± 0.8 (5)
Kepone (50 mg/kg)	Yes	15.77 ± 1.86[c,d] (9)	42 ± 3[c,d] (9)	74.4 ± 2.3[c,d] (5)
Phenobarbital (80 mg/kg)	No	24.86 ± 1.34 (9)	15 ± 1 (9)	5.4 ± 0.5 (5)
Phenobarbital (80 mg/kg)	Yes	25.59 ± 1.37 (9)	29 ± 3[c,d] (9)	59.2 ± 11.4[c] (5)

[a] CHCl$_3$ was administered 18 hr after a single dose of vehicle (PO), Mirex (PO), Kepone (PO), or phenobarbital (IP). The animals were killed 24 hr later. Values represent the mean ± SE in (n) rats.
[b] Values significantly different from the control (vehicle alone) group, $p < 0.05$.
[c] Values significantly different from the corresponding group not given CHCl$_3$, $p < 0.05$.
[d] Values significantly different from the group receiving vehicle + CHCl$_3$, $p < 0.05$.

administered alone. The CHCl$_3$ challenge dose produced a moderate degree of kidney damage in vehicle-pretreated rats, as indicated by an increase in the percentage of abnormal tubules and a small but statistically significant increase in BUN. Mirex pretreatment did not potentiate the nephrotoxic action of the CHCl$_3$ challenge dose, whereas CHCl$_3$-induced renal damage was significantly increased in rats pretreated with Kepone. Thus treatment with Kepone + CHCl$_3$ resulted in an ~40% depression in slice PAH accumulation, an ~83% increase in BUN, and an ~75% increase in abnormal tubules as compared to the CHCl$_3$ challenge dose alone. No necrotic tubules were observed in the vehicle + CHCl$_3$ or Kepone + CHCl$_3$ groups. Interestingly, a single 80 mg/kg dose of phenobarbital (PB) did not produce any indication of kidney damage when administered alone and did not appear to produce an appreciable potentiation of CHCl$_3$ nephrotoxicity in rats (Table II).

Potentiation of CHCl$_3$-induced kidney injury was also observed in rats pretreated with multiple doses of Mirex (10 mg/kg/day) or phenobarbital (80 mg/kg/day) (Table III). CHCl$_3$ administration (0.5 ml/kg, PO) to rats pretreated with any of these agents resulted in a significant elevation of BUN content when compared to CHCl$_3$ treatment alone. In addition, the data obtained from the histologic analysis of the renal damage produced by these treatment regimens suggested that CHCl$_3$-induced nephrotoxicity was potentiated by Mirex, Kepone, and phenobarbital. Thus, abnormal tubules accounted for approximately 50% of the total in vehicle-pretreated rats challenged with CHCl$_3$ and increased to approximately 70, 90, and 94% in rats treated with Mirex + CHCl$_3$, Kepone + CHCl$_3$, and phenobarbital + CHCl$_3$, respectively. A small component of necrotic tubules (~4%) was observed in rats receiving the combination of PB + CHCl$_3$. Necrotic tubules were not observed in the other treatment groups. Slice PAH accumulation was not altered by any of the treatment regimens.

Based on the functional and histologic data from these particular experiments, the relative ranking of these three chemicals, in order of increasing potentiating ability, appears to be Mirex < phenobarbital < Kepone. The most salient aspect of this ranking was, of course, the disparity in potentiating ability between Mirex and Kepone. These structural analogs differ only in that two chlorine atoms in Mirex are replaced by a carbonyl moiety in Kepone. Yet Mirex did not potentiate CHCl₃-induced nephrotoxicity in mice and rats when administered as a single dose and was a less effective potentiating agent than Kepone when multiple doses were administered to rats. These observations suggest that the carbonyl moiety was a prime determinant of the potentiating capability of Kepone. Furthermore, these data were consistent with the hypothesis that ketones or ketogenic chemicals can increase the susceptibility of laboratory animals to haloalkane-induced nephrotoxicity. While the results with phenobarbital appear to support this hypothesis, a note of caution must be interjected. The degree of potentiation observed with phenobarbital was relatively weak and appeared only after administration of multiple doses of this agent. In addition, other investigators have found phenobarbital to have no effect on CHCl₃-induced nephrotoxicity in mice (Kluwe *et al.*, 1978) or to actually protect against the kidney-damaging properties of this haloalkane in mice (Ilett *et al.*, 1973). Thus additional information is needed to clarify the ability of phenobarbital to increase haloalkane-induced nephrotoxicity.

3. Aliphatic Ketone-Induced Potentiation of CHCl₃ Renal Damage

We attempted to confirm the validity of our hypothesis by evaluating the nephrotoxic response to a challenging dose of CHCl₃ in rats pretreated with *n*-hexane (H) (Hewitt *et al.*, 1980b). 2-Hexanone (2-HX), 1-hexanol, and 2,5-

TABLE III. Effects of Mirex, Kepone, and Phenobarbital Pretreatment (Multiple Doses) on CHCl₃-Induced Nephrotoxicity in Male Sprague-Dawley Rats[a]

Pretreatment	CHCl₃ challenge (0.5 ml/kg, PO)	PAH S/M ratio	BUN, mg/100 ml	Abnormal tubules, %
Vehicle	No	24.42 ± 1.26 (7)	14 ± 1 (13)	3.2 ± 0.5 (5)
Vehicle	Yes	24.43 ± 1.42 (7)	20 ± 2 (13)	49.6 ± 7.0[b,c] (5)
Mirex (10 mg/kg/day)	No	27.93 ± 1.43 (7)	14 ± 1 (13)	2.8 ± 0.7 (5)
Mirex (10 mg/kg/day)	Yes	24.76 ± 2.16 (6)	30 ± 4[c,d] (12)	69.8 ± 3.8[c,d] (5)
Kepone (10 mg/kg/day)	No	28.53 ± 1.40 (7)	14 ± 1 (13)	3.2 ± 0.5 (5)
Kepone (10 mg/kg/day)	Yes	21.92 ± 3.48 (7)	42 ± 4[c,d] (12)	89.6 ± 2.7[c,d] (5)
Phenobarbital (80 mg/kg/day)	No	27.95 ± 1.90 (7)	15 ± 1 (7)	3.8 ± 0.4 (5)
Phenobarbital (80 mg/kg/day)	Yes	24.43 ± 2.78 (7)	37 ± 5[c,d] (7)	94.4 ± 1.1[c,d] (5)

[a] Rats received vehicle (PO), Mirex (PO), Kepone (PO), or phenobarbital (IP) once daily for 3 days. CHCl₃ was administered 18 hr after the last dose. Rats were killed 24 hr after CHCl₃ administration. Values represent the mean ± SE determined in (*n*) rats.
[b] Values significantly different from the control (vehicle alone) group, $p < 0.05$.
[c] Values significantly different from the corresponding group not given CHCl₃, $p < 0.05$.
[d] Values significantly different from the group receiving vehicle + CHCl₃, $p < 0.05$.

hexanedione (2,5-HD) have been identified as metabolites of *n*-hexane in the rat (Bus *et al.*, 1979; Dolara *et al.*, 1978), while 2-hexanol (Couri *et al.*, 1976), 5-hydroxy-2-hexanone, and 2,5-HD have been isolated from guinea pigs treated with *n*-hexane (DiVincenzo *et al.*, 1976). Thus this ketogenic chemical should potentiate $CHCl_3$-induced nephrotoxicity if our hypothesis were correct. In the same study, the potentiating ability of two ketonic metabolites of *n*-hexane, 2-HX and 2,5-HD, also were assessed. Acetone (A) was also included in the study. In these experiments, rats were challenged with $CHCl_3$ (0.5 ml/kg, IP) 18 hr after a single oral dose (15 mmole/kg) of *n*-hexane or one of the three ketones. None of the suspected potentiating agents altered the functional parameters of renal damage when administered alone (Table IV). However, when pretreatment with any of these chemicals was followed by the $CHCl_3$ challenge dose, the response to the $CHCl_3$ was significantly increased. Thus administration of $CHCl_3$ to A-, 2-HX-, or 2,5-HD-pretreated rats resulted in a significant depression of slice PAH accumulation. $CHCl_3$ administration to H-pretreated rats did not significantly reduce the slice accumulation of PAH. However, the active slice accumulation of a model organic cation, tetraethylammonium (TEA) ion, was significantly depressed in rats treated with the combination of H + $CHCl_3$, A + $CHCl_3$, 2-HX + $CHCl_3$, or 2,5-HD + $CHCl_3$. In contrast, elevation of BUN content occurred only when $CHCl_3$ was given to rats pretreated with 2-HX or 2,5-HD. These results indicated that the relative ranking of the chemicals in order of increasing potentiating ability was H ⩽ A < 2,5-HD ≈ 2-HX. This ranking was confirmed by a quantitative histologic analysis of the kidney injury produced by the various treatment regimens (Table IV). The percentage of abnormal tubules was greatest in sections taken from rats treated with 2-HX or 2,5-HD and the $CHCl_3$ challenge dose. The combination of *n*-hexane plus $CHCl_3$ produced the smallest percentage of abnormal tubules, while acetone plus $CHCl_3$ produced a lesion of intermediate severity. Also in keeping with the relative ranking of these chemicals was the observation that necrotic tubules appeared only in the 2-HX + $CHCl_3$ (6% necrotic tubules) and the 2,5-HD + $CHCl_3$ (13% necrotic tubules) groups.

The observation that acetone, *n*-hexane, 2-HX, and 2,5-HD were able to potentiate $CHCl_3$-induced nephrotoxicity provided additional confirmation for the hypothesis that the administration of ketones or ketogenic chemicals increases the susceptibility of the kidney to the toxic actions of haloalkanes. In addition, the marked disparity in the ability of acetone, 2-HX, and 2,5-HD to potentiate $CHCl_3$ toxicity may also provide some insight into the connection between the structural properties of ketonic solvents and their ability to potentiate toxicity. Although acetone, 2-HX, and 2,5-HD were administered at equimolar doses (15 mmole/kg), acetone was found to be a less efficient potentiating agent than 2-HX or 2,5-HD, either of which had approximately equivalent potentiating capacities. The observation that the two six-carbon ketones 2-HX and 2,5-HD were more effective potentiators than acetone, a three-carbon ketone, suggested that the relative potentiating capacity of ketonic solvents may increase with increasing length of the carbon skeleton. The number of ketonic moieties appears to be of less importance in determining potentiating ability, since the six-carbon, monoketonic solvent 2-

TABLE IV. Effects of Pretreatment with Acetone, n-Hexane, 2-Hexanone, or 2,5-Hexanedione on CHCl₃-Induced Nephrotoxicity in Male Sprague-Dawley Rats[a]

Pre-treatment	CHCl₃ challenge (0.5 ml/kg, IP)	PAH S/M ratio	TEA S/M ratio	BUN, mg/100 ml	Abnormal tubules, %
Vehicle	No	23.69 ± 1.00 (15)	24.32 ± 0.79 (15)	17 ± 1 (15)	1.0 ± 0.4 (5)
Vehicle	Yes	22.85 ± 1.14 (15)	23.81 ± 1.02 (15)	16 ± 1 (15)	1.2 ± 0.6 (5)
Acetone	No	21.97 ± 0.91 (6)	26.16 ± 0.64 (6)	13 ± 1 (6)	2.2 ± 0.2 (5)
Acetone	Yes	13.68 ± 1.56[c,d] (6)	15.06 ± 1.66[c,d] (6)	30 ± 3 (6)	18.0 ± 2.6 (5)
n-Hexane	No	22.86 ± 1.39 (8)	24.80 ± 1.62 (8)	15 ± 1 (8)	4.8 ± 1.4 (5)
n-Hexane	Yes	17.48 ± 2.24 (8)	18.06 ± 2.12[c,d] (8)	24 ± 4 (8)	13.2 ± 3.2 (5)
2-Hexanone	No	26.02 ± 1.87 (6)	27.07 ± 0.06 (6)	18 ± 3 (6)	9.6 ± 3.3 (5)
2-Hexanone	Yes	14.15 ± 3.05[c,d] (6)	15.96 ± 1.87[c,d] (6)	71 ± 15[c,d] (6)	27.4 ± 5.5[d] (5)
2,5-Hexane-dione	No	22.81 ± 1.43 (10)	22.80 ± 1.17 (10)	16 ± 2 (10)	22.6 ± 4.5[b] (5)
2,5-Hexane-dione.	Yes	11.80 ± 3.11[c,d] (5)	11.59 ± 2.47[c,d] (5)	47 ± 14[c,d] (5)	38.8 ± 11.2[d] (4)

[a] CHCl₃ was administered 18 hr after a single oral dose (15 mmole/kg) of vehicle, acetone, n-hexane, 2-hexanone, or 2,5-hexanedione. The animals were killed 24 hr later. Values represent the mean ± SE determined in (n) rats. Data were compiled from Hewitt et al. (1980b) and are used with the permission of Academic Press.
[b] Values significantly different from the control (vehicle alone) group, $p < 0.05$.
[c] Values significantly different from the corresponding group not given CHCl₃, $p < 0.05$.
[d] Values significantly different from the group receiving vehicle + CHCl₃, $p < 0.05$.

HX potentiated CHCl₃ nephrotoxicity to approximately the same extent as 2,5-HD, a six-carbon diketonic solvent.

The possibility that a relationship might exist between potentiating capacity and the carbon chain length of ketonic solvents was tested by determining the renal damage produced by CHCl₃ (0.75 ml/kg IP) in rats pretreated with ketones ranging in chain length from three to seven carbons. The ketones tested included acetone (2-propanone), 2-butanone (2-BU), 2-pentanone (2-PN), 2-hexanone, and 2-heptanone (2-HP). Treatment of rats with a single dose (15 mmole/kg, PO) of A, 2-BU, 2-PN, 2-HX, or 2-HP alone did not produce an appreciable degree of kidney injury (Table V). The response to the CHCl₃ challenge dose was increased in rats pretreated with 2-PN, 2-HX, or 2-HP. Thus significant increases in BUN content were observed in rats receiving combinations of 2-PN + CHCl₃, 2-HX + CHCl₃ or 2-HP + CHCl₃. Renal cortical slice accumulation of PAH was significantly depressed in rats treated with both 2-HP and CHCl₃ and appeared to be decreased in the 2-PN + CHCl₃ group. In contrast, no apparent potentiation of CHCl₃-induced nephrotoxicity was observed in rats pretreated with A or 2-BU. Interestingly, when the individual data for slice PAH accumulation or BUN content in the groups receiving a ketone + 0.75 ml/kg CHCl₃ were plotted against the carbon chain length of the ketones used, it appeared that a linear relationship existed. Linear regression analysis of these data demonstrated that a statistically significant linear correlation existed between depression of slice PAH accumulation and the increase in ketone chain length ($y = -2.28x + 28.26, r = 0.47, p$

< 0.05). Similarly, a significant linear correlation was observed between the increase in BUN content and the increase in ketone chain length ($y = 8.84x + 2.16$, $r = 0.44$, $p < 0.05$). Thus these results support the concept that the potentiating ability of the ketone increases as the length of the carbon skeleton increases.

Unfortunately, these observations on possible structure–activity relationships among ketonic potentiating agents are complicated by the fact that the actual identity of the compound(s) mediating the potentiation of $CHCl_3$-induced kidney injury remains in question. For example, 2-HX is biotransformed to one or more metabolites (DiVincenzo et al., 1976) and it is conceivable that the potentiation of toxicity could arise from the action of 2-HX itself or one of its metabolites. Similarly, it remains possible that metabolites of the other ketones examined may contribute to the ability of these agents to exacerbate $CHCl_3$ toxicity. Consequently, the contribution of the various metabolites of these ketones to the potentiation of $CHCl_3$ toxicity must be evaluated before a relationship between carbon skeleton length and potentiating ability of ketones can be firmly established.

In summary, the experiments reviewed in this chapter document that the nephrotoxicity produced by $CHCl_3$ can be enhanced by one ketogenic chemical and seven ketones. These results support the hypothesis that the administration or generation of abnormal amounts of ketonic substances increases the susceptibility of the kidney to the toxic effects of haloalkanes. This hypothesis is of particular

TABLE V. Effects of Pretreatment with Various Ketones on $CHCl_3$-Induced Nephrotoxicity in Male Sprague-Dawley Rats[a]

Pretreatment	$CHCl_3$ challenge (0.75 ml/kg, IP)	PAH S/M ratio	BUN, mg/100 ml
Vehicle	No	26.13 ± 1.16 (16)	18 ± 1 (16)
Vehicle	Yes	20.53 ± 0.69 (12)	18 ± 1 (12)
Acetone	No	27.99 ± 2.08 (10)	20 ± 1 (10)
Acetone	Yes	21.19 ± 2.36 (6)	23 ± 2 (6)
2-Butanone	No	26.56 ± 0.92 (11)	19 ± 1 (11)
2-Butanone	Yes	18.52 ± 1.85[b] (6)	30 ± 4 (6)
2-Pentanone	No	27.11 ± 0.61 (13)	19 ± 1 (13)
2-Pentanone	Yes	16.06 ± 1.31[b] (5)	79 ± 16[b,c] (5)
2-Hexanone	No	24.99 ± 0.96 (18)	21 ± 2 (18)
2-Hexanone	Yes	19.70 ± 4.34 (5)	47 ± 13[b,c] (5)
2-Heptanone	No	27.55 ± 0.93 (15)	19 ± 1 (15)
2-Heptanone	Yes	8.30 ± 1.88[b,c] (4)	54 ± 4[b,c] (3)

[a] $CHCl_3$ was administered 18 hr after a single oral dose (15 mmole/kg) of vehicle, acetone, 2-butanone, 2-pentanone, 2-hexanone, or 2-heptanone. The animals were killed 24 hr later. Values represent the mean ± SE determined in (n) rats.
[b] Values significantly different from the corresponding group not given $CHCl_3$, $p < 0.05$.
[c] Values significantly different from the group receiving vehicle + $CHCl_3$, $p < 0.05$.

importance in that it suggests that it may be possible to develop structure–activity relationships useful in predicting the ability of chemicals to interact with haloalkanes, thereby resulting in a potentiated degree of kidney injury.

Acknowledgments

We gratefully acknowledge the expert technical assistance of Johanne Couture, Monique Morisset, Maxine Little, and Mildred Floyd. This work was supported in part by Grant No. 6605-1441-40, Grant No. 6605-1731-52, Contract No. OSU 77-00100 from Health and Welfare—Canada, a grant from the Commission de la Santé et de la Sécurité du Travail, Gouvernement du Québec, and Grant No. OH00986-01 from the U.S. Public Health Service. H. M. is a visiting scientist from Takeda Chemical Industries, Osaka, Japan.

References

Bus, J. S., White, E. L., and Barrow, C. S., 1979, Disposition of n-hexane in rats after single and repeated inhalation exposure, *Toxicol. Appl. Pharmacol.* **48**:A167.

Couri, D., Abdel-Rahman, M. S., and Hetland, L. B., 1976, Biotransformation of hexane and methyl n-butyl ketone, *Toxicol. Appl. Pharmacol.* **37**:124.

DiVincenzo, G. D., Kaplan, C. J., and Dedinas, J., 1976, Characterization of the metabolites of methyl n-butyl ketone in guinea pig serum and their clearance, *Toxicol. Appl. Pharmacol.* **36**:511.

Dolara, P., Franconi, F., and Basosi, D., 1978, Urinary excretion of some n-hexane metabolites, *Pharmacol. Res. Commun.* **10**:503.

Hewitt, W. R., Miyajima, H., Côté, M. G., and Plaa, G. L., 1979, Acute alteration of chloroform-induced hepato- and nephrotoxicity by mirex and Kepone, *Toxicol. Appl. Pharmacol.* **48**:509.

Hewitt, W. R., Miyajima, H., Côté, M. G., and Plaa, G. L., 1980a, Modification of haloalkane-induced hepatotoxicity by exogenous ketone and metabolic ketosis, *Fed. Proc.* **39**:3118.

Hewitt, W. R., Miyahima, H., Côté, M. G., and Plaa, G. L., 1980b, Acute alteration of chloroform-induced hepato- and nephrotoxicity by n-hexane, methyl n-butyl ketone and 2,5-hexanedione, *Toxicol. Appl. Pharmacol.* **53**:230.

Ilett, K. F., Reid, W. D., Sipes, I. G., and Krishna, G., 1973, Chloroform toxicity in mice: Correlation of renal and hepatic necrosis with covalent binding of metabolites to tissue macromolecules, *Exp. Mol. Pathol.* **19**:215.

Klaassen, C. D., and Plaa, G. L., 1966, Relative effects of various chlorinated hydrocarbons on liver and kidney function in mice, *Toxicol. Appl. Pharmacol.* **9**:139.

Kluwe, W. M., and Hook, J. B., 1978, Polybrominated biphenyl-induced potentiation of chloroform toxicity, *Toxicol. Appl. Pharmacol.* **45**:861.

Kluwe, W. M., McCormack, K. M., and Hook, J. B., 1978, Selective modification of the renal and hepatic toxicities of chloroform by induction of drug-metabolizing enzyme systems in kidney and liver, *J. Pharmacol. Exp. Ther.* **207**:566.

Kluwe, W. M., Hermann, C. L., and Hook, J. B., 1979, Effects of dietary polychlorinated biphenyls and polybrominated biphenyls on the renal and hepatic toxicities of several chlorinated hydrocarbon solvents in mice, *J. Toxicol. Environ. Health* **5**:605.

Watrous, W. M., and Plaa, G. L., 1971, The potentiation of CHCl₃-induced nephrototoxicity by some aliphatic alcohols in mice, *Pharmacologist* **13**:227.

V

IMMUNOLOGIC MECHANISMS AND TOXIC NEPHROPATHIES

GIUSEPPE A. ANDRES, Section Editor

31

Drug-Induced Renal Lesions
Immunopathologic Mechanisms

GIUSEPPE A. ANDRES

In the symposium on Drug Effects on the Kidney held in Sidney in 1979, practically all the papers delt with various aspects of nephrotoxicity. To my knowledge, there was not a session dedicated to renal injury related to drug-induced hypersensitivity reactions. The chapters in this section address this topic, a task made very difficult by the ignorance that still persists in this area of nephropathology. During the last decade, but especially during the last 3–4 years, drug-induced hypersensitivity has received increasing attention due to its importance in clinical medicine and the accelerated pace of progress in basic research. There is evidence, based on experimental models, human disease, or both, that the four main mechanisms of hypersensitivity, namely (1) antibody reacting with renal antigens, (2) immune complex formed *in situ* or in the circulation, (3) cell-mediated hypersensitivity, and (4) immediate IgE-type hypersensitivity, may by activated, independently or synergistically, by drugs and that the kidney may be a target organ.

The work of Tan's laboratory (Tan, 1974) has shown that the induction of autoimmune-like disease by certain drugs, such as hydralazine and procainamide, probably depends on many factors, including (in the case of lupus-like syndrome) the complexing of hydralazine with deoxyribonucleoprotein. This drug–nucleoprotein complex has been shown *in vitro* to resist destruction by proteolytic digestion, and *in vivo* this may be a mechanism for enhancing the immunogenic

GIUSEPPE A. ANDRES • Departments of Microbiology, Pathology and Medicine, School of Medicine, State University of New York at Buffalo, Buffalo, New York 14214.

potential of circulating nucleoprotein. A further factor may be the phenotypic pattern of acetyltransferase activity in a particular individual, since in the case of hydralazine the development of autoantibodies and disease activity are related to the low activity of this acetylating enzyme.

The possibility that drugs may act as haptens, binding to renal protein components and thereby inducing an antibody response which generates renal injury, has been studied by Border, Wilson, and their collaborators (Border *et al.*, 1974). They have shown that methacillin may occasionally be responsible for formation of anti-tubular basement membrane (TBM) antibody and tubulointerstitial nephritis. One of the most challenging problems in the area of antibody-mediated injury is that concerning the etiology of Goodpasture's disease. Several reports have stressed the association between exposure to volatile hydrocarbon and development of anti-basement membrane antibodies. Wilson has studied the effect of exposure of laboratory animals to hydrocarbons and has monitored the effect of this environmental toxicant by a sensitive radioimmunoassay for anti-basement membrane antibody; he describes the results of these studies in Chapter 33.

Clinical or experimental observations, or both, suggest that cell-mediated hypersensitivity may be occasionally involved in the pathogenesis of immunologically mediated renal diseases. There is only a preliminary evidence that drugs may evoke a cell-mediated hypersensitivity reaction in the kidney. McCluskey and his collaborators (Bhan *et al.*, 1978) have provided, together with the laboratories of Unanue and Cotran (Schriener *et al.*, 1978), significant contributions in this area. In Chapter 34 McCluskey and Bhan discuss the state of the art in this field and present some new data concerning the identification of T-cell subgroups in kidney tissue of patients with drug-associated interstitial nephritis.

In Chapter 35 Border attempts an even more difficult task: to analyze the role of immediate IgE-type hypersensitivity in renal diseases. This is a difficult assignment because the evidence that immediate IgE-type hypersensitivity plays a role in certain forms of drug-induced tubulointerstitial nephritis is mainly based on clinical observations. A reproducible experimental model has not yet been developed.

Finally, there is considerable evidence that heavy metals, especially mercury and gold, are implicated in the pathogenesis of certain forms of immune complex glomerulonephritis in man and in animals. The demonstration, provided by Bariety *et al.* (1971), that $HgCl_2$ produced GN in the rat has generated a number of studies which have shown that antibodies to basement membrane and collagen matrix antigens, and immune complexes containing these antigens, cooperate in the genesis of this renal disease. In Chapter 37 Albini *et al.* discuss the histopathology and the immunopathogenesis of the systemic autoimmune disease that develops in rabbits and rats exposed to $HgCl_2$ (Roman-Franco *et al.*, 1976; Sapin *et al.*, 1977). In Chapter 38, Druet and collaborators, who have provided important contributions in the definition of the genetic susceptibility of a certain strain of rats (Brown Norway) to develop this autoimmune disease (Druet *et al.*, 1977), discuss this topic and present the results of new studies indicating that $HgCl_2$ induces a polyclonal B-cell activation. Milgrom and his collaborators have shown that administration of gold may induce formation of circulating antigen–antibody complexes in man (Palosuo *et al.*, 1976). In Chapter 36 Palosuo and Milgrom discuss the possible role of immune complexes in gold nephropathy. Garvey and her collaborators have addressed their attention to the development of a sensitive radioimmunoassay for metallothionein (Chang *et al.*, 1980). This assay is based on the use of antibody to purified cadmium-binding protein. The results have contributed to the hypothesis that metallobinding protein may play a role in the pathogenesis of dysfunction associated with exposure to heavy metals. In Chapter 39 Garvey discusses this new technique.

The increased concerns caused by environmental pollution and by the widespread use of drugs makes it important to deepen our understanding on the effect of these agents on the immune system. I hope that the subsequent chapters may contribute to this goal.

References

Bariety, J., Druet, P., Laliberte, F., and Sapin, C., 1971, Glomerulonephritis with γ- and β1C-globulin deposits induced in rats by mercuric chloride, *Am. J. Pathol.* **65**:293.

Bhan, A. K., Schneeberger, E. E., Collins, A. B., and McCluskey, R. T., 1978, Evidence for a pathogenic role of cell-mediated immune mechanism in expimental glomerulonephritis, *J. Exp. Med.* **148**:246.

Border, W. A., Lehman, D. H., Egan, J. D., Sass, H. J., Globe, J. E., and Wilson, C. B., 1974, Anti-tubular basement membrane antibodies in methicillin-associated interstitial nephritis, *N. Engl. J. Med.* **291**:381.

Chang, C. C., Vander Mallie, R. J., and Garvey, J. S., 1980, A radioimmunoassay for human metallothionein, *Toxicol. Appl. Pharmacol.* **55**:94.

Druet, E., Sapin, C., Günther, E., Feingold, N., and Druet, P., 1977, Mercuric chloride-induced anti-glomerular basement membrane antibodies in the rat, Genetic control, *Eur. J. Immunol.* **7**:348.

Palosuo, T., Provost, T. T., and Milgrom, F., 1976, Gold nephropathy: Serologic data suggesting an immune complex disease, *Clin. Exp. Immunol.* **25**:311.

Roman-Franco, A. A., Turiello, M., Albini, B., Ossi, E., and Andres, G. A., 1976, Anti-basement membrane antibody and immune complexes in rabbits injected with mercuric chloride (HgCl$_2$), *Kidney Int.* **10**:549.

Sapin, C., Druet, E., and Druet, P., 1977, Induction of anti-glomerular basement membrane antibodies in the Brown-Norway rat by mercuric chloride, *Clin. Exp. Immunol.* **28**:173.

Schreiner, G. F., Cotran, R. S., Pardo, V., and Unanue, E. R., 1978, A mononuclear cell component in experimental immunological glomerulonephritis, *J. Exp. Med.* **147**:369.

Tan, E. M., 1974, Drug-induced autoimmune disease, *Fed. Proc.* **33**:1894.

32

Antihistone Antibodies Induced by Procainamide and Hydralazine

JOSEPH P. PORTANOVA, ROBERT L. RUBIN, and ENG M. TAN

1. Introduction

It is well documented that symptoms resembling those of systemic lupus erythematosus (SLE) as well as antinuclear antibodies (ANAs) may appear in patients undergoing prolonged drug therapy (Weinstein, 1980; Blomgren, 1973; Tan, 1974). Several drugs differing in molecular structure and metabolic pathways of biotransformation have been implicated (Lee and Chase, 1975).

Prospective studies have shown that patients receiving hydralazine (Hy) or procainamide (Pr) are particularly prone to developing both lupus symptoms and ANA. Approximately 10% of patients treated with Hy or Pr will develop symptoms of SLE, such as arthralgia, myalgia, arthritis, fever, pericarditis, and pleuritis (Perry, 1973; Blomgren *et al.*, 1972). Clinical renal involvement and central nervous system disease occur rarely. Classically, symptoms subside within weeks after discontinuation of drug therapy, whereas ANA may persist for over 10 years

JOSEPH P. PORTANOVA, ROBERT L. RUBIN, and ENG M. TAN • Division of Rheumatic Diseases, Department of Medicine, School of Medicine, University of Colorado Health Sciences Center, Denver, Colorado 80262.

(Perry, 1973; Blomgren *et al.*, 1972). In certain patients, symptoms may also persist for several years after drug treatment is terminated (Alarcon-Segovia *et al.*, 1967).

Antinuclear antibodies have been detected in 30% of Hy-treated patients (Condemi *et al.*, 1967) and in at least 50% of patients receiving Pr (Blomgren *et al.*, 1967; Woosley *et al.*, 1978). Virtually all symptomatic patients have ANA (Perry, 1973; Woosley *et al.*, 1978). Antinuclear antibodies may also occur in patients who never develop lupus symptoms throughout the course of drug therapy (Molina *et al.*, 1969).

A recent comparison of ANA in SLE and drug-induced LE revealed different antibody specificities in the sera of patients with these diseases (Fritzler and Tan, 1978). Antibodies in SLE sera were directed to a variety of nuclear antigens, including single-stranded DNA, double-stranded DNA, histones, and nonhistone proteins such as Sm, nuclear ribonucleoprotein, and Sjogren's syndrome antigen B. In contrast, antibody specificities in drug-induced LE were restricted to histones and single-stranded DNA.

The observations of Fritzler and Tan (1978) regarding antihistone antibodies in drug-induced lupus sera were derived mainly from patients treated with Pr. In the present chapter, we extend these observations by describing our findings concerning the specificity of antihistone antibodies induced by Pr as well as Hy. Information about antibody specificity for histones has been obtained by a three-step immunofluorescent assay using histone-reconstituted mouse kidney sections (Tan *et al.*, 1976) and by a solid-phase radioimmunoassay using polystyrene tubes coated with chromatin or isolated histones (Rubin *et al.*, 1982). The results indicate that Pr-induced antibodies react preferentially with the H2A–H2B complex, whereas antibodies induced by Hy react poorly with this histone complex. Furthermore, preliminary evidence suggests that Hy-induced antihistone antibodies react strongly with uncomplexed histones H2A and H3; Pr-induced antibodies are substantially less reactive with these histones.

2. Methods and Results

Procainamide and Hy sera were examined for antihistone antibodies using the three-step immunofluorescent technique (Tan *et al.*, 1976). In this assay, nuclei of acetone-fixed mouse kidney sections are depleted of histones and most nonhistone proteins by extraction with 0.1 N HCl. The DNA is retained in the nucleus. The acid-extracted nuclei are then reconstituted with a solution of total histones to provide a histone–DNA substrate in the absence of other nuclear antigens. In the three-step immunofluorescent method, a serum containing antibodies to histones demonstrates ANA with acetone-fixed kidney sections, does not react with acid-extracted tissue, and does react with histone-reconstituted tissue (Fig. 1).

Table I shows reactions of sera from Pr-treated patients in the immuno-fluorescent assay. As a group, these sera reacted strongly on acetone-fixed mouse kidney sections. Eight of ten sera had IgM ANA titers ≥512. The GMTs (geometric mean titer) of Pr-induced IgM and IgG ANA were 830 and 1259, respectively. No

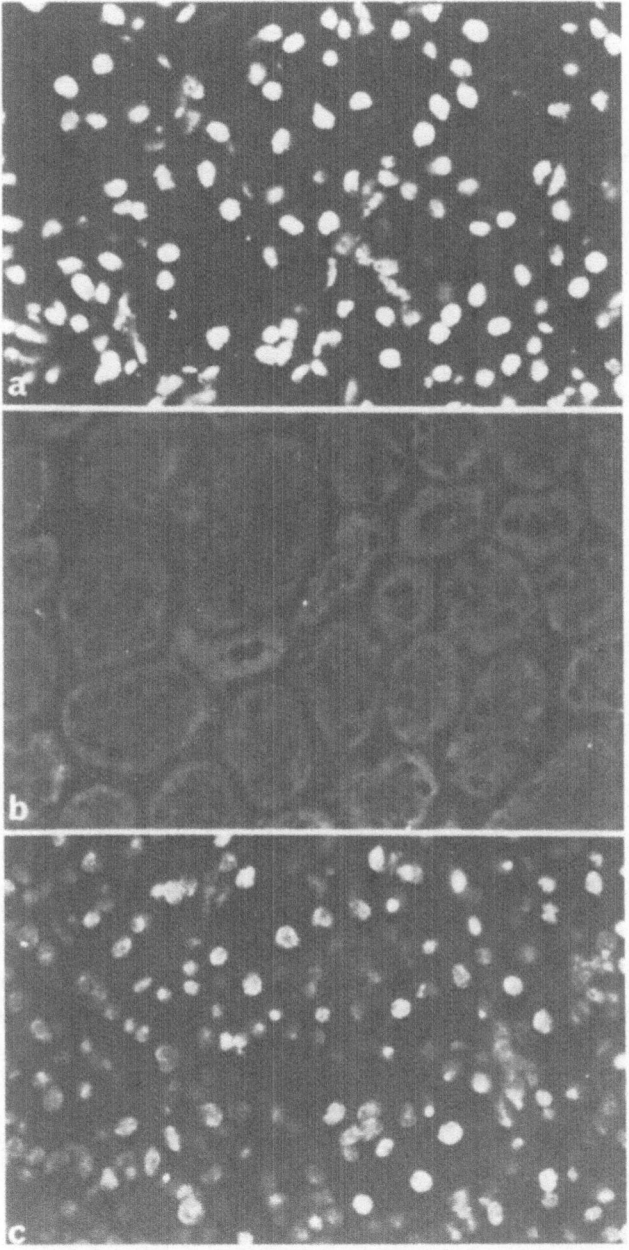

FIGURE 1. Three-step immunofluorescent technique for the detection of antihistone antibodies. (a) Acetone-fixed mouse kidney section reacted with serum of patient with drug-induced antinuclear antibodies. (b) Tissue section extracted with 0.1 N HCl and reacted with patient's serum. No nuclear staining was observed. (c) Tissue section extracted with 0.1 N HCl, reconstituted with histones, and reacted with patient's serum. A clumpy pattern of nuclear staining was characteristic of a positive test for antihistone antibodies.

Table I. Procainamide-Induced Antinuclear Antibodies

Patient	Ig class	Titer on mouse kidney		
		Acetone fixed	0.1 N HCl extracted	Histone reconstituted
Ta	IgM	8192	<4	4096
	IgG	4096	<4	512
No	IgM	4096	<4	2048
	IgG	1024	<4	<4
Ga	IgM	2048	<4	2048
	IgG	2048	<4	1024
Er	IgM	2048	<4	1024
	IgG	4096	<4	512
Wh	IgM	512	<4	16
	IgG	4096	<4	2048
Gu	IgM	512	<4	256
	IgG	2048	<4	256
Mo	IgM	512	<4	256
	IgG	1024	<4	256
Om	IgM	512	<4	64
	IgG	32	<4	<4
Pr	IgM	256	<4	<4
	IgG	512	<4	64
Ho	IgM	64	<4	<4
	IgG	2048	<4	2048

serum reacted with acid-extracted nuclei. However, all sera reacted with histone-reconstituted nuclei, demonstrating antihistone antibodies of the IgM and/or IgG class.

The ANA titers of Hy sera on acetone-fixed mouse kidney sections were considerably lower than those of Pr sera (Table II). Only two of ten sera had titers ≥ 512 for IgM and IgG ANA. The GMTs of Hy-induced IgM (174) and IgG (151) were significantly less than those of Pr sera, $p < 0.05$ and <0.01, respectively. Furthermore, no Hy serum reacted with acid-extracted nuclei or with nuclei reconstituted with histones. These results indicate a clear difference in the ANA induced by these two drugs.

In further experiments, a solid-phase radioimmunoassay was used to define more precisely the antibody specificities in the Pr and Hy sera. Sera were tested at a dilution of 1:100 on polystyrene tubes that were coated with calf thymus chromatin or isolated calf thymus histones. The detecting reagents were ^{125}I-labeled antibodies to human IgM and IgG. Initial studies involved tests of Pr and Hy sera using tubes coated with chromatin. As shown in Fig. 2, antibodies induced by Pr were substantially more reactive with chromatin than were antibodies induced by Hy. The mean antichromatin activity in Pr sera was significantly greater than that in Hy sera for both IgM (log mean cpm = 3.4 ± 0.4 vs 2.6 ± 0.5) and IgG (log mean cpm = 3.1 ± 0.6 vs 2.0 ± 0.4), $p < 0.05$ and <0.01, respectively. Thus, a heightened reactivity of Pr sera was observed in tests with chromatin as determined by radioimmunoassay and with mouse kidney nuclei as determined by indirect immunofluorescence.

Sera were then tested directly for antibodies to histones by solid-phase radioimmunoassay using polystyrene tubes coated with histone H1 or with mixtures of histones H2A and H2B or histones H3 and H4. Histone mixtures were prepared under conditions of high ionic strength which favor the formation of histone complexes H2A–H2B and H3–H4 (Isenberg, 1974). The results are shown in Table III. The predominant antihistone activity in the Pr sera was to the H2A–H2B complex. Anti-H2A–H2B activity in Pr sera was significantly greater than that in Hy sera for both IgM (mean cpm = 2348 vs 478) and IgG (mean cpm = 1212 vs. 151), $p < 0.01$ and <0.01, respectively. The Pr sera also demonstrated significantly higher anti-H1 activity than Hy sera for IgM, $p < 0.05$, but not for IgG. Finally, there was no difference between the anti-H3–H4 activity in Pr and Hy sera for both the IgM and IgG class.

The elevated anti-H2A–H2B activity in Pr sera prompted us to determine

TABLE II. Hydralazine-Induced Antinuclear Antibodies

Patient	Ig class	Titer on mouse kidney		
		Acetone fixed	0.1 N HCl extracted	Histone reconstituted
Bo	IgM	1024	<4	<4
	IgG	256	<4	<4
Ma	IgM	512	<4	<4
	IgG	256	<4	<4
Mi	IgM	256	<4	<4
	IgG	128	<4	<4
Wh	IgM	256	<4	<4
	IgG	128	<4	<4
Jo	IgM	256	<4	<4
	IgG	128	<4	<4
Ly	IgM	256	<4	<4
	IgG	128	<4	<4
Ba	IgM	128	<4	<4
	IgG	128	<4	<4
Kl	IgM	128	<4	<4
	IgG	4	<4	<4
Ra	IgM	64	<4	<4
	IgG	2048	<4	<4
We	IgM	16	<4	<4
	IgG	512	<4	<4

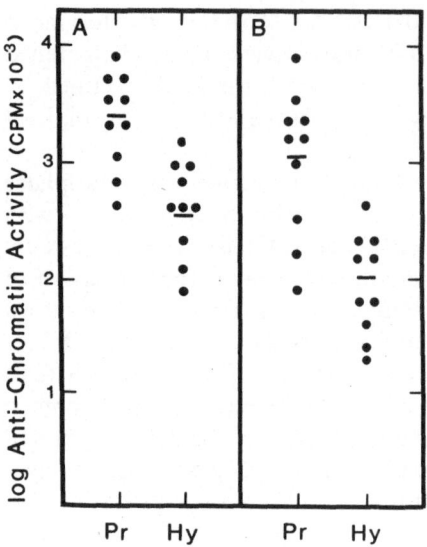

FIGURE 2. Activities of procainamide (Pr)- and hydralazine (Hy)-induced antibodies to calf thymus chromatin. Antibodies were detected with ^{125}I-labeled anti-human IgM (A) or anti-human IgG (B) by solid-phase radioimmunoassay using chromatin-coated tubes. Mean log antichromatin activities are indicated by horizontal bars.

whether this activity was related to the capacity of Pr but not Hy sera to react with histone-reconstituted nuclei in the three-step immunofluorescent test. The relationship between anti-H2A–H2B activity and histone-reconstitution immuno-fluorescence titers of Pr sera is shown in Fig. 3. The Pr sera with anti-H2A–H2B activity $\geq$800 cpm for IgM (Fig. 3A) and $\geq$650 cpm for IgG (Fig. 3B) reacted with histone-reconstituted nuclei to a titer $\geq$4. Linear regression analysis revealed a significant correlation between activity to the H2A–H2B complex and the titer on histone-reconstituted nuclei for both IgM and IgG, $p < 0.001$ and <0.01, respectively. The Pr sera with IgM anti-H2A–H2B activity $<$800 cpm did not react with histone-reconstituted nuclei. Similarly, no reactions on histone-reconstituted nuclei were obtained with Pr sera that had IgG anti-H2A–H2B activity $<$650 cpm.

Anti-H2A–H2B activities in Hy sera are also plotted in Fig. 3. It can be seen that the negligible anti-H2A–H2B activity in most sera correlated with a lack of reactivity with histone-reconstituted nuclei. However, there were four sera that had

TABLE III. Activities of Antihistone Antibodies Induced by Procainamide (Pr) and Hydralazine (Hy)

Ig class	Drug	Mean antibody activity to histones, cpm		
		H1	H2A–H2B	H3–H4
IgM	Pr	741[a]	2348[b]	805
	Hy	141[a]	478[b]	750
IgG	Pr	225	1212[b]	0
	Hy	167	151[b]	0

[a] $p < 0.05$.
[b] $p < 0.01$.

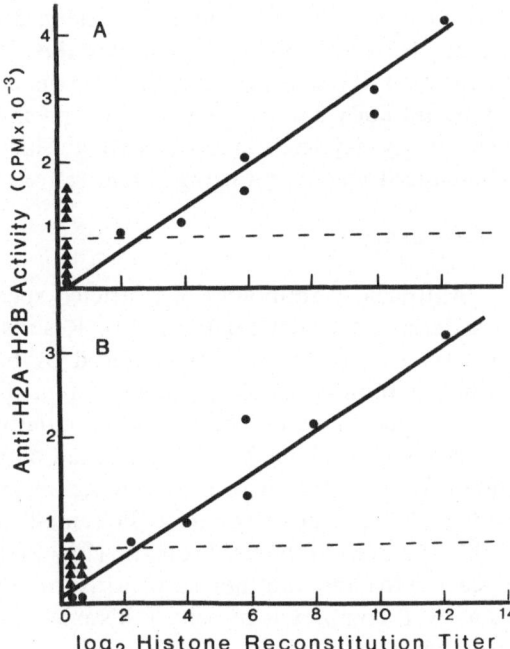

FIGURE 3. Relationship between anti-H2A–H2B activity in solid-phase assay and titer on histone-reconstituted nuclei in the immunofluorescent assay. Sera were from patients treated with procainamide (●) or hydralazine (▲). The correlation coefficient for IgG antibodies (B) was 0.86, $p < 0.01$, and for IgM antibodies (A) was 0.96, $p < 0.001$. The horizontal broken line separates those procainamide sera that were positive in the immunofluorescent assay from those that were negative.

IgM anti-H2A-H2B activities >800 cpm (Fig. 3A) and one serum with IgG anti-H2A-H2B activity >650 cpm (Fig. 3B). Therefore, insufficient levels of anti-H2A-H2B activity may account for the failure of most but not all Hy sera to react in the immunofluorescent assay.

Significant differences in the reactivities of Pr and Hy sera for individual histones were revealed by solid-phase radioimmunoassay. Antibody determinations were performed on tubes coated with individual histones H2A, H2B, or H3 as well as with tubes coated with the H2A–H2B complex. The results are presented in Table IV. Differences in antibody specificity were most readily apparent for antihistone antibodies of IgM class. The Pr-induced IgM antibodies reacted strongly with the H2A–H2B complex but not with individual histones H2A, H2B,

TABLE IV. Activities of Antihistone Antibodies Induced by Procainamide (Pr) and Hydralazine (Hy)

Ig class	Drug	Mean antibody activity to histones, cpm			
		H2A	H2B	H2A–H2B	H3
IgM	Pr	20	2	1475	65
	Hy	2114	193	473	2435
IgG	Pr	1462	230	3705	178
	Hy	108	82	346	167

or H3. Conversely, IgM antibodies induced by Hy reacted well with individual histones H2A and H3 but poorly with the H2A–H2B complex. This preferential reactivity of Hy sera for individual versus complexed histones is evidence for a unique antibody population that is not present in substantial amounts in Pr sera. Interestingly, Hy sera did not contain antihistone antibodies of the IgG class with an enhanced reactivity for individual histones H2A and H3.

3. Discussion

Antinuclear antibodies of restricted specificity for histones have previously been described in drug-induced LE. Results of the present study indicate that (1) antibodies to histones are induced by both Pr and Hy and (2) antihistone antibodies induced by Pr are of a different specificity from those induced by Hy.

The predominant antihistone specificity in Pr sera is for the H2A–H2B complex as determined by solid-phase radioimmunoassay (Tables III and IV). This finding is consistent with a heightened reactivity of Pr sera for substrates that contained this histone complex (Felsenfeld, 1978), such as mouse kidney nuclei (Table I) and calf thymus chromatin (Fig. 2). Furthermore, the capacity of Pr sera to react in the immunofluorescent assay for antihistone antibodies suggests that the H2A–H2B complex is present in histone-reconstituted nuclei in an antigenically active form.

Of interest, most Pr sera have IgM ANA in relatively high titer. This finding is in contrast to sera containing antibodies to other nuclear antigens, such as Sm and nuclear ribonucleoprotein. The IgG ANA in these sera are generally in higher titer than IgM ANA. The substantial IgM antibody activity in Pr sera is suggestive of an ongoing primary immune response, possibly to a T-independent histone antigen.

The failure of Hy sera to react in the immunofluorescent assay was not due to an absence of antihistone antibodies in these sera. Preliminary studies by radioimmunoassay demonstrate substantial antibody activity in Hy sera for individual histones but not for histones present in complexed form. Evidently, antibodies directed to antigenic determinants on individual histones H2A and H3 are not readily demonstrable on histone-reconstituted nuclei by the immunofluorescent assay. The low reactivity of Hy-induced antibodies for nuclei (Table II) and chromatin (Fig. 2) is also consistent with a preferential reactivity for individual histones rather than for histone complexes.

The mechanism by which drugs induce antihistone antibodies is not known. Evidence for an interaction of Hy and Pr with DNA (Eldredge et al., 1974) or histone–DNA complexes (Tan, 1968) has been obtained in vitro. Other studies (Kirtland et al., 1980) have demonstrated an inhibitory effect of drugs on the function of suppressor T lymphocytes. The extent to which these interactions contribute to the induction of antihistone antibodies in drug-induced LE remains to be determined.

Acknowledgment

This work was supported in part by National Institutes of Health grants #AM20705 and #AI07166.

References

Alarcon-Segovia, D., Wakin, K. G., Worthington, J. W., and Ward, L. E., 1967, Clinical and experimental studies on the hydralazine syndrome and its relationship to systemic lupus erythematosus, *Medicine* **46**:1.

Blomgren, S. E., 1973, Drug-induced lupus erythematosus, *Semin. Hematol.* **10**:345.

Blomgren, S. E., Condemi, J. J., Bignall, M. C., and Vaughan, J. H., 1967, Antinuclear antibody induced by procainamide: A prospective study, *N. Engl. J. Med.* **281**:64.

Blomgren, S. E., Condemi, J. J., and Vaughn, J. H., 1972, Procainamide-induced lupus erythematosus: Clinical and laboratory observations, *Am. J. Med.* **52**:338.

Condemi, J. J., Moore-Jones, D., Vaughan, J. H., and Perry, H. M., 1967, Antinuclear antibodies following hydralazine toxicity, *N. Engl. J. Med.* **276**:486.

Eldredge, N. T., van B. Robertson, W., and Miller, J. J., 1974, The interaction of lupus-inducing drugs with deoxyribonucleic acid, *Clin. Immunol. Immunopathol.* **3**:268.

Felsenfeld, G., 1978, Chromatin, *Nature* **271**:115.

Fritzler, M. J., and Tan, E. M., 1978, Antibodies to histones in drug-induced and idiopathic lupus erythematosus, *J. Clin. Invest.* **21**:560.

Isenberg, I., 1974, Histones, *Annu. Rev. Biochem.* **48**:159.

Kirtland, H. H., Mohler, D. N., and Horwitz, D. A., 1980, Methyldopa inhibition of suppressor-lymphocyte function, *N. Engl. J. Med.* **302**:825.

Lee, S. L., and Chase, P. H., 1975, Drug-induced systemic lupus erythematosus: A critical review, *Semin. Arth. Rheum.* **5**:83.

Molina, J., Dubois, E. L., Bilitch, M., Bland, S. L., and Friou, G. J., 1969, Procainamide-induced serologic changes in asymptomatic patients, *Arth. Rheum.* **12**:608.

Perry, H. M., Jr., 1973, Late toxicity to hydralazine resembling systemic lupus erythematosus or rheumatoid arthritis, *Am. J. Med.* **54**:58.

Rubin, R. L., Joslin, F. G., and Tan, E. M., 1982, A solid-phase radioimmunoassay for anti-histone antibodies in human sera: comparison with an immunofluorescence assay, *Scand. J. Immunol.* **15**:63.

Tan, E. M., 1968, The influence of hydralazine on antigen–antibody reactions, *Arth. Rheum.* **11**:511.

Tan, E. M., 1974, Drug-induced autoimmune diseases, *Fed. Proc.* **33**:1894.

Tan, E. M., Robinson, J., and Robitaille, P., 1976, Studies on antibodies to histones by immunofluorescence, *Scand. J. Immunol.* **5**:811.

Weinstein, A., 1980, Drug-induced systemic lupus erythematosus, *Prog. Clin. Immunol.* **4**:1.

Woosley, R. L., Drayer, D. E., Reidenberg, M. M., Nies, A. S., Carr, K., and Oates, J. A., 1978, Effect of acetylator phenotype on the rate at which procainamide induces antinuclear antibodies and the lupus syndrome, *N. Engl. J. Med.* **298**:1157.

33

Drug- and Toxin-Induced Nephritides
Anti-Kidney Antibody and Immune Complex Mediation

CURTIS B. WILSON

1. Introduction

Current evidence suggests that drugs and toxins can damage the kidney through humoral immunologic mechanisms. These pathways of nephritogenic humoral immune injury will be described in this chapter, and several ways in which drugs and toxins potentially play a role will be outlined. Particular attention will be given to hydrocarbons as a toxic agent that has been reported to induce at least one kind of anti-kidney antibody response, namely that directed toward the renal basement membranes.

2. Antibody Mechanisms of Immune Renal Injury

To begin unraveling the role of drugs and toxins as inducers of or contributors to nephritogenic humoral immune responses in the glomerulus and tubulointer-

---

CURTIS B. WILSON • Department of Immunopathology, Research Institute of Scripps Clinic, La Jolla, California 92037.

stitial tissues, it is necessary to have a basic understanding of the ways that antibodies can damage the kidney. As recently reviewed, studies first in animals and over the past 25 years in humans have defined two basically different ways in which antibodies can incite renal damage (Wilson and Dixon, 1981). For discussion, these processes are most easily divided according to the physical nature and location of the antigens involved in the particular nephritogenic immune attack in question. These antigens are first categorized as soluble (in the body fluids) or insoluble (tissue-fixed), and the soluble antigens are then subdivided into either intravascular or extravascular groups, according to the location where they are available to interact with antibody. Antibodies reacting with soluble antigens in the vascular compartment cause formation and deposition of immune complexes within vessels. This mechanism, recognized by studies of serum sickness in animals (Dixon, 1963), is well known. A growing number of antigens has been implicated in nephritogenic circulating immune complex formation. These antigens include drugs and other exogenous materials, predominantly from infectious agents, as well as endogenous elements such as nuclear proteins and antigens from neoplasms (Wilson and Dixon, 1981).

Local immune complex formation, as in the Arthus reaction (Cochrane and Janoff, 1974) or such disorders as thyroiditis (Clagett et al., 1974), can occur when antigens in extravascular fluids surrounding the site of injection or production react with antibody. This mechanism appears to operate in some forms of tubulointerstitial nephritis. Rabbits injected with renal tubular antigens form immune complex deposits in the potential extravascular space between the tubular cell plasma membrane and the tubular basement membrane (TBM) (Unanue et al., 1967). Rats immunized against the Tamm-Horsfall protein develop tubulo-interstitial nephritis apparently by a similar mechanism (Hoyer, 1980).

As with the soluble antigens, the insoluble or tissue-fixed antigens can be subdivided into two categories. In one, structural components of the kidney serve as antigens; in the other, foreign or extrarenal materials that become trapped or "planted" in the kidney are the antigenic source. Of the kidney's nephritogenic structural antigens, the glomerular basement membrane (GBM) and TBM are the best understood from research both in experimental models and in human immunologic renal disease [reviewed in Wilson and Dixon (1981)]. In humans, anti-GBM antibodies cause the majority of cases of Goodpasture's syndrome. Anti-GBM antibodies can also cause glomerulonephritis, most often rapidly progressive, with extensive glomerular crescent formation on histologic examination. The glomerulonephritis can also be milder and sometimes self-remitting. Occasionally, lung injury is seen in the absence of clinically overt glomerular damage, with a clinical presentation of idiopathic pulmonary hemosiderosis.

The potential for other structural, nonclassic GBM or TBM antigens to serve as targets for nephritogenic immunologic attack is just beginning to be appreciated (Wilson, 1979, 1981; Couser and Salant, 1980). The same is true of planted antigens, as we see from evidence for these processes now developing in animal models. Brief mention of these alternative systems of immune injury is warranted to familiarize the reader with them as new concepts. By routine immunofluorescence study, the latter mechanisms may not be distinguishable from those of classic granular immune complex deposition, because of the irregular distribution of the reactive insoluble antigens. Whenever granular glomerular deposits of immuno-globulin are encountered in patients with drug or toxic exposure, physicians seeking the cause should consider alternatives to the circulating immune complex mechanism that would include antibody reactions with irregularly deposited nonclassic GBM glomerular capillary wall antigens or randomly trapped foreign antigens as well.

At least two models of nonclassic GBM glomerular capillary wall antigens have been studied in animals. New Zealand White rabbits may develop glomerulonephritis spontaneously (Verroust et al., 1974). The glomerular lesion is characterized by irregular granular and segmental deposits of IgG and C3 along the GBM. By electron microscopy electron-dense deposits are found in a nearly continuous, sawtooth-like pattern at the subepithelial aspect of the GBM. The IgG can be recovered by acid elution of washed kidney homogenates. When the eluate is tested by indirect immunofluorescence on normal rabbit kidney sections, reactivity with the glomerular capillary wall results in an irregular, somewhat fuzzy pattern of deposition. When evaluated at the ultrastructural level with immunoperoxidase techniques, one finds that the predominant site of reactivity is in the area of the epithelial cell foot process attachment to the GBM (Neale and Wilson, 1978). The reactivity cannot be related to the classic GBM antigen by radioimmunoassay or absorption experiments. Thus, these observations indicate that antibodies associated with glomerulonephritis can react with other than the classic nephritogenic GBM antigen.

In a second, more complicated model, antibodies have been identified that are reactive with yet another glomerular capillary wall component. This is the Heymann nephritis model induced in rats by injecting suspensions of rat kidney tissue cleared of large particulate elements (Heymann et al., 1959). The antigen is termed Fx1A (Edgington et al., 1967). The lesion produced is morphologically similar to that of human membranous glomerulonephritis and has fine granular accumulations of IgG and C3 along the GBM, with subepithelial electron-dense deposits. The lesion, once thought to be caused by glomerular deposition of tubular brush border antigen–antibody complexes (Edgington et al., 1967), is now known, as shown by recent elution studies, to have antibodies reactive with glomerular capillary wall antigens as well (Neale and Wilson, 1979). The sites of eluate reactivity are small, granular areas, most easily visualized along the subepithelial aspect of the GBM on ultramicroscopic examination. Studies using heterologous anti-Fx1A antibodies in rats have shown binding to similar sites after selective perfusion that excluded any possible deposition of circulating immune complexes (Couser et al., 1978; Van Damme et al., 1978). Other additional contribution of circulating immune complex formation in the active model remains unresolved. We have identified occasional eluates from human glomerulonephritic kidneys that react with human kidney sections in a pattern similar to that of the models just discussed, indicating the likelihood of uncovering human counterparts of these nephritogenic immune systems and emphasizing the need to consider such mechanisms when evaluating the roles of drugs and toxins in immune glomerular injury.

The planted antigen mechanism of immune glomerular injury was first identified in the autologous phase of experimental anti-GBM antibody-induced glomerular injury. This phase occurs when the recipient of heterologous anti-GBM antibody produces antibodies reactive with the heterologous immunoglobulin bound to its glomeruli. Circulating immune complex deposits that have become trapped in glomeruli can also serve as a source of planted antigen (or antibody) for continued interaction with antibody (or antigen) from the circulation (Wilson and Dixon, 1981). The moment-to-moment composition of immune complex deposits is determined by the continuing dynamic equilibrium of the immune complex formation. The deposits can actually be dissolved by creating a deliberate state of huge antigen excess (Wilson and Dixon, 1971). Other elements of previously trapped immunologic reactants also could potentially serve as planted antigens for interaction with rheumatoid factors, anti-idiotypic antibodies, and immunoconglutinins. Materials that for physicochemical reasons are capable of binding to the glomerular capillary wall and potentially other renal structures could be involved. We have shown that the plant lectin concanavalin A can bind to the carbohydrate of the glomerular capillary wall and serve as a nephritogenic planted antigen for interaction with passively administered or actively formed anti-concanavalin A antibody (Golbus and Wilson, 1979). The polyanionic charged sites of the glomerular capillary wall have been shown to interact with cationic dyes and drugs such as protamine (Seiler et al., 1977), again indicating the potential for trapping of potentially antigenic material. Other substances, such as DNA, have been suggested to bind to the GBM (Izui et al., 1976), and material taken out of the circulation by the mesangium can also be a target of nephritogenic immune attack (Mauer et al., 1972).

3. Relationship of Drugs and Toxins to Nephritogenic Immune Responses

As seen from the foregoing, a drug or toxin could potentially engender immune glomerular damage in a great many ways. The association of renal injury with drugs or toxins is not infrequent; however, in many instances the mechanisms involved remain speculative. Probably the most straightforward mechanism is for the drug or toxin to serve as a circulating antigen that forms immune complexes with subsequent vascular deposition. Foreign serum proteins from antitoxins or anti-lymphocyte globulins act as good examples of such reactions leading to serum sickness. Repeated injections of toxoids have been associated with apparent immune complex-induced glomerulonephritis in one patient (Boulton-Jones et al., 1974). Antibiotics also could be involved. For example, we have detected carbinicillin-containing circulating immune complexes in one individual with carbinicillin-associated glomerulonephritis (Neale and Wilson, unpublished observations). Illicit drug use is complicated by glomerular injury with the immunopathologic chracteristics of immune complex-induced glomerulonephritis (Rao et al., 1977). The drug or its contaminants may lead to immune complex formation or can potentially interact with the kidney in other ways. In some instances, glomerulonephritis can be related to immune complex formation with antigens from infections such as subacute bacterial endocarditis or hepatitis, common in such patients. Drugs can also induce a systemic lupus erythematosus-like disease having multiple features of circulating immune complex disorders (see Chapter 32, this volume).

Such diverse therapeutic agents as mercury, tridione, gold, and penicillamine have been temporally associated with glomerulonephritis and irregular granular deposits of immunoglobulin and complement, consistent with but not diagnostic of circulating immune complex disease [reviewed in Wilson and Dixon (1981)]. Experimental and clinical studies have not clearly implicated these drugs in the suspected immune complex deposits, opening the door for wide-ranging speculation about the ways in which such agents might generate nephritogenic immune reactions. In diseases such as rheumatoid arthritis, in which immune complexes are ever present, there is the possibility that a drug such as gold may increase immune complex formation or impair handling of immune complexes, thereby enhancing glomerular localization. The drug could also combine with tissue components to act as a hapten, to release immunogenic tissue antigens, or to induce neoantigens with subsequent stimulation of immune complex disease or local nephritogenic antigen-antibody reactions. For example, autoantibody to "ubiquitous tissue antigen" has been noted in one patient with gold nephropathy (Palosuo *et al.*, 1976).

Since the nephritogenicity of immune complexes depends on their relative antigen–antibody ratios, the effect of drugs on ongoing, potentially nephritogenic immune reactions should be considered. The drug itself could promote the release of antigens, thereby enhancing glomerular injury, as has been noted in thyroiditis of rabbits and humans. In one such patient studied in our laboratory, two episodes of serum sickness-like glomerulonephritis followed administration of radioactive iodine to manage hyperthyroidism in a patient with thyroiditis (Ploth *et al.*, 1978). Later, thyroglobulin-containing immune complexes were found in the glomeruli. The effect of immunosuppression must also be considered as a possibility for shifting the relative antigen–antibody ratios to favor nephritogenic immune complex formation. Immunosuppressive therapy could additionally influence handling of circulating immune complexes and certainly carries with it an increased likelihood of infection and associated immune complex formation. Drugs or toxins may also damage tissues, leading to diseases that in turn could be complicated by immune complex injury. An example of this would be alcoholic cirrhosis, which was reported as accompanied by glomerulonephritis characterized by striking IgA accumulation, predominantly in the mesangial area of the glomerulus (Berger *et al.*, 1977).

Drugs and toxins may directly induce antibody responses to insoluble structural renal antigens and could serve as planted antigens as well. The clearest example is the induction by methicillin of anti-TBM antibodies along with methicillin-related tubulointerstitial nephritis in some individuals. A patient with such tubulointerstitial nephritis whom we studied by immunofluorescence (Border *et al.*, 1974) had linear deposits of IgG along his TBM. Later detection of circulating anti-TBM antibodies confirmed the significance of the immunofluorescent findings. Additional studies showed that dimethoxyphenylpenicilloyl (DPO), a breakdown product of methicillin, was bound to his TBM. This suggested that the DPO might have functioned as a hapten, which when bound to the TBM could induce antibodies to the TBM "carrier" portion of the DPO–TBM hapten–carrier conjugate. Nevertheless, the occurrence of anti-TBM antibodies in methicillin-associated tubulointerstitial nephritis is low, indicating that anti-TBM antibodies are not the common immunopathologic mechanism of this condition. DPO is not found in the TBM of all patients receiving methicillin, and genetic differences in the immune response may also contribute to the outcome. Anti-TBM antibodies have been identified in phenytoin toxicity, with many similarities to the methicillin mechanisms just noted (Hyman *et al.,* 1978). As will be discussed in more detail, other toxins, and particularly hydrocarbons, have been associated with anti-GBM antibody production, although not clearly implicated as its cause.

Therapeutic procedures such as radiation may cause anti-GBM antibody formation, considering that anti-GBM antibody-induced diseases developed in three patients from our series who had been treated for Hodgkin's disease. The potential for lymphoid stroma to induce anti-GBM antibodies was previously noted when heterologous anti-human lymphocyte antibodies used therapeutically for renal transplant patients were found to be contaminated by anti-GBM antibodies related to inclusion of lymphoid stroma in the inoculum (Wilson *et al.,* 1971). This observation emphasizes the possibility of nephrotoxic immune injury by administration of therapeutic antibody preparations inadvertently contaminated with kidney-reactive antibodies. Drugs and toxic agents could also act by making normally sequestered antigen accessible to antibody. This possibility is suggested by the recent observation that oxygen therapy enchanced the binding of passively administered anti-basement membrane antibody to the lung basement membrane of rabbits (Jennings *et al.,* 1981).

The possibility that drugs can interact with other nonclassic basement membrane glomerular capillary wall or tubulointerstitial antigens, thereby exposing or releasing antigens or inducing neoantigens, leading to nephritogenic immune reactions, is also real. The potential for drugs or toxic materials to become trapped or planted in the glomerulus or other renal sites, thereby serving as sites for nephritogenic antibody attack, is another attractive possibility for investigations into the mechanisms whereby such agents can lead to immune injury.

4. Noxious Stimuli in the Induction of Anti-GBM Antibodies

Factors responsible for the induction of autoimmune anti-GBM antibody production are poorly understood. Urine (and serum) contains antigens cross-reactive with the GBM and capable of inducing anti-GBM antibodies in experimental animals (Lerner and Dixon, 1968). GBM antigens unique to some individuals can be responsible for induction of anti-GBM antibodies, e.g., after transplantation of GBM antigen-positive kidneys into GBM antigen-negative individuals, as sometimes occurs in the Alport's syndrome type of hereditary nephritis (Wilson, 1980). Preceding immunologic or other forms of injury to the kidney sometimes antedates the onset of anti-GBM antibody production, as does influenza A2 infection in occasional patients [reviewed in Wilson and Dixon (1981)]. As noted earlier, methicillin and phenytoin therapies have also been associated occasionally with anti-TBM antibody formation.

After a few scattered reports linked hydrocarbon exposure with Goodpasture's syndrome, Beirne and Brennan (1972) detailed exposures in six of eight patients with Goodpasture's syndrome and/or evidence of anti-basement membrane antibodies to solvents used in degreasing, painting, and refinishing, or contained in jet fuel. Subsequently, these investigators reported hydrocarbon solvent exposure in 11 of 16 patients with immunopathologic evidence of anti-basement membrane disease, presenting either as Goodpasture's syndrome or rapidly progressive glomerulonephritis (Beirne et al., 1977). Hydrocarbon exposure may also worsen poststreptococcal glomerulonephritis (Ravnskov, 1978) and, in one study, was noted more frequently in patients with proliferative glomerulonephritis than in a variety of other renal disorders (Zimmerman et al., 1975). Moreover, membranous glomerulopathy has been related to hydrocarbon solvent exposures of four patients (Ehrenreich et al., 1977). Both benzine and N,N'-diacetylbenezidine have been reported as the cause of glomerulonephritis in experimental animals; however, immunopathologic studies are lacking (Klavis and Drommer, 1970; Harman, 1971).

In our own experience, based on information available to us from physicians caring for 407 patients with anti-basement membrane disease (253 having a Goodpasture's presentation), a history of unusual hydrocarbon exposure, although sometimes present, is not especially common. Only about 4% of the patients with Goodpasture's syndrome had major exposures to hydrocarbon fuel, and an additional 8% had lesser exposures to fuel, paint, cleaning solvents, or toluene. Less complete histories were available from the patients presenting with kidney involvement only, but hydrocarbon exposure of any sort was noted in only 6%. It should be stressed that detailed histories obtained by a trained toxicologist were rarely available in this group, and no adequate control population is available to judge the relative frequency of similar exposures. Based on the prevalance of hydrocarbons in our environment, the frequency with which anti-basement membrane antibodies may be induced by hydrocarbon exposure must be small.

We have begun to examine this question by testing for anti-basement membrane antibodies in rabbits, rats, and mice that have been exposed to 50–3000 parts per million of unleaded gasoline vapors 5 days per week for periods of 6 weeks to 2 years. Radioimmunoassays employing human nephritogenic GBM antigens have not detected any significant reactivity in sera from these animals. Additional studies in progress include immunofluorescence and elution studies of renal and lung tissue, as well as further evaluation of circulating antibodies by indirect immunofluorescence. In addition, no difference in anti-basement membrane antibody production was noted in groups of rabbits purposefully immunized with basement membrane fractions, whether concomitantly exposed to hydrocarbon solvent vapors or not.

The historical data available, as well as preliminary results from animal exposure experiments, do not suggest that hydrocarbon solvent exposure, in particular gasoline, is a major inducer of anti-GBM antibodies. It is not possible to confirm or refute its possible inductive role in isolated cases, however. Theoretically, hydrocarbons could enhance the pathogenicity of anti-basement membrane antibodies formed for other reasons by exposing normally sequestered antigens, perhaps in the lung, thereby making the underlying anti-basement membrane antibody process clinically apparent. Until more and firmer information is available, hydrocarbon solvents will continue to be included in the list of potential inducers of anti-basement membrane antibody responses.

5. Conclusions

In summary, drugs and toxins can be involved with the various recognized forms of humoral immune renal injury. They can serve as antigens for immune complex formation. They may also alter an ongoing immune complex disease or induce one. These agents may otherwise play a role in inducing antibodies reactive with kidney antigens, either the well-known basement membrane antigens or potentially with the newly recognized nonclassic basement membrane antigens. Drugs and toxins potentially also serve as haptens, planted antigens, inducers of neoantigens, or exposers of sequestered antigens, in all cases capable of leading to nephritogenic immune responses. The roles of drugs and toxins, as they may induce renal injury via the immediate hypersensitivity and cellular limbs of the immune system, are topics of other chapters. Elements of these systems, particularly the cellular immune system, could act in concert with the humoral mechanisms described throughout this chapter.

Acknowledgments

This is publication No. 2459 from the Department of Immunopathology, Research Institute of Scripps Clinic, La Jolla, California. This work was supported in part by United States Public Health Service Grants AM-20043, AM-18626, and Al-07007, and Biomedical Research Support Grant RRO-5514.

References

Beirne, G. J., and Brennan, J. T., 1972, Glomerulonephritis associated with hydrocarbon solvents: Mediated by antiglomerular basement membrane antibody, *Arch. Environ. Health* **25**:365.

Beirne, G. J., Wagnild, J. P., Zimmerman, S. W., Macken, P. D., and Burkholder, P. M., 1977, Idiopathic crescentic glomerulonephritis, *Medicine* **56**:349.

Berger, J., Yaneva, H., and Nabarra, B., 1977, Glomerular changes in patients with cirrhosis of the liver, *Adv. Nephrol.* **7**:3.

Border, W. A., Lehman, D. H., Egan, J. D., Sass, H. J., Glode, J. E., and Wilson, C. B., 1974, Anti-tubular basement-membrane antibodies in methicillin-associated interstitial nephritis, *N. Engl. J. Med.* **291**:381.

Boulton-Jones, J. M., Sissons, J. G. P., Naish, P. F., Evans, D. J., and Peters, D. K., 1974, Self-induced glomerulonephritis, *Br. Med. J.* **3**:387.

Clagett, J. A., Wilson, C. B., and Weigle, W. O., 1974, Interstitial immune complex thyroiditis in mice. The role of autoantibody to thyroglobulin, J. Exp. Med. 140:1439.

Cochrane, C. G., and Janoff, A., 1974, The Arthus reaction: A model of neutrophil and complement-mediated injury, in: The Inflammatory Process, 2nd ed. (B. W. Zweifach, L. Grant, and R. T. McCluskey, eds.), Academic Press, New York, Volume III, p. 85.

Couser, W. G., and Salant, D. J., 1980, In situ immune complex formation and glomerular injury, Editorial, Kidney Int. 17:1.

Couser, W. G., Steinmuller, D. R., Stilmant, M. M., Salant, D. J., and Lowenstein, L. M., 1978, Experimental glomerulonephritis in the isolated perfused rat kidney, J. Clin. Invest. 62:1275.

Dixon, F. J., 1963, The role of antigen–antibody complexes in disease, Harvey Lect. 58:21.

Edgington, T. S., Glassock, R. J., and Dixon, F. J., 1967, Autologous immune complex pathogenesis of experimental allergic glomerulonephritis, Science 155:1432.

Ehrenreich, T., Yunis, S. L., and Churg, J., 1977, Membranous nephropathy following exposure to volatile hydrocarbons, Environ. Res. 14:35.

Golbus, S. M., and Wilson, C. B., 1979, Experimental glomerulonephritis induced by in situ formation of immune complexes in glomerular capillary wall, Kidney Int. 16:148.

Harman, J. W., 1971, Chronic glomerulonephritis and the nephrotic syndrome induced in rats with N,N'-diacetylbenzidine, J. Pathol. 104:119.

Heymann, W., Hackel, D. B., Harwood, S., Wilson, S. G. F., and Hunter, J. L. P., 1959, Production of nephrotic syndrome in rats by Freund's adjuvants and rat kidney suspensions, Proc. Soc. Exp. Biol. Med. 100:660.

Hoyer, J. R., 1980, Tubulointerstitial immune complex nephritis in rats immunized with Tamm-Horsfall protein, Kidney Int. 17:284.

Hyman, L. R., Ballow, M., and Knieser, M. R., 1978, Diphenylhydantoin interstitial nephritis. Roles of cellular and humoral immunologic injury, J. Pediatr. 92:915.

Izui, S., Lambert, P. H., and Miescher, P. A., 1976, In vitro demonstration of a particular affinity of glomerular basement membrane and collagen for DNA. A possible basis for a local formation of DNA–anti-DNA complexes in systemic lupus erythematosus, J. Exp. Med. 144:428.

Jennings, L., Andres, G., and Brentjens, J., 1981, Experimental Goodpasture's disease (GPD) in the rabbit, Kidney Int. 19:183 (abstract).

Klavis, G., and Drommer, W., 1970, Goodpasture-syndrom und benzineinwirkung, Arch. Toxikol. 26:40.

Lerner, R. A., and Dixon, F. J., 1968, The induction of acute glomerulonephritis in rabbits with soluble antigens isolated from normal homologous and autologous urine, J. Immunol. 100:1277.

Mauer, S. M., Fish, A. J., Blau, E. B., and Michael, A. F., 1972, The glomerular mesangium. 1. Kinetic studies of macromolecular uptake in normal and nephrotic rats, J. Clin. Invest. 51:1092.

Neale, T. J., and Wilson, C. B., 1978, Non-GBM glomerular antigen in spontaneous nephritis in rabbits, Kidney Int. 14:715 (abstract).

Neale, T. J., and Wilson, C. B., 1979, Fixed glomerular antigen in Heymann's nephritis: Eluted antibody reactivity with normal rat glomeruli, Kidney Int. 16:799 (abstract).

Palosuo, T., Provost, T. T., and Milgrom, F., 1976, Gold nephropathy: Serologic data suggesting an immune complex disease, Clin. Exp. Immunol. 25:311.

Ploth, D. W., Fitz, A., Schnetzler, D., Seidenfeld, J., and Wilson, C. B., 1978, Thyroglobulin–antithyroglobulin immune complex glomerulonephritis complicating radioiodine therapy, Clin. Immunol. Immunopathol. 9:327.

Rao, T. K. S., Nicastri, A. D., and Friedman, E. A., 1977, Renal consequences of narcotic abuse, Adv. Nephrol. 7:261.

Ravnskov, U., 1978, Exposure to organic solvents—A missing link in poststreptococcal glomerulonephritis?, Acta. Med. Scand. 203:351.

Seiler, M. W., Rennke, H. G., Venkatachalam, M. A., and Cotran, R. S., 1977, Pathogenesis of polycation-induced alterations ("fusion") of glomerular epithelium, Lab. Invest. 36:48.

Unanue, E. R., Dixon, F. J., and Feldman, J. D., 1967, Experimental allergic glomerulonephritis induced in the rabbit with homologous renal antigens, J. Exp. Med. 125:163.

Van Damme, B. J. C., Fleuren, G. J., Bakker, W. W., Vernier, R. L., and Hoedemaeker, Ph. J., 1978, Experimental glomerulonephritis in the rat induced by antibodies directed against tubular antigens. V. Fixed glomerular antigens in the pathogenesis of heterologous immune complex glomerulonephritis, *Lab. Invest.* **38**:502.

Verroust, P. J., Wilson, C. B., and Dixon, F. J., 1974, Lack of nephritogenicity of systemic activation of the alternate complement pathway, *Kidney Int.* **6**:157.

Wilson, C. B., 1979, Immune reactions with antigens in or of the glomerulus, in: *Immunopathology* (F. Milgrom and B. Albini, eds.), Karger, Basel, p. 127.

Wilson, C. B., 1980, Individual and strain differences in renal basement membrane antigens, *Transpl. Proc.* **12**(Suppl. 1):69.

Wilson, C. B., Nephritogenic antibody mechanisms involving antigens within the glomerulus, 1981, *Immunol. Rev.* **55**:257.

Wilson, C. B., and Dixon, F. J., 1971, Quantitation of acute and chronic serum sickness in the rabbit, *J. Exp. Med.* **134**:7s.

Wilson, C. B., and Dixon, F. J., 1981, The renal response to immunological injury, in: *The Kidney*, 2nd ed. (B. M. Brenner, and F. C. Rector, Jr., eds.), Saunders, Philadelphia, Pennsylvania, Volume 1, p. 1237.

Wilson, C. B., Dixon, F. J., Fortner, J. G., and Cerilli, J., 1971, Glomerular basement membrane-reactive antibodies in anti-lymphocyte globulin, *J. Clin. Invest.* **50**:1525.

Zimmerman, S. W., Groehler, K., and Beirne, G. J., 1975, Hydrocarbon exposure and chronic glomerulonephritis, *Lancet* **2**:199.

34

Immunologic Mechanisms in Drug-Induced Acute Interstitial Nephritis

ROBERT T. McCLUSKEY and ATUL K. BHAN

1. General Features of Drug-Induced Acute Interstitial Nephritis

It is well established that certain therapeutic agents can cause acute interstitial nephritis (AIN),* which is characterized histologically by interstitial edema, leukocytic infiltration, and tubular cell damage (Baldwin *et al.*, 1968; McCluskey and Colvin, 1978). The infiltrate usually consists principally of mononuclear cells, most of which have the appearance of lymphocytes, but eosinophils are usually present, often in large numbers, and plasma cells, as well as small numbers of basophils, are frequently found. Tubules are often invaded by leukocytes and severe tubular damage may be seen. The glomeruli and blood vessels appear

---

*The term acute interstitial nephritis (AIN) is used here for convenience, although since tubules are also affected, the condition might more properly be called acute tubulointerstitial nephritis. Tubulointerstitial nephritis (TIN) is used here as a more general term, to refer to any form of tubulointerstitial nephritis, whether acute or chronic.

ROBERT T. McCLUSKEY and ATUL K. BHAN • Department of Pathology, Massachusetts General Hospital and Harvard Medical School, Boston, Massachusetts 02117.

normal. In most cases there are no characteristic immunofluorescence findings and distinctive ultrastructural features are not seen.

The drugs that have been most frequently incriminated are methicillin, ampicillin, rifampicin, phenindione, and sulfonamides (Mery and Morel-Maroger, 1976). With certain of these agents, especially methicillin and related antibiotics, there is considerable evidence that the renal lesions result from immunologic mechanisms rather than toxic effects. Thus, only a small percentage of patients taking the drug develop AIN and the reactions are not clearly dose-related. Furthermore, many of the patients with renal lesions have evidence of immunologically mediated reactivity to the drug, in the form of fever, rash, arthralgia, and eosinophilia. The lesions generally subside rapidly after discontinuance of the drug.

In many cases the clinical and histologic features are quite characteristic and the diagnosis of drug-induced immunologically mediated AIN can be made with confidence. In many patients, however, it is impossible to determine whether the renal damage is due to a toxic effect of a drug, an immunologic reaction initiated by a drug, an infectious process, or some other, unknown cause. Generally, agents causing direct injury to tubular cells produce less interstitial edema and less leukocytic accumulation (and in particular fewer eosinophils) than do immunologically mediated reactions, and certain toxic agents produce characteristic ultrastructural tubular changes, but the criteria for distinguishing between these two forms of renal damage are not precise. Furthermore, it is possible that in some instances both direct toxic and immunologic mechanisms are involved.

Certain infections, in particular scarlet fever and diphtheria (Councilman 1898; Kimmelstiel, 1930), were sometimes observed to be associated with AIN (apparently not due to bacterial invasion of the kidney). The lesions were described as showing conspicuous interstitial mononuclear cell infiltration. Although it is possible that some patients who develop AIN while on antibiotic treatment for an infectious disease are suffering from a complication of the infection rather than of the therapy, it is unlikely that this accounts for very many cases. For one thing, AIN has been observed in patients treated for a variety of infections, most of which were not associated with renal damage in the days before antibiotics. Furthermore, it is clear that methicillin, at least, can induce AIN in the absence of infection, since certain patients receiving this antibiotic prophylactically have been observed to develop typical renal lesions (Olsen and Asklund, 1976).

2. Immunologic Mechanisms in Tubulointerstitial Nephritis (TIN): General Considerations

The immunologic mechanisms responsible for drug-induced AIN have not been elucidated. Progress has been hampered by the lack of a suitable experimental model. The closest model is one described by Domoto *et al.* (1977), who injected fluorescein isothiocyanate (FITC) into the aorta of rats and found at 7 days mild AIN with mononuclear cells and occasional eosinophils. FITC was seen in the basal portion of the tubules on day 7. Immunoglobulins were not found. The authors suggested that FITC bound to kidney proteins and remained localized there long enough to be a target for an immunologic reaction, probably a cell-mediated reaction.

In addition, there are several models of immunologically mediated tubulo-interstitial nephritis (TIN), which although not drug-induced, have suggested mechanisms that may be involved in drug-induced lesions. TIN can result either from antibody- or cell-mediated mechanisms and either autologous or exogenous antigens can be involved (Table I) (Andres and McCluskey, 1975).

2.1. Antibody-Mediated TIN

Two types of antibody deposits can be identified in renal tissue by immunofluorescence: (1) immune complexes are recognized as granular deposits of immunoglobulins and complement components along tubular basement membranes (TBM) or in the interstitium; and (2) anti-TBM antibodies are seen as continuous linear deposits of IgG along the TBM. Tubulointerstitial immune complex deposits are often accompanied by interstitial collections of mononuclear cells, interstitial fibrosis, and tubular damage. Anti-TBM antibodies are usually associated with severe TIN, which in the guinea pig is characterized by a predominantly mononuclear cell infiltrate, often with multinucleated giant cells (the nature of infiltrating mononuclear cells is discussed below). Antibodies directed against cell surface components may also cause damage through complement-mediated cell lysis (Andres and McCluskey, 1975) or antibody-dependent cell-mediated cytotoxicity (ADCC) (Phillips and Neilson, 1981); these forms cannot usually be identified by immunofluorescence.

2.2. Cell-Mediated Mechanisms in TIN

Several years ago it was shown that delayed type hypersensitivity (DTH) reactions can be elicited in the cortex by the direct intrarenal injection of an exogenous antigen, bovine gamma globulin (BGG), in appropriately sensitized

TABLE I. Experimental Models of Immunologically Mediated Tubulointerstitial
Renal Lesions

Antibody-mediated
  Tubulointerstitial immune complex deposits
  a. Containing autologous antigens (Klassen *et al.*, 1971)
  b. Containing exogenous antigens (Brentjens *et al.*, 1974)
  Antitibular basement membrane (TBM) antibody disease (Steblay and Rudofsky, 1971)
  Cytotoxic damage to tubular cells due to autoantibodies against surface antigens (Andres and
    McCluskey, 1975)
  Antibody-dependent cell-mediated cytotoxicity (ADCC) to tubular cells (Phillips and Neilson, 1981)
Cell-mediated
  Exogenous antigens; DTH type reaction (Van Zwieten *et al.*, 1977)
  Autologous antigens; DTH or possibly cytolytic reactions (Sugisaki *et al.*, 1980)
  Hapten-autologous component conjugate; DTH or cytolytic reactions (Domoto *et al.*, 1977)

---

guinea pigs or rats (Van Zwieten *et al.,* 1977). The reactions exhibited a mononuclear cell infiltrate, with focal tubular destruction. Reactions were regularly produced with heat-aggregated BGG, whereas soluble BGG was generally ineffective, presumably because it rapidly diffused from the injection site. In contrast, soluble BGG was equally as effective as aggregated material in triggering DTH reactions in the skin. The findings indicate that in the kidney only those antigens that are aggregated, particulate (such as bacteria), or fixed (such as structural proteins or drugs covalently bound to structural components) are able to trigger DTH reactions.

A model of cell-mediated TIN resulting from sensitization to autologous antigens has recently been reported (Sugisaki *et al.,* 1980). Lewis rats were injected with homogenates of syngeneic renal tissue in adjuvants; many of the animals developed severe TIN, which was most intense at 10–14 days. The kidneys showed severe, irregular infiltrates of mononuclear cells, most of which appeared to be large lymphocytes. By immunofluorescence no tubulointerstitial deposits of immunoglobulins or C3 were detected. Transfer of lymph node or spleen cells, but not of serum, from kidney-injected donors resulted in focal interstitial infiltrates and tubular cell damage at 24–48 hr. The responsible autologous antigen(s) were not identified.

3. Analysis of Cell-Mediated Reactions *in Vivo*

Although studies of experimental models have provided criteria (by immunofluorescence findings) for recognition of immune complex deposits or anti-TBM antibodies in humans, they have so far failed to provide ways that permit identification of other forms of antibody-mediated reactions (IgE-mediated or ADCC), or of cell-mediated reactions.

Cell-mediated immunologic reactions are defined here as reactions triggered by interactions of sensitized T cells and antigen (McCluskey and Bhan, 1977). Two major categories are DTH reactions and those resulting from the action of cytolytic T cells. In the mouse it has been shown that these two types of reactions depend on different subsets of T lymphocytes. DTH reactions are triggered by contact of sensitized cells of the Ly 1 set (inducer/helper) (Cantor, 1979) with antigen, appropriately presented by macrophages or other accessory cells. The lymphocytes are stimulated to produce and release a group of lymphokines that mediate the reaction, in large part through recruitment of various leukocytes. In contrast, cytolytic T cells belong to the Ly 23 subset and exert their effects through direct contact with target cells (Cantor, 1979).

DTH reactions are themselves quite heterogeneous; the variability depends on a number of factors, including the species studied, the nature of the antigen, the tissue involved, and the method used for immunization. In the mouse, DTH reactions elicited in the foot pad are generally quite rich in granulocytes. In the guinea pig, DTH reactions elicited in animals sensitized with incomplete Freund's adjuvant or without adjuvants often contain numerous basophils (Colvin and Dvorak, 1979). In several species, DTH reactions produced against nondiffusible, insoluble antigens assume the appearance of granulomas. All of these reactions are presumably initiated by inducer-type lymphocytes. The nature of the cells in an infiltrate becomes even more complicated if cytolytic T cells, natural killer cells, or cells involved in ADCC participate in a reaction.

Although a predominantly mononuclear cell infiltrate is considered to indicate a cell-mediated reaction, nonimmunologic and even certain antibody-mediated reactions may be associated with such an infiltrate (McCluskey and Bhan, 1977). Clearly, recognition of DTH reactions cannot be based on histologic findings. Furthermore, the demonstration that a host exhibits cell-mediated reactivity against a particular antigen, either through elicitation of a DTH reaction or by *in vitro* analogs (such as MIF production), does not provide direct evidence that cell-mediated mechanisms are playing a pathogenetic role. The most reliable evidence for a pathogenetic role of DTH reactivity is provided by the demonstration that lesions develop in recipients of T lymphocytes (of the inducer type) and not in recipients of antibodies. Such studies obviously cannot be performed in humans.

3.1. Identification of Mononuclear Cells in Tissues

An important question is whether newer methods for the identification of mononuclear cells in tissue sections will lead to more meaningful interpretation of inflammatory infiltrates. Methods for the identification of cells in tissue sections are particularly important to the study of drug-induced AIN, since they can be applied to small renal biopsy specimens, which provide too few cells for suspension studies. Furthermore, since there is no entirely satisfactory experimental model, the only approach to drug-induced AIN is to study the human condition. Several years ago techniques were introduced that permitted demonstration of cells with Fc and C3 receptors in frozen tissue sections, or in cell suspensions, through rosetting techniques (Jaffe and Green, 1977). Sheep erythrocytes coated with IgG (IgGEA) or IgM and complement (IgMEAC) are used to identify cells with Fc or C3 receptors, respectively. Under most conditions, mononuclear cells with Fc receptors identified in tissue sections were thought to be mononuclear phagocytes (monocytes and macrophages) and cells with C3 receptor B lymphocytes.

These rosetting methods have been used to study the infiltrating cells in guinea pigs with anti-TBM disease (McCluskey *et al.*, 1981). This model is of interest since, although the lesions are apparently mediated by antibodies—the disease is transferrable with serum and not with cells—the infiltrate is in its earliest stages almost exclusively mononuclear in nature. Many of the cells were found to have Fc receptors, which provides some evidence for the presence of mononuclear phagocytes. However, many of the cells have the appearance of lymphocytes; these did not react with IgMEAC, which suggested that they were either T cells or "null" cells. It has been suggested that the Fc-receptor-bearing cells, which probably include cells other than mononuclear phagocytes, are responsible for tubular cell injury via ADCC (Phillips and Neilson, 1981).

Although the introduction of rosetting techniques represented an advance in the study of mononuclear cells in infiltrates, there are serious problems with the use of these methods. For one thing, both C3 and Fc receptors are present on a number of cell types and cannot therefore identify individual cell types with certainty (although, as noted, under conditions generally used in tissue sections to identify mononuclear cells, IgGEA forms rosettes principally with mononuclear phagocytes and IgMEAC with B cells). Furthermore, there are no rosetting techniques that can reliably identify T cells in tissue sections. Aside from receptor studies, certain heteroantisera against T-cell marker or B-cell antigen receptors (IgM or IgD) have been employed with immunohistochemical techniques to identify cells in frozen sections.

3.2. Use of Monoclonal Antibodies to Identify Human Mononuclear Cells

Lately, the outlook for identification of mononuclear cells (and in particular of human cells) has brightened considerably because of the development of monoclonal antibodies that are specific for surface antigens of T cells, T-cell subsets, or B cells (Reinherz and Schlossman, 1980; Stashenko *et al.*, 1980; Bhan *et al.*, 1981). These antibodies, as well as certain heteroantisera or monoclonal antibodies against heavy chain isotypes (anti IgM, IgD), can be used with immunofluorescence or with immunoperoxidase (PAP) techniques to identify individual cells in frozen sections and have been used to characterize cells in the human thymus and lymph nodes (Bhan *et al.*, 1980; Poppema *et al.*, 1981). The reactivities of the most useful monoclonal antibodies are as follows: antibodies designated as anti T1 or T3 react with all peripheral T cells; anti T4 reacts with T cells having inducer or helper function; anti T5 and T8 react with T cells possessing cytotoxic or suppressor function (Reinherz and Schlossman, 1980). There are at present no monoclonal antibodies that distinguish human suppressor from cytotoxic cells. Although identification of T cells with receptors for IgG (T$_G$ cells) has been used as evidence of suppressor cells (Husby *et al.*, 1981), this approach is not entirely reliable.

To date, relatively few studies employing monoclonal antibodies have been performed to analyze the nature of mononuclear cells in inflammatory infiltrates and in particular in cases of interstitial nephritis. Husby *et al.* (1981) have reported that the majority of mononuclear cells in the infiltrates in various forms of chronic interstitial nephritis (apparently not drug-related) were T lymphocytes and an appreciable number were T$_G$ cells. They concluded that cell-mediated immunity played a major role in "interstitial nephritis" and that there was local activation of suppressor cells. In preliminary studies we have found that the majority of cells in early rejecting renal allografts are T cells (Bhan *et al.*, 1982). Studies on drug-induced AIN are discussed below. What is needed now are studies that delineate the composition of the infiltrate in prototypic reactions, such as tuberculin or contact reactions and in unmodified allograft reactions (using skin grafts in volunteers). Only then will it be known if there are distinctive patterns that permit recognition of DTH or other cell-mediated reactions. If mononuclear antibodies specific for human natural killer cells, monocytes, and the various cells engaged in ADCC become available, analysis of infiltrates should become even more meaningful.

4. Immunologic Mechanisms in Drug-Induced AIN

The evidence favoring the interpretation that drug-induced AIN is immunologically mediated has been summarized earlier. The responsible immunologic mechanisms have not been completely elucidated. Apparently several mechanisms participate and these appear to vary in importance from one case to another. Whatever the type of immunologic reaction involved, it seems likely, on the basis of general considerations of drug hypersensitivity (Parker, 1980), that a hapten derivative of the drug becomes covalently bound to certain sites in renal tissue, and that the reactivity is directed against the hapten–host conjugate. In some cases of methicillin or penicillin-associated interstitial nephritis, a derivative of the antibiotic is demonstrable by immunofluorescence in renal tissue, along the TBM or in the interstitium. Assuming that renal fixation is important, two general possibilities can be considered: (1) The drug hapten binds to renal tissues in all patients receiving the drug, and the development of lesions depends on an unusual immune response (either humoral or cell-mediated); or (2) the binding of hapten occurs only in those patients who develop interstitial nephritis. There is conflicting evidence concerning this point. Border *et al.* (1974) were unable to find methicillin antigen in the kidneys of animals given methicillin; however, they did not provide information concerning the dosage or time of administration. In contrast, Colvin *et al.* (1974) described binding of anti-penicillin antibodies to the TBM and interstitium in autopsy specimens of histologically normal kidneys from patients who received large amounts of penicillin shortly before death. Further studies are needed to clarify this issue. However, it is possible that haptens may be bound to cell surfaces and not be readily detectable by immunofluorescence.

Certain observations suggest that cell-mediated mechanisms are more important in drug-induced AIN than antibody-mediated mechanisms. In most cases, a predominantly mononuclear cell infiltrate is found without associated immunoglobulin deposits; and DTH reactions to intracutaneous injection of the drug or *in vitro* evidence of cell-mediated reactivity have been described in a few patients (Baldwin *et al.,* 1968). Furthermore, in preliminary studies employing monoclonal antibodies we have found in three cases of AIN (oxacillin-, methicillin-, and penicillin-related) that there are numerous T cells; T4 cells appear to be more numerous than T8 cells, which is a pattern that might be expected of DTH reactions.

On the other hand, the presence of plasma cells and eosinophils in the infiltrate and the frequent presence of circulating antibodies against the drug have been used as arguments favoring antibody-mediated components. There is, however, no direct evidence for the participation of antibody-mediated mechanisms in most cases. Since the specificities of the antibodies produced by the plasma cells in the infiltrates are unknown, the role played by these cells is unclear. Immunofluorescence studies have not revealed granular tubulointerstitial deposits of immunoglobulins and complement in drug-induced AIN and thus have not provided evidence that immune complexes are responsible for the lesions. It should be noted, however, that in certain models of immune complex disease, deposits demonstrable by immunofluorescence disappear more quickly than signs of tissue damage.

A few cases of drug-induced AIN have been associated with anti-TBM antibodies, either bound to renal TBM or in the circulation (Border *et al.,* 1974). In the majority of cases, however, evidence of anti-TBM antibodies is lacking (Galpin *et al.,* 1978; Ooi *et al.,* 1975). The findings indicate, however, that in an occasional patient a true autoimmune response is initiated, almost certainly in addition to a response directed against conjugates of drug and host components. Autoantibodies against other renal antigens, such as surface antigens, have not been described in patients with AIN, but these might escape detection by traditional immunofluorescence techniques.

Although it is difficult to imagine why the kidney should be a selective target for an anaphylactic type reaction (normally there are relatively few interstitial mast cells), several observations have led to the suggestion that IgE antibodies participate in the pathogenesis of the lesions. Elevated serum IgE levels are found in some patients. In one case of AIN associated with phenobarbital therapy many of the plasma cells were said to stain for IgE (Faarup and Christensen, 1974). Furthermore, numerous eosinophils are present in the infiltrates in many cases, whereas neutrophils are often scanty. However, although anaphylactic type reactions lead to the generation of the greatest number of factors that selectively attract eosinophils, factors chemotactic for eosinophils are also produced and released in DTH reactions (Weller and Goetzl, 1979). Furthermore, it is known that certain DTH reactions, when accompanied by an antibody (presumably IgG) component, exhibit infiltrates that contain numerous eosinophils as well as mononuclear cells.

5. Concluding Remarks

Although there is abundant evidence that certain therapeutic agents, in particular methicillin or other penicillin analogs, can produce immunologically mediated AIN, the responsible mechanisms have not been defined. More than one mechanism may participate, and these may vary from case to case. In rare cases, anti-TBM antibodies appear to contribute to the damage. The roles of IgE antibodies and ADCC remain to be elucidated. In the absence of a suitable experimental model the most rewarding approach to elucidation of pathogenesis appears to be analysis of the infiltrating cells through the use of appropriate monoclonal antibodies; preliminary studies are consistent with the interpretation that cell-mediated mechanisms of the DTH type are involved.

Acknowledgment

This work was supported in part by National Institutes of Health grant RO1 AM 18729.

References

Andres, G. A., and McCluskey, R. T., 1975, Tubular and interstitial renal disease due to immunologic mechanisms, *Kidney Int.* **7**:271.

Baldwin, D. S., Levine, B. B., McCluskey, R. T., and Gallo, G. R., 1968, Renal failure and interstitial nephritis due to penicillin and methicillin, *N. Engl. J. Med.* **279**:1245.

Bhan, A. K., Reinherz, E. L., Poppema, S., McCluskey, R. T., and Schlossman, S. F., 1980, Location of T cell and major histocompatibility complex antigens in the human thymus, *J. Exp. Med.* **152**:771.

Bhan, A. K., Nadler, L. M., Skashenko, P., McCluskey, R. T., and Schlossman, S. F., 1981, Stages of B cell differentiation in human lymphoid tissue, *J. Exp. Med.* **154**:737.

Bhan, A. K., Colvin, R. B., Cosimi, A. B., and McCluskey, R. T., 1982, Nature of cellular infiltrate in renal allograft rejection. *Kidney Int.* **21**:293 (abstract).

Border, W. A., Lehman, D. H., Egan, J. D., Sass, H. J., Glade, J. E., and Wilson, C. B., 1974, Anti-tubular basement membrane antibodies in methicillin associated interstitial nephritis, *N. Engl. J. Med.* **291**:381.

Brentjens, J. R., O'Donnell, D. W., Pawlowski, I. B., and Andres, G. A., 1974, Extra-glomerular lesions associated with deposition of circulating antigen–antibody complexes in kidneys of rabbits with chronic serum sickness, *Clin. Immunol. Immunopathol.* **3**:112.

Cantor, H., 1979, Control of the immune system by inhibitor and inducer T lymphocytes, *Ann. Rev. Med.* **30**:269.

Colvin, R. B., and Dvorak, H. F., 1979, Role of granulocytes in cell-mediated immunity, in: *Mechanisms of Immunopathology* (S. Cohen, P. A. Ward, and R. T. McCluskey, eds.), Wiley, New York, p. 69.

Colvin, R. B., Burton, N. E., and Hyslop, N. E., 1974, Penicillin associated interstitial nephritis, *Ann. Intern. Med.* **81**:404.

Councilman, W. T., 1898, Acute interstitial nephritis, *J. Exp. Med.* **3**:393.

Domoto, D. T., Askenase, P. W., and Kashgarian, M., 1977, Tubulointerstitial nephritis (TIN) due to fluorescein isothiocyanate (FITC). A possible hapten-immunologically mediated reaction, *Kidney Int.* **12**:512 (abstract).

Faarup, P., and Christensen, E., 1974, IgE containing plasma cells in acute tubulointerstitial nephropathy, *Lancet* **2**:718.

Galpin, J. E., Shinaberger, J. H., Stanley, T. M., Blumenbrantz, M. J., Bayer, A. S., Friedman, G. S., Montgomerie, J. Z., Guzr, L. B., Coburn, J. W., and Glassock, R. J., 1978, Acute interstitial nephritis due to methicillin, *Am. J. Med.* **65**:756.

Husby, G., Tung, K. S. K., and Williams, R. C., 1981, Characterization of renal tissue lymphocytes in patients with interstitial nephritis, *Am. J. Med.* **70**:31.

Jaffe, E. S., and Green, I., 1977, Neoplasms of the immune system in: *Mechanisms of Tumor Immunity* (I. S. Green, S. Cohen, and R. T. McCluskey, eds.), Wiley, New York, p. 251.

Kimmelstiel, P., 1938, Acute hematogenous interstitial nephritis, *Am. J. Pathol.* **14**:737.

Klassen, J., McCluskey, R. T., and Milgrom, F., 1971, Nonglomerular renal disease produced in rabbits by immunization with homologous kidney, *Am. J. Pathol.* **63**:333.

McCluskey, R. T., and Bhan, A. K., 1977, Cell-mediated reactions in vivo, in: *Mechanisms of Tumor Immunity* (I. S. Green, S. Cohen, and R. T. McCluskey, eds.), Wiley, New York, p. 1.

McCluskey, R. T., and Colvin, R. B., 1978, Immunological aspects of renal tubular and interstitial diseases, *Ann. Rev. Med.* **29**:191.

McCluskey, R. T., Bhan, A. K., and Colvin, R. B., 1982, Experimental and human anti-tubular basement membrane (TBM) nephritis, in *NIH Symposium,* in press.

Mery, J. P., and Morel-Maroger, L., 1976, Acute interstitial nephritis: A hypersensitivity reaction to drugs, in: *Proceedings of the 6th International Congress of Nephrology* (S. Giovannetti, V. Bonomini, and G. D'Amico, eds.), Basel, p. 524.

Olsen, S., and Asklund, M., 1976, Interstitial nephritis with acute renal failure following cardiac surgery and treatment with methicillin, *Acta Med. Scand.* **199**:305.

Ooi, B. S., 1975, Acute interstitial nephritis. A clinical and pathologic study based on renal biopsies, *Am. J. Med.* **59**:614.

Parker, C. W., 1980, Drug allergy, in: *Clinical Immunology,* Volume II (C. W. Parker, ed.), Saunders, Philadelphia, Pennsylvania, p. 1219.

Phillips, S. M., and Neilson, E. G., 1981, Antibody-dependent cellular cytotoxicity (ADCC): A new effector mechanism in interstitial nephritis, *Kidney Int.* **19**:189 (abstract).

Poppema, S., Bhan, A. K., Reinherz, E. L., McCluskey, R. T., and Schlossman, S. F., 1981, Distribution of T cell subsets in human lymph nodes, *J. Exp. Med.* **153**:30.

Reinherz, E. L., and Schlossman, S. F., 1980, The differentiation and function of human T lymphocytes, *Cell* **19**:821.

Stashenko, P., Nadler, R., Hardy, S., and Schlossman, S. F., 1980, Characterization of a human B lymphocyte-specific antigen, *J. Immunol.* **125**:1678.

Steblay, R. W., and Rudofsky, U. H., 1971, Renal tubular disease and autoantibodies against tubular basement induced in guinea pigs, *J. Immunol.* **107**:589.

Sugisaki, T., Yoshida, T., McCluskey, R. T., Andres, G. A., and Klassen, J., 1980, Autoimmune cell-mediated tubulointerstitial nephritis induced in Lewis rats by renal antigens, *Clin. Immunol. Immunopathol.* **15**:33.

Van Zwieten, M. J., Leber, P. D., Bhan, A. K., and McCluskey, R. T., 1977, Experimental cell-mediated interstitial nephritis induced with exogenous antigens, *J. Immunol.* **118**:589.

Weller, P. F., and Goetzl, E. J., 1979, The regulatory and effector roles of eosinophils, in: *Advances in Immunology,* Volume 27 (H. G. Kunkel and F. J. Dixon, eds.), Academic Press, New York, p. 339.

35

Drug-Induced Nephritides
Immediate Hypersensitivity Mechanism

WAYNE A. BORDER

Tubular injury in the form of acute interstitial nephritis (AIN) or acute renal failure is a well-recognized complication of the administration of a variety of drugs or diagnostic agents, such as radiocontrast dye (Linton *et al.,* 1980). The acute onset of AIN as well as the frequently associated findings of fever, rash, eosinophilia, and elevated IgE levels have strongly suggested an allergic or hypersensitivity reaction. The underlying immunologic mechanism(s) of these reactions has not been established and it will be the purpose of this chapter to examine the possible role of immediate hypersensitivity.

Immunologically mediated tissue damage has been broadly separated into four distinct mechanisms by Coombs and Gell (1975): type I, anaphylactic or immediate hypersensitivity; type II, cytotoxic; type III, immune complex; and type IV, delayed hypersensitivity.

Immediate hypersensitivity is defined as a reaction occurring within minutes of exposure to an antigen in any member of a species (anaphylaxis) or in a uniquely predisposed member (atopy). Any antigen is potentially able to elicit such a response, if the animal is properly sensitized; however, such antigens are usually proteins or drugs acting as haptens attached to proteins. The classic proof of immediate hypersensitivity was the ability to transfer the reaction by the

WAYNE A. BORDER • University of California at Los Angeles School of Medicine, Harbor-UCLA Medical Center, Torrance, California 90509.

intradermal injection of serum into a nonsensitized host (Prausnitz-Kustner test). The basis for this transfer was the presence of reagenic antibody, now identified in humans to be IgE, that fixed to specific receptors or became absorbed, on the surface of mast cells and basophils. The immunobiology of IgE has been studied in detail by Ishizaka and Ishizaka (1978) and the molecular basis of immediate hypersensitivity is now well understood.

Upon a sensitized individual's second exposure to antigen, the antigen reacts with cell-fixed IgE, resulting in an architectural distortion in the cell membrane. Within the mast cell and/or basophil cell a complicated enzymatic cascade is activated that results in release of intracellular granules. Once discharged, the granules undergo dissolution and their contents are released into the surrounding tissue and/or blood. The granules contain powerful mediators of inflammation: histamine, serotonin, bradykinin, slow reacting substance of anaphylaxis (SRS-A), and eosinophil chemotactic factor of anaphylaxis (ECF-A). Interaction of the mediators with smooth muscle and endothelial cells produces the clinical findings characteristic of anaphylaxis: respiratory tract—bronchial obstruction, laryngeal edema; gastrointestinal tract—nausea, vomiting, diarrhea; cardiovascular system—hypotension, shock; and skin—wheals, angioedema. The drug cromolyn sodium acts to inhibit the release of mediators by stabilizing the cell membrane, thus preventing degranulation.

Secondary mediators may play a role in type I reactions. The same enzymatic activity producing degranulation can release prostaglandin A and F, which cause smooth muscle contraction and increased capillary permeability. Damage to endothelial cells can activate Hageman factor, leading to kinin generation (bradykinin has been shown to be released directly from human mast cells). Accumulation of eosinophils is a hallmark of allergic reactions and is in response to release of ECF-A. The purpose of eosinophil migration has been a mystery, but it is now known that eosinophils can engulf immune complexes and also release arylsulfatase B and histaminase, two enzymes that degrade SRS-A and histamine; eosinophils are thus able to limit an allergic reaction by removing antigenic material and destroying released mediators.

In the laboratory one can design *in vitro* systems that allow clear division among the four types of immune mechanisms; however, this division is somewhat artificial. In the whole animal the responsible mechanisms can be clouded by the secondary activation of the humoral amplification systems of complement, coagulation, kinin, and fibrinolysis. Thus, a type III reaction, mediated by immune complexes, can mimic anaphylaxis by activation of complement and release of anaphylotoxins C3a and C5a. In clinical practice matters may be even more complex, with one or more immune mechanisms being active simultaneously. Penicillin allergy is a perfect example of this, where anaphylaxis (type I), hemolytic anemia (type II), serum sickness (type III), and skin rash (type IV) may occur all at the same time or sequentially (Wells, 1978).

Mediators of inflammation can also be released by direct, nonimmunologic contact with drugs or chemicals giving rise to a pseudoallergic reaction that may be indistinguishable from immediate hypersensitivity. The administration of radio-contrast dye, often for intravenous pyelography, produces a 5–8% incidence of anaphylactoid reactions, ranging from mild to severe, consisting in varying degrees of nausea, vomiting, urticaria, faintness, headache, bronchospasm, and, occasionally, pulmonary edema and cardiac arrest (Till *et al.,* 1980). The agents involved are usually triiodinated benzoic compounds that are capable of a number of direct actions. *In vitro,* upon exposure to mast cells, histamine release occurs and following intravenous administration to humans, increased plasma levels of histamine are detectable; however, reactions do not correlate with histamine levels. Several workers have shown that radiocontrast dye can directly activate complement either via the classical pathway or through a still unknown pathway involving a nondirect enzymatic action. It remains to be defined whether significant complement activation occurs *in vivo* and whether it correlates clinically with the appearance of reactions. Isolated case reports have suggested a role for immune complex formation and even direct interaction with IgE antibody; however, other experiments have shown that the kinetics of histamine release induced by these agents is quite different from that produced by IgE antibody. Thus, most experts do not believe that antibody plays a major role in the clinical reactions in humans.

Similar anaphylactoid reactions occur after the ingestion of aspirin. There is convincing evidence that these reactions, like those due to radiocontrast agents, are also nonimmunologic in origin and they are classified as pseudoallergic. Part of the evidence for a pseudoallergic reaction being nonimmunologic is that, unlike in true allergy, reactions can occur without previous contact with the agent and the frequency of reactions does not increase with repeated exposures. Although acute renal failure may follow such a reaction, the evidence suggests that it is secondary to hemodynamic alterations and not to a direct effect on the kidney (aspirin has been reported also as a cause of AIN). It is well to be aware of the existence of pseudoallergic reactions, before one casually assumes a role for immediate hypersensitivity, based solely upon clinical findings that resemble an allergic or hypersensitivity reaction.

In some cases of AIN in humans there is clear evidence of an immunologic pathogenesis (van Ypersele de Strihou, 1979; Andres *et al.,* 1979). In rare cases antitubular basement membrane antibodies are formed or, more commonly, as in Goodpasture's syndrome, anti-glomerular basement membrane antibodies cross-react with tubular basement membrane (type II). In patients with severe immune complex disease, such as lupus erythematosus, immune complex deposits (type III) are easily identified along the tubular basement membrane. Animal models of AIN induced by anti-basement membrane antibody and immune complex deposition have been described and seem to reproduce the clinical events in humans. In addition, an experimental model of AIN mediated by delayed hypersensitivity (type IV) has also been studied extensively (Van Zwieten *et al.,* 1977). To date there is no animal model of AIN that has been described that involves a proven role for immediate hypersensitivity (type I). However, certain clinical features of AIN in humans suggest the possibility of a role for immediate hypersensitivity and this information must be considered before a final conclusion can be reached (Andres *et al.,* 1979).

By 1980, over 300 cases of drug-induced AIN had been reported (Linton *et al.,* 1980). The drugs most frequently involved are listed in Table I. It is thought, based

TABLE I. Drugs Most Frequently Associated with Acute Interstitial Nephritis[a]

Common	Uncommon
Methicillin	Penicillin G
Ampicillin	Other semisynthetic penicillins
Cephalosporin	Thiazide
Rifampin	Furosemide
Phenindione	Nonsteroidal anti-inflammatory drugs
Allopurinol	Carbenicillin
Phenytoin	Cephalothin

[a] Data compiled from Linton et al. (1980) and van Ypersele de Strihou (1979).

on renal biopsy studies, that about 10% of all cases of acute renal failure in humans are due to AIN; however, the incidence of AIN in all patients taking the same drug or the prevalence of AIN in all patients taking drugs is unknown. The epidemiologic facts strongly suggest an immunologic or allergic pathogenesis for AIN: (1) Reactions occur only in a small fraction of persons exposed. (2) Reactions occur with a decreased latent period with each new exposure. (3) Reactions are elicited by minimal exposure. (4) Reactions are associated with other signs of allergic disease. In unusual cases patients have been rechallenged with the same drug and have developed AIN on each occasion followed by a complete recovery. The duration of drug exposure before AIN is detected is highly variable, but ranges from 5 days to 5 weeks. Although allergic symptomatology is found in many patients with AIN, it also is often absent and cannot be relied upon for diagnosis. As estimated from the literature (Linton et al., 1980; van Ypersele de Strihou, 1979), the percent of AIN patients with allergic findings are shown in Table II and are: arthralgias 22%, rash 43%, elevated serum IgE 64%, fever 67%, eosinophilia 77%, and eosinophiluria 79%. A note of clinical caution is in order. Although eosinophiluria is the most common finding, it is also the least sensitive, because it usually occurs late in the course; proteinuria and azotemia usually precede it and should alert the clinician. The triad of fever, rash, and arthralgia is present in less than 40% of patients.

Notably absent in patients with AIN are signs and symptoms of anaphylaxis, especially evidence of urticaria, angioedema, and smooth muscle spasm. Such

TABLE II. Allergic Phenomena Occurring in Patients with Drug-Induced Acute Interstitial Nephritis

	Percent patients positive
Arthralgia	22
Rash	43
Elevated serum IgE	64
Fever	67
Eosinophilia	77
Eosinophiluria	79

[a] Data compiled from Linton et al. (1980) and van Ypersele de Strihou (1979).

findings can occur as a consequence of pseudoallergic reactions, but their absence strongly suggests that IgE-induced release of vasoactive mediators is not occurring. Eosinophilia is not a specific event in immediate hypersensitivity and eosinophils do not possess IgE on their surface. The histology of AIN is also nonspecific and does not indicate a specific immune mechanism. At this point the clinical picture of AIN is consistent with types II, III, or IV mechanism of immune injury and this is reinforced by the noted duration of drug exposure prior to onset of AIN.

One finding to be explained, however, is the presence of elevated serum levels of IgE reported in 64% of a small number of patients tested. IgE levels can now be measured accurately and are normally not affected by bacterial or viral infection (Johansson and Foucard, 1978). Levels of IgE are known to be increased in patients with mononucleosis, parasitic infections, asthma, hay fever, and other atopic disorders. In some disorders, such as pollen allergy, IgE levels correlate both with disease activity and response to immunotherapy. An important advance in the study of immediate hypersensitivity has been development of the radioallergo-sorbent test (RAST), which allows the measurement of antigen-specific IgE. The RAST is performed by mixing the test serum with the allergen coupled to an immunosorbent. After reacting and thorough washing, radiolabeled anti-IgE is added and after a repeat washing step and centrifugation, the radioactivity in the pellet is determined as a measure of the specific IgE antibody. The RAST test can detect as little as 1 ng of IgE antibody.

The RAST test has been compared with the classic Prausnitz-Kustner test for detecting reagenic antibody and both tests show a high degree of correlation (Johansson and Foucard, 1978). RAST also correlates with intradermal skin tests performed with specific allergens (Johansson and Foucard, 1978). Thus, the use of RAST has become the preferred way to test for IgE and the potential for immediate hypersensitivity.

In the majority of cases of AIN, specific tests for immediate hypersensitivity to the offending drug have not been performed. Ooi *et al.* (1979) measured IgE levels in five patients with AIN and found them to be elevated in three; sequential determinations of IgE tended to show an increase with progression of AIN. This report does not explain the apparent normal levels found in the remaining two patients and why, if the IgE acted as reagenic antibody, there were electron-dense deposits found along some tubular basement membranes. The obvious conclusion is that IgE participated in an immune complex (type III) mechanism of injury, although the authors consider these findings as evidence of "drug hypersensitivity." A sophisticated evaluation of an individual patient was conducted recently in a case of cephradine-induced AIN (Wiles *et al.,* 1979). A RAST test failed to show antibody to the inciting agent but did show a positive reaction with the penicilloyl group of penicillin G. The patient's serum caused the release of histamine from leukocytes *in vitro* due to the presence of an IgG antibody. Skin tests performed with a variety of antigens were not positive. In this patient the clear presence of an IgG reagenic antibody was established, suggesting a possible role in production of AIN.

It is clear from examination of the available experimental and clinical data that there is little evidence to support a role for immediate hypersensitivity as an important pathogenic mechanism for AIN. However, very few studies have directly attempted to define the immune mechanism(s) operative in AIN. With the advent of sophisticated tests such as RAST, it is now possible to specifically search for antigen-specific IgE antibody in groups of patients with AIN. A similar approach has been taken in patients with penicillin allergy and has proven highly successful (Chandra *et al.*, 1980). Furthermore, the use of the RAST test and/or intradermal tests might allow the identification of patients at risk to develop AIN.

References

Andres, G. A., Albini, B., Brentjens, J., Milgrom, M., Noble, B., Ossi, E., and Szymanski, C., 1979, Immunologically mediated tubulointerstitial nephritis in experimental animals and in man, in: *Immunopathology* (F. Milgrom and B. Albini, eds.), Karger, Basel, p. 123.

Chandra, R. K., Goglekar, S. A., and Tomas, E., 1980, Penicillin in allergy: Antipenicillin IgE antibodies and immediate hypersensitivity skin reactions employing major and minor determinants of penicillin, *Arch. Dis. Child.* **55**:857.

Coombs, R. R. A., and Gell, P. G. H., 1975, Classification of allergic reactions responsible for clinical hypersensitivity and disease, in: *Clinical Aspects of Immunology* (P. G. H. Gell, R. R. A. Coombs, and P. J. Lachmann, eds.), Blackwell, London, p. 761.

Ishizaka, K., and Ishizaka, T., 1978, Immunology of IgE-mediated hypersensitivity, in: *Allergy, Principles and Practice* (E. Middleton, C. E. Reed, and E. F. Ellis, eds.), Mosby, St. Louis, Missouri, Volume 2, p. 52.

Johansson, S. G. O., and Foucard, T., 1978, IgE in immunity and disease, in: *Allergy, Principles and Practice* (E. Middleton, C. E. Reed, and E. F. Ellis, eds.), Mosby, St. Louis, Missouri, Volume 2, p. 551.

Linton, A. L., Clark, W. F., Driedger, A. A., Turnbull, D. I., and Lindsay, R. M., 1980, Acute interstitial nephritis due to drugs, *Ann. Int. Med.* **93**:735.

Ooi, B. S., Pesce, A. J., First, M. R., Pollak, V. E., Bernstein, I. L., and Wellington, J., 1979, IgE levels in interstitial nephritis, *Lancet* **1**:1254.

Till, G., Voigtländer, V., and Rother, U., 1980, Complement and pseudo-allergic reactions to drugs, in: *Cytotoxic and Complement-Mediated Reactions* (P. Dukor, P. Kallós, H. D. Schlumberger, and G. B. West, eds.), Karger, Basel, p. 105.

Van Ypersele de Strihou, C., 1979, Acute oliguric interstitial nephritis, *Kidney Int.* **16**:751.

Van Zwieten, M. J., Leber, P. D., Bhan, A. K., and McCluskey, R. T., 1977, Experimental cell-mediated interstitial nephritis induced with exogenous antigens, *J. Immunol.* **118**:589.

Wells, J. V., 1978, Immune mechanisms in tissue damage, in: *Basic and Clinical Immunology*, 2nd ed. (H. H. Fudenberg, D. P. Stites, J. L. Caldwell, and J. V. Wells, eds.), Lange Medical Publications, Los Gatos, California, p. 267.

Wiles, C. M., Assem, E. S. K., Cohen, S. L., and Fisher, C., 1979, Cephradine-induced interstitial nephritis, *Clin. Exp. Immunol.* **36**:342.

36

Gold-Induced Autoimmune Reactions

TIMO PALOSUO and FELIX MILGROM

Despite widespread clinical use of gold in the treatment of rheumatoid arthritis (RA) for over 50 years (Anon, 1979), little is known about the mechanism of the therapeutic action of gold and of its toxicity. The efficacy of gold in RA has been well documented, and despite its occasionally serious side effects, such as proteinuria, nephrotic syndrome, exfoliative dermatitis, aplastic anemia, agranulocytosis, and thrombocytopenic purpura, gold is today accepted in most countries as a first-line treatment for RA. In the 1970s gold salts were also introduced and successfully used in the treatment of pemphigus (Penneys *et al.,* 1973).

When the still-current mode of gold therapy of RA was established in the early 1930s, it was noted that the only effective gold compounds were those containing a sulfur radical (Forestier, 1935). Attention has recently been paid to the possible role of the thiol group in gold preparations, since it has been recognized that d-penicillamine, a widely used alternative for gold in RA, contains a similar thiol group and has very similar therapeutic action and even side effects (Jellum *et al.,* 1977). Although conflicting views as to the mechanisms of the therapeutic effects of gold have been presented, many authors seem to consider gold salts as mildly immunosuppressive agents, capable of inhibiting the complement system,

TIMO PALOSUO • National Public Health Institute, Helsinki, Finland; and Department of Microbiology, School of Medicine, State University of New York at Buffalo, Buffalo, New York 14214. **FELIX MILGROM** • Department of Microbiology, School of Medicine, State University of New York at Buffalo, Buffalo, New York 14214.

decreasing lysosomal enzyme activity and phagocytosis, and depressing the titers of rheumatoid factors in patients with RA (Hurd, 1977).

Autoimmune phenomena induced by gold therapy have been suspected, especially in "toxic" reactions occurring during the treatment. In gold nephropathy, immunoglobulins and complement have regularly been demonstrated in a characteristic fashion along the glomerular capillary walls, but metallic gold has not been found in these deposits (Törnroth and Skrifvars, 1974; Davies *et al.*, 1977). It may be suggested that gold, which has been shown to accumulate in inflamed synovial tissues and in the renal tubular epithelial cells (Davies *et al.*, 1977; Vernon-Roberts *et al.*, 1976), releases, from the affected tissues, components that themselves are, or are rendered, immunogenic. This leads to autoantibody response and formation of pathogenetic immune complexes. Recently, it has been suggested that gold-related pulmonary fibrosis, a rare complication of chrysotherapy, might also have an autoimmune pathogenesis (Smith and Ball, 1980). Penicillamine seems to be frequently associated with autoimmune reactions: many types of "autoimmune diseases," including systemic lupus erythematosus, myasthenia gravis, Goodpasture's syndrome, and dermatomyositis, and variety of autoantibodies, such as anti-acetylcholin receptor, and antimitochondrial and antinuclear antibodies have been reported to develop in patients treated with this drug (O'Brien, 1980; Russel and Lindstrom, 1978; Kirby *et al.*, 1979; Zilko *et al.*, 1977).

We have been involved in studies on possible autoimmune reactions induced by gold therapy on two lines of investigation, dealing with pemphigus and RA.

In 1976, we (Palosuo *et al.*, 1976a) described a multitude of immunologic phenomena preceding the onset of significant proteinuria during gold treatment of a patient suffering from pemphigus. In a pre-complication serum of this 41-year-old Caucasian male, who had had pemphigus for 11 years, precipitating antibodies combining with boiling-resistant and ethanol-insoluble antigens referred to as BE antigens (Milgrom, 1980) of human kidney and liver, as well as against plain saline extracts of the same tissues, were found. The serum specimen also contained precipitating antibodies against an antigen present in sodium deoxycholate extracts of human and rat liver, previously termed ubiquitous tissue antigen (UTA), to denote its lack of species and tissue specificity. The serologically active moiety of UTA was $\alpha(1 \rightarrow 6)$ linked glucopyranose, or dextran (Palosuo and Milgrom, 1981). Even a third antigen–antibody system was observed in sera of this patient. Several serum specimens contained an undefined thermostable "serum antigen," against which another serum sample contained precipitating antibodies. Before, and at the time of the complication, the patient also had circulating immune complexes, detected by the platelet aggregation test, and he developed a positive Waaler-Rose test. All these serologic activities disappeared when gold was discontinued, although significant proteinuria persisted for 3 months.

Following the above-described findings in pemphigus, we became interested in the role of anti-dextran antibodies in RA patients receiving gold. A further impetus for this was that we had found dextran and anti-dextran antibodies (termed by us UTA and anti-UTA at that time), frequently in the form of immune complexes, in sera of patients with renal diseases (Palosuo *et al.*, 1976b), most notably in membranous nephropathy, which is the characteristic histopathologic presentation of gold nephropathy (Törnroth and Skrifvars, 1974). Dextran (UTA) was also identified as the antigen of immune deposits in the kidney of a patient with chronic glomerulonephritis (Milgrom *et al.,* 1976). In a prospective study (Palosuo *et al.,* 1978) we monitored RA patients receiving gold therapy for several months in order to learn whether they form anti-dextran antibodies. Precipitating anti-dextran antibodies were demonstrated in three of 84 (3.6%) patients with acute RA of short duration, but the incidence of these antibodies increased to as much as 17% shortly after gold treatment was initiated, i.e., usually within 1 month. Interestingly enough, anti-dextran antibodies were demonstrable in almost half of the patients (47%) who had developed proteinuria during gold (or d-penicillamine) therapy. These antibodies often disappeared from sera of RA patients after discontinuation of the therapy (Palosuo, unpublished observations). In contrast, anti-dextran antibodies were found in only 0.3% of apparently healthy middle-aged people.

Whether the observed anti-dextran antibodies represented antibodies to exogenous or endogenous antigens remains to be shown. As a rule, dextrans are considered bacterial polysaccharide products, and chains consisting of more than a few $\alpha(1 \rightarrow 6)$ linked glucopyranoses are not believed to exist in normal structures of the human body. However, bacteria capable of producing dextrans reside in the oral cavity and in the intestine, and dextrans are frequently contaminants of various food products, especially sugars. Anti-dextran antibodies are also known to cross-react with certain other polysaccharides, such an pneumococcal polysaccharides and teichoic acids, and apparently also slightly with dextrins and glycogens (Palosuo and Milgrom, 1981), but the original stimulus for anti-dextran antibody production in the above-discussed patients remains obscure. Immune response to dextrans in man may be of further interest since it was recently reported that RA patients who possess the HLA DRw3 alloantigen have a 32-fold higher risk of developing proteinuria during gold therapy than patients who lack this antigen (Wooley *et al.,* 1980). The possibility that the same genetic factors that seem to govern the susceptibility to the renal complication of gold treatment may be involved also in the regulation of immune response to dextrans (and possibly other T-independent polysaccharide antigens) may be worth considering.

The formation of autoantibodies related to gold—in contrast to penicillamine therapy—may thus be a rare phenomenon. However, as exemplified by our findings in a pemphigus patient, antibodies to tissue components may develop and they may be associated with gold nephropathy. The presence in sera of such antibodies may be very short-lived, however, and therefore difficult to demonstrate. Antibodies combining with dextran seem to develop frequently during gold treatment of RA patients, especially when proteinuria develops as a complication of the therapy. The mechanism of the formation of the dextran-reactive antibodies and their biologic significance has, however, remained unknown.

Acknowledgment

This work was supported by Research Grant AM-17317 from the National Institute of Arthritis, Metabolism, and Digestive Diseases.

References

Anon, 1979, Fifty years of gold in rheumatoid arthritis, *Br. Med. J.* **1**:289.

Davies, D. J., Dowling, J., and Xipell, J. M., 1977, Gold nephropathy, *Pathology* **9**:281.

Forestier, J., 1935, Rheumatoid arthritis and its treatment by gold salts, *J. Lab. Clin. Med.* **20**:827.

Hurd, E. R., 1977, Drugs affecting the immune response, in: *Immunology in Medicine* (E. J. Holborow and W. G. Reeves, eds.), Academic Press, London, p. 1080.

Jellum, E., Aaseth, J., and Munthe, E., 1977, Is the mechanism of action during treatment of rheumatoid arthritis with penicillamine and gold thiomalate the same? *Proc. R. Soc. Med.* **70**(Suppl. 3):136.

Kirby, J. D., Dieppe, P. A., Huskisson, E. C., and Smith, B., 1979, d-Penicillamine and immune complex deposition, *Ann. Rheum. Dis.* **38**:344.

Milgrom, F., 1980, Antigens of vertebrate cells, tissues, and body fluids, *Transpl. Proc.* **12**(suppl. 1):2.

Milgrom, F., Campbell, W., and Andres, G. A., 1976, Antigen in immune complex nephritis, *Immunology* **30**:277.

O'Brien, W. M., 1980, Toxicity of d-penicillamine in rheumatoid arthritis *Ann. Intern. Med.* **92**:120.

Palosuo, T., and Milgrom, F., 1981, Appearance of dextrans and anti-dextran antibodies in human sera, *Int. Arch. Allergy Appl. Immunol.,* **65**:153.

Palosuo, T., Provost, T. T., and Milgrom, F., 1976a, Gold nephropathy: Serologic data suggesting an immune complex disease, *Clin. Exp. Immunol.* **25**:311.

Palosuo, T., Andres, G. A., and Milgrom, F., 1976b, A ubiquitous tissue antigen and its corresponding antibody in sera of patients with glomerulonephritides, *Int. Arch. Allergy Appl. Immunol.* **52**:129.

Palosuo, T., Kajander, A., von Essen, R., and Milgrom, F., 1978, Precipitating autoantibody to a ubiquitous tissue antigen: Association with rheumatoid arthritis treated with sodium aurothiomalate or d-penicillamine, *Clin. Immunol. Immunopathol.* **10**:355.

Penneys, N. S., Eaglstein, W. H., Indgin, S., and Frost, P., 1973, Gold sodium thiomalate treatment of pemphigus, *Arch Dermatol.* **108**:56.

Russel, A. S., and Lindstrom, J. M., 1978, Penicillamine-induced myasthenia gravis associated with antibodies to acetyl-choline receptor, *Neurology* **28**:847.

Smith, W., and Ball, G. V., 1980, Lung injury due to gold treatment, *Arth. Rheum.* **23**:351.

Törnroth, T., and Skrifvars, B., 1974, Gold nephropathy prototype of membranous glomerulonephritis, *Am. J. Pathol.* **75**:573.

Vernon-Roberts, B., Doré, J. L., Jessop, J. D., and Henderson, W. J., 1976, Selective concentration and localization of gold in macrophages of synovial and other tissues during and after chrysotherapy in rheumatoid patients, *Ann. Rheum. Dis.* **35**:477.

Wooley, P. H., Griffin, J., Panayi, G. S., Batchelor, J. R., Welsh, K. I., and Gibson, T. J., 1980, HLA DR antigens and toxic reaction to sodium aurothiomalate and d-penicillamine in patients with rheumatoid arthritis, *N. Engl. J. Med.* **303**:300.

Zilko, P. J., Dawkins, R. L., and Cohen, M. L., 1977, Penicillamine treatment of rheumatoid arthritis: Relationship of proteinuria and autoantibodies to immune status, *Proc. R. Soc. Med.* **70**(Suppl. 3):118.

37

Mercuric Chloride-Induced Immunologically Mediated Diseases in Experimental Animals

B. ALBINI, I. GLURICH, and GIUSEPPE A. ANDRES

1. Introduction

Large doses of cadmium, gold, or mercury salts produce acute toxic lesions, e.g., renal tubular necrosis (Gritzka and Trump, 1968). Administration of low doses of heavy metals may induce immunologically mediated diseases. Thus, membranous nephropathy has been described in humans in association with administration of mercury-containing compounds (Mendema *et al.*, 1963). During the last decade, several animal models of chronic mercuric chloride-induced, immunologically mediated diseases have been established (Bariety *et al.*, 1971; Roman-Franco *et al.*, 1976, 1978; Sapin *et al.*, 1977; Albini and Andres, 1982). In these models, pathogenesis involves autoimmunization to connective tissue components and development of immune complex-mediated pathology.

B. ALBINI, I. GLURICH, and GIUSEPPE A. ANDRES • Departments of Microbiology, Pathology, and Medicine, State University of New York at Buffalo, Buffalo, New York 14214.

2. Disease in the Rabbit

The mercuric chloride-induced disease in the rabbit has been described in detail by Albini and Andres (1982). Suffice it here to summarize the most salient features of this disease (Roman-Franco *et al.,* 1976, 1978). The rabbits receive intramuscularly 2 mg mercuric chloride per kg body weight twice a week. Four stages may be distinguished in the ensuing disease.

1. An acute toxic phase with the major clinical manifestation of profuse diarrhea is seen in the first week after initiation of injections and subsides after 3–7 days.

2. The autoimmune stage: Antibodies to connective tissue components are produced and bind, *in vivo,* to endo- and perimysium of skeletal muscles, glomerular basement membranes (GBM), tubular basement membranes (TBM), vascular media basement membrane-like substance, elastica interna, capillaries of the heart, and reticulin of the spleen. This stage is seen 2–5 weeks after first injection with mercuric chloride. No clinical signs of illness are seen, tissues show normal structure, and proteinuria is absent or minimal. There is no or only negligible binding of rabbit C3 to the structures reacting with antibodies to rabbit IgG in immunofluorescent staining procedures. In indirect immunofluorescence tests, sera of these rabbits show weak to moderate reaction with basement membranes, elasticae, and reticulin of normal rabbit tissue; antibodies to basement membranes can be eluted from deposits in tissues by citrate buffer (pH 3.2) elution (Woodroffe and Wilson, 1977). Eluates and sera tested on kidney sections from other species reveal a partial species restriction; there is moderate reactivity with human, weak reactivity with rat, but no reactivity with mouse, guinea pig, and chicken basement membranes. At the beginning of this stage, around days 14 to 21 after initiation of injection, C1q-binding assays document a peak of reactive material (immune complexes or collagen-like material; unpublished observations). At the end of this stage, Raji-cell and anti-antibody inhibition assays detect low concentrations of circulating immune complexes.

3. Four to ten weeks after the first injection, there is a gradual superimposition of granular deposits of IgG and C3 on the linear deposits of IgG in the glomerulus (seen in stage 2) and in many other vascular regions. Electron-dense deposits are seen scattered on the subepithelial side of the GBM. There is proteinuria, and the animals lose weight and are inactive. Histopathology reveals characteristic membranous nephropathy with increasingly heavy deposits in the GBM, ultimately forming broad, ribbonlike deposits. Eluates of these kidneys react with basement membranes, reproducing the pattern seen with sera and eluates of animals in stage 2. After partial elution of tissue sections, granular deposits can be reconstituted by antibodies to basement membranes. In this stage, high concentrations of immune complexes may be detected in the circulation by Raji cell and anti-antibody, but not with C1q, assays.

4. After discontinuation of mercuric chloride injections, immune deposits remain in the kidney for at least 28 weeks. However, they change their pattern. Immune deposits in these animals are seen mainly in the mesangium. There is only minimal and intermittent proteinuria and the animals recover. Histopatho-

logically, mild widening of the mesangial matrix and proliferation of mesangial cells are seen.

Stages 2 and 3 represent a superimposition of immune complex-mediated glomerulonephritis on basement membrane antibody-mediated pathology, and stages 3 and 4 exemplify the transition of membranous to mesangial glomerulopathy. The antigen involved probably is a connective tissue component widely distributed throughout the body. Its precise nature is not yet known (Albini and Andres, 1982). Interestingly, mercury is not detectable in basement membranes and thus with the antigens involved in the subsequent immunologically mediated disease (Roman-Franco *et al.,* 1978).

3. Injection of Mercuric Chloride into Mice

Eighteen C3H.NB mice obtained as a gift from Dr. M. Zaleski of the Department of Microbiology, State University of New York at Buffalo, Buffalo, New York, were injected intramuscularly with 20 μg of mercuric chloride per 10 g body weight twice a week for a period of 4 months. During the last 5 weeks, five mice developed proteinuria (mean 24-hr urinary protein 34.2 ± 7.2 mg in this group compared to the mean 24-hr urinary protein of untreated age and sex matched controls 9.3 ± 1.8). Fifteen animals had diffuse to granular mesangial deposits strikingly stronger than those seen in untreated controls. Two mice showed immunoglobulin deposits in the TBM and one mouse showed granular immune deposits in the GBM. Media of vessels always had linear or linear and granular immune deposits of IgG. The C3 deposits were moderate in the mice with IgG deposits in TBM and GBM, but only minimal in nine of the mice with mesangial deposits, and equivocal in two other mice. Thus, mercuric chloride injections in the mice induce a predominantly mesangial glomerulopathy, with proteinuria occurring only in few cases. Two of 16 sera tested in indirect immunofluorescence tests on mouse kidney showed minimal reaction with mesangium and the media of medium-sized and large vessels.

4. Disease in Brown Norway Rats

Mercuric chloride-induced disease of the brown Norway rat has been studied extensively by Bariety *et al.* (1971) and, more recently, by Druet and his group (Bernaudin *et al.,* 1979; Druet *et al.,* 1978; Hinglais *et al.,* 1979; Sapin *et al.,* 1977). These studies are discussed in detail in Chapter 38. In the present section this disease is illustrated with data from experiments performed in Buffalo on 95 brown Norway rats. The brown Norway rats used in the experiments were of two different ages: 30 were 11 months old and 65 were 3 months old at the beginning of administration of mercuric chloride. This compound was administered in doses of 80 μg/100 g body weight to 20 11-month-old rats (group A) and to 30 3-month-old rats (group B); 25 3-month-old rats received 20 μg mercuric chloride/100 g body weight. Ten 11-month-old and ten 3-month-old rats served as uninjected controls.

The experimental rats were injected twice a week intramuscularly. Urine and blood were obtained once per week. Kidney biopsies were performed every 2–4 weeks in 3–5 animals of each group.

4.1. Group A

The protein excretion over an observation period of 49 weeks is summarized in Table I. There are four periods of high frequency of proteinuria: 5–7, 11–15, 21–25, and 29–33 weeks after first injection. Six animals never developed proteinuria. Other rats showed two or three distinct periods of proteinuria.

Direct immunofluorescence first revealed minimal linear deposits of IgG in the GBM and in the mesangium 3 weeks after initiation of the experiment. Subsequently, immune deposits were found in all rats tested in the glomerular mesangium and in few of them in the basement membrane of renal collecting ducts. This staining pattern prevailed in the time period of 5–11 weeks. Later, until week 25, all biopsies examined showed extensive mesangial deposits of IgG, IgG deposits in the media of vessels, and weak to moderate linear staining of the GBM in some

TABLE I. Proteinuria in Brown Norway Rats of Group A($n = 20$)

Weeks[a]	Protein in urine, mean ± SD	Frequency of proteinuria in rats,[b] %
−1	19.4 ± 12.6	0
3	26.4 ± 29.0	0
5	36.4 ± 24.0	42
7	45.6 ± 42.1	28
9	31.4 ± 21.3	14
11	39.2 ± 31.4	38
13	40.6 ± 28.6	23
15	55.8 ± 54.5	36
17	27.7 ± 32.3	9
19	25.5 ± 13.8	9
21	49.3 ± 49.6	40
23	52.6 ± 62.1	33
25	47.5 ± 54.6	29
27	68.6 ± 105.3	14
29	55.7 ± 38.9	43
31	55.2 ± 28.3	50
33	26.6 ± 9.9	0
35	37.5 ± 19.2	33
37	28.8 ± 9.1	0
39	37.1 ± 8.9	0
41	30.6 ± 13.6	0
43	60.3 ± 38.7	40
45	32.6 ± 2.2	20
47	32.2 ± 9.9	0
49	59.05 ± 20.3	20

[a] Weeks after first administration of mercuric chloride.
[b] Proteinuria defined as protein greater than mean + 2SD of untreated controls.

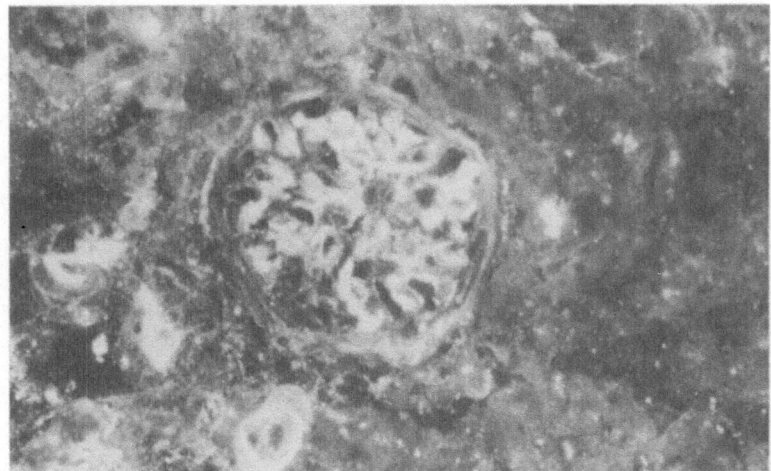

FIGURE 1. Direct immunofluorescent staining for rat IgG on kidney of a rat from group A(#2033) 35 weeks after initiation of experiment. × 250.

tissue specimens. From week 29 on, granular staining was detectable in the mesangium and the GBM, the media of vessels, and sometimes in the TBM and Bowman's capsule (Fig. 1). Indirect immunofluorescence tests with sera from these rats showed reactivity on normal tissue duplicating the pattern seen in direct immunofluorescent staining. All rats had IgG deposits in the peri- and endomysium of skeletal muscles, first in linear and later in a linear and granular pattern (Fig. 2).

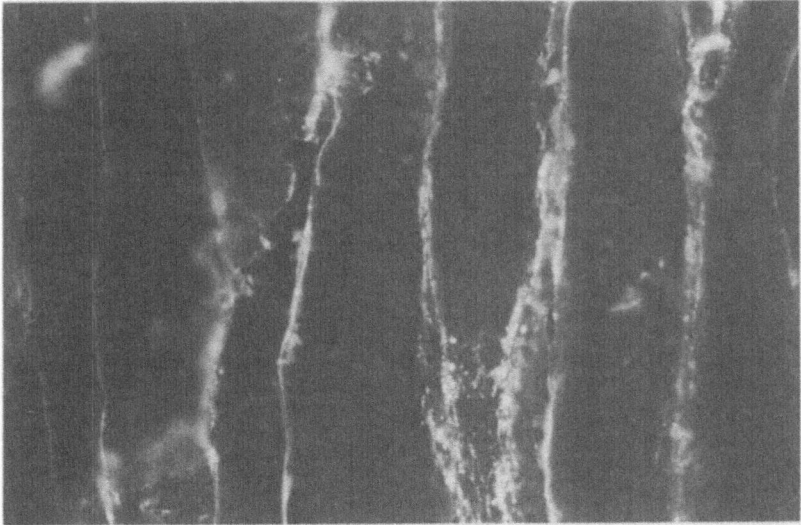

FIGURE 2. Direct immunofluorescent staining for rat IgG on skeletal muscle of a rat from group A (#2044). × 400.

Rats with antibody to TBM showed focal deposits of IgG in capillaries of the heart, in lung alveoli, in the skin, liver, ovaries, uvea and iris of the eye, and in the peritoneum. In the late stage of the disease, diffuse granular immune deposits were seen in various organs, and renal lesions were extensive and severe.

4.2. Group B

The protein excretion of rats in this group over an observation period of 29 weeks is summarized in Table II. In this group, four periods of proteinuria could be observed. During three weeks after the first injection, strong linear staining of GBM for IgG was seen (Fig. 3). In some rats, strong linear staining was also seen in the TBM and sometimes in the adventitia of vessels. The first granular staining appeared in the GBM after 3–4 four weeks. Four to 5 weeks after the first injection, almost all biopsy specimens contained distinct granular immune deposits in the GBM. Granular deposits became detectable also in peritubular capillaries (Fig. 4); staining of TBM at this time was only weak. Nine rats with heavy proteinuria died during this time period. At 9 weeks after initiation of experiment, rats also showed diffuse and granular staining in the mesangial region superimposed in linear staining in GBM (Fig. 5). At 23–29 weeks, mesangial deposits were prominent, and in some rats the only ones seen. In the majority of rats, however, linear and granular deposits of IgG in the GBM and in the media of vessels still were detectable. Eluates prepared by citrate buffer extraction (pH 3.2) showed moderate to strong reactivity with normal rat GBM and TBM (stronger in rats killed or deceased early in the disease), and strong reactivity with the adventitia of vessels (Fig. 6). Preliminary experiments with the elution technique using glycin buffers at decreasing pH, as proposed by Bartolotti (1977), suggest that the eluates contain antibodies with distinct affinities for GBM, TBM, and vessels (Table III).

TABLE II. Proteinuria in Brown Norway Rats of Group B ($n = 30$)

Weeks[a]	Protein in urine, mean ± SD	Frequency of proteinuria in rats,[a] %
1	30.2 ± 7.6	0
3	73.9 ± 101.8	39
5	43.3 ± 28.5	38
7	68.0 ± 54.9	38
9	59.3 ± 49.5	53
11	29.6 ± 12.6	14
13	45.3 ± 19.1	45
15	66.5 ± 46.6	62
17	28.4 ± 11.1	9
19	39.6 ± 9.3	18
21	51.2 ± 39.4	27
23	35.2 ± 18.6	27
25	19.5 ± 7.1	0
27	24.8 ± 8.0	0
29	62.4 ± 53.0	44

[a] See footnotes to Table I.

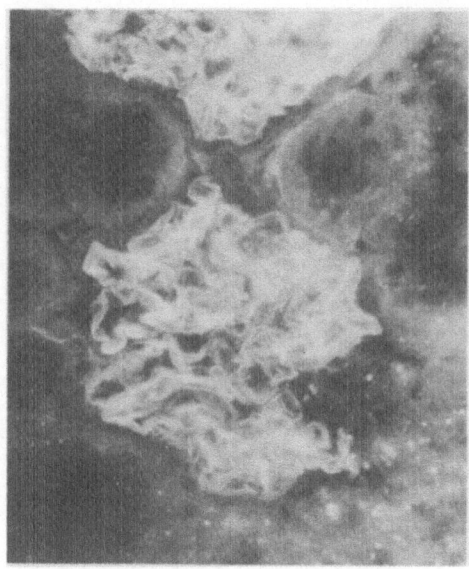

FIGURE 3. Direct immunofluorescent staining for rat IgG on kidney of a rat of group B (#2503). × 250.

4.3. Group C

The protein excretion of rats in this group over an observation period of 29 weeks is summarized in Table IV. Proteinuria in this group was rather infrequent and developed later than in group B. Three weeks after initiation of the experiment, direct immunofluorescence tests showed weak linear IgG deposits in the GBM and in the mesangium. Six weeks later, almost exclusive mesangial staining, and subsequently, staining of the media of vessels became detectable. This remained the predominant pattern seen in direct immunofluorescent staining procedures (Fig. 7). Granular staining, although infrequent, was seen from week 17 on. Eluates

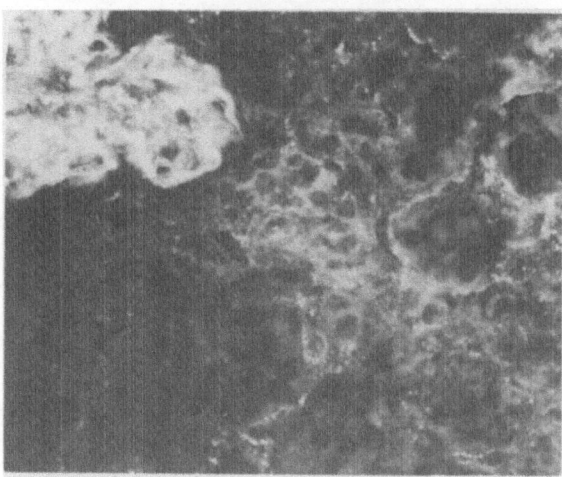

FIGURE 4. Direct immunofluorescent staining for rat IgG on kidney of a rat of group B (#2585). × 250.

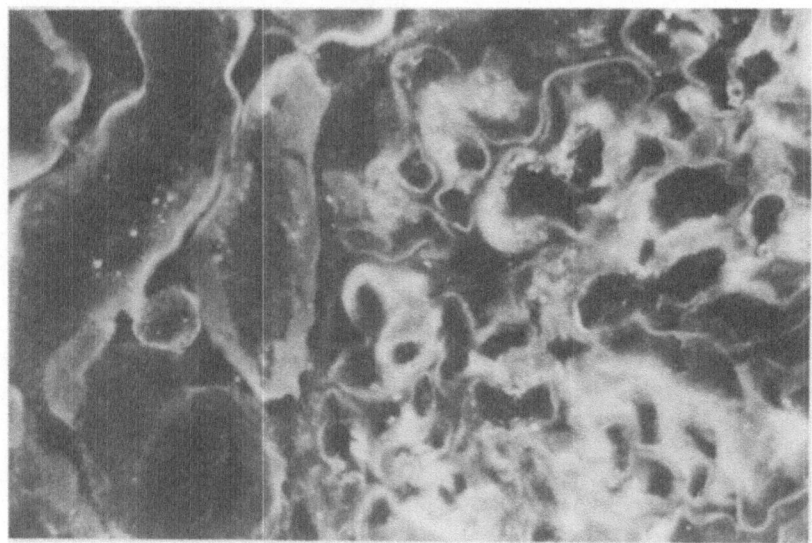

FIGURE 5. Direct immunofluorescent staining for rat IgG on kidney of a rat of group B (#2586). ×630.

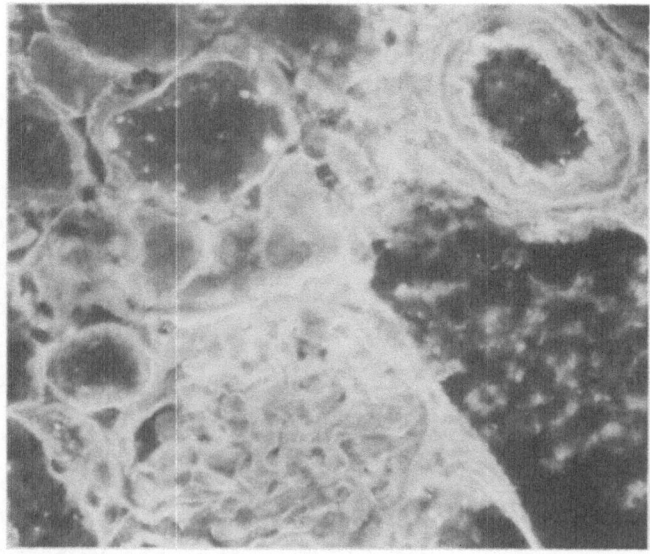

FIGURE 6. Indirect immunofluorescent staining of normal rat kidney with eluate (#2586) from rats of group B. × 250.

TABLE III. Reactivity of Eluates Obtained from Four Kidneys of Proteinuric Rats of Group B on Normal Kidney Tissue

pH of elution with glycin-HCl buffer	Reactivity with:		
	GBM	TBM	Vessel
3.2	++	++	++
3.0	0	+	++
2.8	0	+	+
2.6	0	+	+

obtained from kidneys of rats from group C reacted only with mesangium and the media of vessels of untreated rats. In organs other than the kidneys, only the media of vessels showed linear deposits of IgG.

In animals of all groups, C3 was definitively detectable only in granular but not in linear deposits. It is important to note that in all three groups there was no correlation between extent and intensity of immune deposits and levels of proteinuria. All rats showed deposits of IgG in the peri- and endomysium of skeletal muscles. Reactivity with connective tissue components was restricted to medium and large vessels (media) of other organs in rats that showed only mesangial deposition in the kidney. Rats with antibodies to GBM and/or TBM showed reactivity with reticulin and capillaries in other organs. The experiments reported here confirm the notion of a transition from anti-basement membrane antibody-mediated to immune complex-mediated disease, as first observed in

TABLE IV. Proteinuria in Brown Norway Rats of Group C ($n = 25$)

Weeks[a]	Protein in urine, mean ± SD	Frequency of proteinuria in rats,[a] %
1	28.3 ± 9.6	0
3	18.9 ± 6.8	0
5	18.8 ± 7.8	0
7	34.4 ± 14.8	10
9	28.9 ± 20.8	14
11	30.3 ± 13.4	14
13	32.8 ± 8.9	0
15	56.4 ± 45.7	35
17	27.4 ± 8.9	0
19	40.1 ± 15.5	33
21	49.8 ± 40.1	22
23	30.6 ± 13.5	17
25	19.1 ± 6.9	0
27	30.3 ± 14.4	12
29	29.3 ± 12.5	13

[a] See footnotes to Table I.

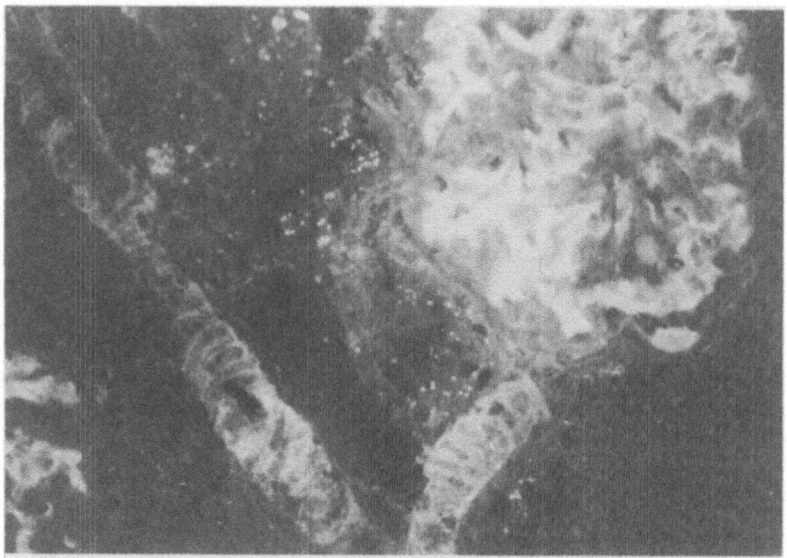

FIGURE 7. Direct immunofluorescent staining for rat IgG on kidney of rat of group C (#2508). ×400.

mercury-induced nephropathy in the rabbit (Roman-Franco *et al.,* 1976, 1978). The rat disease, however, is characterized by marked proteinuria also during the stage of antibody-mediated injury—a feature different from the findings in the rabbit. Furthermore, the disease in the rat is characterized by a series of intermittent proteinuric attacks.

The nature of the autoantigen involved is as yet unknown in all animal models studied. It seems that in the rat model there are at least two systems of antigens, a mesangial–vascular media system (group C) and a GBM–TBM–adventitia system (group B). The former is seen predominantly in rats injected with extremely low doses of mercuric chloride and the latter in those injected with moderately high doses.

From preliminary studies, it seems rather unlikely that some of the well-defined connective tissue antigens are involved. Antibodies to such components (e.g., fibronectin, collagen IV) are found only in a negligible portion of sera, never in the kidney eluates, and never in more than borderline titers (Martin, Terranova, Albini, and Andres, unpublished observations). Search for other connective tissue antigens now seems warranted.

The disease induced by chronic administration of low doses of mercuric chloride is characterized by slightly different patterns in various species, and in the same species there is some variability according to the age of the animal and the dose administered. Thus differences in the characteristics of the disease not only are determined genetically (see Chapter 38 by Druet *et al.*) but also may depend on other constitutional and environmental factors.

Acknowledgments

We thank Carol Soroka, Joy Kennedy, and Joy Niesen for efficient technical help. The studies reported here were supported by National Institutes of Health grant AI 10334.

References

Albini, B., and Andres, G., 1982, Autoimmune disease induced in rabbits by administration of mercuric chloride. Evidence suggesting a role for antigens of the connective tissue matrix, in: *Fogarty International Center Series,* in press.

Bariety, J., Druet, P., Laliberte, F., and Sapin, C., 1971, Glomerulonephritis with gamma and beta 1 C globulin deposits induced in rats by mercuric chloride, *Am. J. Pathol.* **65**:293.

Bartolotti, S. R., 1977, Quantitative elution studies in experimental immune complex and nephrotoxic nephritis, *Clin. Exp. Immunol.* **29**:334.

Bernaudin, J. F., Druet, E., Belair, M. F., Pinchon, M. C., Sapin, C., and Druet, P., 1979, Extrarenal immune complex type deposits induced by mercuric chloride in the Brown Norway rat, *Clin. Exp. Immunol.* **38**:265.

Druet, P., Druet, E., Potdevin, F., and Sapin, C., 1978, Immune type glomerulonephritis induced by HgCl₂ in the Brown-Norway rat, *Ann. Immunol. (Inst. Pasteur)* **129C**:777.

Gritzka, T. L., and Trump, B. F., 1968, Renal tubular lesions caused by mercuric chloride, *Am. J. Pathol.* **52**:1225.

Hinglais, N., Druet, P., Grossetete, J., Sapin, C., and Bariety, J., 1979, Ultrastructural study of nephritis induced in Brown-Norway rats by mercuric chloride, *Lab. Invest.* **41**:150.

Mendema, E., Arends, A., van Zeijst, J., Vermeer, G., von der Hem, G. K., and von der Slikke, L. B., 1963, Mercury and the kidney, *Lancet* **i**:1266.

Roman-Franco, A. A., Turiello, M., Albini, B., Ossi, E., and Andres, G. A., 1976, Anti-basement membrane antibody (ABMAb) and immune complexes (IC) in rabbits injected with mercuric chloride, *Kidney Int.* **10**:549.

Roman-Franco, A. A., Turiello, M., Albini, B., Ossi, E., Milgrom, F., and Andres, G. A., 1978, Anti-basement membrane antibodies and antigen–antibody complexes in rabbits injected with mercuric chloride, *Clin. Immunol. Immunopathol.* **9**:464.

Sapin, C., Druet, E., and Druet, P., 1977, Induction of anti-glomerular basement membrane antibodies in the Brown Norway rat by mercuric chloride, *Clin. Exp. Immunol.* **28**:173.

Woodroffe, A. J., and Wilson, C. B., 1977, An evaluation of elution techniques in the study of immune complex glomerulonephritis, *J. Immunol.* **118**:1788.

38

Genetic Control of Mercury-Induced Immune Response in the Rat

PHILIPPE DRUET, CATHERINE SAPIN,
ELVIRA DRUET, and FRANÇOIS HIRSCH

1. Introduction

Mercurials have been associated with immune-type glomerulonephritis (GN) in man (Mandema *et al.*, 1963) based on the finding of immune-type deposits. However the antigen(s) involved and the mechanism(s) responsible for this disease have not been elucidated. Similar glomerular deposits may occur in patients treated with gold salts or d-penicillamin and some drugs are known to induce lupus-like syndromes in humans. Thus it appears that a number of drugs and environmental or occupational toxins are responsible for immunologically mediated disorders in humans.

In order to elucidate the pathogenetic mechanism of mercury, we described a model of mercury-induced membranous nephropathy in outbred Wistar rats (Bariety *et al.*, 1971). However, only 30% of the rats developed the disease. Genetic control of the immune response and the role of immune response (Ir) genes linked

PHILIPPE DRUET, CATHERINE SAPIN, ELVIRA DRUET, and FRANÇOIS HIRSCH •
Laboratory of Morphology and Renal Immunopathology, Broussais Hospital, Paris, France.

to the major histocompatibility complex (MHC) have been established since the pioneering work of Benacerraf and McDevitt (1972). Genes linked to the MHC have also been shown to be involved in the genetic control of susceptibility to experimental allergic encephalomyelitis (Gasser et al., 1973) and autologous immune complex nephritis (Stenglein et al., 1975) in rats. We therefore decided to evaluate the role of MHC-linked genes in the susceptibility of various inbred rat strains to mercury-induced GN.

2. Susceptibility of Rats Bearing the RT1$^{n \ or \ l}$ Haplotype to Mercury-Induced Autoimmune Disease

2.1. Mercury-Induced Autoimmune Disease in the Brown Norway Rat

The mercury autoimmune disease was extensively studied in Brown Norway (BN) rats (RT1^n),* which were found to be the most susceptible strain (Sapin et al., 1977; P. Druet et al., 1978). Brown Norway rats develop a biphasic disease when injected three times a week for 2 months with HgCl$_2$ doses ranging from 200 to 5 μg/100 g body weight.

2.1.1. Clinical and Immunopathologic Findings

Glomerular and extrarenal linear IgG deposits were detected 8 days after the first HgCl$_2$ injection. IgG deposition was intense on day 15 and diminished or disappeared by day 60; linear staining for C3 was weaker (Sapin et al., 1977; Bernaudin et al., 1979). Granular renal and extrarenal deposits that stained with anti-IgG and anti-C3 conjugates were found by day 15 and were prominent after 1 month (P. Druet et al., 1978; Bernaudin et al., 1979).

Heavy proteinuria and nephrotic syndrome, both transient, appeared between days 12 and 20 in most rats (P. Druet et al., 1978). No major abnormalities could be seen upon light microscopy examination of kidney samples, except for the well-documented early HgCl$_2$-induced tubular necrosis (Hinglais et al., 1979). Detachment of endothelial cells and monocytic infiltration were the most prominent features observed by electron microscopy during the second week of the disease (Hinglais et al., 1979). Subepithelial and mesangial electron-dense deposits were only found after the second week (Hinglais et al., 1979).

Proteinuria in these animals was not influenced by decomplementation with cobra venom factor, while linear C3 deposits were no longer observed, indicating the non-complement dependence of proteinuria (Capron et al., 1982). Thus, anti-glomerular basement membrane (GBM) antibodies, circulating immune complex (IC)-like material (Bellon et al., 1982), monocytic infiltration (Hinglais et al., 1979), and intravascular coagulation with intracapillary fibrin deposits (Michaud et al., 1981), which have all been shown to occur in the course of the disease, may participate in the appearance of proteinuria.

---

*RT1$^{l,n,\cdots}$ refers to haplotypes of the rat major histocompatibility system.

2.1.2. Serologic Findings

Circulating anti-GBM antibodies of anti-laminin and anti-procollagen IV specificities were transiently detected between day 14 and day 30 by radioimmuno-assay and enzyme immunoassay (Bellon *et al.*, 1982). Circulating IC-like material was also present at the same period of time, usually by day 11, as assessed by the Raji cell and the Clq binding assays (Bellon *et al.*, 1982). Complement activation through the classical pathway could also be demonstrated to occur, transiently (Capron *et al.*, 1980), at the same time that circulating IC-like material was detected. In early experiments, a basement membrane antigen was suggested to be present within the renal immune deposits (P. Druet *et al.*, 1978); however, the antigen involved in the formation of circulating IC has not yet been characterized. The fact that anti-GBM antibodies and circulating IC are transiently present together with the self-limited character of the disease although $HgCl_2$ injections were pursued suggests that at one point immunologic events induced by $HgCl_2$ are suppressed (Bellon *et al.*, 1982).

2.2. Failure to Induce Mercury Autoimmune Disease in Rats Bearing the $RT1^l$ Haplotype

In contrast to BN rats ($RT1^n$), where the disease was constantly observed after $HgCl_2$ injections, all tested strains bearing the $RT1^l$ haplotype were found to be resistant. Thus, Lewis (LEW), BS, AS, and F.344 ($RT1^l$) rats when injected with 200 μg $HgCl_2$ did not develop any evidence of glomerular disease on day 15 or 60 despite the occurrence of acute renal tubular necrosis (E. Druet *et al.*, 1977, 1979). Resistance of F.344 and LEW rats to the $HgCl_2$ autoimmune disease was still observed when up to 400 μg $HgCl_2$ was injected. LEW rats did not develop circulating anti-GBM antibodies, and complement level remained within a normal range in these animals. Resistance of LEW rats was not due to the absence of antigenic determinants that are present in the basement membrane of BN rats, since circulating IgG with anti-GBM activity and kidney acid eluates obtained from BN rats injected with $HgCl_2$ revealed the basement membrane from both BN and LEW rats when tested by indirect immunofluorescence (E. Druet *et al.*, 1977). Furthermore, resistance of LEW rats was not due to a difference in the half-life or organ distribution of $HgCl_2$ when compared to BN rats, as shown by turnover studies using $^{203}HgCl_2$ (unpublished data).

3. Susceptibility of Segregants between BN and LEW Rats to Mercury-Induced Autoimmune Disease

In order to examine the susceptibility of segregants to the induction of anti-GBM antibodies and of IC-type deposits, (LEW × BN) F_1 and F_2 hybrids, (LEW × BN) F_1 × LEW and (LEW × BN) F_1 × BN backcrosses, and congenic LEW.1N rats were injected with 200 or 100 μg $HgCl_2$ three times a week for 2 months. The RT1 haplotype was determined in these animals using the dextran hemagglutination test with specific reciprocal alloantisera (E. Druet *et al.*, 1977). Kidney samples obtained on days 15 and 60 were stained with an anti-rat IgG conjugate. A linear pattern of fixation was considered as demonstrative of the presence of anti-GBM antibodies. A granular pattern of fixation was considered as the hallmark of an IC-type disease.

3.1. Susceptibility of Segregants between LEW and BN Rats to the Induction of Anti-Glomerular Basement Membrane Antibodies by HgCl₂

3.1.1. F₁ Hybrids

All F_1 hybrids, whether injected with 200 or 100 μg $HgCl_2$, either male or female, exhibited linear IgG deposits on day 15, indicating that susceptibility was inherited as a dominant and autosomal trait (Table I).

3.1.2. Segregants Bearing the RT1$^{(1/1)}$ Haplotype

Segregants bearing the RT1$^{(1/1)}$ haplotype did not exhibit anyglomerular deposit on day 15 or 60 (Table I).

3.1.3. Segregants Bearing the RT1^{n} Haplotype

Linear IgG deposits along the GBM were only found among segregants bearing the RT1$^{n(1/n \text{ or } n/n)}$ haplotype; however several of these segregants were negative on day 15 and remained negative on day 60 (Table I).

These results demonstrate the role of genes, or of a cluster of genes, linked to the MHC in the susceptibility of segregants to the induction of anti-GBM antibodies by $HgCl_2$. The absence of linear IgG deposits in several segregants

TABLE I. Immunofluorescence Findings in Segregants between BN and LEW Rats Injected with HgCl₂[a]

Segregants	RT1	Number positive/number tested	
		Linear IgG deposits	Immune-complex type deposits
(LEW×BN) F₁	1/n	17/17	11/11
(LEW×BN) F₂	1/1	0/22	0/16
	1/n	19/31	37/38
	n/n	8/14	15/16
(LEW×BN) F₁ × LEW	1/1	0/27	0/33
	1/n	9/20	30/30
(LEW×BN) F₁ × BN	n/n	NT	25/25
	1/n	NT	34/34
LEW.1N	n/n	0/8	NT

[a] Segregants injected with 200 μg or 100 μg per 100 g body weight three times a week for 2 months. Linear IgG deposits and immune-complex-type deposits were looked for by immunofluorescence on days 15 and 60.

bearing the $RT1^n$ haplotype suggests that non-MHC-linked genes are also involved. When the data obtained were compared to those expected in a polygenic model, results were compatible with a two- or three-gene model (E. Druet *et al.*, 1977).

3.1.4. Congenic LEW.1N Rats

Congenic LEW.1N rats ($RT1^n$) did not develop linear IgG deposits, thus confirming the role of non-MHC-linked genes in induction of anti-GBM antibodies by $HgCl_2$. Susceptibility therefore depends on two or three codominant autosomal genes, one of which is RT1-linked (E. Druet *et al.*, 1977).

3.2. Susceptibility of Segregants between LEW and BN Rats to the Induction of Immune Complex-Type Deposits by $HgCl_2$

Since many segregants injected with 200 μg died, only rats injected with 100 μg were tested for the occurrence of IC-type deposits on day 60 (Sapin *et al.*, 1981).

3.2.1. F_1 Hybrids

All F_1 hybrids injected with $HgCl_2$, either male or female, exhibited granular IgG deposits in the glomerulus as well as in vessel walls, in the kidney, and in the spleen (Table I).

3.2.2. Segregants Bearing the $RT1^{(1/1)}$ Haplotype

None of the segregants bearing the $RT1^{(1/1)}$ haplotype showed any granular deposits in the kidney or in the spleen (Table I).

3.2.3. Segregants Bearing the $RT1^n$ Haplotype

All the segregants bearing the $RT1^{n(1/n \text{ or } n/n)}$ haplotype, except for two F_2 hybrids (one 1/n and one n/n), had granular IgG deposits in the kidney and/or in the spleen (Table I). However, differences in the localization and in the amount of granular deposits were observed among the segregants. Some segregants had only few scattered granular deposits in the kidney. Others had granular deposits in vessel walls, in the spleen, and in the kidney, but not in the glomerulus. Others had diffuse granular IgG deposits in vessel walls, in the spleen, and in the kidney as well as in the glomerulus. Since such important differences in IgG deposition could be observed, it is likely that the two negative F_2 hybrids had too few IgG deposits to be detected.

The susceptibility to the second phase of $HgCl_2$-induced disease is inherited as an autosomal and dominant trait and depends on one major gene, or cluster of genes, MHC-linked. Differences observed between segregants suggest that other genes are probably involved (Sapin *et al.*, 1981). Two other conclusions may be drawn from these data. First, the two phases probably do no depend on each other, because several segregants had IC-type deposits without evidence for the presence of anti-GBM antibodies. Second, spleen is, in this model, a more susceptible target to IC deposition than is the kidney.

3.3. Role of Immunocompetent Cells in Susceptibility to Mercury-Induced Autoimmune disease

Experiments with segregants presented above suggested that the resistance of LEW rats could be overcome provided they had the suitable immunocompetent cells. In order to test this hypothesis, LEW rats were lethally irradiated and reconstituted with either bone narrow cells or spleen cells from (LEW × BN) F_1 hybrids. Animals were tested for acquisition of $RT1^n$ haplotype 1 month after reconstitution. Chimeric animals were injected with 200 μg $HgCl_2$ three times a week for 2 months and kidney specimens were examined using direct immuno-fluorescence on days 15 and 60. Controls included (1) LEW rats irradiated but not reconstituted, (2) LEW rats reconstituted with syngenic immunocompetent cells, and (3) irradiated LEW rats that spontaneously returned to the initial $RT1^1$ haplotype after reconstitution with immunocompetent cells from F_1 hybrids (Sapin et al., 1980).

Linear IgG deposits on day 15 and granular deposits on day 60 were only observed in rats reconstituted with either bone narrow or spleen cells and which remained chimeric ($RT1^{1/n}$).

These data demonstrate that products of the genes involved in controlling susceptibility to $HgCl_2$ autoimmune disease are expressed on immunocompetent cells (Sapin et al., 1980).

4. Evidence for HgCl$_2$-Induced Polyclonal Activation in the Brown Norway Rat

The mechanism by which $HgCl_2$ induces an autoimmune disease in the BN rat could be either a modification of a self-component, such as basement membrane, or/and a direct triggering of immunocompetent cells. We investigated the latter possibility, since we observed an increase in total serum IgE in BN rats injected with $HgCl_2$.

4.1. Serum IgE in Brown Norway and Lewis Rats Injected with HgCl$_2$

Nonspecific serum IgE was sequentially measured in both LEW and BN rats injected with $HgCl_2$ and with control solution. IgE level was determined by radial-immunodiffusion and by bioassay (Prouvost-Danon et al., 1981). Baseline IgE in BN rats was lower than 4 μg/ml and lower than 2 μg/ml in LEW rats.

Total serum IgE increased in BN rats, but not in LEW rats, after $HgCl_2$ injections. IgE level rose from day 5 to reach its maximum value by day 15. The maximal value obtained could reach 2000 μg/ml (Fig. 1). No specificity could be demonstrated for IgE induced by $HgCl_2$ in BN rats: no IgE was deposited along the GBM on day 15 and serum from BN rats at the peak of the IgE response did not

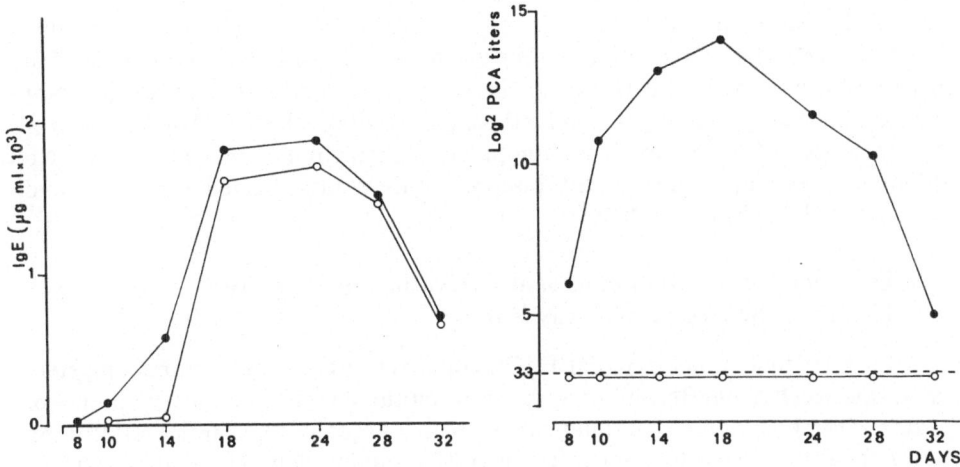

FIGURE 1. Total serum IgE level (left panel) and specific anti-ovalbumin IgE response (right panel) in ten BN rats immunized with ovalbumin from day 0 and injected with HgCl$_2$ from day 4. Tests were performed on pooled sera before (●) and after (○) absorption with ovalbumin.

stain normal BN kidneys by indirect immunofluorescence using an anti-rat ε antiserum; furthermore, there was no decrease in IgE level after absorption of BN rat serum obtained at the peak of the IgE response, with rat GBM or with insolubilized normal rat serum, whether pretreated or not with HgCl$_2$ (Prouvost-Danon *et al.*, 1981).

These data are reminiscent of those described in rats after parasitic infection. Parasites induce an increase of nonspecific IgE by potentiating the IgE response to various ongoing antigens, as seen in good IgE responders such as BN rats. We therefore compared the specific anti-ovalbumin IgE response in BN and LEW rats injected with HgCl$_2$. The specific anti-ovalbumin IgE response was potentiated in BN but not in LEW rats. However, it only represented a trace amount of the total serum IgE as assessed by absorption experiments using insolubilized ovalbumin (Fig. 1).

Preliminary results from our laboratory suggest that the HgCl$_2$-induced IgE response is under polygenic control.

4.2. Evidence for *in Vivo* Polyclonal Activation Induced by HgCl$_2$ in the Brown Norway Rat

Spleen and lymph nodes from BN rats injected with HgCl$_2$ were enlarged on day 12, with a significant increase of the number of spleen cells as compared to control BN rats. No similar modification was found in LEW rats injected with HgCl$_2$. At the same time, total serum IgM level significantly increased in BN rats injected with HgCl$_2$ (Hirsch *et al.*, 1982).

We then investigated the anti-DNA and anti-TNP (2,4,6-trinitrophenyl) activities in BN and LEW rats injected with HgCl$_2$, since these antibody specificities are considered as markers of *in vivo* polyclonal activation. The single-stranded DNA (SS-DNA) binding activity of serum was significantly increased on day 7 in BN rats injected with $HgCl_2$ but not in LEW rats. There was also a significant increase in both IgM and IgG anti-TNP plaque-forming cells (PFC) in the spleen of BN rats injected with $HgCl_2$ as compared to control BN rats (Table II). No difference was found between PFC obtained in LEW rats injected with $HgCl_2$ and in controls (Hirsch *et al.*, 1982).

4.3. Evidence for *in Vitro* Polyclonal Activation Induced by $HgCl_2$ in the Brown Norway Rat

In order to demonstrate that $HgCl_2$ could act directly on immunocompetent cells, spleen cells from BN and LEW rats were cultured in the presence or absence of $HgCl_2$. The $HgCl_2$ induced a significant increase of IgM and IgG anti-TNP PFC on day 4 of culture. In contrast, no effect of $HgCl_2$ could be demonstrated on cultured spleen cells from LEW rats (Hirsch *et al.*, 1982).

These data support the hypothesis that $HgCl_2$ acts on immunocompetent cells so as to induce polyclonal activation. Recent experiments by Weening (1980) using PVG/c rats demonstrated that $HgCl_2$ reduced the generation of suppressive cells by concanavalin A. Our observations in the BN rat are compatible with an effect of $HgCl_2$ on B cells and/or on T cells.

5. Susceptibility of Various Inbred Rat Strains to $HgCl_2$-Induced Autoimmune Glomerulonephritis

Since BN rats ($RT1^n$) differed from rats bearing the $RT1^l$ haplotype in their susceptibility to the induction of $HgCl_2$-induced GN, we investigated the ability of various inbred rat strains to develop after $HgCl_2$ injections an immune-type GN (Table III) and an antibody response toward GBM, brush border from convoluted proximal tubule, nuclear antigens, and SS-DNA.

None of the rats bearing the $RT1^l$ haplotype developed an immune-type GN or autoantibodies. Several strains—PVG/c and AUG ($RT1^c$), DA and AVN

TABLE II. Anti-TNP Response in Rats Injected with $HgCl_2$ Compared to Control

Strain	$HgCl_2$	PFC per spleen $\pm$ SD on day 12 of intoxication[a]	
		IgM	IgG
BN	+	$44,400 \pm 5,200^b$ (5)	$32,400 \pm 1,200^b$ (5)
BN	−	$9,300 \pm 5,100$ (4)	$1,200 \pm 400$ (4)
LEW	+	$14,200 \pm 2,900$ (4)	$3,100 \pm 900$ (4)
LEW	−	$18,000 \pm 300$ (4)	$3,100 \pm 1,000$ (4)

[a] Direct (IgM) and indirect (IgG) plaque-forming cells (PFC). Numbers in parentheses indicate the number of rats tested.
[b] $p < 0.01$ as compared to control.

TABLE III. Susceptibility of Inbred Rat Strains to Immune Complex Type (ICT) Nephritis Induced by $HgCl_2$: Immunofluorescence Findings[a]

Strain	RT1	ICT deposits, no. positive/no. tested	Strain	RT1	ICT deposits, no. positive/no. tested
LEW	l	0/11	LEW.1D	d	6/6
F.344	l	0/6	AS$_2$	f	8/8
BS	l	0/5	OKA	k	5/5
AS	l	0/8	WAG	u	0/9
PVG/c	c	30/30	LOU	u	0/5
AUG	c	5/5	LEW.1U	u	0/5
DA	a	16/16	WF	u	1/10
AVN	a	3/16	WF[b]	u	5/5
BD$\underline{V}$	d	9/9			

[a] Rats were injected with 200 µg $HgCl_2$/100 g body weight.
[b] Injected with 400 µg.

(RT1^a), AS$_2$ (RT1^f), OKA (RT1^k), BD$\underline{V}$ and LEW.1D (RT1^d), and BUF (RT1^b)—developed an IC-type GN with granular IgG deposits in the glomerular capillary wall and/or in the mesangium; granular deposits were also found in the vessel walls and in the tubular basement membrane from PVG/c and BUF rats (E. Druet *et al.*, 1979). No anti-GBM or anti-brush border antibodies were found in any strain. Antinuclear antibodies and anti-SS DNA antibodies were found on day 60 to a significant titer as compared to control rats in PVG/c, DA, BD$\underline{V}$, OKA, and BUF rats. Individual high values were also observed among AUG and OKA rats. The AS$_2$ and LEW.1D rats had no detectable anti-SS-DNA antibodies, although they develop an IC-type GN. Anti-SS DNA antibodies were recovered after kidney acid elution in PVG/c rats (E. Druet *et al.*, 1979). Weening *et al.* (1978), who first described the disease in PVG/c rats, found circulating antinucleoprotein antibodies.

Four strains (WAG, WF, LOU, and LEW.1U), all bearing the RT1^u haplotype, were also tested for their susceptibility to develop similar autoimmune abnormalities after $HgCl_2$ injections. All strains were found to be resistant. However, WR rats, when injected with a higher $HgCl^2$ dose (400 µg), developed a membranous nephropathy with heavy granular IgG deposits, without detectable anti-brush border or anti-SS DNA antibodies.

These data show the peculiarity of BN rats after $HgCl_2$ administration as compared to 17 other strains. It is noteworthy that the disease induced in BN rats closely resembles that obtained in rabbits after $HgCl_2$ injections (Roman-Franco *et al.*, 1978).

Our results in several strains of rat bearing the same RT1 haplotype together with those obtained in BD$\underline{V}$ and congenic LEW.1D suggest a role for MHC linked gene(s) in the susceptibility of rats to $HgCl_2$-induced IC-type nephritis. Anti-SS-DNA antibodies were only found among those rats that develop an IC-type nephritis and were eluted from PVG/c rat kidneys, suggesting a pathogenetic role for these antibodies. The absence of anti-SS-DNA antibodies in AVN, AS$_2$, and congenic LEW.1D rats indicates that these antibodies are not necessarily required for the nephritis to occur.

6. Conclusion

Several conclusions can be drawn from these studies.

Brown Norway rats developed a unique biphasic autoimmune disease when injected with $HgCl_2$. The susceptibility of BN rats to both phases of the disease depends upon a few genes, including MHC-linked genes, as demonstrated in segregants between BN (susceptible) and LEW (resistant) rats injected with $HgCl_2$. The compound $HgCl_2$ behaves as an *in vivo* and *in vitro* polyclonal activator in BN but not in LEW rats.

Genes linked to the MHC are also involved in the susceptibility of several rat strains to an immune-type but not anti-GBM-mediated GN. Anti-SS-DNA antibodies participate in the pathogenesis of the latter immune-type GN in some strains of rats.

Thus, under similar experimental conditions, the phenotypic expression of the disease differs with the strain that is tested.

Acknowledgments

We wish to thank Dr. M. Kazatchkine for reviewing the manuscipt and Laurence Meurisse for typing it.

Several inbred rat strains were provided by Dr. E. Gunther (Max-Planck Institüt für Immunbiologie, Freiburg, West Germany) and by Dr. H. Bazin (Université Catholique de Louvain, Brussels, Belgium).

This work was supported by DGRST (78/7/2625) and INSERM (CRL 805023).

References

Bariety, J., Druet, P., Laliberté, F., and Sapin, C., 1971, Glomerulonephritis with δ and Blc globular deposits induced in rats by mercuric chloride, *Am. J. Pathol.* **65**:293.

Bellon, B., Capron, M., Druet, E., Verroust, P., Vial, M. C., Sapin, C., Girard, J. F., Foidart, J. M., Mahieu, P., and Druet, P., 1982, Mercuric chloride induced auto-immune disease in BN rats. Sequential search for anti-basement membrane antibodies and circulating immune complexes, *Eur. J. Clin. Invest.* **12**:127.

Benacerraf, B., and McDevitt, H. O., 1972, The histocompatibility linked immune response genes, *Science* **175**:273.

Bernaudin, J. F., Druet, E., Belair, M. F., and Pinchon, M. C., Sapin, C., and Druet, P., 1979, Extrarenal immune complex type deposits induced by mercuric chloride in the Brown-Norway rat, *Clin. Exp. Immunol.* **38**:265.

Capron, M., Ayed, K., Druet, E., Sapin, C., Mandet, C., Druet, P., and Girard, J. F., 1980, Complement studied in BN rats with mercuric chloride induced immune glomerulonephritis, *Ann. Immunol. (Inst. Pasteur)* **131D**:43.

Capron, M., Bascou, C., Vial, M. C., Grossetete, J., Hinglais, N., Girard, J. F., and Druet, P., 1982, Effects of decomplementation on mercuric chloride induced glomerulonephritis in Brown-Norway rats, *Clin. Exp. Immunol.,* in press.

Druet, E., Sapin, C., Günther, E., Feingold, N., and Druet, P., 1977, Mercuric chloride induced anti-glomerular basement membrane antibodies in the rat. Genetic control, *Eur. J. Immunol.* **7:**348.

Druet, E., Fournié, G., Mandet, C., Sapin, C., Günther, E., and Druet, P., 1979, Genetic control of susceptibility to immune complex type nephritis induced by $HgCl_2$ in rats, *Transpl. Proc.* **11:**1600.

Druet, P., Druet, E., Potdevin, F., and Sapin, C., 1978, Immune type glomerulonephritis induced by $HgCl_2$ in the Brown-Norway rat, *Ann. Immunol. (Inst. Pasteur)* **129C:**777.

Gasser, D. L., Newlin, C. M., Palm, J., and Gonatas, N. K., 1973, Genetic control of susceptibility to experimental allergic encephalomyelitis in rats, *Science* **181:**872.

Hinglais, N., Druet, P., Grossetete, J., Sapin, C., and Bariety, J., 1979, Ultrastructural study of nephritis induced in Brown-Norway rats by mercuric chloride, *Lab. Invest.* **41:**150.

Hirsch, F., Couderc, J., Sapin, C., Fournié, G., and Druet, P., 1982, Polyclonal effect of $HgCl_2$ in the rat. Its possible role in mercury induced glomerulonephritis, *Eur. J. Immunol.,* in press.

Mandema, E., Arends, A., Van Zeijst, J., Vermeer, G., Van Der Hem, G. K., and Van Der Slikke, L. B., 1963, Mercury and the kidney, *Lancet* **i:**1266.

Michaud, A., Sapin, C., Aïach, M., Druet, P., and Forestier, F., 1981, Clotting abnormalities during the course of immune type glomerulonephritis induced by $HgCl_2$ in the Brown-Norway (BN) rat, in: VIIIth International Congress on Thrombosis and Haemostasis, Toronto, July 1981.

Prouvost-Danon, A., Abadie, A., Sapin, C., Bazin, H., and Druet, P., 1981, Induction of IgE synthesis and potentiation of anti-ovalbumin IgE antibody response by $HgCl_2$ in the rat, 1981, *J. Immunol.* **126:**699.

Roman-Franco, A. A., Turiello, M., Albini, B., Ossi, E., Milgrom, F., and Andres, G. A., 1978, Anti-basement membrane antibodies and antigen–antibody complexes in rabbits injected with mercuric chloride, *Clin. Immunol. Immunopathol.* **9:**464.

Sapin, C., Druet, E., and Druet, P., 1977, Induction of anti-glomerular basement membrane antibodies in the Brown-Norway rat by mercuric chloride, *Clin. Exp. Immunol.* **28:**173.

Sapin, C., Druet, P., and Mandet, C., 1980, Induction of susceptibility to $HgCl_2$ immune glomerulonephritis in the Lewis rat by immunocompetent cells from susceptible F_1 hybrids, *Eur. J. Immunol.* **10:**371.

Sapin, C., Mandet, C., Druet, E., Günther, E., and Druet, P., 1981, Immune complex type disease induced by $HgCl_2$ in the BN rat: genetic control of susceptibility, *Transpl. Proc.* **13:**1404.

Stenglein, B., Thoenes, G. H., and Günther, E., 1975, Genetically controlled autologous immune complex glomerulonephritis in rats, *J. Immunol.* **115:**895.

Weening, J. J., 1980, Mercury induced immune complex glomerulopathy. An experimental study, Thesis, Groningen, The Netherlands.

Weening, J. J., Fleuren, G. J., and Hoedemaeker, P. J., 1978, Demonstration of anti-nuclear antibodies in mercuric chloride induced glomerulopathy in the rat, *Lab. Invest.* **39:**405.

39

The Application of a Radioimmunoassay for Sensitive Detection of Metallothionein (Thionein) in Physiologic Fluids of Humans and Rats

JUSTINE S. GARVEY

1. Introduction

Recently a radioimmunoassay (RIA) has been developed in my laboratory for the sensitive detection of metallothionein (Mt) or its precursor, thionein (Th). These metal-binding proteins are of considerable interest to toxicologists (see this volume, Chapters 25 and 26). The assay (VanderMallie and Garvey, 1979; Chang *et al.*, 1980a) is a competitive, double antibody (Ab) type of RIA, set up in tubes and involving three stages of reaction at 4°C. For standardization, the first stage is incubation of primary Ab (specific Ab vs Mt) with known amounts of unlabeled

JUSTINE S. GARVEY • Department of Biology, Syracuse University, Syracuse, New York 13210.

Mt; the second stage is a continued incubation after addition of $[^{125}I]$-Mt; and the third stage is addition of secondary Ab and continued incubation. Centrifugation is then used to separate the bound $[^{125}I]$-Mt from the free $[^{125}I]$-Mt and the radioactivity in the precipitate is determined by gamma counting. Along with the assay tubes with varying amounts of known unlabeled Mt, there is an assay for "maximum bound" (B_0), i.e., a tube that lacks any nonlabeled competing Mt, and also an assay for "nonspecific bound" (NSB), i.e., a tube that lacks the primary Ab. Using the precipitated counts (cpm) in the relationship (cpm − NSB)/(B_0 − NSB), a plot is prepared relating % bound (Y) to log Mt protein. This customarily sigmoid curve is called a standard curve. In the case of the RIA for Mt it characteristically has an essentially linear region which can be well-represented by a linear-log regression relating Y and log Mt. This regression is then used to quantitate samples containing unknown amounts of Mt. The standardizing data in this RIA are also readily amenable to logit-log analysis relating logit Y {where Y is now fraction bound; logit Y is $\log[Y/(1 − Y)]$} and log Mt. Such a regression extends the region of utility of the data to relatively high or low Mt concentrations. In contrast to the RIA for Mt, many substances are characterized by assay data that are not amenable to linear-log or logit-log regressions; polygonal curves are sometimes used, and more complicated cases exist (Vogt *et al.*, 1978).

In the utilization of the RIA for the quantitation of unknowns, the latter are added in the first stage of the reaction in place of the nonlabeled competing Mt; otherwise the assay proceeds as described above. In the following sections I will briefly discuss some recent experiments involving use of the RIA in studies involving Mt and various metals.

2. Identity of Human and Rat Species of Liver Mt

Figure 1 shows experimental standard curves from which linear-log regressions were developed for determination of rat liver Mt and of human liver Mt (Chang *et al.*, 1980a). Two points are noteworthy here: sensitivity of the RIA for Mt is 100–200 pg; and there is an apparent parallelism of the human and rat standard curves in the linear region. The linear-log regressions then developed from the data confirm that the two slopes are not significantly different ($p > 0.05$). This indicates an identity of the two reacting protein species, these two species of Mt having complete cross-reactivity.

3. Confirmation That a Copper-Binding Protein Is a Copper-Thionein

Figure 2 shows an example of a logit-log regression analysis of RIA data which was developed in a collaborative study with Dr. Dennis Winge and associates, involving the quantitation of Cu-binding protein to determine if it actually was a Cu-Th (Winge *et al.*, 1981). The logit-log analysis permitted use of the RIA data at the extremes of the sigmoid standard curves. The regressions shown are for assays of a known (Cd, Zn)-Th (labeled Cd), and two samples of Cu-binding proteins

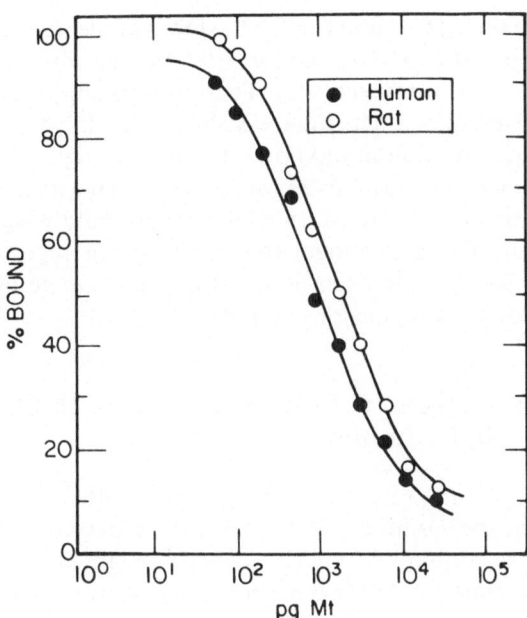

FIGURE 1. Standard curves in the RIA for human and rat liver Mt. Percent bound is shown as a function of log Mt concentration (in pg). The sigmoid character of the response is evident. The data in the essentially linear region are later used to develop linear-log regressions for subsequent quantitation of unknowns. These regressions are characterized by slopes of -35.3 and -37.4, respectively; these slopes are not significantly different ($p > 0.05$). The two species of Mt thus appear to have complete cross-reactivity. [Reprinted with permission from Chang *et al.* (1980a).]

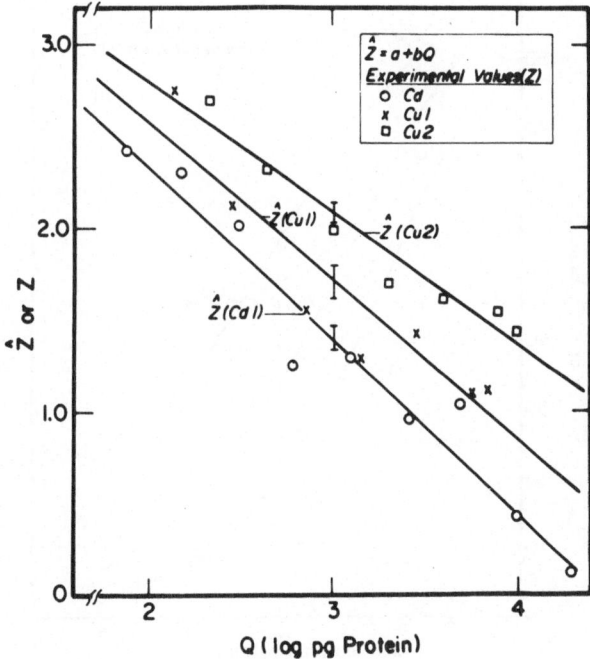

FIGURE 2. Analysis of RIA standardizing data involving a Cu-thionein and a (Cd, Zn)-thionein: logit-log regressions are shown for the assay data associated with a (Cd, Zn)-thionein ($\bigcirc$, labeled Cd) and two preparations of a Cu-binding protein ($\times$, labeled Cu1; $\square$, labeled Cu2). $\hat{Z}$ is log $100W$, where W is logit Y, with Y expressed as fraction bound protein; Mt concentration is in pg, and Q is log Mt. The regressions for the three $\hat{Z}$'s are (for Cd, Cu1, and Cu2, respectively) $4.230 - 0.941Q$, $4.326 - 0.871Q$, $4.222 - 0.714Q$. In the same sequence, the correlation coefficients are $-0.976, -0.931, -0.972$. For the three slope differences (Cu2–Cd, Cu2–Cu1, and Cu1–Cd) the null hypothesis (that the slopes are equal) cannot be rejected ($p > 0.1, p > 0.3$, $p > 0.5$ in the listed sequence). [Reprinted with permission from Winge *et al.* (1981).]

(labeled Cu1 and Cu2). The slopes of the three curves are not significantly different ($p > 0.1$, $p > 0.3$, and $p > 0.5$ for the three differences Cu2–Cd, Cu2–Cu1, and Cu1–Cd, respectively). These results support the conclusion that the Cu-binding protein is a Mt. Figure 3 shows, for illustrative purposes only (not used in the quantitation of unknowns), the logit-log relations of Fig..2 when transformed to linear-log form (these curves are drawn with a flexcurve through five points only in each case). Transformed curves are mildly sigmoid compared to the linear form (restricted in range) that a linear-log regression would have taken, and they illustrate the extension of the useful range of the standard curve data that the logit-log regressions permit.

4. Detection of Mt in Humans with Occupational Exposure to Cadmium

In rats treated subcutaneously with 2 mg Cd/kg body weight, Mt was detected in the serum at 29 hr; it was also detected at 48 hr after the first of two daily injections of 0.5 mg Cd/kg body weight. These findings, along with the reaction of identity of rat Mt and human Mt mentioned in the preceding section, indicated the potential use of the RIA for the assay of endogenous human Mt. The levels of Mt in plasma and urine of groups of humans occupationally exposed to Cd and of such levels in groups not so exposed have been studied in collaboration with Dr. Robert

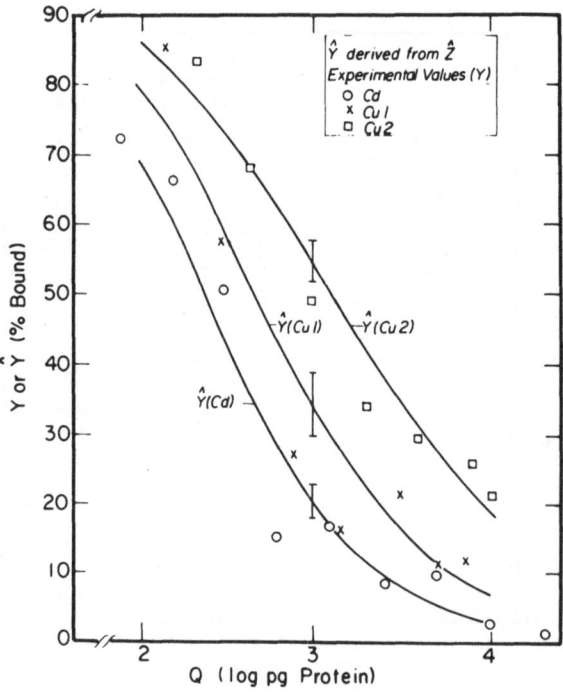

FIGURE 3. Analysis of RIA standardizing data involving a Cu-thionein and a (Cd, Zn)-thionein: linear-log representations of percent bound (Y) vs log Mt concentration are derived from the regressions of Fig. 2. Curves are presented for illustrative purposes only, and are based on values calculated for Q (log Mt in pg) equal to 2, 2.5, 3, 3.5, and 4, these connected by flexcurve. [Reprinted with permission from Winge *et al.* (1981).]

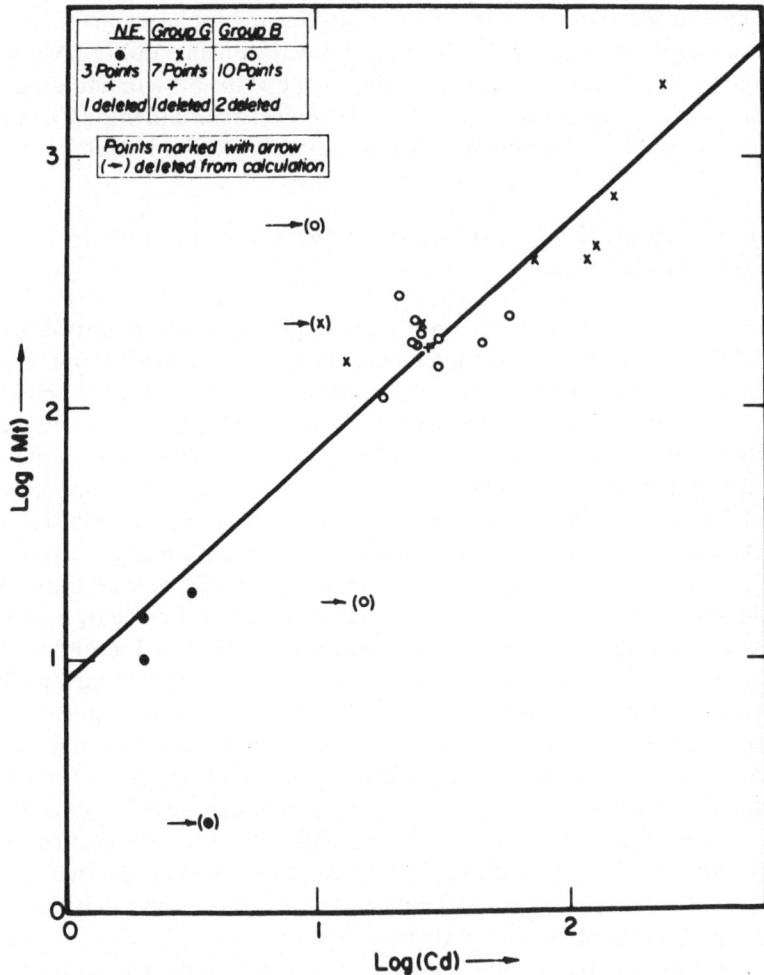

FIGURE 4. Analysis of RIA data relating Mt and Cd concentrations in urine of Cd-exposed and nonexposed humans: log Mt as a function of log Cd. The Mt and Cd concentrations are expressed in $\mu g/g$ creatinine in urine. N.E. indicates nonexposed; exposed subjects are from two differing occupational environments (G,B). The regression is $\log Mt = 0.9157 + 0.9058 \log Cd$; $r = 0.962$. It is based on 20 values (of the total of 24), with four variant values indicated by arrows. [Based on data from Chang *et al.* (1980b).]

Lauwerys and associates (Chang *et al.*, 1980b). The study included data from assays of body fluids of 32 subjects, including two groups with differing occupational exposure and two groups of nonexposed subjects; one of the latter was a group of eight whose Mt concentrations in urine were at or below the limit of accurate detection in the assay. The results suggested that Mt concentration in urine as measured by the RIA was more related to the Cd level in urine than to renal status (as shown by various biologic variables). Despite there being no demonstrable difference in the distribution of Mt levels regardless of renal parameter examined, there was a significant correlation between log Cd in urine and log Mt in urine ($r = 0.94$; $p < 0.05$). This log–log relationship is one of the principal findings of the experiment. The data evidenced no simple relation between either log Cd or log Mt and years of exposure, a matter to which attention is next addressed.

5. Commentary on the Experiment on Mt and Cd Levels in Cd-Exposed Humans

Some speculations resulting from additional processing of the data reported in the experiment of Section 4 are intruded here. They concern possible potential uses of the RIA based on various interpretations of the data. If we omit the mentioned eight controls with vanishing Mt levels and restrict the analysis to the remaining 24 subjects (including four nonexposed subjects, eight from one occupationally exposed group, and 12 from another such group), one observes that four of the total number of subjects have Mt–Cd relations considerably variant from the remaining 20 subjects. Shown in Fig. 4 is a plot of log Mt as a function of log Cd, with the indicated log–log regression (log Mt $= 0.9157 + 0.9058$ log Cd; $r = 0.962$) derived for the mentioned 20 values only. The four variant points are indicated by arrows. The slope implies that log(Mt/Cd) is almost independent of log Cd. Figure 5 shows the regression [log(Mt/Cd) $= 0.9134 - 0.0926$ log Cd; $r = -0.339$] relating log(Mt/Cd) and log Cd. The slope and correlation coefficient support the implication from Fig. 4; again the four variant points are indicated by arrows. As mentioned in the previous section, a plot of log Mt or log Cd as a function of years of exposure yields no clear relation, but a plot of log(Mt/Cd) vs years of exposure X yields (Fig. 6) a log-linear regression [log(Mt/Cd) $= 0.7193 + 0.004123X$; $r = 0.351$] with little slope and small correlation. This result might be interpreted as evidence that there is a "normal" Mt/Cd ratio of about 6 (the mean value is 6.013) and that this ratio is relatively independent of years of exposure. This might imply, using 6200 for the molecular weight of thionein and a maximum of seven bound Cd atoms per Th molecule, that a "normal" condition in urine for Cd-exposed subjects is (approximately) 3–4 free Cd atoms for each Mt molecule (assumed saturated). However, the actual range of Mt/Cd values is 3–12, and one would appreciate more information on the concentrations of Zn, Cu, and Hg in the urine before making more elaborate speculations. The same is true when considering the four variant points (these points are 3.4, 4.8, 6.0, and 6.4 standard deviations from the mean of the other 20 points). A tentative hypothesis regarding these four points might be that they reflect either unusual conditions (body function) or the presence of significant amounts of bound metals (or Mt-inducing metals and/or hormones) other than Cd.

It is conceivable that the use of the RIA in conjunction with determined concentrations of the most prominent of the metals bound by Th (Zn, Cd, Cu, Hg) might gradually lead to establishing useful "normal" values of certain parameters and thus permit identification of "abnormal" values, these subsequently to be correlated with specific functional or toxicologic conditions.

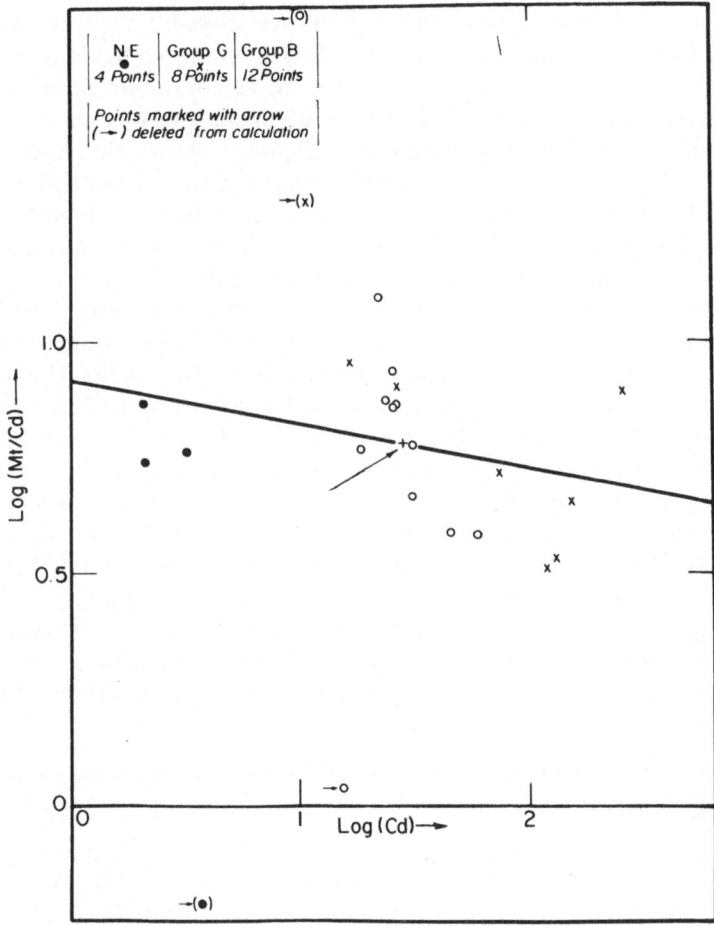

FIGURE 5. Analysis of RIA data relating Mt and Cd concentrations in urine of Cd-exposed and nonexposed humans: log(Mt/Cd) as a function of log Cd. Units of concentration are as in Fig. 4. The derived regression is based on the 20 values of Fig. 4, and the four variant values are indicated by arrows. The regression is log(Mt/Cd) = 0.9134 − 0.0926 log Cd; $r = -0.339$. [Based on data from Chang *et al.* (1980b).]

6. Levels of Circulating Mt in Pretreated Rats and in Rats Treated with Nontoxic Amounts of Zinc and Cadmium

An experiment was recently performed in my laboratory using rats to establish the capability of the RIA to quantify low concentration of circulating Mt (Garvey and Chang, 1981). This was in part stimulated by the fact that the detection of Mt in plasma had not been quantitatively observed (Kägi and Nordberg, 1979). From single and multiple injection studies it was concluded to use 0.8 mg Cd/kg body weight and 5 mg Zn/kg body weight as the treatment levels (adequate to allow accurate measurements of Mt in serum and yet not be toxic), and to make three intraperitoneal (IP) injections of these quantities spaced 3 days apart (Cd was injected as $CdCl_2$ in saline, Zn as $ZnSO_4$ in saline, and controls received only physiologic saline). The animals were bled at 18- or 24-hr intervals (a few bleedings were also done at 6- or 12-hr intervals) over a 13-day period. The data were analysed using logit-log regressions. Figure 7 shows the results for the Cd-treated and Zn-treated rats ($n = 5$ in both cases) and for the controls ($n = 2$). The indicated standard errors of the means include those associated with use of the regressions (the error is a function of distance from the mean value of the regression). The mean control levels are shown for the 0–3, 3–6, 6–9, and 9–13 day periods. For the entire 13-day period the mean control level was 1.79 ± 0.41 ng Mt/ml serum (significantly different from zero; $p < 0.001$). The mean control levels for the four periods are not statistically distinguishable ($p > 0.25$). The regression developed for the Zn injections and the controls had a better correlation coefficient (-0.99) than did that for the Cd injections (-0.97) and the resultant smaller values of standard errors associated with measurements following Zn injections permitted assigning significance ($p < 0.01$) to the differences between the second and third relative maxima and their respective control levels as well as to the difference between the relative maxima at 4 and 7 days. In the case of the Cd injections, although the initial maximum was significantly different from the control level ($p < 0.02$), and the successive relative maxima similarly distinguishable from control levels ($p < 0.05$), the reduced level of significance primarily reflects the reduced accuracy of the regression used compared to that developed for the Zn

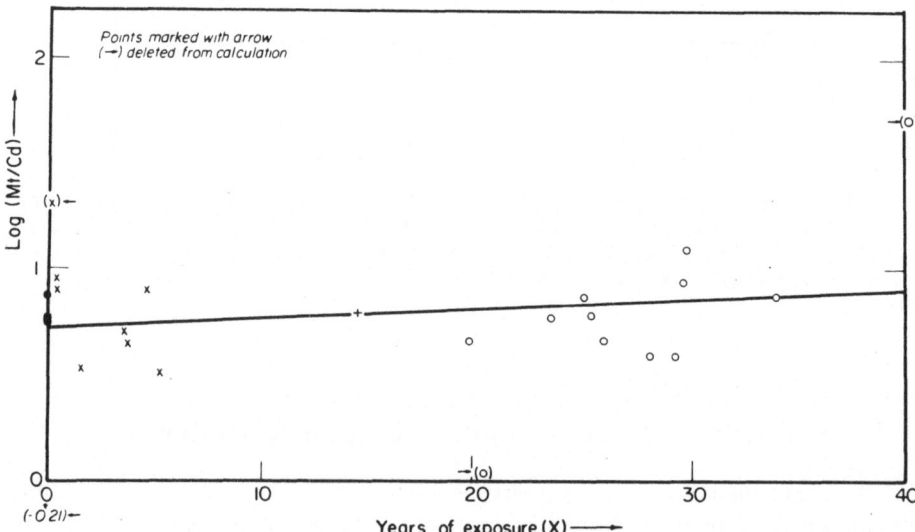

FIGURE 6. Analysis of data relating Mt and Cd concentrations in urine of Cd-exposed and nonexposed humans: log(Mt/Cd) as a function of years of exposure. Concentration units are those of Fig. 4. The derived regression is based on the 20 values used in Figs. 4 and 5; the four variant values are indicated by arrows. The regression is log(Mt/Cd) = 0.7193 + 0.004123X; X is years of exposure; $r = 0.351$. Mean log(Mt/Cd) is 0.7791 ±0.0347. [Based on data from Chang *et al.* (1980b).]

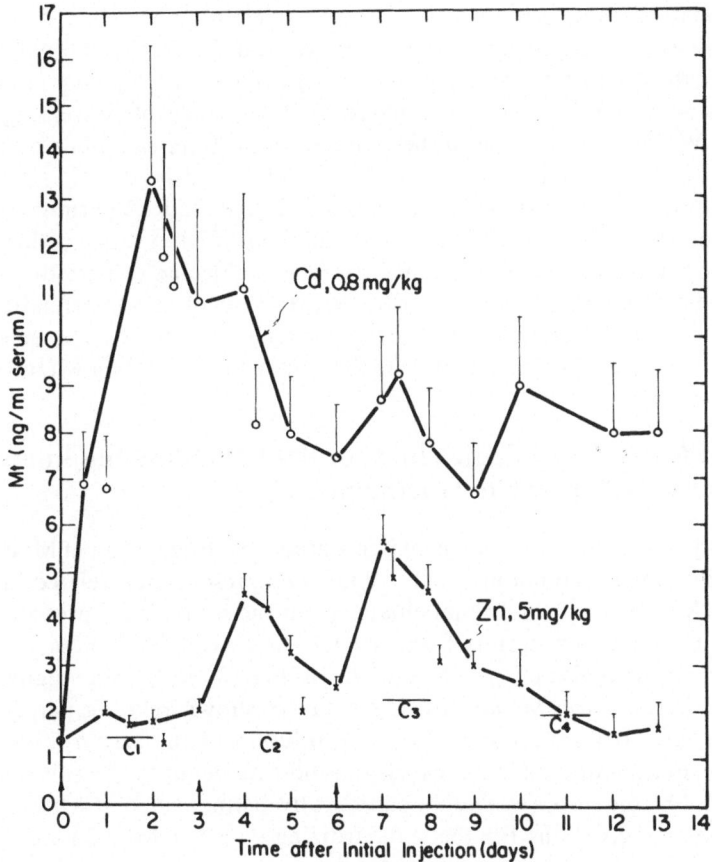

FIGURE 7. Concentrations of metallothionein (thionein) in serum following IP injections of zinc and cadmium into rats. The polygonal curves show circulating Mt (Th) concentrations (ng Mt/ml serum) as a function of time (days) after the first of a series of three IP injections (made at intervals of 3 days) of either Zn or Cd into male Sprague-Dawley rats (275–325 g). The Zn injections ($ZnSO_4$ in saline) were each of 5 mg Zn/kg body weight; Cd injections ($CdCl_2$ in saline) were each of 0.8 mg Cd/kg body weight; $n = 5$ in both cases. Controls ($n = 2$) were injected with physiologic saline only on the same schedule. The arrows on the abscissa indicate injection. Bleedings were made at 18- or 24-hr intervals (with a few explorative cases of 6- or 12-hr bleedings; the 6-hr values and one 12-hr value are left as isolates in view of their apparently depressed values). The mean response levels for the controls in the 0–3, 3–6, 6–9, and 9–13-day periods are indicated by the symbols C1-C4. The initial value of circulating Mt concentration in the entire cadre of rats was 1.35 ± 0.25 ng Mt/ml serum. The mean response level of the controls over the 13-day period was 1.79 ± 0.41 ng Mt/ml serum. The smaller standard errors of the mean values for the Zn-injected and control rats is primarily a consequence of a superior correlation for the standardizing data (the logit-log regression developed had $r = -0.99$) than was the case for that of the Cd-injected rats (the logit-log regression developed had $r = -0.97$). Significance levels of differences between mean values reported in the text are considered conservative; they followed a procedure of summing variances rather than pooling, and the use of effective degrees of freedom rather than available degrees of freedom (which usually reduced available degrees of freedom by 50%). [Data from Garvey and Chang (1981).]

injections. This fact also precludes any statement of significance of differences between relative maxima in the case of the Cd injections.

In passing, it may be noted that there was a difference in the nature of the response to the Zn injections relative to that following the Cd injections (a sequence of increasing relative maxima compared to a sequence of decreasing relative maxima). While it is tempting to speculate on this difference, much of it may be dose-related.

The principal conclusion of the experiment is that the RIA permits circulating Mt concentrations to be readily detected and quantified in rats existing in usual laboratory conditions and in the same rats when subjected to periodic injections and periodic bleedings. In general, when regressions of good correlation (0.98– 0.99) are derivable from the data the accuracy of the method is adequate to establish significance for measurements that differ by about 2 ng Mt/ml serum.

7. RIA of Mt in Physiologic Fluids of BN Rats Showing Immune Complexes after HgCl$_2$ Treatment

Finally, I mention an experiment in progress on the detection of Mt in BN rats treated with HgCl$_2$ and about which Albini *et al.* present their related findings in Chapter 37. The serum and renal eluates from these rats were provided by Dr. Albini and preliminary results from the RIA are available. These preliminary results are presented in Figs. 8 and 9. In Fig. 8, linear-log regressions of standardizing data for Mt and for a serum are shown relating percent bound protein to log Mt (circles) and to a serial dilution of the serum (crosses). The parallelism of the slopes of these regressions indicates identity of a substance in the serum with Mt (the slopes can not be statistically distinguished; $p > 0.05$). Figure 9 shows similar regressions for the standardizing data for Mt ($\bigcirc$) and for serial dilutions of three renal eluates ($\square, \times, \triangle$). Again the parallelism of the regressions indicates identity of a substance in the eluates with Mt ($p > 0.15$, $p > 0.50$, and $p > 0.50$ for the significance of differences between the slope of the regression for the standardizing data and each of the eluates in the sequence listed above).

8. Summary Comments

In sum, the RIA for Mt has been used to develop data that have allowed the subsequent use of either linear-log or logit-log regressions in establishing an identity of rat liver Mt and human liver Mt, in establishing that the Cu-binding protein isolated by Winge and associates is a Cu-thionein, in establishing that in Cd-exposed humans there is a log-log relation between Cd concentration in urine and Mt concentration in urine, and in establishing that in Hg-treated BN rats there is an identity between Mt and the protein in the serum and renal eluates from these rats. The RIA has detected and accurately quantified levels of circulating Mt and it has been sufficiently accurate to allow assigning statistical significance to differences of Mt concentration of about 2 ng/ml serum in the detection of

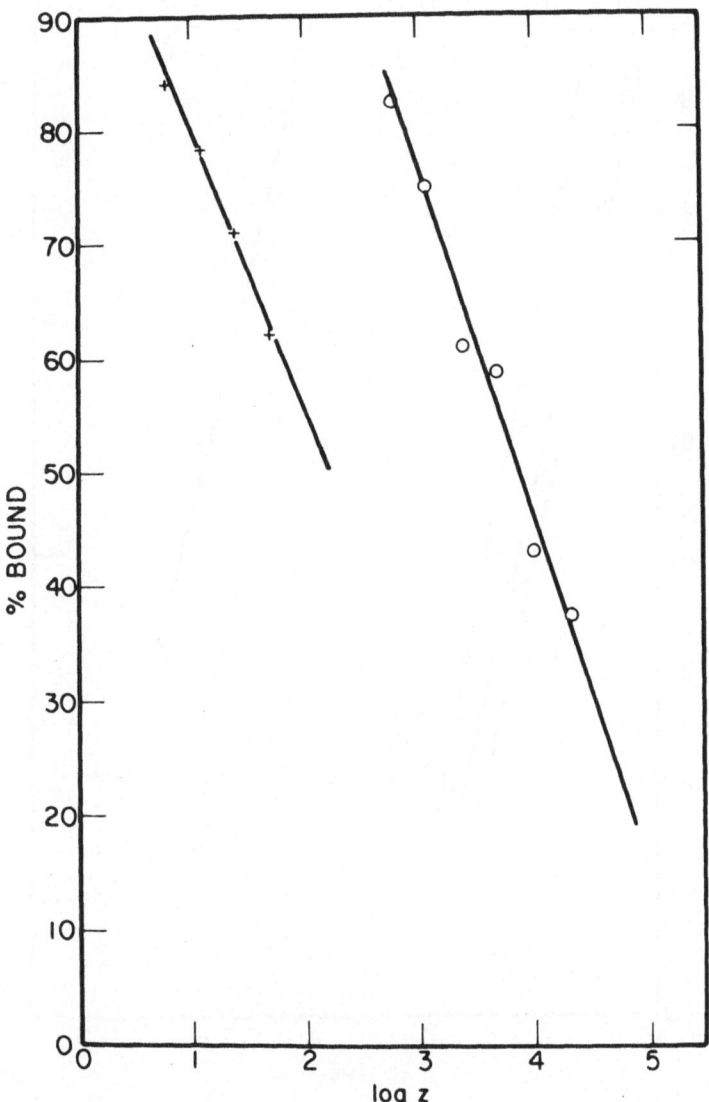

FIGURE 8. Analysis of RIA data of serum from HgCl₂-treated BN rats. Circles indicate standardizing data relating percent bound (Y) to log Mt concentration (z in pg Mt). Crosses indicate similar data derived from serial dilutions of rat serum (z in units 10^{-5} liter whole serum). The linear-log regressions are (○) $Y = 169.265 - 30.939 \log z$; $r = -0.994$; $n = 6$. (+) $Y = 104.300 - 24.482 \log z$; $r = -0.996$; $n = 4$. The slopes are not statistically distinguishable ($p > 0.05$), and the null hypothesis (that the slopes are equal) cannot be rejected.

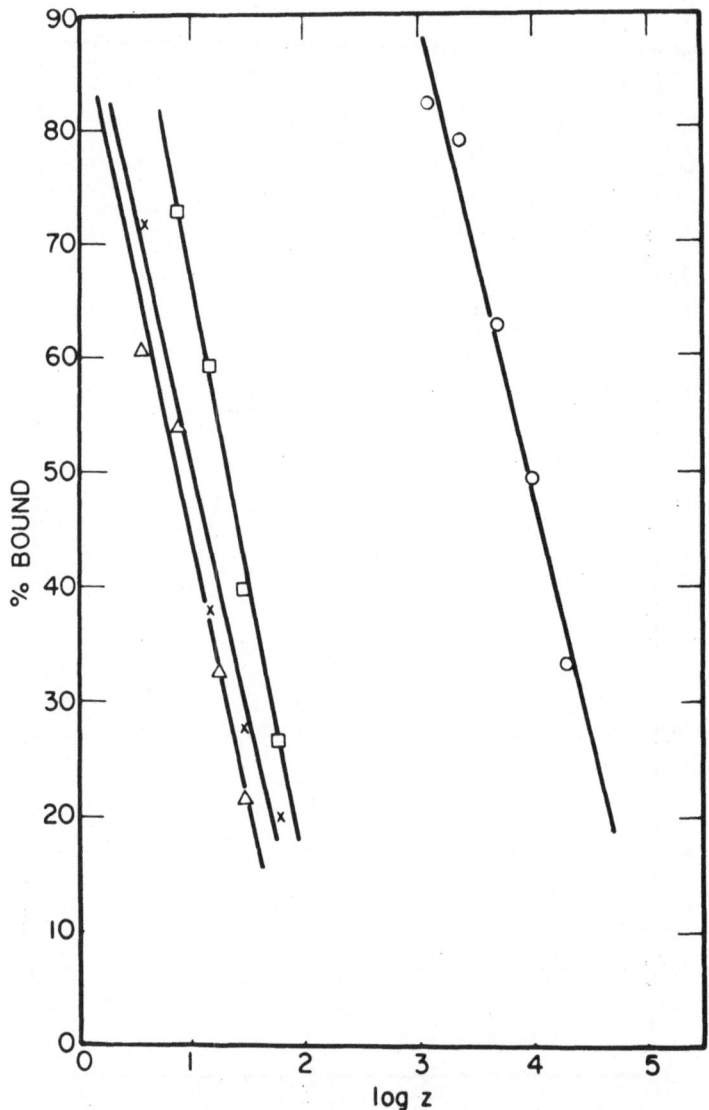

FIGURE 9. Analysis of RIA data of renal eluates from $HgCl_2$-treated BN rats. Circles indicate standardizing data relating percent bound (Y) to log Mt concentration (z in pg Mt). The symbols □, ×, △ indicate data from serial dilutions of three renal eluates (z in units of 10^{-4} liter whole serum). The linear-log regressions are as follows: (○) $Y = 218.485 - 42.504 \log z$; $r = -0.983$; $n = 5$. □: $Y = 119.973 - 52.323 \log z$; $r = -0.997$; $n = 4$. Significance level of slope difference (□-○): $p > 0.15$. (×) $Y = 94.849 - 43.665 \log z$; $r = -0.984$; $n = 4$. Significance level of slope difference (×-○): $p > 0.50$. (△) $Y = 90.171 - 45.962 \log z$; $r = -0.978$; $n = 4$. Significance level of slope difference (△-○): $p > 0.50$. In all three cases above the null hypothesis cannot be rejected. For a discussion of analyses similar to those of Figs. 8 and 9 see Hunter and Greenwood (1964).

circulating Mt. Finally, there appear to be additional potential uses for the RIA in various aspects of Mt-related toxicologic problems, these likely to be in conjunction with determinations of concentrations in tissues and fluids of the cadre of typical metals bound by thionein (Zn, Cu, Cd, Hg).

Acknowledgment

Support of this research by National Institutes of Health grant ES 01629 is acknowledged.

References

Chang, C. C., Vander Mallie, R. J., and Garvey, J. S., 1980a, A radioimmunoassay for human metallothionein, *Toxicol. Appl. Pharmacol.* **55**:94.

Chang, C. C., Lauwerys, R., Bernard, A., Roels, H., Buchet, J. P., and Garvey, J. S., 1980b, Metallothionein in cadmium-exposed workers, *Environ. Res.* **23**:422.

Garvey, J. S., and Chang, C. C., 1981, Detection of circulating metallothionein in rats injected with zinc or cadmium, *Science* **214**:805.

Hunter, W. M., and Greenwood, F. C., 1964, A radio-immunoelectrophoretic assay for human growth hormone, *Biochem. J.* **91**:43.

Kägi, J. H. R., and Nordberg, M. (eds.), 1979, *Proceedings of the First International Meeting on Metallothionein and Other Low Molecular Weight Metal-binding Proteins, Experientia Suppl.* **34**:133.

VanderMallie, R. J., and Garvey, J. S., 1979, Radioimmunoassay of metallothioneins, *J. Biol. Chem.* **254**:8416.

Vogt, W., Sandel, P., Langfelder, Ch., and Knedel, M., 1978, Performance of various mathematical methods for computer-aided processing of radioimmunoassay results, *Clin. Chim. Acta* **87**:101.

Winge, D. R., Geller, B. L., and Garvey, J. S., 1981, Isolation of Cu-thionein from rat liver, *Arch. Biochem. Biophys.* **208**:160.

Index